RADAR
DESIGN
PRINCIPLES

RADAR DESIGN PRINCIPLES

Signal Processing and the Environment

FRED E. NATHANSON

Manager
Washington Division
Technology Service Corporation
Silver Spring, Maryland

McGRAW-HILL BOOK COMPANY   New York   St. Louis
San Francisco   Düsseldorf   Johannesburg   Kuala Lumpur
London   Panama   Paris   São Paulo   Singapore
Sydney   Tokyo   Toronto

RADAR DESIGN PRINCIPLES

07–046047–7

10 VBVB 0987654

PREFACE

The need for a broad coverage of the radar waveform and processing problem became obvious to the author during several years of participation on radar system design and evaluation teams for the Navy, Army, and Air Force. In these design studies there was rarely a conventional "text book" problem in meeting the detection range and accuracy requirements. The requirements usually included a loosely defined "all weather capability" and the ability to rapidly and automatically detect low flying aircraft or missiles. The wide variety of proposed waveforms and processors seemed to be adequate on the surface. However, on closer inspection, many of these techniques had questionable performance in ordinary radar environments (windblown precipitation or strong reflections from regions near the radar, etc.). In addition the nature of the target reflections (scintillation, etc.) was not always given sufficient attention. As a result the quoted capabilities of these radars were often illusory.

The emphasis of this book is on the choice of transmit waveform and the types of receivers, detectors, etc., that may be designed to cope with the "total radar environment" rather than any single performance specification. The total environment as defined here includes the unwanted reflections from the sea, land areas, precipitation, and chaff, as well as thermal and jamming noise.

This book will neither suggest or even imply the existence of a single optimum waveform or processor. However, it is hoped that the later chapters will suggest desirable if not optimum types of processing for specific radar environments. In complex environments or with multifunction radars the current trend toward multiple modes of transmission may be the best solution if the radar can somehow sense the environment and adapt to it.

While the book is devoted to the study of radar signals, the problem of sonar processing in a reverberation-limited environment is almost identical. With proper scaling the analyses given here are applicable.

The book is essentially divided into three parts. The first four chapters contain an introduction to signal processing, a review of the radar equations for the detection of targets in the presence of noise and natural and man-made interference, the statistics of target detection, and the techniques for obtaining automatic detection. The next three chapters are a thorough survey and analysis of old and new material on the reflectivity of both material and man-made targets. Then spectral and wavelength properties are discussed in detail as they are shown to have a substantial effect on the choice of processing technique. The final seven chapters contain descriptions of the various signal processing techniques that are widely used or proposed for future radar systems. After a general discussion of processing concepts, specific techniques are discussed for the detection of moving targets by use of the doppler effect (CW, MTI, and Pulse Doppler) and the pulse compression techniques (phase-coding, frequency-coding, and linear FM). In most of these signal processing chapters there is a discussion of the theory of operation, diagrams of typical processors including digital implementations, equations for performance analysis, and advantages and disadvantages of each technique. The final chapter describes some hybrid processors with special emphasis on digital techniques.

While this book is not intended as a textbook it can easily be used as such with appropriate derivations and examples that are available in the references. Some of the draft chapters were used for an intensive course in radar and clutter given by the author and Dr. E. C. Watters, at the University of Alabama in Huntsville in 1967 and 1968. The manuscript has also been used for the second semester of a graduate course in radar at the Evening College of the Johns Hopkins University.

Fred E. Nathanson

ACKNOWLEDGMENTS

The development of signal processing as a specific part of radar system design is the result of the contributions of many individuals and organizations. The source of many of the ideas described in this book is not always clear, and the concepts often evolved at the same time in different laboratories and different countries of the world. The cited references, while not always the original source of the various advances, were often chosen for clarity of presentation or wide availability rather than for historical significance. My apologies to those not receiving credit for their original work.

This author is deeply indebted to many of his co-workers and those in other organizations. Primary credit is due to J. P. Reilly of the Applied Physics Laboratory of the Johns Hopkins University (APL) who wrote Chapter 9, a substantial portion of Chapter 5, and material in some sections of Chapters 1, 3, 6, 7, and 14. He also was of great assistance in the organization and editing of the entire book. Dr. P. J. Luke of APL helped considerably with some of the analysis and wrote sections in Chapters 8, 10, and 11. Notes by Dr. E. C. Watters of Westinghouse Electronics were used in Chapters 1 and 8. A considerable portion of unpublished material was also obtained from Dr. L. Fehlner, I. Katz, V. W. Pidgeon, J. L. Queen, S. A. Taylor, A. Chwastyk, D. M. White, and T. A. Wild of the Applied Physics Laboratory, M. Ares of General Electric Co., and Dr. L. Slobodin (Sec. 8.7) of Lockheed Electronics.

I wish to thank M. Davidson and C. Towle of APL for devoting considerable time and much insight into editing and commenting on the manuscript. Many of the illustrations are the fine work of S. J. Kundin. Special thanks to Mrs. S. Arnold, D. Shaw, and P. Goodman for typing the barely legible manuscript material.

Thanks to Dr. J. B. Garrison and J. L. Queen for their guidance into the radar processing field and for the many hours of challenging discussion on the merits of various waveforms and processors. Also to Dr. R. E. Gibson and Dr. A. Kossiakoff, the Directors of the Applied Physics Laboratory for their encouragement and support.

Some of the material was originally developed for the Naval Ordnance Systems Command, under contract NOrd 62-0604c, and the author wishes to thank the Navy for permission to publish that material. The author is indebted to the editors of the IEEE for permission to publish many figures from the Proceedings and Transactions.

Finally my utmost appreciation to my wife and daughter for their encouragement, understanding, and patience during the preparation of the manuscript and to my ten-year-old son who expects me to build him a radar now that I have written a book about it.

Fred E. Nathanson

CONTENTS

1 RADAR AND ITS COMPOSITE ENVIRONMENT

1.1 BACKGROUND AND PURPOSE OF SIGNAL PROCESSING

The term *radar signal processing* encompasses the choice of transmit waveforms for various radars, detection theory, performance evaluation, and the circuitry between the antenna and the displays or data processing computers. The relationship of *signal processing* to radar design is analogous to *modulation theory* in communication systems. Both fields continually emphasize communicating a maximum of information in a specified bandwidth and minimizing the effects of interference. The somewhat slow evolution of signal processing as a subject can be related to the time lags between the telegraph, voice communication, and color television.

Although P. M. Woodward's book [407][1] "Probability and Information Theory with Applications to Radar" in 1953 laid the basic ground rules, the term *radar signal processing* was not used until the late 1950's.

---

[1] Numbers in brackets indicate works listed in Bibliography and References at the back of the book.

During World War II there were numerous studies on how to design radar receivers in order to optimize the signal-to-noise ratio for pulse and continuous wave transmissions. These transmitted signals were basically simple, and most of the effort was to relate performance to the limitations of the components available at the time. For about ten years after 1945 most of the effort was on larger power transmitters and antennas and receiver-mixers with lower noise figures. When the practical peak transmitted power was well into the megawatts, the merit of further increases became questionable from the financial aspect if not from technical limitations. The pulse length of these high-powered radars was being constantly increased because of the ever present desire for longer detection and tracking ranges. The coarseness of the resulting range measurement led to the requirement for what is now commonly referred to as pulse compression. The development of the power amplifier chain (klystron amplifiers, etc.) gave the radar designer the opportunity to transmit complex waveforms at microwave frequencies. This led to the development of the "chirp" system and to some similar efforts in coding of the transmissions by phase reversal, whereby better resolution and measurements of range could be obtained without significant change in the detection range of the radar. I prefer to think of this as the beginning of signal processing as a subject in itself.

At about the same time, diode mixers gave way to the maser, the parametric amplifier, and in some cases the traveling wave tube. The promise of vastly increased sensitivity seemed to open the way for truly long-range systems. Unfortunately, the displays of these sensitive high-power radars became *cluttered* by rain, land objects, sea reflections, clouds, birds, etc. The increased sensitivity also made it possible for an enemy to jam the radars with low-power wide-band noise or pulses at approximately the transmit frequency. These problems led to experiments and theoretical studies on radar reflections from various environmental reflectors. It was soon realized that the reflectivity of natural objects varied by a factor of over 10^8 with frequency, incidence angle, polarization, etc. This made any single set of measurements of little general value. At the same time that moving target indicator (MTI) systems were being expanded to include multiple cancellation techniques, pulse doppler systems appeared to take advantage of the resolution of pulse radars, chirp systems were designed using various forms of linear and nonlinear frequency modulation, and frequency coding was added to numerous forms of phase coding.

As the range of the radars increased and their resolution became finer, the operator viewing the conventional radar scope was faced with more information than he could handle, especially in air defense radar networks. The rapid reaction time required by military applications led to attempts to implement automatic detection in surface radars for detecting both aircraft and missiles.

Proliferation of signal processing techniques and the resulting hardware have too often preceded the analysis of their overall effectiveness. To a great extent, this has been caused by an insufficient understanding of the statistics of the radar environment and an absence of standard terminology (e.g., *subclutter visibility* or *interference rejection*). By 1962, with the help of such radar texts as M. I. Skolnik's "Introduction to Radar Systems" [354], the target detection range of a completely designed radar could be predicted to within about 50 percent when limited by receiver noise and to perhaps within a factor of two when limited by simple countermeasures. As late as 1968, however, the estimated range performance of a known technique in an environment of rain, chaff, sea, or land clutter often varied by a factor of three or more; and the performance of untried *improvements* even exceeded this factor.

While the calculations of radar performance may never achieve the accuracy expected in other fields of engineering, it has become necessary for the radar engineer to be able to predict performance in the total radar environment and to present both the system designer and his *customer* with the means for comparing the multitude of radar waveforms and receiver configurations. The goal of this book will be to accomplish the above requirements. In order to do this, it is necessary to summarize what is known about radar environment and to present quantitative estimates of the amplitude and the statistical distributions of radar targets and *clutter*. This book will not be a panacea for the radar designer in that there will not be a unique waveform proposed nor a single receiver configuration suggested that is applicable to a wide range of radars. There will be considerable emphasis on the choice of transmit frequency for a given environment, but it is likely that radar carrier frequencies will continue to span over four orders of magnitude and that the general technology will also cover sonic and optical regions. It is safe to predict that the trend in processing will be toward digital implementations. In most of the processing chapters (8 through 14) examples will be given of digital processors that either extend the processing capabilities or at least promise more reliable hardware.

1.2 RADAR AND THE RADAR EQUATION

It is clear that volumes could be and have been devoted to the theory and design of radar. It is assumed throughout this text that the reader is familiar with the basic concepts of radar and with the elements of probability theory. If this is not the case a quick survey of the first few chapters of "Introduction to Radar Systems" [354], "Radar System Analysis" [20], or "Modern Radar" [38] should suffice. In this section a few of the basic relations are reviewed in order to establish the terminology to be used throughout the book.

The emphasis in this book will be on radars radiating a pulsed sinusoid of duration τ. This duration is related to distance units by $c\tau/2$, where c is the velocity of propagation of electromagnetic waves. The factor of two accounts for the two-way path and appears throughout the radar equations. If the pulse is a sample of a sine wave without modulation, τ is often called the range resolution in time units and $c\tau/2$ the range resolution in distance units. This is illustrated on Fig. 1-1. To a first approximation the echo power from all targets within the radar beam over a distance $c\tau/2$ will add. Inasmuch as the target phases are random, on the average the power returned is the sum of the power reflected by the individual targets. Note that in the last sentence the statement reads "on the average the power . . ." not the "average power." The precise total power reflected is a function of the power backscattered from each target and the relative phases of each reflected signal of amplitude σ_k. More specifically, the voltage return from a collection of N nonmoving

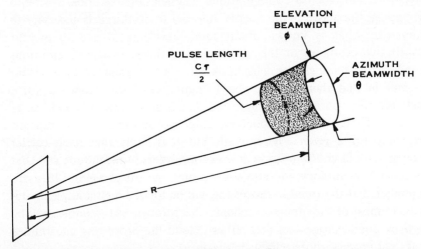

figure 1-1 *Geometry for pulse radar and volume reflectors.*

targets in a single resolution cell may be written

$$E(t) = \sum_{k=1}^{N} \sigma_k \cos(\omega t + \theta_k) \tag{1-1}$$

where θ_k is the phase angle of the return signal from the kth target. The instantaneous power return can be written

$$\text{``}P(t)\text{ on the average''} \approx \sum_{k=1}^{N} \sigma_k^2/2 \tag{1-2}$$

Since the instantaneous power at the radar varies in a statistical manner, the tools of probability theory are required for its study. A much more precise formulation and discussion of distributed targets or *clutter* problems will be given in Chaps. 2, 6, and 7.

Doppler Shift

One of the principal techniques used to separate real targets from background clutter (clutter is the undesired echo signal from precipitation, chaff, sea, or ground) is the use of the doppler shift phenomenon. It relies on the fact that although most targets of interest have a radial range rate with respect to the radar most clutter has near zero range rate for ground-based radars. (Airborne and shipboard radars present more difficult problems.) If target and clutter radial velocities do not differ, doppler discrimination does not work.

The doppler shift for a CW signal is given by

$$f_d = \frac{2\dot{R}}{c} f = \frac{2v_r}{\lambda} \tag{1-3}$$

where c is propagation velocity, f is transmit frequency, and v_r or $\dot{R}$ is range rate. For $f = 10^9$ Hz, $c = 3 \times 10^8$ m/sec, and $\dot{R} = 300$ m/sec, the doppler shift $f_d = 2$ kHz. It is apparent that either a continuous signal or more than one pulse must be used in a radar to take advantage of doppler shift. (A single pulse usually has a bandwidth measured in hundreds of kilohertz or megahertz.) It should be noted that relativistic effects have been neglected. For typical monostatic radar applications no problem arises. For radars used for instrumentation of rocket flights, the error made by neglecting relativistic effects might become important [174]. It is thus possible to discriminate between targets in the same range resolution cell if the targets possess different range rates. A quantitative assessment of the discrimination depends upon both the vagaries of nature (how stationary is a tree with its leaves moving in the wind?) (see Chap. 7) and equipment limitations (what spurious signals are generated in the radar?).

Antenna-gain Beamwidth Relations

If power P_T were to be radiated from an omnidirectional antenna, the power density (power per unit area) at a range R would be given by $P_T/4\pi R^2$ since $4\pi R^2$ is the surface area of a sphere with radius R. Such an omnidirectional antenna is physically unrealizable, but it does serve as a reference to which real antennas may be compared. For example, the gain G_T of a transmitting antenna is the ratio of its maximum radiation intensity to the intensity that would be realized with a lossless omnidirectional antenna if both were driven at equal power levels.

To obtain some appreciation for the relationships between antenna gain, antenna size, and directivity, consider a linear array of $2N + 1$ omnidirectional radiators separated by one-half wavelength with each element radiating the same in-phase signal

$$E = \cos \omega t \qquad\qquad (1\text{-}4)$$

At a great distance from the antenna, the resultant radiation pattern is

$$E(\theta) = A \sum_{k=-N}^{N} \cos\left[\omega\left(t - \frac{k\lambda \sin\theta}{2c}\right)\right] = A \sum_{k=-N}^{N} \cos(\omega t - k\pi \sin\theta) \qquad (1\text{-}5)$$

where A is a constant depending on range, c is the propagation velocity, θ is the angle measured from broadside, and λ is the wavelength. This particular sum may be evaluated to yield

$$E(\theta) = A \frac{\sin[(2N + 1)(\pi/2) \sin\theta]}{(2N + 1) \sin[(\pi/2) \sin\theta]} \qquad (1\text{-}6)$$

It can be seen that for an antenna array many wavelengths long, the radiated energy is concentrated in a narrow beam. For small θ, $\sin \theta$ can be approximated by θ, and the half-power beamwidth (3 dB one-way beamwidth) is approximately

$$\varphi_1 = \frac{1.77}{(2N + 1)} \text{ radians} = \frac{102}{(2N + 1)} \text{ degrees} \tag{1-7}$$

In addition, it can be determined from Eq. (1-6) that the first sidelobes are down 13.2 dB from the mainlobe. The narrower the antenna beam becomes, the greater the gain G_T becomes. For a lossless antenna, the gain may be computed once the antenna pattern is known. For purposes of preliminary system design, it has been found that a complete description of the antenna pattern function is not essential.

A concept of special interest is that of effective antenna aperture. If an antenna intercepts a portion of a wave with a given power density (viz., P(watts)$/m^2$), the power available at the antenna terminals is the power intercepted by the effective area of the antenna. The relationship between effective area A_e and gain is [220, 44]

$$A_e = \frac{G\lambda^2}{4\pi} \tag{1-8}$$

Transmitted and Received Power

If a pulse of peak power $\hat{P}_T$ is radiated, the peak power density at a target at range R is

$$\frac{\hat{P}_T G_T}{4\pi R^2}$$

If it is assumed that the target reradiates the intercepted power, the peak power density at the radar is

$$\frac{\hat{P}_T G_T \sigma_t}{(4\pi)^2 R^4} \tag{1-9}$$

where σ_t is the radar cross-section area defined in Chap. 5. The peak power received by the radar is then

$$\hat{P}_r = \frac{\hat{P}_T G_T \sigma_t}{(4\pi)^2 R^4} A_e \tag{1-10}$$

where A_e is the effective receiving area of the antenna. If the relationship between effective area and receive antenna gain G_R is used

$$\hat{P}_r = \frac{\hat{P}_T G_T G_R \lambda^2 \sigma_t}{(4\pi)^3 R^4} \tag{1-11}$$

It can be shown [345] that the ability to *detect* the pulse reflected by the target depends upon the pulse energy rather than its peak power. Thus, including system noise power density, the range of the pulse radar may be described by

$$R^4 \approx \frac{\hat{P}_T \tau G_T G_R \lambda^2 \sigma_t}{(4\pi)^3 K T_s (S/N)} \tag{1-12}$$

where τ is the pulse length, (S/N) is required signal-to-noise, K is Boltzmann's constant, and T_s is the system noise temperature.

The preceding simplified relationships neglected such things as atmospheric attenuation, solar and galactic noise, clutter, jamming, and system losses. Further discussion of the radar range equations in a normal, a jammed, and a clutter environment are given in Chap. 2.

1.3 FUNCTIONS OF VARIOUS TYPES OF RADAR

The rather simple algebraic equation that is given for radar detection range is often misleading in that it does not emphasize either the multiplicity of functions that are expected from the modern radar or the performance of the radar in adverse environments. Before any treatise on signal processing can begin, it is necessary to discuss some criteria for measuring the *quality* of performance of a radar system. It is well to start by paraphrasing those postulated in a paper by Siebert [345] and to expand on them as appropriate. The relative importance of these criteria depends on the particular radar problem.

1. *Reliability of detection* includes not only the maximum detection range but also the probability or percentage of the time that the desired targets will be detected at any range. Since detection is inherently a statistical problem, this measure of performance must also include the probability of mistaking unwanted targets or noise for the true target.

2. *Accuracy* is measured with respect to target parameter estimates. These parameters include target range; angular coordinates; range and angular rates; and, in more recent radars, range and angular accelerations.

3. A third quality criterion is the extent to which the accuracy parameters can be measured without *ambiguity* or, alternately, the difficulty encountered in resolving any ambiguities that may be present.

4. *Resolution* is the degree to which two or more targets may be separated in one or more spatial coordinates, in radial velocity, or in acceleration. In the simplest sense, resolution measures the ability to distinguish between the radar echoes from similar aircraft in a formation or to distinguish a missile from possible decoys. In the more sophisticated sense resolution in ground mapping radars includes the separation of a multitude of *targets* with widely divergent radar echoing areas without *self-clutter* or cross-talk between the various reflectors [227].

To these four quality factors a fifth must be added: discrimination capability of the radar. Discrimination capability can be thought of as the resolution of echoes from different classes of targets.

5. *Discrimination* is the ability to detect or to track a target echo in the presence of environmental echoes (clutter). It is convenient to include here the discrimination of a missile or an aircraft from man-made dipoles (chaff) or decoys, target identification from radar signatures, and the ability to separate the echoes from a reentry body from its *wake*.

In a military radar system, another measure of performance must be defined: the relative electronic countermeasure or jamming immunity. This has been summarized by Schlesinger [335].

6. *Countermeasure immunity* is (*a*) "selection of a transmitted signal to give the enemy the least possible information from reconnaissance (Elint*) compatible with the requirements of receiver signal processing." (*b*) "Selection of those processing techniques to make the best use of the identifying characteristics of the desired signal, while also making as much use as possible of the known characteristics of the interfering noise or signals." (*c*) "In some cases, the information received by two or more receivers, or derived by two or more complete systems at different locations or utilizing different principles, parameters or operation may be compared (correlated) to provide useful discrimination between desired and undesired signals."

Finally a term must be added to describe a radar's performance in the presence of friendly interference or RFI. The increasing importance of

*Electronic Intelligence.

this item results from the proliferation of both military and civil radar systems.

7. *Immunity to radio frequency interference (RFI)* measures the ability of a radar system to perform its mission in close proximity to other radar systems. It includes both the ability to inhibit detection or display of the transmitted signals (direct or reflected) from another radar and the ability to detect the desired targets in the presence of another radar's signals. RFI immunity is sometimes called *electromagnetic compatibility*.

Perhaps it would have been desirable to have written this entire book using these *quality* factors as an outline, but this author feels that such a book would only be suitable for a particular class of radars and that a separate volume would then be needed for each class of radar. Instead, this book outlines the functions of the various radars in this chapter and reviews the *detection* of targets in the first part of the book. The discussion of clutter and false target discrimination, which is emphasized in later chapters, will be preceded by a detailed discussion of the appropriate radar backscatter characteristics of targets and clutter. The goal is to provide a background of those properties of radar target-clutter echoes that allow *discrimination* to take place. Measures of *ambiguity* and *resolution* are introduced in Chap. 8 on signal processing concepts and waveform design and are expanded upon in the remaining chapters. The immunity of a system to *electronic countermeasures* (ECM) and RFI are not discussed as such but are implied in any discussion of signal processing techniques. Chapter 2 presents the basic ECM equations, and Chap. 4 discusses automatic detection. The ground rules for determining performance can be found in Schlesinger's book [335].

The next four sections introduce the many types of radars. Some of the more common types are air surveillance, traffic control, meteorological, and mapping. Sections 1.4 to 1.6 will help establish the function and the total environment for each type. Even within the types of radars sensitivity to weather effects, sea clutter echoes, etc. is dependent on the transmitter frequency, the radar location, and the expected range coverage.

1.4 TARGET DETECTION RADARS FOR AIRCRAFT, MISSILES, AND SATELLITES

The radar processing techniques described in this book emphasize the detection, discrimination, and resolution of man-made targets rather than those targets important to mapping and meteorological radars. This emphasis is consistent with the general trend of radar research which in recent years has concentrated on improving the capability of radars searching for usually distant targets of small radar cross section. The book also will emphasize the detection of airborne or space targets such as aircraft, missiles, and satellites.

The problem of detection of small airborne or space targets is generally characterized by the large volume of space that is to be searched and by the multitude of competing environmental reflectors, both natural and man made. The advent of satellites and small[1] missiles would have made the military radar problem almost impossible if it were not for velocity discrimination techniques. It is perhaps this combination of small radar cross section and a complex environment that justifies the heavy emphasis placed here on this type of radar. In most of this book, the discussion of radar processing applies to a radar located on the surface of the earth. These discussions may be applied to satellite, missile, or airplane radars without much difficulty except for the effect of the radar backscatter from the surface of the earth, which is discussed in detail in Chap. 7 and in [254, 227].

The detection and tracking of low-altitude aircraft or missiles require a specialized analysis. There are at least three unique problems that are discussed in later sections.

1. The vertical *lobing* effect of low-frequency radars due to forward scatter causes nulls in the antenna patterns due to reflections from the earth (discussed in this chapter in the sections on surface targets and forward scatter).

2. The target echoes must compete with the backscatter or clutter from surface features of the earth.

3. The tracking radar may attempt to track the target's reflected signals or to track the clutter itself [315, 110].

Several other factors must be considered before a discussion of the numerous signal processing techniques for surface radars can begin. In order to determine the applicability of the various processing techniques discussed in the text, it is useful to make a check list to determine if the system specifications constrain the receiver design or if some parts of the radar design have been *frozen* and limit the choice. As an example, a check list for an air surveillance radar might include the following:

[1] Small in the sense of having a low radar cross section.

1. Can the transmitter support complex waveforms?

2. Is the transmitter suitable for pulsed or continuous (CW) transmissions?

3. Is there an unavoidable bandwidth limitation in the transmitter, receiver, or antenna?

4. Has the transmitter carrier frequency been chosen? Is frequency shifting from pulse to pulse practical?

5. How much time is allotted to scan the volume of interest? How much time per beam position?

6. Is there a requirement to detect crossing targets (zero radial velocity) or stationary targets?

7. Is there an *all weather* requirement? How much rain, etc. should be used for the design case?

8. Will the radar be subject to jamming or chaff?

9. Is the radar likely to detect undesired targets, birds, insects, and enemy decoys, etc., and interpret them as true targets?

10. Is automatic detection of the target a requirement, or does an operator make the decisions?

11. Will nearby radars cause interference (RFI)?

12. Are the transmit and receive antenna polarizations fixed? Can the transmit and receive polarizations be switched from pulse to pulse?

13. What is the positional accuracy that is needed in range, velocity, and angle? How much smoothing time can be allowed?

14. Is there more than one target to be expected within the beamwidth of the antenna?

Ideally the choice of signal processing technique should be established at the earliest possible time in the design of the radar system so that arbitrary decisions on transmitters, antennas, carrier frequencies, etc., do not lead to unnecessarily complex processing. As an example, a change in one frequency band (typically a factor of 1.5 or 2 to 1) may have little effect on the detection range of a radar on a clear day, but the higher frequency radar may require an additional factor of 10 in rejection of unwanted weather echoes since weather backscatter generally varies as the fourth power of the carrier frequency.

In order to avoid continual repetition throughout the book several general assumptions will be made for subsequent discussions of radar signal processing techniques and surveillance radars.

1. The bandwidth of the radar transmission is assumed to be small compared to the carrier frequency.

2. The *target* is assumed to be physically small compared to the volume defined by the pulse length and the antenna beamwidths at the target range.

3. The targets are assumed to have either a zero or constant radial velocity, allowing target acceleration effects to be neglected. This radial velocity is small enough compared to the speed of light to neglect relativity effects, and the doppler frequency shift is small compared to the carrier frequency.

4. The compression of the envelope of the target echoes due to the radial velocity of high-speed targets is neglected.

5. *Positive* doppler frequencies correspond to inbound targets, and *negative* doppler frequencies to outbound targets.

6. The receiver implementations that are shown fall into the general class of *real time processors*, meaning that the radar output, whether it be a *detection* or an estimate of a particular parameter, will occur within a fraction of a second after reception of the target echoes. This does not preclude the increasingly prevalent practice of storing the input data on tape, in magnetic cores, or with digital logic, which in general means the insertion of the storage elements into the appropriate block diagrams. Block diagrams for digital processors will be shown for moving target indicator (MTI), pulse doppler, and phase-coded pulse-compression systems.

7. The variations in electronic gain required for different processors are neglected since the cost of amplification is negligible compared to other parts of the processors unless the signal bandwidth is in excess of 100 MHz.

The environment for the surveillance radar is emphasized in much of the discussion of signal processing techniques. Some of the reasons for this emphasis are

1. While many radar engineers can design radars and predict their performance to an acceptable degree in the absence of weather, sea, or land clutter, radar design and analysis in adverse environments have left much to be desired.

2. It requires relatively little chaff, interference, or jamming power to confuse many large and powerful surveillance radars.

3. Both natural and man-made environments create tremendous demands on the dynamic range of the receiver of the surveillance radar to avoid undesired nonlinearities.

4. In the missile era, the demand for rapid identification of potential enemy targets has led to increasing requirements for automatic or

semiautomatic detection. Later chapters show that the design effort for an automatic processing system must be different from that for radars designed for display purposes only. Inadequate dynamic range is often the problem in many current radars rather than inadequate signal-to-noise or signal-to-clutter ratios. As we enter the age of the commercial supersonic aircraft, the demand for *quick reaction* and automatic systems will also extend to air traffic control radars.

5. As the radar cross sections of missile targets are decreased by using favorable geometric designs or by using radar absorbing materials (RAM), the target echoes will reduce to those of small natural scatterers. For example, the theoretical radar cross section of an object shaped like a cone-sphere may be less than that of a single metallic half-wave dipole at microwave frequencies. The use of high-resolution radar for detecting these targets is a subject in itself often calling for combinations of several of the techniques discussed in later chapters.

6. Finally, in this era of intense competition for large radar contracts, the proposals for new radars are required to be quite explicit for all environments. Both the system engineer and the potential customer need to know how to make performance computations and apply *figures of merit* under all environments.

Chapters 2, 3, and 4 on the review of radar range performance equations, statistical relationships for various detection processes, and automatic detection by nonlinear, sequential, and adaptive techniques provide the basis for computation of detection capability in the presence of noiselike interference. Chapter 5 on radar targets then allows the choice of the correct statistical model for the target cross-section term.

Later chapters emphasize that the *noise* of the radar is often the backscatter from objects other than the desired targets. When these echoes are larger than the effective receiver noise power, the ratio of target cross sections σ_t to an equivalent clutter cross section σ_c dominates the radar performance computations. Chapters 6 and 7 on atmospheric effects, weather, chaff, and sea and land clutter are included in this book to allow a better estimate of the appropriate values of the clutter backscatter. Figure 1-2 illustrates in a simplified way the magnitude of the problem for a typical, but fictitious, narrow beamwidth pulsed air surveillance radar with a C-band (5,600 MHz) carrier frequency. The presentation is somewhat unusual in that most of the radar's parameters have been held constant. The echoes from the environmental factors are plotted versus radar range; the left ordinate shows the equivalent radar

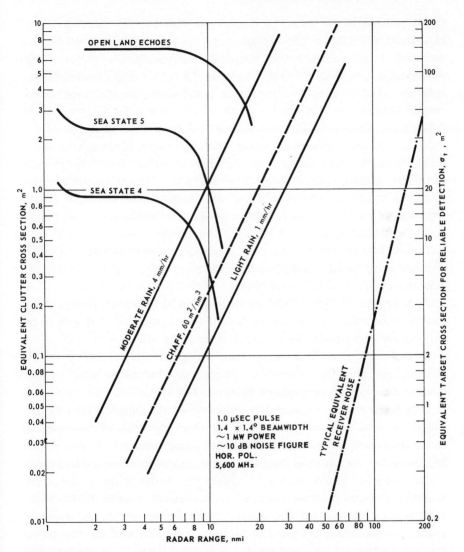

figure 1-2 *Environment for typical C-band surveillance radar.*

cross section. The slopes of the various lines illustrate the different range dependencies of different kinds of environmental factors to be discussed in later chapters. The rapid drop of the sea and land clutter curves illustrates the rapid reduction of backscatter echoes at the radar horizon in a rather arbitrary way. Also shown on the graph is the equivalent receiver noise power in the bandwidth of the transmitted pulse. This is the amount of receiver noise power that is equal to the target power into the radar that is returned from a particular range.[1] The right-hand ordinate is the typical target cross section σ_t that can be detected at the various ranges assuming that the signal-to-mean clutter power ratio is typically[2] 20 to 1 (13 dB). The other major assumption is that the clutter does not appear at ambiguous ranges (*second-time-around* echoes, etc.). A similar graph of environmental effects, but at 3,000 MHz, appears in Figure 4-1. The effects of forward scatter of the radar waves have been neglected in both figures. In normalizing these graphs to radar cross section, backscatter following an R^{-4} law,[3] such as targets or land or sea clutter echoes at low grazing angles (small angles from the horizon), appears as a constant cross section; and echo power from uniform rain and chaff appears to increase as the square of range.

There are two significant features of this graph.

1. The surface radar, especially in a military environment, is rarely receiver noise limited except at high elevation angles on a clear day (at frequencies of 3,000 MHz and above).

2. Even the relatively short 1-μsec pulse (150 m in radar range) in the example needs further processing for the echoes from small targets to be sufficiently above the mean backscatter from the *clutter*.

In summary, some form of clutter, and perhaps electronic counter-measure rejection, is required for detection of small (~ 1 m^2) radar cross-section targets to ranges of 100 nmi (185 km).

In the processing chapters, the reduction of the clutter echoes from a single uncoded transmitted pulse as compared to those of the target is called the *improvement factor I*. The basic equations used to draw graphs of this nature are given in Chap. 2; quantitative values for the undesired clutter are given in Chaps. 6 and 7. Since the clutter shown in the figure generally has considerable extent, it is generally desirable to minimize the antenna beamwidths and usually the pulse length. Another appropriate generalization is that the clutter signals, as well as the target echoes, increase with increasing transmitted power. In designing a noise-limited surface radar system, the system engineer can either increase the transmit energy, reduce the receiver noise density, or

---

[1] If the target echo, which usually varies as R^{-4}, is held constant, then noise power, which is usually constant with R, *effectively* increases as R^4.

[2] The choice of this number is arbitrary, but it has been found to be widely useful in evaluating surveillance radars.

[3] The use of R^{-4} rather than the usual R^{-3} is a generalization of the data in Chap. 7.

increase the antenna size. In the cluttered environment, increased antenna size is usually most desirable.

1.5 SURFACE TARGET DETECTION

Although radars for the detection of targets on the surface of the earth do not make up a single class, it is useful to discuss them in one section. The radars for detection of surface ships, submarines, vehicles, and human targets face common problems caused by the forward scatter of radar signals and the shadowing effect of the surface features of the earth. A forward-scattered electromagnetic wave[1] adds to the direct wave between the radar and the target which causes either vector addition or subtraction of the two radar waves at the target depending upon the relative phase of the reflected wave. The return signal from the target may also follow two paths with a similar interference effect [181, 354]. This subject is studied later in Sec. 1.10 on forward scatter; a more detailed study of the calculations for surface targets can be found in Durlach [99].

An additional factor that compounds the processing problem is that of the backscatter from the surface of the earth and the "cultural" features surrounding the target. The amplitude and spectral characteristics of this *clutter* are discussed in detail in Chap. 7 on sea and land backscatter. At this time, it will suffice to say that surface clutter usually dominates over receiver noise in any attempt to detect men or vehicles on land and that sea clutter is dominant in most attempts to detect ships in all but the calmest sea conditions. Since detection of the ship target at sea allows a somewhat more straightforward discussion, it will be considered first.

Sea Targets

For a low-frequency (< 1 GHz) surface radar attempting to detect ships at sea, forward-scatter interference is the most severe problem. The surface of the water acts as a mirror, and the phase of the reflected electromagnetic wave will be almost in antiphase with that of the direct signal for small grazing angles (almost tangent to the surface). As early as

---

[1] The term *scattered* will be used throughout the book to denote all energy that is not absorbed; *forward scatter* will denote all reflections away from the radar, and *backscatter* will denote all energy (called *clutter* if it is not the desired signal) redirected toward the originating radar.

World War II, it was shown that the received power P_r from a point reflector for low-frequency radars (0.2-2.0 GHz) near the surface could be approximated by [354, p. 503]

$$P_r \doteq \frac{4\pi G^2 P_t \sigma_t (h_a h_t)^4}{\lambda^2 R^8} \qquad (1\text{-}13)$$

where G^2 is the two-way antenna gain, h_a is the radar antenna height above the water, h_t is the target height, λ is the carrier wavelength, and σ_t is the target radar cross section. Note that the received power is a strong function of the carrier frequency, target height, and range for a given radar. As a result, radars designed for detection of targets near or on the water generally have a carrier frequency above 3,000 MHz, where the effect of forward scatter is of less importance except in calm seas.[1] At these frequencies and for ranges of less than 10-15 km, the usual radar equations can be applied. Unfortunately, the backscatter from the sea at low grazing angles also increases rapidly with carrier frequency, and the sea clutter echoes usually far exceed receiver noise, as was shown in Fig. 1-1. The sea clutter on a radar display generally appears to have long persistence and a coarser grain than receiver noise [82]. Target detection equations are given in Chaps. 2 and 3 for which sea clutter echoes are considered to be "colored" or partially correlated noise.

The backscatter from ships of 100 ft length or longer is generally greater than from typical sea clutter echoes; however, this is not generally the case for buoys, snorkels, and periscopes. Since the height h_t of these targets is only a few feet, the forward scatter interference tends to reduce the reflected signals even above 3,000 MHz. The obvious, though not always successful, technique for detecting these targets is to use a radar having a very narrow beamwidth and a very short pulse length. Unfortunately, this tends to resolve the ocean waves; and a careful study of the nature of sea backscatter is necessary for proper choice of parameters.

Land Targets

The detection of men and land vehicles by a field radar primarily involves the major problems resulting from clutter echoes from the

---

[1] A discussion of the merits of 3,000- and 9,300-MHz radars for detecting marine targets is contained in a paper by Harrison [161].

terrain and cultural features, the attenuation resulting from natural or man-made obstacles and wooded areas, and forward scatter phenomena. While doppler radars have been successful in detecting and measuring the velocities of vehicles with proper siting of the radar at close ranges (i.e., police traffic radars), there is little in the open literature on the general problem. If the radar has sufficient sensitivity, the detection range of a man or vehicle can be estimated by using the radar cross sections given in [354] together with the land-clutter reflectivity data of Chap. 7 as inputs to the clutter range equations of Chap. 2. The radar must have a higher transmit power than for the *free space* situation since the attenuation by woods can be significant. Saxton and Lane [333] summarized the results of several experiments between 100- and 3,000-MHz carrier frequency. They found that for either antenna polarization, attenuation by trees with leaves and in that frequency region is given by

$$A = 0.25 f^{3/4} \text{ (dB/m)} \tag{1-14}$$

where f is the carrier frequency in gigahertz. At higher microwave frequencies, 100 m of heavily wooded area will prevent detection of all but the largest objects. See also Tamir [370] for attenuation in forests.

1.6 METEOROLOGICAL RADAR PROCESSING

Weather radars, both large land-based systems and long-range airborne systems, are finding increasing application. Determination of the paths of thunderstorms, tornadoes, heavy rain, hail, or snow and even of the location of clear air turbulence by radar is helping to avoid discomfort and disaster in many parts of the world. The advent of the supersonic transport demands that advanced radars with ranges of 200 to 300 nmi be carried in commercial aircraft. As the requirements increase, the use of advanced signal processing techniques for meteorological radars becomes mandatory just as it has in most other types of radars. The radar meteorologist needs information on weather velocities and turbulence as well as quantitative measures of precipitation.

The intensity of radar returns from rain, snow, and clouds has been reported in many places, including the numerous Radar Meteorological Conferences of the American Meteorological Society, and in such texts as those of Atlas [12] and Battan [25]. A summary of attenuation and reflectivity of various meteorological phenomena is given in Chaps. 5 and 6. Chapter 6 also gives some recent experimental results on the spectrum and carrier frequency effects of the backscatter from rain and snow. While there is no other section of this book specifically on processing for meteorologists, Chaps. 3, 11, and 14 should yield insight into processors that can be used to extract more information from weather echoes.

The basic radar range equations for weather radars are shown in Chaps. 2 and 6 to differ from those for target detection. For a given antenna area A_e, the received power from rain or snow is proportional to the fourth power of the carrier frequency if the attenuation term is negligible. The decibel attenuation in rain, which varies about inversely as $\lambda^{2.8}$, can be neglected for wavelengths longer than 10 cm (S band). At 3 cm attenuation can become appreciable. Long-range weather radars usually transmit at carrier wavelengths between 5 and 10 cm; airborne weather radars usually use wavelengths of 5 cm or less. It should be noted that the experimental results for attenuation at 3 cm reported in Chap. 6 show it to be considerably higher than earlier theoretical predictions. This may be a reason for using wavelengths near 5 cm if considerable penetration of storms is desired.

Katz [206] and others have shown that clear air echoes do not have the λ^{-4} backscatter relationship of weather echoes. Katz suggests a *polychromatic radar* (multiple carrier frequencies) approach to discriminate between weather effects, clear air turbulence (CAT) echoes that often occur in thin layers, and various birds and insects. The scattering cross sections of some of these anomalous effects are discussed in Chap. 5. The polychromatic radar can also aid in studies of precipitation by helping determine drop diameter, liquid water content, etc. with the longer wavelength radars giving more penetration and the shorter wavelengths giving better data on weaker echoes.

Measures of the mean velocity and the internal turbulence of airborne reflectors are necessary to predict the probable path and the severity of storms. Since this generally requires range, angle, and velocity resolution, the weather radar designer desires a coherent-pulse or pulse-doppler processor; but he usually cannot justify the cost of a complete processor. In addition, data processing of the large quantity of range and velocity information can become unwieldy unless the analysis is confined to a small volume. While the full processors described in Chaps. 11 and 14 are applicable to weather radar, a brief mention will be given in this section of some of the *compromise* systems that have been utilized to measure the mean velocity and spectrum of weather echoes.

One technique to determine the standard deviation of the spectrum of weather echoes (but not the mean velocity) is to measure the rate of crossings of the average value of the envelope of fluctuating radar returns. A device using this technique, the R Meter, is described by Rogers [323] and Atlas [12]. With a stable radar and a nonfluctuating target, the envelope of the echoes would have a constant amplitude; and the number of crossings about the mean value would be zero. When the reflecting objects are a collection of random scatterers that are in motion with respect to one another, there is a fluctuation about the mean value at a rate that is proportional to the velocity spectrum width. The relationship for a square-law detector can be approximated by [12, 32]

$$\sigma_v \approx \frac{N\lambda}{2.4} \quad \text{to} \quad \frac{N\lambda}{2.6} \tag{1-15}$$

where N is the mean number of positive crossings per second, σ_v is the standard deviation of the velocity spectrum, and λ is the carrier wavelength. While a coherent radar is not required, the carrier frequency must be stable enough to prevent any erroneous widening of the measured spectra. An erroneous velocity spread of as much as 5 m/sec can be indicated for a 1-μsec pulse and a transmitter FM deviation of 5 MHz at a 100-kHz rate and at 10-GHz carrier frequency [12].

The velocity-azimuth display (VAD) is a method for determining the mean velocity of precipitation. It is based on pulse doppler techniques [12]. A pencil-beam radar is pointed at a fixed elevation angle ($\sim 35°$) above the horizon and is slowly rotated in azimuth. The processor measures the doppler frequency at a fixed range and many azimuths and thus determines the radial velocity at each azimuth. The mean wind speed and some of the statistics of the spectrum can be determined from one full rotation if some assumptions are made about the uniformity of the mean wind speed in the area and wind shear effects. To determine the winds at different altitudes, the echoes at different ranges can be sampled. The mean radial velocities of the winds V_0 which are obtained as a function of angle (from the doppler relationship) are computed by

$$V_0 = V_f \sin\alpha + V_h \cos\alpha \cos\beta \tag{1-16}$$

where V_0 is the radial velocity at each value of α and β, V_f is the fall velocity of the precipitation, V_h is the horizontal wind velocity, α is the elevation angle, and β is the azimuth angle. It was predicted that with this technique the mean wind speed could be measured to better than ± 0.5 m/sec. The effect of the rapid variations of wind shear and the non-uniformities in local winds indicate that this accuracy may be difficult to obtain except with narrow beamwidth radars and fairly rapid rotation rates.

Another technique, used by this author and his co-workers to obtain the spectra of weather, chaff, and sea clutter echoes, computes the correlation coefficients of precipitation echoes for various time displacements with an autocorrelator. It is based on the fact that turbulent precipitation echoes produce a fluctuation in radar return signals at a rate that is proportional to the short-term radial motions of the individual reflecting particles. A more complete implementation that also measures the average radial velocity will be discussed in Chap. 14. Under most conditions the familiar doppler equation can be rewritten as

$$\sigma_f = \frac{2\sigma_v}{\lambda} \tag{1-17}$$

where σ_f is the standard deviation of the doppler echoes in hertz and σ_v is the rms radial velocity of the particles.

A complete pulse doppler processor can measure these properties at a given location, but it requires a rather complex implementation. Similarly a spectrum analyzer can determine the complete power spectrum of the echoes, but it is rather wasteful of transmission time unless all the return signals are recorded in some manner (such as on tape) and analyzed repeatedly. There is another relationship that leads to a simple processor if the precipitation echo signals are large compared to the receiver noise.[1] The Wiener-Kintchine theorem states that the power spectrum of a signal is the Fourier transform of its autocorrelation function. If the spectrum of the received echoes is approximately gaussian, the Fourier transform method shows that, if the average doppler component is removed (see Sec. 14.9)

---

[1] It can be shown [224, chap. 12] that an autocorrelator is not very efficient if the signal-to-noise ratio is less than unity.

$$\sigma_v = \frac{0.095\,\lambda}{\tau_{0.5}} \qquad\qquad (1\text{-}18)$$

where $\tau_{0.5}$ is the time when the autocorrelation function of the echoes drops to 0.5 of its maximum amplitude (at $\tau = 0$). This relationship is quite useful in modern clutter studies, since frequently it is easier to compute autocorrelation functions than to compute the Fourier transform.

1.7 CRITERIA FOR CHOICE OF SIGNAL PROCESSING TECHNIQUES

The desire for ever greater detection and tracking ranges for weapon-system radars forced the peak power of the radars of the post World War II era well into the megawatt region. Even then, the detection ranges were not considered adequate for short-pulse radar transmissions. When longer pulses were transmitted, target resolution and accuracy became unacceptable. While efforts to increase detection range were also substantial in the field of low noise receivers, it became apparent that external noise and clutter in the military *all weather* environments would negate the improvements in noise figure. At about that time Siebert [345] and others pointed out that the detection range for a given radar and target was only dependent on the ratio of the received signal energy to noise power spectral density and was *independent of the waveform.* The efforts at most radar laboratories then switched from attempts to construct higher power transmitters to attempts to use pulses that were of longer duration than the range resolution and accuracy requirements would allow. These pulses were then internally coded in some way to regain the resolution. The next few paragraphs show that there is a conflict in obtaining range resolution and accuracy and simultaneously obtaining velocity resolution and accuracy with simple waveforms. Range resolution is defined here as the ability to separate two targets of similar reflectivity. For a pulsed sinusoid, resolution can be approximated by $\Delta R \approx (1/B)$, where ΔR is the range resolution in time units and B is the transmission 3-dB bandwidth of a pulsed sinusoidal signal.

The range accuracy or ability to measure the distance from the radar to the target is described by Woodward [407]. For a single-pulse transmission when the target velocity is known and acceleration is neglected,

$$\sigma_\tau = \frac{1}{\beta (2E/N_0)^{1/2}} \tag{1-19}$$

where σ_τ = the standard deviation of range error in time units.

$2E/N_0$ = the signal energy to noise power per hertz of the double-sided spectrum assuming white and gaussian noise. For near optimum receivers, and considering only the real part of the noise, this is numerically equal to S/N the peak signal-to-noise power at the output of the receiver matched filter.

β = root-mean-square (angular) bandwidth of the signal envelope about the mean frequency. β^2 is the normalized second moment of the spectrum about the mean (taken here to be zero frequency) [354, p. 467].

Neglecting the pulse shape for a moment, it can be shown that for uncoded pulses $B \approx 1/\tau$ and $\beta \approx 1/\tau$, where τ is the 3-dB duration of the pulse. Thus, both range resolution and accuracy requirements generally are in direct opposition to detectability requirements which vary as $\hat{P}_t \tau$ (peak power times pulse length) for simple pulses.

In its simplest form, velocity resolution, i.e., the ability to separate two targets separated in doppler, can be expressed by $\Delta V \approx 1/T$, where ΔV is doppler resolution in frequency units and T is time duration of the waveform. As one would expect, the longer the duration of the signal, the easier it is to accurately measure the doppler shift. It has been shown that the accuracy with which doppler frequency can be determined (when range is known) is given by [354, p. 473]

$$\sigma_d = \frac{1}{t_e (2E/N_0)^{1/2}} \tag{1-20}$$

where σ_d is the standard deviation of the doppler frequency measurement and t_e is the effective time duration of the waveform.

Thus, the requirement for velocity resolution and accuracy implies long duration waveforms whereas that for range resolution and accuracy implies short waveforms. This apparent paradox will be expanded on in Chap. 8 on signal processing concepts and waveform design.

1.8 ANTENNA CONSIDERATIONS

In military radar systems prior to the early 1960's, the conflicting requirements for detection, resolution, and accuracy were often resolved by having two or more different radars, each of which was most suited to a particular function. These would include search radars, height finders, track radars, and sometimes gunfire control and missile guidance radars. In civil radars for airports there is often a functional division into general air surveillance radars, precision approach radars, and meteorological radars. To circumvent this need for separate radars there are many current programs aimed at the design of multifunction radars. It is worthwhile to point out some of the compromises that are necessary in the design of multifunction radars.

It can be seen from Eq. (1-10) that long-range target detection is heavily dependent on having a large effective receiving aperture area; however, large aperture areas imply narrow beamwidths as can be seen from the following relationships for a rectangular aperture.

$$G = \frac{4\pi A_e}{\lambda^2} \cong \frac{4\pi}{\theta_1 \phi_1} \tag{1-21}$$

where G is the antenna gain, A_e is the effective antenna aperture, λ is the transmit wavelength, and $\theta_1 \phi_1$ are one-way 3-dB beamwidths[1] (radians). These terms will be elaborated upon in Chap. 2. It is useful to describe the cross-sectional area of the radar beam as

$$\theta_1 \phi_1 \approx \frac{\lambda^2}{A_e} \tag{1-22}$$

Thus, beamwidth decreases as $\sqrt{A_e}$ increases. If good angular resolution is also desired, the wavelength is usually kept small. When the radar requirement includes surveillance of the entire hemisphere, the number of beam positions to be searched, N_B, with some overlap of the beam, is then

$$N_B > \frac{2\pi}{\theta_1 \phi_1} \approx \frac{2\pi A_e}{\lambda^2} \approx \frac{G}{2} \tag{1-23}$$

For $A_e = 10$ m^2, the number of beams in the hemisphere is about 6,000 at a wavelength $\lambda = 10$ cm (S band). If the desired scan time is 10 sec, there are about 600 beam positions to be observed per second or 1.6 msec per beam position. The significance for signal processing is that pulse compression techniques, MTI, or short bursts of coherent pulses

---

[1] With uniform illumination, "4-dB beamwidth along the principal axes" is a more precise definition.

can yield good performance in a 1.6-msec period and that CW or pulse doppler techniques are generally more time consuming. The relationship of performance to *dwell time* per beam will be analyzed in the chapters on the individual techniques.

A related problem with narrow beam surveillance radars is the difficulty of achieving reliable detection with only one or two pulses per beamwidth and complex targets. It will be shown in Chap. 3 on statistical relationships for various detection processes and in Chap. 5 on radar targets that greater total energy is needed to have a high probability of detecting a slowly fluctuating target in a single pulse than if several pulses are transmitted such that *independent* target reflections are obtained. This has been verified with both automatic detection systems and operator displays.

For frequency or phase scanned array antennas the usable spectrum of the transmitted waveform is limited because of the beam dispersion as bandwidth is increased. This is generally not a significant problem with conventional dish antennas. With a frequency scanned antenna [354, sec. 7] the transmission bandwidth limit is obvious since a change in carrier frequency is deliberately used to steer the beam in elevation or azimuth. However, with a phase-scanned antenna the limitations are somewhat more subtle. These limitations can be separated into at least two classes: the phase approximation to time delay and the transient effects. The phase approximation results from the widespread use of array phase shifters rather than time delays that are limited to a maximum of about $360°$ variation (modulo 2π). When arrays whose linear dimensions are many wavelengths are steered more than a few beamwidths from the boresight position, the modulo 2π phase approximation to the desired wavefront is only accurate over a narrow band of carrier frequencies. The reduction in antenna gain (one way) for a parallel-fed antenna as a function of scan (steering) angle from boresight, for short-pulse or compressed-pulse systems, can be approximated by [353]

$$ L \approx 0.72 \left(\frac{U}{\tau \alpha f_0} \right)^2 \quad \text{in dB, for} \quad \left(\frac{U}{\tau \alpha f_0} \right) \le 2 \tag{1-24} $$

where U = sine of the steering angle from boresight at which L is being measured

τ = $-$ 3-dB pulsewidth at the matched-filter output, sec

α = sine of the one-way beamwidth at boresight and the center frequency, measured in the plane through boresight and the steering angle

f_0 = center frequency of the transmission, Hz

It can also be shown [65] that at a $60°$ scan angle a reasonable maximum transmission bandwidth occurs where

$$\frac{\text{Transmission bandwidth}}{\text{Carrier frequency}} = \frac{\text{boresight beamwidth, degrees}}{100} \qquad (1\text{-}25)$$

At lesser scan angles the antenna gain will be reduced by less than 1 dB (two way) from the gain at the single frequency for which the phase approximation was made if this criterion is satisfied. The highest antenna sidelobe will remain 10 dB below the main beam.

The second *transientlike* effect introduces an antenna bandwidth limitation when a linear array is excited from its center or from one end. If a rectangular pulse of RF energy is fed to one end of an antenna array, the signal will immediately begin propagating into space. Until the array becomes fully energized, the radiated energy will originate from only a portion of the full aperture and will have a much broader beamwidth than the *normal* antenna pattern, which is based on continuous wave excitation. During the transient period, a considerable portion of the energy is radiated into what would normally be considered the sidelobes of the antenna. If the pulse duration τ is long compared to the fill time (propagation time) across the array t_f, most of the total energy is radiated into the normal or steady-state pattern; and there is little increase in the antenna sidelobes.[1] Obviously there is a similar transient effect at the trailing edge of the pulse. The length of the steady-state portion is $t_{ss} = (\tau - 2t_f)$.

An example of the transient effects on one-way energy patterns is shown in Fig. 1-3 for a center-fed array that was tapered by Taylor weighting for 30-dB steady-state sidelobes [172]. The *steady-state* pattern for a long pulse with $(t_{ss}/t_f) = 1,666$ is shown in Fig. 1-3A, and the pattern for short or coded pulses with $(t_{ss}/t_f) = 2$ is shown in Fig. 1-3B. While the pattern for the short pulse is considerably degraded, some of the transient energy is at frequencies that may be eliminated by filtering in the receiver.

Although the two bandwidth limiting effects were discussed separately they can be considered together. Adams [2] has studied the combined

---

[1] The fill time for the center-fed array is one-half that of the end-fed array.

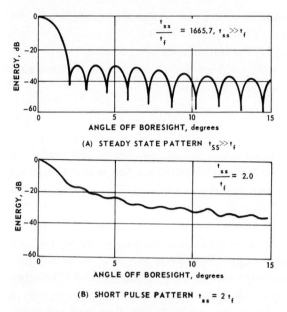

figure 1-3 *Transient energy patterns for center-fed antenna arrays. (After Hill [172].)*

effect and shows that the parallel-fed antenna has broader bandwidth at boresight and that the series-fed antenna degrades somewhat less at high scan angles. Antenna tapering increases the allowable bandwidth of both types of antennas.

1.9 RADAR GRAZING ANGLE FOR 4/3 EARTH APPROXIMATION

In radars designed for detecting or tracking targets near or on the surface of the earth, it is not always adequate to neglect the bending of electromagnetic waves passing through the atmosphere. To account for the curved propagation path in radar computations, the *4/3 earth approximation* is used. In this approximation the radius of the earth is considered to be 4/3 the actual radius and the curved path is considered straight [44, p. 49].[1] This approximation is of great importance for surface radars observing targets near the optical or geometric horizon. It is also of importance in determining the maximum detection range of airborne early-warning (AEW) radars. The 4/3 earth approximation is

[1]The 4/3 radius of the earth is taken to be 4,587 nmi. See also Sec. 6.7 and [310, sec. 4-1].

also essential in attempting to compute the backscatter from the surface of the earth. This section consists of a derivation of the true grazing angle ψ of the radiation in terms of the elevation of the radar above the surface of the earth h and of the depression angle of the radar beam α.

Accurate computations for the detection and tracking of such targets as ships, land vehicles, or low-flying aircraft can only be made with a rather complicated calculation of the true grazing angle and the forward scatter of electromagnetic waves incident on the earth's surface. A thorough analysis of the subject has been made by Durlach [99].

The simplified equations are based on a straight-line propagation path. This path is in reality a curved line. The curved path can be transformed into a straight line by substituting an equivalent earth radius $r_e = 4/3\, r_0$ in place of the true earth radius r_0. Reference [310] gives a good discussion of this transformation and its equivalence to the physical picture. The important results are summarized here.

Figure 1-4a shows a sketch of a propagating ray, which is a curved line because of the refractive index rate of change with height. The symbols are defined as follows:

Δt = time to traverse P-Q

ℓ = ground range = $r_0\theta$

$\Delta h/\Delta \ell$ = change in height with distance

r_0 = actual earth radius

α = depression angle of the ray with respect to local horizontal

We also define a variable s such that

$$n = \left(\frac{r_0}{r}\right)^{s+1} \tag{1-26}$$

where n is the refractive index and r is a radial distance from the center of the earth to the point at which n is measured.

Consider a new sketch with the following conditions (Fig. 1-4b):

1. $s' = -1$ (implies constant index of refraction)
2. $\alpha' = \alpha$
3. $r_0'\theta' = r_0\theta$
4. $r_0' = -r_0/s$ (experimentally, $s \cong -3/4$ on the average)[1]

---

[1]The value of $1/s$ varies with time and location. References [310] and [354] explain that values for $1/s$ may vary over the United States from -1.25 to -1.45 for altitudes less than 1 km. Values considerably greater than -1.3 have been reported

(Footnote continued on next page)

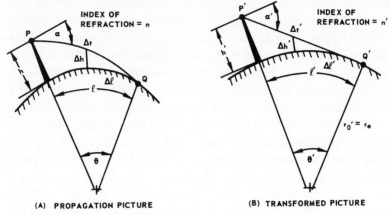

(A) PROPAGATION PICTURE   **(B) TRANSFORMED PICTURE**

figure 1-4 *Propagation paths for electromagnetic energy. (A) Actual path; (B) straight-line path from equivalent earth transformation.*

Because of condition (1) there is no bending of the rays. Condition (2) implies both situations have equivalent depression angles. Condition (3) imposes the constraint that in both situations the point at which the ray strikes the earth occurs at the same ground range. Condition (4) requires an equivalent earth radius of $r_0' = r_e = 4/3 \, r_0$. With these conditions, two important relationships are proven in [334]

$$\Delta t = \Delta t'$$

$$\frac{\Delta h}{\Delta \ell} = \frac{\Delta h'}{\Delta \ell'} \tag{1-27}$$

The first relationship establishes the fact that the time for the energy to traverse both paths is the same. The second indicates that the grazing angles are the same for both pictures (the angle between the local horizontal at the target and the radar beam direction).

The exact expression for the depression angle α of a radar beam illuminating a point Q is

$$\alpha = \sin^{-1}\left(\frac{2r_e h + h^2 + R^2}{2R(r_e + h)}\right) \tag{1-28}$$

near England [181]. It is also pointed out that $1/s = -4/3$ is an average that is almost independent of frequency up to very high radar frequencies. For the visible region, $1/s \cong -1.20$.

Equation (1-28) applies to the configuration shown in Fig. 1-4. P is the location of the radar, Q is the location of the target, α is the depression angle of the radar, h is the antenna height, r_e is the radius of earth (4/3 actual radius) and R is the slant range of target. In most instances Eq. (1-28) may be simplified because $h/r_e \ll 1$ and $h^2/2r_e r \ll (h/r + R/2r_e)$. This simplification yields

$$\alpha \cong \sin^{-1}\left(\frac{h}{R} + \frac{R}{2r_e}\right) \tag{1-29}$$

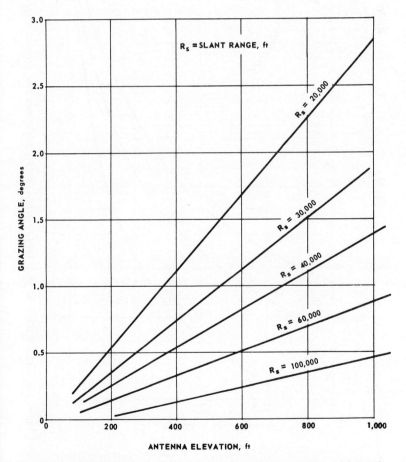

figure 1-5 *Grazing angle versus antenna height for low grazing angles and medium antenna heights.*

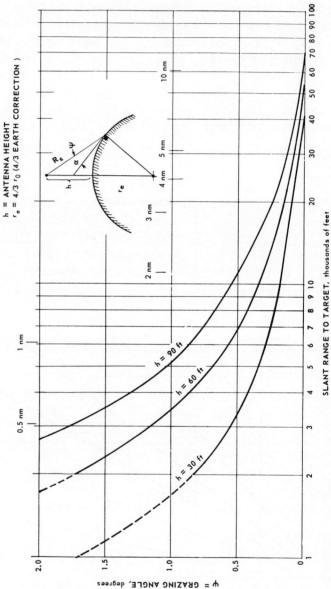

figure 1-6 *Grazing angle vs. slant range for low grazing angles and low antenna heights.*

32

The exact expression for the grazing angle ψ is

$$\psi = \sin^{-1}\left[\frac{h}{R}\left(1 + \frac{h}{2r_e}\right) - \frac{R}{2r_e}\right]$$

(1-30)

$$\psi \cong \sin^{-1}\left(\frac{h}{R} - \frac{R}{2r_e}\right)$$

Figures 1-5 and 1-6 show Eq. (1-30) in graphical form. By comparing Eqs. (1-29) and (1-30), we can see that the sines of the grazing and depression angles differ by approximately $2R/r_e$.

1.10 FORWARD SCATTER EFFECTS

A radar target situated near the earth's surface intercepts energy from a ground-based radar along two paths (Fig. 1-7). One path is the direct line between the source and the target; the other path is a line reflected from the earth's surface. This multipath propagation results in two effects that are generally deleterious to radar performance: (1) an interference pattern between the two rays results, which causes a wide variation in the reflected power versus elevation angle (lobing), and (2) false target

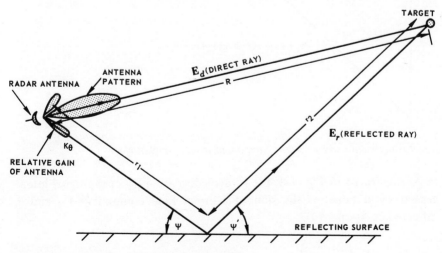

figure 1-7 *Propagation over a reflecting surface.*

images are created, which degrade the angular accuracy of the radar [110]. This section summarizes how the surface properties of the earth alter the magnitude and phase of the scattered signals. Since multipath effects are most severe for a specularly reflected wave (angle of incidence equal to angle of reflection), the results emphasize specular reflection. Backscatter toward the radar is discussed in Chap. 7.

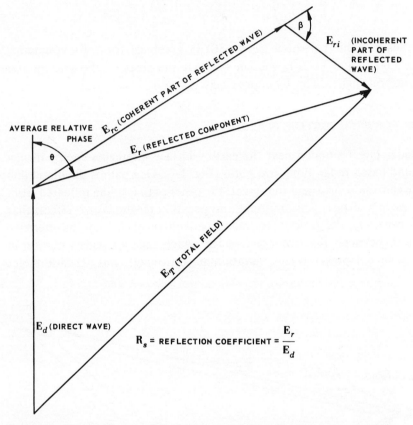

figure 1-8 *Vector diagram of the field intensity for propagation composed of a direct component and a reflected component from a rough surface.*

As illustrated in Fig. 1-8, the total electric field E_T of the signal intercepted by a target is the sum of a direct field component E_d and a reflected component E_r

$$E_T = E_d \left(1 + \frac{E_r}{E_d} \right)$$

$$(1\text{-}31)$$

The vector ratio E_r/E_d is normally expressed in terms of a reflection coefficient R_s which is the ratio E_r/E_d for an isotropic radiator. Equation (1-31) can be expressed in terms of the reflection coefficient

$$E_T = E_d(1 + K_\theta^{1/2} R_s) \qquad (1\text{-}32)$$

where $R_s = E_r/E_d$ for an isotropic radiator and K_θ is the power gain of the antenna at the angle of the reflected wave relative to the gain at the target angle. (Refer to Fig. 1-7.) The relative phase of the reflection coefficient R_s is usually separated into two factors: one which accounts for the relative phase of E_r due to the path length difference $R - (r_1 + r_2)$ and one which accounts for the phase due to the electrical properties of the reflecting surface, that is,

$$R_s = (\overline{R}_s + r_s e^{i\beta}) e^{i\phi} e^{i\alpha}$$

where ϕ = phase due to the electrical properties of the reflecting surface of the reflected ray relative to the ray incident on the surface

$\alpha = (2\pi/\lambda)[R - (r_1 + r_2)]$ = phase of the reflected ray relative to the direct ray at the target from the target-radar geometry

$\overline{R}_s$ = magnitude of coherent part of reflection coefficient

r_s = magnitude of incoherent part of reflection coefficient

β = phase of the incoherent part relative to the coherent part (β is uniformly distributed between 0 and 2π)

The vector diagram of Fig. 1-8 illustrates that the reflected wave E_r is composed of a coherent part E_{rc} and an incoherent, or random, part E_{ri}. Because of the random nature of the incoherent component, the reflection coefficient must be described statistically.

The average power in the total field is proportional to[1]

$$P \sim \overline{E_T^2} = E_d^2(1 + K_\theta \overline{R_s^2}) \qquad (1\text{-}33)$$

(the bar indicates average values). Notice that this expression allows us to specify the amount of power in the reflected beam relative to that in the main beam. $\overline{R_s^2}$ is broken into a coherent power term $\overline{R}_s^2$ and an incoherent power term σ_{Rs}^2.

---

[1] This result makes use of the assumption that $\overline{\cos \alpha} = 0$ because the surface is fluctuating.

The relative coherent reflected power $\bar{R}_s{}^2$ is derived for a rough surface; it can be related to the reflection coefficient for a smooth surface R_0 and a surface parameter $\bar{\rho}$ [33] by[1]

$$\bar{R}_s{}^2 = \bar{\rho}^2 (DR_0)^2 \tag{1-34}$$

where R_s is the reflection coefficient for specular reflection from a rough surface, R_0 is the reflection coefficient for specular reflection from a smooth surface, and D is the divergence factor, which accounts for the fact that the reflecting surface is spherical (viz., the earth). The quantity $\bar{\rho}$ is defined by

$$\bar{\rho} = e^{-g/2} \tag{1-35}$$

where $g/2 = (1/\lambda^2)\, 8\pi^2\, \sigma_h{}^2\, \sin^2 \psi$
  λ = radar wavelength
  $\sigma_h{}^2$ = variance of the surface height about the mean height
  ψ = grazing angle

The value of σ_h required to evaluate Eq. (1-34) may be related to significant wave height (and in turn to the hydrographic sea state) through the Burling relationship[2] $H_{1/3} = 4.0\,\sigma_h$. Equation (1-35) is plotted for several values of sea state and radar wavelength in Fig. 1-9.

The only additional information needed to specify the coherent reflected wave is the value of the average relative phase angle ϕ; Figs. 4-8 and 4-10 of [233] illustrate the variation of the phase of the reflection coefficient as a function of grazing angle for smooth seas. Experimental data on the influence of waves is given by Beard and Katz [27, 29, 30, 28].

The power in the incoherent portion of the reflection wave $\sigma_{Rs}{}^2$ cannot be completely defined because of the present lack of knowledge

---

[1] Equation (1-34) is derived from the assumption that the height distribution and the spatial correlation function are both gaussian and that the reflection is in the specular direction. The author further assumed a uniformly rough surface of infinite conductivity. Although the sea is not a perfect conductor, the error introduced is not serious since the scattered field is determined more by the surface roughness than by the surface conductivity. For small grazing angles, Eq. (1-34) must be modified by a shadowing function.

[2] See Sec. 7.1.

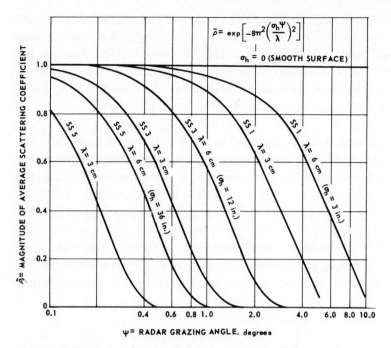

$$\bar{\rho}= \exp\left[-8\pi^2\left(\frac{\sigma_h \Psi}{\lambda}\right)^2\right]$$

$\sigma_h = 0$ (SMOOTH SURFACE)

$\bar{\rho}=$ MAGNITUDE OF AVERAGE SCATTERING COEFFICIENT

1.0

0.8

0.6

0.4

0.2

0

0.1   0.4   0.6 0.8 1.0   2.0   4.0   6.0 8.0 10.0

Ψ= RADAR GRAZING ANGLE, degrees

SS 5 λ= 3 cm

SS 5 λ= 6 cm

SS 3 λ= 3 cm (σ_h = 36 in.)

SS 3 λ= 6 cm (σ_h = 12 in.)

SS 1 λ= 3 cm

SS 1 λ= 6 cm (σ_h = 3 in.)

figure 1-9 *Typical surface coefficient for scattering from seas of various sea states SS (for horizontal or vertical polarization).*

about the spatial correlation function of the sea's surface (see Chap. 7). The appendix of [211] contains plots from which D may be evaluated. The value of the smooth surface reflection coefficient R_0 may be determined by Figs. 4 through 6 of [233]. Beckmann, [32, 34] has shown that for small depression angles the value of R_s must be multiplied by a *shadowing function* to account for the reduction in scattered power caused by the shadowing effect that the *hills* of the surface have on the *valleys.*

The previous theoretical consideration used a number of simplifications that are not strictly correct, namely

1. Perfectly conducting reflector
2. Gaussian height distribution
3. Gaussian spatial correlation function
4. Uniform surface conditions
5. Constant grazing angle within the illuminated area

In spite of these simplifications, the theory compares favorably with experimental findings, as illustrated in Fig. 1-10 [203], which compares experimental values of the coherent term $\bar{\rho} = \overline{R}_s/DR_0$ with those determined from Eq. (1-33). Although the experimental data for the incoherent term cannot be quantitatively compared with theory, notice that the peak value of the experimental curve (corresponding to $\sigma_h \psi/\lambda$ = 100 mils) occurs at approximately the value predicted ($\sigma_h \psi/\lambda$ = 80 mils).

The calculation of the effects of forward scatter on the detection and tracking of low-altitude or surface targets is quite complex. Numerous approximations [44, 20] are needed to accurately determine the detection ranges. As an example consider an L-band radar (1.3 GHz) situated 60 ft above sea level searching for targets over a smooth ocean. For targets at a range of 16 nmi with altitudes of 50, 100, and 200 ft the received power computed from free-space equations must be modified by about − 17, − 7, and + 9 dB, respectively. If the sea becomes rough, the coherent reflection will decrease and the free-space equations will be

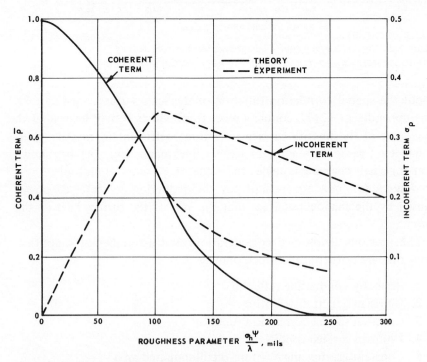

figure 1-10 *Coherent and incoherent reflection components for rough ocean surfaces (for horizontal or vertical polarization). (From Katz [203].)*

more appropriate; however, the backscatter from rough seas is much greater than for calm seas, and the target echoes may be obscured by sea clutter. Some examples of the received power from ship targets near the horizon are shown in Chap. 5. Analysis leading to more rigorous calculations is given by Durlach [99] and Heffner [163].

The tracking radar has similar problems. If the surface is smooth, the radar may try to track the target or its image. If the surface is rough, the tracking error may be dominated by the clutter signals [315]. One series of X-band tests over land ($\rho \approx 0.4$) showed elevation tracking errors of about ± 2 mils for targets near zero elevation [110]. A graph of elevation error versus grazing angle is given by Barton [20, p. 330] from C-band tracking radar tests. The approximate rms elevation error can be written

$$\sigma_\phi = \frac{\rho\phi}{(8G_s)^{1/2}} \qquad\qquad (1\text{-}36)$$

where ρ is the coefficient of reflectivity (0.3-0.4 for typical land), G_s is the power ratio of mainlobe (sum pattern) to sidelobe gain for the error pattern at an angle of $2E$ below the beam axis, E is the elevation angle of target, and ϕ is the elevation beamwidth.

2 REVIEW OF
RADAR RANGE PERFORMANCE
COMPUTATIONS

This chapter will review the basic forms of the radar equation and the variations needed to calculate the performance of a radar that is attempting to detect targets masked by the basic thermal noise of the radar receiver. Subsequently this development will be extended to cover a general environment, which includes the multitude of superfluous returns that often exceed the basic receiver noise. These returns will include electronic countermeasures, friendly interference, sky noise, and the backscatter from atmospheric effects, land, seas, and chaff.

2.1 GENERAL RADAR RANGE EQUATION

The general radar equation for pulse radars shown below includes a larger number of parameters than is commonly used in current radar texts

For general references see Barton [20], Berkowitz [38], Blake [44], and Skolnik [354].

[354, 20, 38]. The separation of certain loss terms from the basic parameters will facilitate rewriting the equations for the various external environments. The loss terms in Eq. (2-1) are less than unity or are subtracted from the numerator when the equation is expressed in decibel form. The backscattered power received from a target of radar cross section σ_t at the first mixer or preamplifier can be written

$$\hat{P}_r = \frac{\hat{P}_T G_T L_T G_R L_R \lambda^2 L_P L_a \sigma_t}{(4\pi)^3 R^4} \tag{2-1}$$

where $\hat{P}_r$ is peak received power

Before elaborating on the terms of this equation, it is useful to convert the received power $\hat{P}_r$ to the minimum detectable received power S_{min} and to rewrite the equation in terms of the detection range R for a pulse radar. Then

$$R^4 = \frac{\hat{P}_T G_T L_T G_R L_R \lambda^2 L_P L_a \sigma_t}{(4\pi)^3 S_{min}} \tag{2-2}$$

where R = the detection range of the desired target with the statistics of detection to be defined later.

$\hat{P}_T$ = the peak transmit power (the average power during the pulse) at what will arbitrarily be defined as the output of the transmitter unit.

G_T = the transmit power gain of the antenna with respect to an omnidirectional radiator. This is a dimensionless quantity equal to $4\pi A_e/\lambda^2$. This term generally refers to the centerline of the antenna beam.

A_e = the effective aperture of the antenna, which is equal to the projected area in the direction of the target times the efficiency. This includes the fractional losses due to spillover and tapering of the aperture to reduce the sidelobes. Then considering the losses, $A_e = \epsilon A$ where A is the physical projection of the antenna area and $\epsilon \approx 0.4$ to 0.7.

λ = the wavelength of the radiation.

L_T = the losses between the transmitter output and free space including transmit-receive duplexers, power dividers, waveguide or coax, radomes, and any other losses not included in

A_e. It is preferred to include in this term any losses which do not affect the beamwidth of the radar.

G_R = the receive power gain of the radar defined in a similar manner to the transmit gain.

L_R = the receive antenna losses defined in a similar manner to the transmit losses. However, in the case of ultrasensitive receivers, the losses may be included in the effective system noise temperature T_s. These will be defined in conjunction with the minimum detectable signal.

L_P = the beam shape and scanning and pattern factor losses,[1] which include several factors to compensate for the antenna gain terms being calculated on the centerline or *nose* of the beam. A typical target for a search radar will tend to have an arbitrary position with respect to this line, and the ground and sea reflections will create a lobing pattern resulting from the relative phase of the free space and reflected signals. An analogous loss occurs in conical scan tracking radars where the beam is *nutated* or scanned such that the full gain of the antenna is not continuously pointed at the target. This is sometimes called the crossover loss [20]. A typical value for search radars is 1.6 dB for each dimension of scan and for 50 percent detection probability. Similar losses caused by finite range gates and doppler filters will be covered in the chapters on processing techniques.

The pattern factors caused by forward scatter are quite complex. They are dependent upon the grazing angle with respect to the land or sea and the height of terrain objects (trees, hills, buildings) or ocean waves at the point of reflection. The reflections may add or subtract to the main beam antenna gain relative to free space depending upon the reflection coefficient and polarization. Blake [44, pp. 26-35] gives the general computation technique. Chapter 7 on sea clutter effects gives additional experimental data for rough sea conditions and some of the basic properties of the reflected signal. There will be no further discussion of lobing effects except as they affect detection of targets near the horizon. (See Sec. 5.6.)

---

[1] Including F of Blake's notation. See Barton [20, pp. 141-152], Blake [44, pp. 27-31], Skolnik [354, pp. 501-506], and Durlach [99].

The description of beam shape losses will be slightly expanded in the section on detection statistics. A more thorough treatment will be found in Barton [20, 23].

L_a = the two-way pattern absorption or propagation losses of the medium. These are calculated separately since they are usually a function of the target range, the elevation of the target, and the type of interference. It is a common practice to compute the free-space range and then adjust for these losses, as in Fig. 2-1, rather than complicate the radar equation. L_a is often expressed as exp $-$ ($2\alpha R$), where α is the attenuation constant of the medium and the factor 2 is for a two-way path.

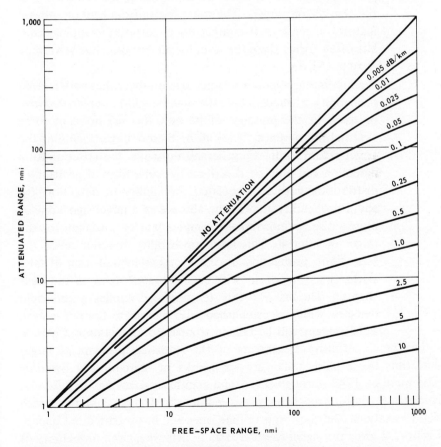

figure 2-1 *Free-space radar range vs. attenuated radar range for one-way attenuation rates of α dB/km. (From Skolnik [354].)*

σ_t = the radar cross-sectional area of the object that is being detected and is equal to

$$\sigma_t = \frac{\left(\begin{array}{c}\text{power reflected toward the receiving}\\ \text{aperture/unit solid angle}\end{array}\right)}{(\text{incident power density}/4\pi)}$$

The definition has a simple physical significance only for the case of a large metallic sphere, where the physical cross section is equal to the radar cross section (when the diameter is large with respect to the radar wavelength). (See Chap. 5.) The use of the term cross section will be expanded to include the amplitude and spectral distributions of the reflectivity in the chapter on targets. The radar return from land or cultural features appears as the target for an airborne mapping radar but often forms the *noise* level for an Airborne Early-Warning System (AEW).

S_{min} = the minimum detectable target signal power that with a given probability of success the radar can be said to *detect, acquire,* or *track* in the presence of its own thermal noise or some external interference. Since all of these factors (including the target return itself) are generally noiselike, the criterion for a detection can only be described by some form of probability distribution with an associated probability of detection P_D and a probability that in the absence of a target signal one or more noise or interference samples will be mistaken for the target of interest. This latter probability is often called the false alarm probability P_F or P_N. The related term of false alarm time T_F or n' will be elaborated upon in the next chapter. The use of this terminology implies a threshold detector which is established with respect to the noise level. This concept will be related to operator detection in Chap. 3.

The most definitive treatments of the statistical problem of target detection by a pulse radar are the works of Marcum and Swerling presented in 1947 through 1954 and republished by the IRE [244]. Numerous graphs have been presented in the technical journals with slight variations; the most appropriate ones will be presented in Chap. 3. Figure 2-2 gives the signal-to-noise ratio to achieve a given probability of detection of a target signal in a single look. The false alarm number n' is

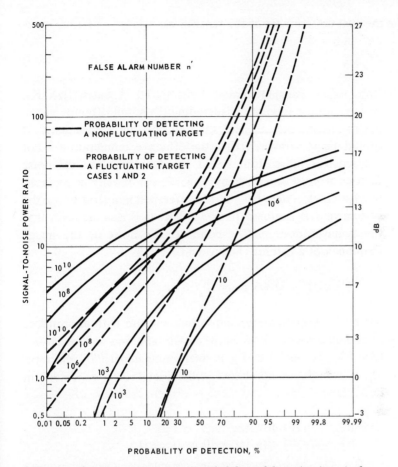

figure 2-2 *Signal-to-noise ratio vs. probability of detection in a single look. (After Fehlner [113] and Marcum and Swerling [244].)*

the parameter. The solid lines apply when the target echo is a constant sinusoid and the noise is gaussian; the dashed lines apply when the target echo is fluctuating. These curves are for single-pulse detection. Multi-pulse detection curves with a more detailed explanation will be given in Chap. 3.

The classic radar equation is based on the assumption that the thermal noise of the receiver is the fundamental limitation on minimum detectable target signal. As more sensitive receivers have been developed the effects of antenna noise and sky noise and the effects of the sun, galaxies, and the earth must be included as described in Berkowitz [38, pt. V, chap. 3]. Since this text is on signal processing, it will suffice to define the thermal noise power of the receiver as

$$N = KT_s B_N \tag{2-3}$$

where K = Boltzmann's constant (equal to 1.38×10^{-23} watts/(Hz)(°K).

T_s = system noise temperature, including the antenna temperature, environmental effects, and the noise of the receiver itself. It should be remembered that the effective temperature T_e of the receiver noise itself is not simply equated to the noise factor (or noise figure) $\overline{F}$ by $T_0 \overline{F}$. T_e can usually be replaced by $T_0(\overline{F} - 1)$, where T_0 is the reference temperature of the system and is assumed to be 290°K. It is also assumed that the receiver *does not* respond to the noise at the image frequency of a heterodyne receiver.[1] Then

$$T_e = (\overline{F} - 1)\, 290°\text{K} \quad \text{or} \quad T_s \doteq T_e + T_A$$

where T_A is the antenna noise temperature with appropriate consideration for losses between the antenna and receiver.[2] (See [38, 20, 44].) If T_A is small compared to T_e, the term T_0 is sometimes combined with K and then expressed in decibels.

$$KT_0 = -204 \text{ dBW/Hz}$$
$$KT_s = -204 \text{ dBW/Hz} + 10 \log(\overline{F} - 1)$$

The noise figure of typical microwave receivers is shown in Fig. 2-3.

B_N = noise bandwidth of the receiver in hertz after amplification but before envelope detection. The actual noise bandwidth varies with the transfer function of the filter or set of filters prior to detection and with the type of waveform. For a simple radar transmitting a rectangular pulse, the *optimum* B_N at IF is

---

[1] See Description of Noise Performance of Amplifiers and Receiving Systems, *Proc. IEEE*, vol. 51, no. 3, pp. 436-442, 1963.

[2] In this equation T_A is often written as T_A/L_R.

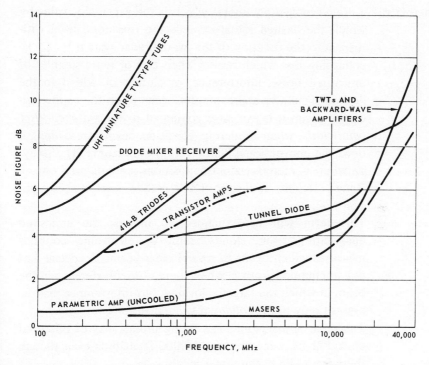

figure 2-3 *Typical noise figure for several types of receivers as a function of frequency.*

$B_N = 1.2/\tau$ to $1.4/\tau$ for a rectangular filter
     passband
$B_N = 0.6/\tau$ to $0.8/\tau$ for synchronously tuned
     filters [354, p. 414] [1]

(2-4)

where τ is the width in seconds of the rectangular transmitted pulse. The value of $1.2/\tau$ is commonly used for operator decisions (Barton [20, p. 21], Blake [44]).

With the preceding assumptions the minimum detectable signal can be written

$$S_{\min} = KT_s B_N \left(\frac{S}{N}\right)\left(\frac{1}{L_c}\right)$$

[1]The loss for not using a matched filter must be included in L_c or L_s.

where (S/N) = a dimensionless parameter which is the acceptable value by which the desired signal exceeds the rms noise level with respect to the function of the overall radar system.[1]

L_c = collapsing loss which results when two or more samples of unwanted noise, interference, or clutter are added to the desired signal. Several examples occur when the bandwidth of the receiver is excessive compared to the *matched-filter bandwidth*, when numerous elevation beams are *collapsed* onto the conventional Plan Position Indicator (PPI) display, or when numerous azimuth positions are *collapsed* onto a Range Height Indicator (RHI). (See Barton [20, pp. 128-132].)

This term will be expanded to include the undesired backscatter from clouds, rain, or electronic counter-measures which occur in several radar beams. A special, but not uncommon, case is when there is no interference in the beam in which the target is located, but interference may be present only in the *collapsed* beam positions.

The radar return from a low flying aircraft may be obscured on a PPI by rain clouds at 10,000 ft altitude even though there is no rain at the target location.

The collapsing ratio ρ, another common term, is defined as

$$\rho = \frac{m + n}{n}$$

where m is the number of samples containing noise alone and n is the number of samples containing signal plus noise. The collapsing loss (in dB) can then be found from the curves on integration loss for n samples integrated inco-herently (Chap. 3). The loss L_c is determined for $n + m$ samples (in dB) and then subtracted from the loss (in dB) for n samples.

Similar collapsing losses for PPI displays as taken from Barton [20] are based on

---

[1] In detection equations (S/N) is usually the required signal-to-noise ratio, and in this text S/N represents the actual power ratio for the stated conditions.

$$\rho = \frac{1/\tau + 2B_v}{2B_v}$$

for insufficient video bandwidth where B_v is the video bandwidth (Hz). In the above cases the increase in the total number of noise samples presented to the detector or operator must be taken into account when evaluating the statistics of the detection process. This type of calculation is expanded upon in Chap. 3. The collapsing loss L_c is expressed here as a fraction less than unity or in decibel form. This loss should be subtracted from the numerator of the radar equations.

Before discussing what constitutes an acceptable value of (S/N) it is useful to write the full radar equation and some of its variations. The *pulse radar equation* is

$$R^4 = \frac{\hat{P}_T G_T L_T G_R L_R \lambda^2 L_P L_a L_c L_s \sigma_t}{(4\pi)^3 KT_s B_N (S/N)} \tag{2-5}$$

for the receiver (and antenna) noise limited case, and where L_s is the signal processing loss, which will be loosely defined as the deviation from a linear receiver with a perfectly matched filter. The numerical value to be used will be expanded upon in Chap. 4 on automatic and adaptive detection and in the various signal processing chapters. This loss will vary in the presence of clutter and jamming because of the change in the spectrum and amplitude distribution of the interference.

Since the noise bandwidth B_N is approximately equal to $1/\tau$ for a system approximating a matched filter, a more descriptive presentation of the detection properties of a radar system is to replace

$$\frac{\hat{P}_T}{B_N} \quad \text{by} \quad \hat{P}_T \tau = E$$

where E is the transmitted energy per pulse in joules. This substitution illustrates that the detection properties of a particular radar (for target obscured by receiver noise) are dependent only upon the transmitted energy and the noise per Hz (KT_s). It will be shown in later sections that the relationship is essentially independent of the pulse length or transmitted waveform. This substitution also allows simpler extension to the CW (continuous wave), FM, and pulse doppler radars, where it can be shown that the average power determines the detection performance if the receiver can be made to approach the *matched* filter.

The radar equation expressed in the above form is somewhat misleading in that it implies a λ^2 dependence on detection range by using the gain of the antenna as a parameter. In a search radar, higher gain with the resulting narrower beamwidth means that less time or energy or fewer pulses can be radiated in any one direction if it is necessary to search a volume which is large compared to the cross-sectional area of any one beam. Another common form of the radar equation substitutes $4\pi A_e/\lambda^2$ for G_R and results in the *preferred search radar equation*

$$R^4 = \frac{\hat{P}_T G_T L_T A_e L_R L_P L_a L_c L_s \sigma_t}{(4\pi)^2 KT_s B_N (S/N)} \tag{2-6}$$

which has no wavelength dependence.

Hall [154] has shown (with different notation) that to search a solid angle Ψ with a given observation time for a single beam[1] T, the minimum total search time (T_{sc}) is

$$T_{sc} = T\frac{\Psi}{\Omega}$$

where Ω = the solid angle of the radar beam

Ψ = the total solid angle to be searched

T_{sc} = time allowed in seconds to search a total solid angle Ψ

The solid angle of the radar beam, however, can be expressed as $\Omega = 4\pi/G_T$. The average power $\bar{P}_T$ is equal to $\hat{P}_T \tau/T$ for a pulse radar, and $B_N \tau \approx 1$ from Eq. (2-4). Then substituting for B_N in Eq. (2-6) gives the *volume search radar equation*

$$R^4 = \frac{\bar{P}_T L_T A_e L_R L_P L_a L_c L_s \sigma_t T_{sc}}{4\pi KT_s (S/N) \Psi} \tag{2-7}$$

---

[1] Assuming continuous transmission.

This equation holds for pulse radars if the two-way propagation time to the target is short enough to use the full energy of the radar in the desired search time without a considerable percentage of *dead time.*

In a similar manner the transmit gain G_T can be converted to aperture area. This form also has misleading effects in countermeasures and in clutter limited situations but applies where there is a firm constraint on antenna size. For a common transmit and receive antenna Eq. (2-6) becomes the *antenna size limited equation*

$$R^4 = \frac{\hat{P}_T L_T A_e^{2} L_R L_P L_a L_c L_s \sigma_t}{4\pi \lambda^2 K T_s B_N (S/N)} \tag{2-8}$$

This form with the inverse λ^2 dependence is useful for aircraft, satellite, and missile radar computations where transmit frequencies of over 10,000 MHz are common. The greater detection range implied for higher frequency radars is often negated by increased atmospheric and weather attenuation and a tendency of receiver noise temperatures to increase with frequency. In addition, the number of beam positions to search a given volume increases.

2.2 RADAR DETECTION WITH NOISE JAMMING OR INTERFERENCE

It is convenient at this point to introduce the detection equations for noise jamming or interference as they follow the basic radar equation in form and do not require the introduction of statistical distributions other than wide-band gaussian noise. The simplest geometry to consider is where the target is carrying a relatively high power noise source to attempt to deny knowledge of *his* location to a search radar or to prevent a tracking radar from *acquiring* him.

The noise power radiated from a broadband jammer (barrage jammer) has a power density at the search radar of

$$\text{Power density at radar} = \frac{\bar{P}_J G_J}{4\pi R_J^{2}} \text{ watts/unit area} \tag{2-9}$$

For general references see Waddell [390], Schlesinger [335], and Barton [24].

where $\overline{P}_J$ = the average power of jammer at its antenna input

G_J = gain of jammer antenna, including losses

R_J = range from jammer to radar

The power entering the radar and added to the receiver noise is dependent on the aperture area of the radar and the relative bandwidth of the jammer B_J and the radar receiver B_N. It is convenient to assume that $B_J \geq B_N$ (otherwise the jammer would be a *spot jammer* and the radar system response would become intimately involved with the type of signal processing). At the receiver input the total noise power is

$$N = KT_s B_N + \frac{\overline{P}_J G_J B_N A_e L_R L_a'}{4\pi R_J^2 B_J} \tag{2-10}$$

where L_a' is the one-way propagation loss. This term can replace $KT_s B_N$ in the previous radar equations, but it is more common to replace $\overline{P}_J/B_J$ by the jammer power per unit bandwidth.[1]

If the barrage jammer is to have any effect on the radar it must be large enough to make

$$\frac{\overline{P}_J G_J A_e L_R L_a'}{4\pi R_J^2 B_J} \gg KT_s$$

that is, the jamming noise density must be greater than the system noise density. With this simplification the radar equation can be written for the jammer at the target range and angle as the *self-screening range equation*

$$R_{SS}^2 = \frac{\hat{P}_T G_T L_T L_P' L_a' L_c' L_s'}{4\pi B_N (S/J)} \left(\frac{\sigma_t}{P_J G_J} \right) \tag{2-11}$$

where R_{SS} = *self-screening* range of the radar, or the range at which a target of cross section σ_t can be detected in jamming noise

[1] Despite the confusion of units, $\overline{P}_J/B_J$ is often called P_J and is usually expressed in watts/MHz if B_N and B_J are in MHz. It is also assumed in all discussions that the total jamming noise does not saturate the receiver.

L'_P = one-way scanning and pattern losses of the radar since the jamming signal also has similar losses

L'_a = one-way atmospheric or weather losses from radar to target

(S/J) = signal-to-jam ratio which is defined in an analogous manner to (S/N)

L'_c = collapsing losses only in the receiver itself since the jammer is assumed to occupy only one beam

L'_s = signal processing losses applicable to jamming (includes image losses, etc.)

P_J = jammer power density (watts/MHz if B_N in MHz)

In operational analysis it has been found convenient to express the self-screening equation as

$$R_{SS} = \alpha \left(\frac{\sigma_t}{P_J G_J} \right)^{\frac{1}{2}} (L'_a)^{\frac{1}{2}}$$

where $(P_J G_J / \sigma_t)$ is expressed in watts/(MHz)(m^2) and the remaining terms form α if B_N is expressed in MHz. The term α is then the range on a 1 m^2 target radiating 1 watt/MHz.

$$\alpha = \left(\frac{\hat{P}_T G_T L_T L'_P L'_c L'_s}{4\pi B_N (S/J)} \right)^{\frac{1}{2}} \tag{2-11a}$$

If the jammer is attempting to screen another target (stand-off jammer) at another range the equation is rewritten and sometimes called the "burn-through" range equation. Neglecting atmospheric losses, the *stand-off jammer equation* is

$$R_{BT} = (\alpha R_j)^{\frac{1}{2}} \left(\frac{\sigma_t}{P_J G_J} \right)^{\frac{1}{4}} \tag{2-12}$$

where R_{BT} is the detection range of the target which is not carrying the jammer. If the jammer is in the radar antenna receive sidelobes, as is the common case for a stand-off jammer, the range equation is written

$$R_{BT} = (\alpha R_j)^{\frac{1}{2}} \left(\frac{\sigma_t}{P_J G_J} \right)^{\frac{1}{4}} \left(\frac{G_R}{G'} \right)^{\frac{1}{4}} \tag{2-13}$$

where (G_R/G') is the ratio of the antenna gain in the direction of the target to the antenna gain in the direction of the jammer.

It can also be shown that the fractional range with barrage jamming is

$$\left(\frac{R_{BT}}{R}\right)^4 = \frac{4\pi KT_s R_J^2}{P_J G_J A_e L_R L_a' + 4\pi KT_s R_J^2} \tag{2-14}$$

or

$$\left(\frac{R_{BT}}{R}\right)^4 = \frac{(4\pi)^2 KT_s R_J^2}{\lambda^2 P_J G_J G_R L_R + (4\pi)^2 KT_s R_J^2} \tag{2-15}$$

If the jammer is in the radar sidelobes, $P_J G_J$ is reduced by the appropriate sidelobe ratio.

It is also useful to derive the volume search radar equation for the self-screening jammer environment in the same manner as Eq. (2-8). If a solid angle Ψ is believed to contain a jamming target and the search time is defined as T_{sc}, it is convenient to assume that

$$T_{sc} = \frac{\hat{P}_\tau \Psi}{\bar{P}\Omega}$$

$$\Omega = \frac{4\pi}{G_T}$$

$$\tau \approx \frac{1}{B_N}$$

Substituting for G_T in Eq. (2-11), the equation for the detection range of a jamming target located within a solid angle Ψ becomes the *volume search equation in jamming*

$$R_{SS}^2 = \frac{\bar{P}_T L_T L_P' L_a' L_c' L_s' T_{sc}}{\Psi(S/J)} \left(\frac{\sigma_t}{\bar{P}_J G_J}\right)$$

where $\bar{P}_J$ is the jammer power density in watts/Hz.

It should be noted that this equation is *independent of both transmit frequency and antenna size* with the assumption that the jammer has a constant power density.[1]

2.3 BEACON AND REPEATER EQUATIONS

The radar echoes from targets are often augmented as for the case of tracking satellites, for air traffic control, and for identification purposes. These augmentors may be passive reflectors, beacon transponders, or repeaters.

The common passive reflectors include Luneburg lenses with partial metallic coatings, corner reflectors, and sometimes directive antennas with appropriate termination. The beacon transponder generates a return signal on the appropriate frequency when it receives an adequate illumination signal from the radar. Since only one-way transmission losses need be considered if the beacon power is high enough to be detected by the radar, the simple *radar beacon illumination* equation is

$$R_B{}^2 = \frac{\hat{P}_T G_T L_T A_B L'_a L'_P}{4\pi S_B} \tag{2-16}$$

where R_B = radar-to-beacon detection range

A_B = receive aperture area of the beacon antenna including losses

L'_a = one-way atmospheric loss

L'_P = antenna pattern losses one way

S_B = minimum detectable signal to beacon

The return signal from the beacon can be calculated by inserting the appropriate parameters in the same equation.

A more complicated equation is needed for the case of the repeater, which amplifies the radar signals and retransmits them to the radar. These repeaters may be used for enhancing the signal at the radar or may be used by an enemy as a jamming technique. Two calculations must be made to cover the two cases when

1. The maximum power output of the repeater is insufficient to be detected by the radar.

---

[1]The same result with different notation was obtained by Barton [21].

2. The noise level of the repeater itself is sufficiently high compared to the signal it receives from the radar [Eq. (2-16)] that the noise output of the repeater is essentially jamming the radar.

If neither of these limitations is reached, the search radar equation [Eq. (2-6)] can be written

$$R^4 = \frac{\hat{P}_T G_T L_T A_e L_R L_P L_a L_c L_s A_B G_B G_E}{(4\pi)^2 KT_s B_N (S/N)} \tag{2-17}$$

if $P_{RB} G_E \leq \hat{P}_{max}$ of the repeater and $P_{RB}/S_B \gg S/N$ in the appropriate bandwidth which depends on the type of processing and

where G_B = retransmit gain including losses of the beacon or repeater antenna[1]

G_E = electronic gain of the repeater including losses

P_{RB} = receive power at the beacon calculated by solving Eq. (2-16) with P_{RB} substituted for S_B

$\hat{P}_{max}$ = maximum peak power output of the repeater transmitter

The detection range for the power-output-limited case of the repeater can be obtained by using the one-way range equation [Eq. (2-16)] and by substituting $\hat{P}_{max}$ for $\hat{P}_T$, G_B for $G_T L_T$, and S_{min} (for the radar) for S_B.

The equations for the repeater-noise-limited case can be solved from the appropriate jamming equations by replacing $KT_s G_E G_B$ with $P_J G_J$ with selection of the proper receiver bandwidth and by replacing σ_t in the same manner as was used in obtaining Eq. (2-17).

2.4 BISTATIC RADAR

While the majority of this text relates mainly to monostatic radars (common location of transmit and receive antennas), it has become increasingly important to study the performance of bistatic radars where there is considerable separation of transmit and receive antennas. The most common application of this technique is the semiactive *homing* missile system where the transmitter and its antenna are located on the ground or in an airplane and the receiver or *seeker* is located in a missile.

---

[1] The total gain for small signals is $G_B G_E$.

This generally allows a larger transmitter and transmit antenna to be used while keeping the size and weight of the receiver system to a minimum. The basic radar detection range for this case is obtained from Eq. (2-5) of Sec. 2.1 and is the *bistatic range equation*

$$(R_T R_R)^2 = \frac{\hat{P}_T G_T L_T G_R L_R \lambda^2 L_P L_{aT} L_{aR} L_c \, \sigma_{tb}}{(4\pi)^3 KT_s B_N S/N} \qquad (2\text{-}18)$$

where $(R_T R_R)$ = *range product* of the system

$\qquad R_T$ = range from the transmitter to the target

$\qquad R_R$ = range from the target to the receiver

$\qquad L_{aT}$ = propagation loss from the transmitter to the target

$\qquad L_{aR}$ = propagation loss from the target to the receiver

$\qquad \sigma_{tb}$ = the bistatic radar cross section of the target

The range computation is quite similar to that for the monostatic radar with the exception of the propagation loss terms and the target cross-section definition. Since in general the propagation paths will not be the same and since the receiver may be at high altitude, both the atmospheric and the propagation losses will be considerably reduced. The bistatic cross section is defined in a similar manner to the monostatic cross section. They will usually have a comparable *average* value with respect to all aspect angles of the target if the bistatic angle[1] is small with respect to 180° and the wavelength is small compared to the dimensions of the target. For a discussion of bistatic cross sections of targets and the unique case of forward scatter (where the bistatic angle $\approx$ 180°) see Skolnik [354] and Chap. 1.

When a self-screening jammer is used, the self-screening range is independent of R_R if proper adjustment is made for the gain of the jammer in the direction of the receiver. Then from Eq. (2-11)[2] the *bistatic self-screening range* is

$$R_{SS}^2 = R_T^2 = \frac{\hat{P}_T G_T L_T L_P' L_c' L_{aT}}{4\pi B_N (S/J)} \left(\frac{\sigma_{tb}}{P_J G_J'} \right) \qquad (2\text{-}19)$$

For general reference see Skolnik [354, pp. 585-594].

[1] The angle between the transmitter and receiver as seen from the target.

[2] Jamming noise power density is assumed to be $\gg KT_s$.

where G'_J is the jammer gain in the direction of the receiver. If the jammer is at a different location than the target, the burn-through range can be written as the *bistatic burn-through range*

$$R_{BT}{}^2 = \left(\frac{\hat{P}_T G_T L_T L'_P L'_c}{4\pi B_N (S/J)} \right) \left(\frac{\sigma_{tb}}{P_J G_J} \right) \left(\frac{L_{aT}}{L_{JR}} \right) \left(\frac{R_{JR}}{R_{TR}} \right)^2 \qquad (2\text{-}20)$$

where R_{BT} = radar transmitter to target range at which the target will be detected with the probabilities associated with (S/J) neglecting receiver noise

L_{JR} = fractional propagation loss between the jammer and the receiver

R_{JR} = range between the jammer and the receiver

R_{TR} = range between the target and the receiver

Substituting α^2 for the first bracket as per Eq. (2-11a)

$$R_{BT} = \alpha \left(\frac{\sigma_{tb}}{P_J G_J} \right)^{\frac{1}{2}} \left(\frac{L_{aT}}{L_{JR}} \right)^{\frac{1}{2}} \left(\frac{R_{JR}}{R_{TR}} \right) \qquad (2\text{-}21)$$

In many cases the ratio R_{JR}/R_{TR} is not known, but it can often be assumed that the four locations are in a straight line as shown in Fig. 2-4. In this case

$$R_{BT} = \frac{\alpha (R_{TR} + R_J)(L_{aT}/L_{JR})^{\frac{1}{2}}}{\alpha (L_{aT}/L_{JR})^{\frac{1}{2}} + R_{TR}(P_J G_J/\sigma_{tb})^{\frac{1}{2}}} \qquad (2\text{-}22)$$

where R_J is the range from transmitter to jammer.

In many cases the factor L_{aT}/L_{JR} can be assumed to be unity because of the short ranges involved and their tendency to cancel each other for typical geometries.

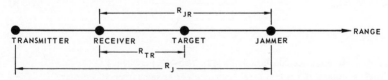

figure 2-4 *Stand-off jammer diagram.*

Again it must be emphasized that if $P_J G_J/\sigma_{tb}$ is expressed in the conventional watts/(MHz)(m^2) the term α should be calculated with B_N in MHz.

2.5 RADAR DETECTION EQUATIONS IN DISTRIBUTED CLUTTER (VOLUME REFLECTORS) FOR PULSE RADARS

If a radar target is situated in a region of space that contains a large number of undesired radar scatterers, such as raindrops, snowflakes, or chaff dipoles, the summation of the reflected radar signals from these objects may well determine the minimum detectable target signal of the radar. Signals reflected from these scatterers form part of the *environmental noise* and are commonly called radar *clutter*. These objects, however, may be the targets of interest, as for example in a weather radar. This section defines the equations that can be used to determine the received power from clutter and the range at which a target will be detected in the presence of clutter signals. Chapter 6 contains the scattering coefficients and other statistics that are required to solve the radar detection equations and describe the spectra of the clutter.

The power density that is incident on a volume of scatterers at a range R (see Fig. 1-1) is

$$P_\alpha = \frac{\hat{P}_T G_T |f(\theta, \phi)|^2 L_T L_a'}{4\pi R^2} \tag{2-23}$$

where P_α = power density at the clutter cell
$G_T |f(\theta, \phi)|^2$ = transmit gain of the antenna with respect to an omnidirectional radiator including the effects of the decreases[1] in gain at angles θ and ϕ from the beam centerline, where

$$G_T = 4\pi \frac{A_e}{\lambda^2}$$

L_a' = the one-way attenuation loss in the transmission medium

Then the reflected power from an elemental volume of scatterers in the far zone of the radar aperture is

---

[1] This factor constitutes part of the pattern loss for targets L_p described earlier.

$$\frac{\hat{P}_T G_T \, |f(\theta,\phi)|^2 \, L_T \, L'_a \Sigma\sigma \, dV}{4\pi R^2}$$

where $\Sigma\sigma$ is the backscattering cross-section coefficient per unit volume (preferably in m^2/m^3) assuming that the scatterers fill the volume dV.

Similarly the echo received by the receive antenna of effective aperture $A_e \, |f(\theta,\phi)|^2$ can be expressed as

$$dP_C \;=\; \frac{\hat{P}_T G_T \, |f(\theta,\phi)|^4 \, L_T \, L_a \, A_e \, L_R \, \Sigma\sigma \, dV}{(4\pi)^2 \, R^4} \tag{2-24}$$

This equation includes the assumptions that there are no *ambiguous* or *second-time*-around *clutter* signals and that the transmitter and receiver are at the same location. For Eq. (2-24)

$$P_C = \text{returned power from the clutter}$$
$$A_e \, |f(\theta,\phi)|^2 = \text{effective aperture as a function of angle from the clutter}$$
$$\text{cell to the radar beam centerline}$$
$$dV = \text{volume element of the clutter cell}$$

Note that for pulse lengths short with respect to R, $dV = R^2(c\tau/2)\,d\Omega$, $c\tau/2$ is the pulse length or cell length in distance units, and $d\Omega$ is the elemental solid angle of the beam.

To obtain the total clutter power for long pulse or continuous wave systems, an integration of Eq. (2-24) should be performed over range. In the chapter on CW radars this will be carried out, but for the present it will be assumed that $c\tau/2 \ll R$.

To eliminate the antenna pattern shape $|f(\theta,\phi)|^4$ term and integrate over the solid angle Ω, Probert-Jones (see Sec. 11.7) [296] has shown that for a gaussian beam pattern and a circular antenna

$$\int_{\Omega} |f(\theta,\phi)|^4 \, d\Omega \;=\; \frac{\pi\,\theta_1\,\phi_1}{8\,\ln 2} \tag{2-25}$$

where θ_1 and ϕ_1 are the one-way, 3-dB beamwidths of the beam pattern. Almost identical results are given in Chap. 10 for a uniformly illuminated circular aperture including the clutter power from the first few sidelobes.

The factor for rectangular apertures should be somewhat smaller than this value when using the usual definitions for beamwidth. Substituting in Eq. (2-24) yields the *clutter return power for single aperture radar*

$$P_C = \left[\frac{\hat{P}_T G_T L_T A_e L_R L_a}{(4\pi)^2 R^2}\right]\left[\frac{\Sigma\sigma(c\tau/2)\pi\theta_1\phi_1}{8\ln 2}\right] \tag{2-26}$$

or by substituting for A_e

$$P_C = \left[\frac{\hat{P}_T G_T L_T G_R L_R \lambda^2 L_a}{(4\pi)^3 R^2}\right]\left[\frac{\Sigma\sigma(c\tau/2)\pi\theta_1\phi_1}{8\ln 2}\right] \tag{2-27}$$

A note of caution: The above equations are only valid for equal transmit and receive beamwidths and antennas with a high ratio of mainbeam to sidelobe gain. If the antenna beamwidth or the gains of the antennas are not known, a good approximation can be obtained by substituting

$$G = \frac{\pi^2}{\theta_1\phi_1}$$

The above equations are adequate for measuring the rain, snow, or chaff backscattered power using the values of $\Sigma\sigma$ from Chap. 6. In most cases these reflectors are *clutter* and act as noise with a similar amplitude distribution to receiver noise. If the backscatter noise P_C is small compared to the system noise $KT_S B_N$, the effect on detection range is negligible. In the usual case of interest $P_C \gg KT_S B_N$. Then the received power from the target can be expressed from Eq. (2-1) of Sec. 2.1 as

$$P_R = \frac{\hat{P}_T G_T L_T G_R L_R \lambda^2 L_a}{(4\pi)^3 R^2}\left(\frac{\sigma_t L_P L_c}{R^2}\right) \tag{2-28}$$

where L_c is the collapsing loss introduced since distributed clutter is often folded onto a two-dimensional display from radar beams other than the one of interest.

To detect a target in a volume of uniformly distributed clutter, the ratio of the target power P_R to that of the clutter return P_C must exceed a value (S/C) in a similar manner to the requirements on signal-to-noise ratio (S/N) or the signal-to-jam ratio (S/J). Then from Eqs. (2-27) and (2-28)

$$\frac{S}{C} = \frac{P_R L_s'}{P_C} = \frac{L_P L_c L_s' \sigma_t (8 \ln 2)}{\pi R^2 \Sigma \sigma (c\tau/2) \theta_1 \phi_1} \tag{2-28a}$$

where L_s' is the signal processing loss for clutter greater than receiver noise. Solving for range gives the *detection range in clutter noise for pulse radars*

$$R^2 = \frac{L_P L_c L_s' \sigma_t \, 2\ln 2}{(\pi/4)\, \theta_1 \phi_1 \,(c\tau/2)\,(S/C)\,\Sigma\sigma} \quad \text{if} \quad P_C \gg KT_s B_N \tag{2-29}$$

The term $(\pi/4)\,\theta_1 \phi_1$ is the usual 3-dB beamwidth definition of the cross section of a circular radar beam and $c\tau/2$ is the pulse length. The remaining undefined term $(2 \ln 2)$[1] is effectively the inverse of the beam pattern loss for the clutter. The fractional pattern loss for the target was defined as L_P, but its numerical value is a complex function of the beam shape, scanning pattern, and the number of pulses in a search or acquisition radar. While the pattern losses L_P for the target will often exceed the factor of $(2 \ln 2)$ for rain in a pencil-beam search radar, a useful approximation for a pulse radar is the *detection range in rain, snow, or chaff*

$$R^2 \doteq \frac{\sigma_t L_c L_s'}{(\pi/4)\, \theta_1 \phi_1 \,(c\tau/2)\,(S/C)\,\Sigma\sigma} \tag{2-30}$$

where σ_t = the radar cross section, m^2
    $c\tau/2$ = the pulse length, m
    $\Sigma\sigma$ = backscatter coefficient of the scatterers, m^2/m^3
    $\theta_1 \phi_1$ = the 3-dB beamwidths in radians for a circular aperture[2]

[1] Numerically equal to 1.39 or 1.4 dB.
[2] A similar approximation for the case where the transmit and receive beamwidths are different is to use the smaller set of the beamwidths assuming that the target is in the smaller beam.

R = the detection range of the target located in a volume filled with clutter, m

(S/C) = the ratio of target echo power to the power received from the clutter cell for detection

L'_s = the signal processing loss based on the spectrum and distribution function of the clutter returns

By substituting $G/(\pi^2)$ for $1/(\theta_1 \phi_1)$

$$R^2 = \frac{4 G \sigma_t L_c L'_s}{\pi^3 (c\tau/2)(S/C)\Sigma\sigma} \tag{2-31}$$

The detection range in the case where the clutter return does not substantially exceed the noise can be found by adding P_C to $KT_s B_N$; substituting this sum in Eq. (2-5), Sec. 2.1; and solving for R.

The single pulse (S/C) required for detection in Rayleigh distributed clutter for matched-filter receivers[1] is usually assumed to be equal to the single pulse (S/N) with respect to detection in noise. This is not generally the case with integration of multiple pulses since the clutter returns are partially correlated from pulse to pulse.

Another problem in defining (S/C), which will be expanded on in Chap. 4 on automatic detection, is caused by the effective clutter bandwidth not always being equal to the noise bandwidth of the receiver. The determination of the target detection statistics from the signal-to-clutter ratio (S/C) and the processing loss L'_s will vary with the transmitted waveform and the implementation of the receiver.

2.6 PULSE RADAR DETECTION EQUATIONS FOR AREA CLUTTER

The development of the range equation for targets in area clutter is similar to that for volume clutter; however, in this case the integral involving the beam pattern shape is not readily evaluated. Therefore, the approach used in this section is somewhat simplified. This development also neglects effects of sidelobe clutter and is valid only in the far field region. Furthermore, because of the simplifications used, the results must be restricted to a pulse radar system.

[1] This is the most common distribution for rain, chaff, or snow.

The power density at a clutter element is given by Eq. (2-32). (Since the pattern shape factor is not used in this development, a closer approximation is the use of the two-way, rather than the one-way, beamwidth.)

$$P_c = \frac{\hat{P}_T G_T L_T L_a'}{4\pi R^2} \tag{2-32}$$

The reflected clutter power is[1]

$$P_{\text{refl}} = \frac{\hat{P}_T G_T L_T L_a' [\sigma_0 A_c]}{4\pi R^2} \tag{2-33}$$

where σ_0 = the backscatter cross section per unit area (m^2/m^2) of a reflecting surface; the mean value unless stated otherwise

A_c = area of the clutter cell

The clutter power at the receiver preamplifier is

$$P_c = \frac{\hat{P}_T G_T L_T L_a L_R A_e}{(4\pi)^2 R^4} (\sigma_0 A_c) \tag{2-34}$$

In considering the clutter area A_c, two cases are of interest depending on whether the pulse length is large or small compared with the projected length in the radial direction of the beam. These two cases lead to different results. In either case the formulation is based on a flat earth approximation, in which case grazing and depression angles are identical. For greater accuracy, the 4/3 earth approximation may be used (Chap. 1).

As suggested by Fig. 2-5, the intercepted clutter area may be approximated by the *clutter cell limited by beamwidths*

$$A_c \cong \pi R^2 \tan \frac{\theta_2}{2} \tan \frac{\phi_2}{2} \csc \Psi \quad \left(\tan \Psi > \frac{\tan \phi_2 R}{c\tau/2} \right) \tag{2-35}$$

or by the *clutter cell limited by pulse length*

$$A_c \cong 2R \; \frac{c\tau}{2} \tan \frac{\theta_2}{2} \sec \Psi \quad \left(\tan \Psi < \frac{\tan \phi_2 R}{c\tau/2} \right) \tag{2-36}$$

[1]The complete equation, including the beam shape factor and antenna sidelobes, is derived in the chapter on CW systems and in [166].

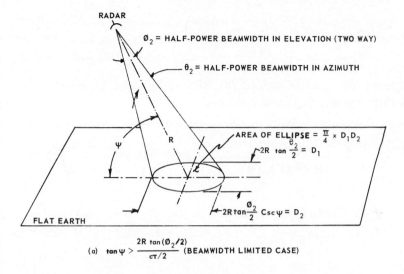

RADAR

$\emptyset_2$ = HALF-POWER BEAMWIDTH IN ELEVATION (TWO WAY)

θ_2 = HALF-POWER BEAMWIDTH IN AZIMUTH

AREA OF ELLIPSE = $\frac{\pi}{4}$ × $D_1 D_2$

$2R \ \tan \frac{\theta_2}{2} = D_1$

$2R \tan \frac{\emptyset_2}{2} \ \text{Csc} \ \psi = D_2$

FLAT EARTH

(a) $\tan \psi > \dfrac{2R \ \tan(\emptyset_2/2)}{c\tau/2}$ (BEAMWIDTH LIMITED CASE)

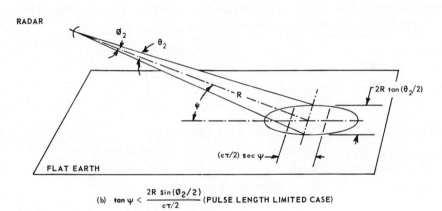

RADAR

$\emptyset_2$

θ_2

$2R \ \tan (\theta_2/2)$

R

$(c\tau/2) \ \text{sec} \ \psi$

FLAT EARTH

(b) $\tan \psi < \dfrac{2R \ \sin(\emptyset_2/2)}{c\tau/2}$ (PULSE LENGTH LIMITED CASE)

figure 2-5 *Area illumination for pulse radar.*

The choice of either Eq. (2-35) or (2-36) depends on the grazing angle Ψ. For small two-way beamwidths θ_2 and ϕ_2 (less than $\approx 10°$) A_c is defined by

$$A_c \cong \frac{\pi R^2}{4} \theta_2 \phi_2 \ \csc \Psi \quad \left(\tan \Psi > \frac{\phi_2 R}{c\tau/2} \right) \quad\quad (2\text{-}37)$$

$$A_c \cong R\,\theta_2 \frac{c\tau}{2} \; \sec \Psi \; , \; \left(\tan \Psi < \frac{\phi_2 R}{c\tau/2} \right) \tag{2-38}$$

Combining Eqs. (2-37) and (2-38) with (2-34), the expression for receiver clutter power in a pulsed radar is

$$P_c = \frac{\hat{P}_T G_T L_T L_a L_R A_e (\pi/4)\,\theta_2 \phi_2 \sigma_0}{(4\pi)^2 R^2 \sin \Psi} \quad \text{for} \quad \left(\tan \Psi > \frac{\phi_2 R}{c\tau/2} \right) \tag{2-39}$$

$$P_c = \frac{\hat{P}_T G_T L_T L_a L_R A_e (c\tau/2)\,\theta_2 \sigma_0}{(4\pi)^2 R^3 \cos \Psi} \quad \text{for} \quad \left(\tan \Psi < \frac{\phi_2 R}{c\tau/2} \right) \tag{2-40}$$

Using a procedure similar to that used in the previous section, the ratio of the equations for P_R and P_c [Eq. (2-39)] results in the following range equation for area clutter-limited cases. This equation applies only when $P_c \gg KT_s B_N$, and the target is in the clutter. The *pulse radar range equation for area clutter* is

$$R^2 = \frac{L'_s L_P L_c (\sin \Psi)\,\sigma_t}{(\pi/4)\,\theta_2 \phi_2 (S/C)\,\sigma_0} \quad \text{for} \quad \left(\tan \Psi > \frac{\phi_2 R}{c\tau/2} \right) \tag{2-41}$$

where L_P = beam pattern losses for the target echo
   L'_s = signal processing loss in a clutter-limited environment
For a circular aperture[1] $G \cong \pi^2/\theta_2 \phi_2\, 2\ln 2$ and using the approximation[2] as explained in the previous section, we have

$$R^2 \cong \frac{4G L'_s L_c (\sin \Psi)\,\sigma_t}{\pi^3 (S/C)\,\sigma_0} \tag{2-42}$$

Another form may be obtained by substituting $\sin \Psi = h/R$, where h is

---

[1] For different transmit and receive antennas the gain corresponding to the smaller beamwidth should be used.
[2] $L_P \approx 1/2 \ln 2$.

the antenna height, and assuming a flat earth

$$R^3 = \frac{4G L_s' L_c h \sigma_t}{\pi^3 (S/C) \sigma_0}$$

(2-43)

The equation for the case where $(\tan \Psi < \theta_2 R/c\tau/2)$ is obtained by taking the ratio of P_R to P_c [Eq. (2-40)] and solving for R. The *pulse radar range equation for area clutter* is

$$R = \frac{L_s' L_P L_c \cos \Psi \sigma_t}{(S/C)(c\tau/2) \theta_2 \sigma_0} \quad \text{for} \quad \left(\tan \Psi < \frac{\phi_2 R}{c\tau/2} \right)$$

(2-44)

As stated previously, the symbols are defined as

G_T = transmit gain of antenna
L_s' = signal processing loss in the clutter-limited case
L_P = beam shape and pattern losses for the target echo
L_c = collapsing loss of processor or display
Ψ = grazing angle of radiation on clutter cell
(S/C) = signal-to-clutter power ratio required for detection
σ_0 = reflectivity per unit cross section of clutter (dimensionless)
θ_2, ϕ_2 = 3-dB beamwidth in azimuth and elevation, respectively (two-way path). In general, $\theta_2 \approx 0.71\theta_1, \phi_2 \approx 0.71\phi_1$.

3 STATISTICAL RELATIONSHIPS FOR VARIOUS DETECTION PROCESSES

3.1 INTRODUCTION AND DEFINITIONS

As alluded to in the previous chapter, the detection of signals in noise or interference is best described in statistical terms although the use of gaussian wide-band noise is not always sufficient to describe performance in a given environment. The first part of this chapter will describe the statistics for detection of steady targets in gaussian and white[1] noise. The discussion will then be extended to target signals that have a fluctuating amplitude with various correlation times. This case is essential to prediction of the performance where the target's returns are immersed in clutter signals. This is followed by a summary of radar

---

For general references see Marcum and Swerling [244], Siebert [345], Fehlner [113], Wainstein and Zubakov [393], Barton [20], Nolan et al. [268], DiFranco and Rubin [96].

[1] See Berkowitz [38, chaps. 2, 3, and 4 by W. R. Bennett].

displays and human operator performance. The techniques of detection by sequential, adaptive, or automatic means are covered in Chap. 4.

There is a multitude of excellent texts and papers on the statistics of radar detection [269, 244, 44, 96]. The purpose of this chapter is to summarize the useful relationships on this topic in engineering terms rather than to delve into the theory. Before proceeding, it is essential to define the terminology to be used here.

1. Signal-to-noise ratio S/N: The ratio of the signal power (when the signal is present) to the average noise power after the matched filter but before detection or postdetection integration.

2. Radar *cell*: A term used to describe a *volume* from which return signals are received. The most common usage is for a pulse radar where the *cell* volume is defined by the pulse length and the half-power beamwidth of the antennas. A *clutter-cell* is a volume of this size containing undesired scatterers. The *target cell* is the same volume containing a target even though it may occupy only a portion of the cell. The number of cells of a simple pulse radar in a given time such as the scan time of the radar, neglecting doppler cells, can be written

$$\begin{pmatrix} \text{Number of} \\ \text{cells in} \\ \text{one scan} \end{pmatrix} = \frac{(\text{PRF})(\text{scan time})(\text{display time/sweep})}{\text{pulse duration}}$$

The PRF is the pulse repetition frequency. It is assumed that the display time per beam is less than the interpulse period.

3. Pulse integration: The summation of target return signals (from the same location) and noise from two or more successive transmit pulses. The purpose is usually to improve the probability of detection or to permit more accurate parameter estimation.

4. Coherent: (*a*) Two or more radar signals are said to be coherent when the relative phases of the signals have a known relationship[1] even though they may be considerably separated in time. (*b*) A term applied to an oscillator that is sufficiently stable to predict the phase at a given time later than the start of the transmission. (*c*) Signals on two or more frequencies may be called coherent if they were derived[2] from the same basic frequency which is stable as specified in (*b*).

---

[1] Phase shifts due to unknown target velocity or acceleration do not alter this definition.

[2] By frequency multiplication or mixing.

5. Coherent integration (predetection integration): The addition of coherent radar signals prior to envelope detection where the signals are summed vectorially (voltagewise). For perfect integration of constant amplitude coherent signals the S/N increases linearly as the number of these signals assuming that the noise is uncorrelated. A predetection matched filter can be considered as a coherent integrator for complex signals. For example, a narrow-band IF filter is a coherent integrator for a single frequency signal. Coherent signals on different frequencies can be added coherently with certain restrictions on the times that this can occur (see Chap. 13) with respect to the relative phases.

6. Correlated: While coherency implies a complete predictability of relative phase (and sometimes amplitude) between a set of radar signals, the term correlated is used in a number of cases where the second, third, fourth, etc., signal of a set is in some way related in phase, amplitude, or power to the first of the set or to each other. The terminology derives from the autocorrelation and cross-correlation functions of random signals.

7. Correlation functions and related distributions: (a) Temporal—describes to what degree the value of a time function $f(t)$ at one time is correlated with another value τ time units later. Usually, unless otherwise specified, the term *autocorrelation function* refers to the temporal function. For a random function $f(t)$

$$R(\tau) = \lim_{T \to \infty} \frac{1}{2T} \int_{-T}^{T} f(t) f(t + \tau) \, dt$$

In practice the finite limits are taken long enough so that a sufficient degree of accuracy is obtained. As stated by the Wiener-Kintchine theorem, the autocorrelation function is simply the Fourier transform of the power spectral density function. Other properties are

$$\rho(\tau) = \frac{R(\tau)}{R(0)}, \text{ the normalized autocorrelation function}$$

$$R(\tau) \leq R(0)$$

$$R(\tau) = R(-\tau)$$

(b) Frequency—describes how well a signal corresponding to a certain transmitted frequency f_0 is correlated[1] with another signal whose

---

[1] Since the correlation in phase of these signals is dependent on the coherency and time delays involved, *correlated* radar returns will generally refer to the power of the returns.

transmitted frequency is different by Δf units. (The two signals are assumed to occur close enough together in time and space so that they are not decorrelated because of the time interval or spatial separation.) The mathematical form is identical to that for the temporal function except that the temporal variable t is replaced by a frequency variable f and the frequency separation Δf replaces τ. (c) Spatial—describes how well the signal reflected from one cell is correlated with another signal reflected from a distant cell. (The two signals are assumed to occur close enough together in time and transmit frequency so that they are not decorrelated because of the time interval or frequency spacing.) The mathematical form is identical with that for the temporal function, except that the temporal variables t and τ are replaced by spatial variables.

8. Probability density function $p(x)$: The probability density function defines the probability of a random variable taking on a value within a small interval. The expression $p(x)\,dx$ gives the probability that a random variable X will lie in the interval $x \leq X \leq x + dx$. Gaussian distribution, often called the *normal distribution*, has the density function

$$ p(x) = \frac{1}{\sqrt{2\pi\,\sigma^2}} \exp\left[-\frac{(x - \bar{x})^2}{2\sigma^2} \right] $$

where $\bar{x}$ is the mean value. The variance (second central moment) is

$$ \sigma^2 = \overline{(x - \bar{x})^2} $$

3.2 TARGET DETECTION BY A PULSED RADAR

J. I. Marcum [244][1] treated the statistical problem of target detection by a pulse radar.[2] His papers have withstood the test of time since their original publication (1947-1948).

---

[1] J. I. Marcum and P. Swerling, Studies of Target Detection by Pulsed Radar, *IRE Trans.*, vol. IT-6, no. 2, April, 1960. (Also published as *Rand Res. Memo* RM 754 by J. I. Marcum, December 1, 1947; *Rand Res. Memo* RM 753 by J. I. Marcum, July 1, 1948; and *Rand Res. Memo* RM 1217 by P. Swerling, March 17, 1954.)

[2] Similar studies were made by Rice [311, 313], Blake [44], Kaplan [201], Hall [154], North [269], and others and with proper interpretation (see Barton [22]) are in close agreement.

Swerling [367] extended Marcum's results to the case of a radar target with an echo of fluctuating strength. Swerling's analytic results are used in this section to extend the numerical data beyond that given in [244].

The text and graphs of these papers, however, are presented in terms so highly mathematical that the results may not appear as useful as they really are. An extremely clear summary of the work of Marcum and the subsequent extensions by Swerling [367] was carried out by Fehlner [113]. The majority of this section is paraphrased from that report.

Marcum's contribution is the definition of the relationship between a threshold value of radar return signal-to-noise power S/N and of the probability that values in excess of the threshold will exist in the presence of both noise and echoes. These threshold crossings will be reported as target detections and will include both false and real reports of targets.

The problem starts with noise. Suppose that the voltage caused by noise alone varies with time as shown in Fig. 3-1. During the period shown, the noise exceeded threshold voltage level A seven times, level B five times, and level C only twice. Obviously, the higher the threshold, the longer will be the average time between occasions when noise alone exceeds the threshold. This time is of considerable concern because if it is too short the system will be too frequently faced with false alarms and if too long excessive radiated energy will be required to achieve reasonable probabilities of target detection. Marcum defined it as the time during which the probability is P_0 that there will not be a false alarm. For purposes of standardization, P_0 is taken to be 0.5. (Not all authors use this definition.)

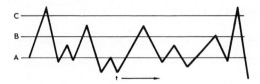

figure 3-1 *Threshold detection of noise.*

P_N is the probability that a false alarm is obtained each time there is an opportunity. The false alarm number n' is the number of independent opportunities for a false alarm in the false alarm time. Then

$$P_0 = (1 - P_N)^{n'} = 0.5 \text{ (Marcum, Eq. 10)}^* \tag{3-1}$$

where P_0 is the probability of no false alarm during the false alarm time, P_N is the probability of a false alarm on each opportunity, and n' is the false alarm number. For a pulse radar (one doppler channel)

$$n' \approx \frac{n}{mN} \approx \frac{T_{FA}}{\tau mN} \tag{3-2}$$

where n is the number of pulse lengths τ in the false alarm time T_{FA}, m is the number of pulses integrated coherently (≥ 1), and N is the number of pulses integrated incoherently (≥ 1).

The probability of a false alarm in the absence of echoes is given by Eq. (3-1). For $n' \gg 1$, a very close approximation is

$$P_N \approx \frac{1}{n'} \log_e \frac{1}{P_0} \text{ (Marcum, Eq. 21)} \tag{3-3}$$

The value of false alarm probability when $P_0 = 0.5$ is

$$P_N \approx \frac{0.69}{n'}$$

The probability P_N that noise alone will exceed a given bias level of voltage is a function of the bias level, the combined law of the detector and integrator, and the characteristics of the noise. The detector referred to here is the envelope detector, the output of which is a given function of the envelope of the carrier wave. The integrator characteristic also affects the statistical problem. The function of the signal voltage that is integrated is called the law of the integrator, e.g., the square of the pulse voltage might be integrated. So long as the same weight is applied to each of the N pulses the integrator is called linear.

The solutions for the bias level are obtained most easily for the *combined square law*, which is usually thought of as a square-law detector coupled with a linear integrator. The bias level for normal

---

*These references are to equations in *Rand Res. Memo* RM 753 from [244], but they are not necessarily quoted verbatim. For example, Marcum used n for false alarm number instead of n'.

distribution of noise voltage is related to P_0 by

$$P_0^{1/n'} = \int_0^{Y_b} \frac{Y^{N-1} e^{-Y}}{(N-1)!} \, dY \quad \text{(Marcum, Eq. 39)} \tag{3-4}$$

and

$$Y = \sum_1^N y \quad \text{(Marcum, Eq. 26)}$$

where Y_b is the bias level normalized to the root-mean-square noise and y is the output of the detector normalized to root-mean-square noise.

A functional block diagram of a pulse radar with automatic detection is shown in Fig. 3-2. The output of the matched filter is fed to the detector and also to a device that measures rms noise,[1] the value of which must not be unduly influenced by echoes. This value is multiplied by a factor Y_b to obtain the bias level. The integrated output of the detector is compared to the bias level in the threshold device. If the output is larger, an alarm is sounded indicating the probable presence of a target. Marcum's analysis applies to only that part of the functional block diagram downstream of the matched filter. If m pulses are integrated coherently, Marcum's data apply to $1/m$ of the pulses, each of which has m times the energy of a single pulse. This analysis applies to integrators that are dumped after the integration period.

The probability of detection, i.e., the probability of sounding the alarm in the presence of signal and noise, can be calculated from Fehlner's equation, an alternate form of Marcum's Eq. 49.

$$P_D(x, Y_b) = e^{-Nx} \sum_{k=0}^{\infty} \frac{(Nx)^K}{K!} \sum_{j=0}^{N-1+K} \frac{e^{-Y_b} Y_b^j}{j!} \tag{3-5}$$

where x is the signal-to-noise power ratio during the pulse (later $x = S/N$).

---

[1] If the root-mean-square noise level is changing as a function of time (because of jamming, etc.), a sufficient number of samples must be averaged before the bias level is determined. This will be covered in Chap. 4 on automatic detection. The range gate is not necessary for simple-pulse radars.

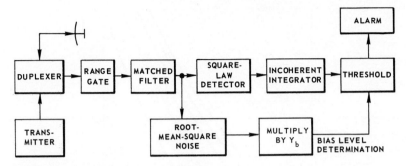

figure 3-2 *Pulse integrator block diagram.*

Since the relationships have been established among the bias level Y_b the false alarm number n' and the number of pulses integrated N by Eq. (3-4), the probability of detection can now be related to n' and N through Eq. (3-5). These relationships are shown in Figs. 3-3 through 3-6, which also show fluctuating target detection curves. The single-pulse case was shown as Fig. 2-2.

3.3 ADDITIONAL RESULTS OF THE MARCUM AND SWERLING ANALYSIS

Marcum made a limited number of calculations for a linear detector and integrator. A comparison with the square-law results shows very small differences. For example, at a detection probability of 0.5, the required signal-to-noise ratio is the same for both laws for $N = 1$ and $N = 70$, where N is the number of pulses integrated. For $N = 10$, the square law requires about 2.5 percent more power ratio than the linear law. For N very large the linear law approaches a requirement of about 4.5 percent more signal-to-noise ratio than the square law. These results are ample justification for preferring the more easily obtained square-law data. Graphs of the linear-rectifier detector *detectability factor* with an explanation of the relationships to (S/N) are given in Blake [44, pp. 8-14].

Another method of detection is based on the probability of signal plus noise exceeding noise alone over a period of time. This time is taken to be the false alarm time. Marcum points out that this criterion is not useful for search radars because of the difficulty of picking out the largest signal over a period of time and because of the difficulty of what to do when many signals are present; however, this criterion is useful for

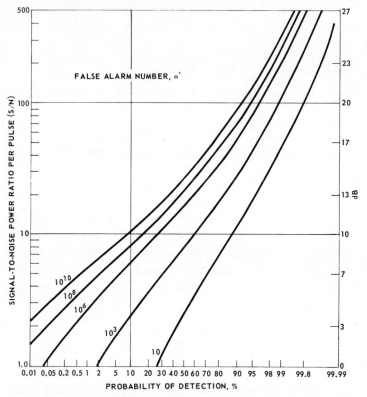

figure 3-3 *Signal-to-noise ratio vs. P_D for a fluctuating target—cases 3 and 4 (N = 1). (After Fehlner [113].)*

determining the probability that an automatic tracking beam will lose its target because of noise or with digital integration.

Marcum proves rigorously that there is a best possible detector law for each signal strength (Marcum, Eq. 217). For small signals the best possible law closely approximates the square law. For large signals, it is the linear law. The numerical results for these extreme cases, however, are not very different, so that if faced with a choice, there is not much reason to prefer one to the other on the basis of detection probability. (See DiFranco and Rubin [96, p. 370].) The linear law, however, has one practical advantage in that a linear detector would be less subject to saturation by large signals.

Swerling extended Marcum's square-law results to four different cases in which targets return echoes of fluctuating strength.[1] Cases 1 and 2

---

[1] A detailed discussion of these cases is given in Chap. 5 on targets.

apply to targets that can be represented as a number of independently fluctuating reflectors of about equal echoing area. The density function is the *chi-square* distribution with two degrees of freedom.

$$w(x, \bar{x}) = \frac{1}{\bar{x}} \exp\left(-\frac{x}{\bar{x}}\right) \qquad x \geq 0 \qquad (3\text{-}6)$$

where $\bar{x}$ is the average signal-to-noise ratio over all target fluctuations. (See Chap. 5.)

Cases 3 and 4 were derived from the density function of Eq. (3-7). It was said to apply to targets that can be represented as one large reflector

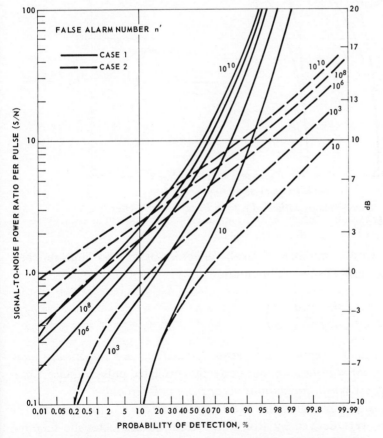

figure 3-4 *Signal-to-noise ratio vs. P_D for a fluctuating target (N = 6). (After Fehlner [113].)*

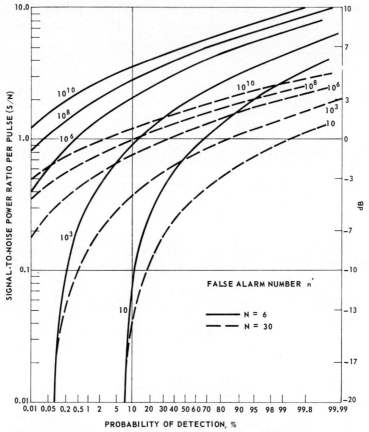

figure 3-5 *Signal-to-noise ratio vs. P_D for a steady target. (After Fehlner [113].)*

together with a number of small reflectors or as one large reflector subject to small changes in orientation.

$$w(x, \overline{x}) = \frac{4x}{\overline{x}^2} \exp\left(-\frac{2x}{\overline{x}}\right) \qquad x \geq 0 \qquad (3\text{-}7)$$

Cases 1 and 3 apply when echo fluctuations are statistically independent from scan to scan, but perfectly correlated pulse to pulse. *Cases 2 and 4 apply when fluctuations occur from pulse to pulse.*

Swerling did not give the exact expression for P_N for Case 4. The equations were derived by Roll [113] from the characteristic function for Case 4 given in [244].

The relationships among P_D, $\bar{x}$, N, and n' are shown in Figs. 3-3 through 3-6. A far more extensive set is available in Fehlner [113], and numerical results are tabulated in the supplement to that report. Similar results are shown in [268] with a slightly different method of computation.

Detection probability P_D is related to the bias level Y_b through integrals, the values of which are very sensitive to small changes in the limits of integration, especially for large N. Accordingly, the bias level Y_b was computed from Eq. (3-4) in double precision on a computer. Various values of n' between 10 and 10^{10} were used. Fehlner's analysis showed that the accuracy of Marcum's data for nonfluctuating targets is at least as good as the accuracy of reading Marcum's graphs. In practice,

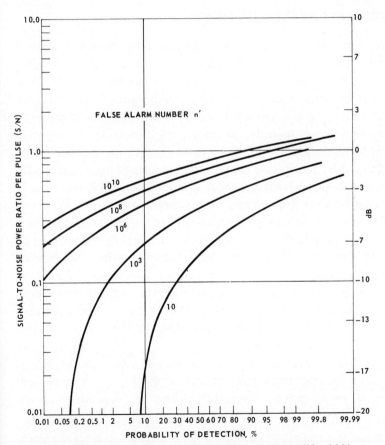

figure 3-6 *Signal-to-noise ratio vs. P_D for a steady target (N = 100).*

the false alarm rate is established, in the absence of a target, by a counter at the output of the threshold circuit of Fig. 3-2.

The preceding discussion has assumed that the integrator was linear with analog inputs. Two other integrator types are

1. A delay line integrator with a feedback gain of less than unity, which permits the older information to decay as new information is added.

2. A digital integrator where the input amplitude is quantized into one or more levels with the integration performed in a digital memory.

It can be shown that the additional loss introduced by these techniques is small while a considerable saving in hardware is obtained. (See Chap. 14.)

3.4 NONCOHERENT INTEGRATION LOSSES

The graphs shown in Figs. 3-3 to 3-6 can be used to obtain the required signal-to-noise ratios per pulse for a given detection criterion. It is often more interesting to know the integration loss or the difference between perfect coherent integration and incoherent techniques. Before this loss can be determined, it is necessary to define the false alarm statistics in more consistent terms in order to resolve differences in the reported curves of integration loss. The primary reason for the discrepancy is that after integration of many noise samples, the number of opportunities for a false alarm is reduced by the number of integrations. If it is desired to keep the time between false alarms T_{fa} constant, the threshold or bias level can be lowered, and a higher probability of a single sample of noise exceeding this threshold P_N can be accepted. Barton [22] has shown that this accounts for most of the differences between earlier publications. The graphs published in this section are for a constant false alarm probability P_N and for the false alarm time increasing with the number of integrations. This allows the estimation of the number of false alarms per beam position to be computed from a knowledge of the number of noise samples in the instrumented range.

It has also been shown [22] that the fluctuation losses[1] of the various Swerling cases are essentially independent of the loss from incoherent integration of many pulses. Since these losses are dependent on the

---

[1] The increase in target echo to achieve the same detection statistics as a steady target of the same radar cross section.

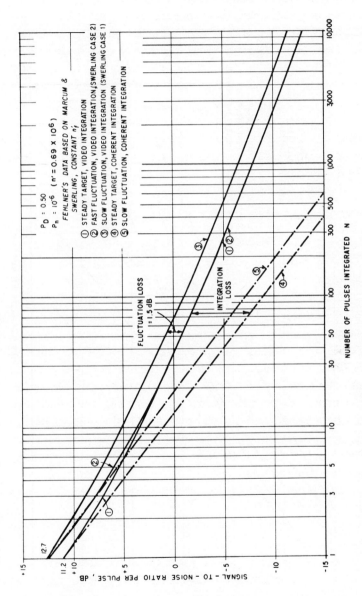

figure 3-7 *Detectability factor (S/N) vs. number of pulses ($P_D = 0.5$).*

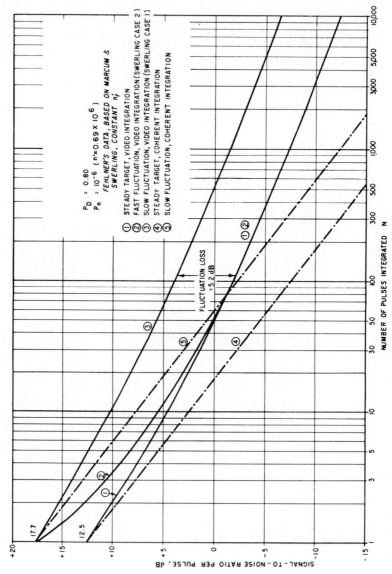

figure 3-8 *Detectability factor (S/N) vs. number of pulses ($P_D = 0.8$).*

$P_D = 0.80$
$P_n = 10^{-6}$ ($n = 0.69 \times 10^6$)

FEHLNER'S DATA, BASED ON MARCUM &
SWERLING, CONSTANT n'_f

① STEADY TARGET, VIDEO INTEGRATION
② FAST FLUCTUATION, VIDEO INTEGRATION (SWERLING CASE 2)
③ SLOW FLUCTUATION, VIDEO INTEGRATION (SWERLING CASE 1)
④ STEADY TARGET, COHERENT INTEGRATION
⑤ SLOW FLUCTUATION, COHERENT INTEGRATION

FLUCTUATION LOSS
= 5.2 dB

NUMBER OF PULSES INTEGRATED N

SIGNAL-TO-NOISE RATIO PER PULSE, dB

various detection and false alarm criteria, the required (S/N) per pulse as a function of the number of pulses can only be drawn for special cases. Figures 3-7 and 3-8 are from Barton [22], who used Fehlner's computations of the Marcum and Swerling equations.

Figure 3-7 shows the required (S/N) per pulse for N pulses for a detection probability P_D of 50 percent and a false alarm probability P_N of 10^{-6}. This is a common case for a low-resolution search radar. Both coherent and incoherent integration are included. The integration loss is the difference in the ordinate between the two cases for a given number of pulses integrated. Figure 3-8 is for a $P_D = 80$ percent and $P_N = 10^{-6}$. This is a common value for acquisition by a tracking radar.

Another comparison is the required signal-to-noise ratio (Fig. 3-9) for the single-pulse case for the various fluctuation models. At desired single scan detection probabilities P_D between 0.2 and 0.4 it makes little difference which target fluctuation model is assumed; however, at a $P_D = 0.9$ an average of 8 dB more S/N is required to detect a slowly fluctuating target for the illustrated case of 10^6 false alarm number (17 dB more for the Weinstock model[1]).

The choice of a particular false alarm number n' has not been emphasized as it is generally less important than the fluctuation model of the target when using search radars and when the number of integrations is small.

The required signal-to-noise ratio versus the false alarm number for P_D equal to 0.5 is shown in Fig. 3-10 [410] where the number of pulses integrated N is the parameter. The target model is the Swerling Case 1.

It can be seen that even for the case of 0.5 detection probability and a fluctuating target, only a 2-dB increase in S/N is required to increase the false alarm number from 10^5 to 10^9 for 10 pulses integrated. At the 10^5 false alarm number a 7 dB less signal-to-noise ratio would be required to detect a constant echo.

Since these curves were also drawn from Fehlner's [113] computations, they can be used to interpolate the previous figures in this chapter.

3.5 POSTDETECTION INTEGRATION WITH PARTIALLY CORRELATED NOISE

Marcum and Swerling [244] showed that for postdetection integration the required signal-to-noise ratio per pulse may be reduced by a factor

---

[1] Discussed in Chap. 5 on targets. This is a chi-square model with one degree of freedom.

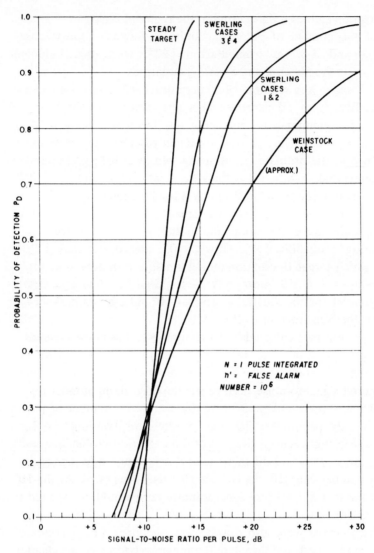

figure 3-9 *Detection probabilities for various target fluctuation models.*

somewhat between $\sqrt{N}$ and N (the number of pulses integrated). These results were obtained for white, gaussian noise; and it was assumed that every pulse processed contained a statistically independent sample of the noise. It is possible to extend the theory to include interference that is time correlated, such as clutter echoes. This extension must account for the fact that the number of independent samples of the interference may

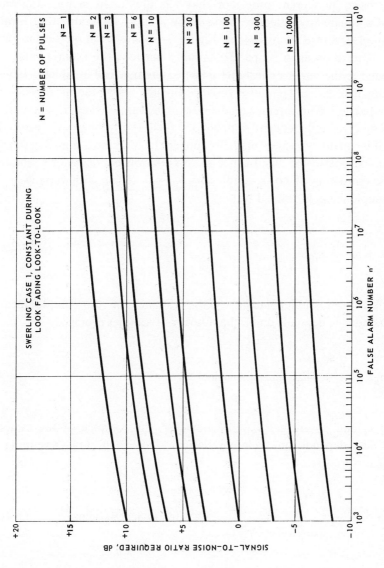

figure 3-10 *Signal-to-noise ratio vs. n' for $P_D = 0.5$. (After Young [410].)*

85

be much smaller than the actual number of pulses processed and the amplitude distribution of the interference.

The basic incoherent processor that was illustrated in Fig. 3-2 consists of a square-law detector followed by a linear integrator that forms the sample average of N sweeps. Figure 3-11 shows sample sweeps one might observe on an A scope. To simplify matters, the average value of the signal is shown as a constant[1] except at a range cell in which a target is assumed to exist. The noise variance in time about the mean value is designated σ_n^2. The variance of the ensemble average is σ_n^2/N_I, where N_I is the number of independent samples obtained at any one range interval. It is this quantity which may be applied to Marcum's and Swerling's results. To obtain the significance of this relationship consider a random process consisting of noise samples x_1, x_2, x_3, ..., x_N with variance σ_x^2. The sample mean is defined by

$$\bar{x} = \frac{1}{N} \sum_{i=1}^{N} x_i \tag{3-8}$$

For statistically independent x_i the variance of the sample mean, designated $\sigma_{\bar{x}}^2$, depends on N, the number of samples averaged

$$\sigma_{\bar{x}}^2 = \frac{\sigma_x^2}{N} \quad \text{(independent samples)} \tag{3-9}$$

When x_i and x_j are correlated,[2] however, the variance of the sample mean also depends on their correlation [224, chap. 11]. Assuming a stationary process,

$$\sigma_{\bar{x}}^2 = \sigma_x^2 \sum_{k=-(N-1)}^{N-1} \frac{N - |k|}{N^2} \rho(k\ell) \quad \text{(dependent samples)} \tag{3-10}$$

---

[1] This assumption will limit the usefulness of the results given here. Tests on *uniform* rain indicate that there is considerable spatial variation of the mean value within one mile. The results are valid if the nonuniformities are compensated for by means of an adaptive threshold (see Chap. 4).

[2] Correlated samples obey the property $\overline{(x_i - \bar{x})(x_j - \bar{x})} \neq 0$.

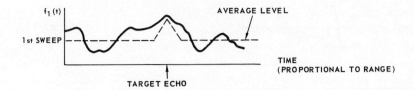

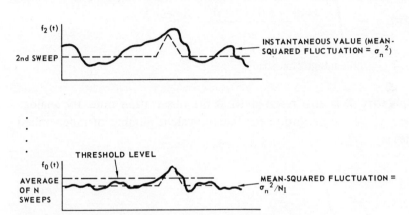

figure 3-11   *Voltage waveforms of an automatic detection process with correlated noise.*

where

$$\rho(k\ell) = \frac{\overline{[x_i - \bar{x}][x_{i+k\ell} - \bar{x}]}}{\sigma_x^2}$$

ℓ = sampling interval

The quantity $\rho(k\ell)$ is the normalized autocorrelation coefficient. As the actual number of samples averaged in a given time T becomes very large, the sample variance becomes that for continuous sampling (continuous integration)[1]

---

[1] Continuous integration does not necessarily minimize $\sigma_{\bar{x}}^2$; however, except for integration intervals quite short compared with the correlation time of the function, the difference between continuous integration and optimum sampling is small. (See Balch et al. [15] and Fine [116].)

$$\sigma_{\bar{x}}^2 = \sigma_x^2 \int_{-T}^{T} \frac{T - |\tau|}{T^2} \rho(\tau) \, d\tau \tag{3-11}$$

where

$$\rho(\tau) = \frac{1}{2T} \int_{-T}^{T} \frac{[x(t) - \bar{x}][x(t + \tau) - \bar{x}] \, dt}{\sigma_x^2}$$

T = total integration time

The quantity T is also referred to as the observation time. By analogy with Eq. (3-9) the definition for the equivalent number of independent samples is

$$\frac{1}{N_I} = \frac{\sigma_{\bar{x}}^2}{\sigma_x^2} = \int_{-T}^{T} \frac{T - |\tau|}{T^2} \rho(\tau) \, d\tau \tag{3-12}$$

Thus, Eq. (3-12) demonstrates that the number of statistically independent samples obtainable in the interval T is dependent on the normalized autocorrelation function $\rho(\tau)$; this in turn depends on the power spectral density function of the clutter.

3.6 INDEPENDENT SAMPLING OF CLUTTER ECHOES

The gaussian power spectral density[1] function has been shown to provide a reasonable description of clutter spectra, namely,

$$S(f) = S_0 e^{-f^2/2\sigma_f^2} \tag{3-13}$$

where σ_f^2 is the second moment of the spectrum. By taking the Fourier

---

[1] Although this section analyzes independent sampling for clutter having gaussian spectra, the results are not very sensitive to the precise spectrum shape. Lhermitte [228], for example, shows that nearly identical results are obtained for triangular or rectangular spectra.

transform of Eq. (3-13), the normalized autocorrelation function, which is also a gaussian function, is obtained

$$\rho(\tau) = e^{-\tau^2/2\sigma_\tau^2} \qquad (3\text{-}14)$$

where σ_τ^2 is the second moment of the correlation function. Equations (3-13) and (3-14) are related by

$$\sigma_\tau = \frac{1}{2\pi\sigma_f} \qquad (3\text{-}15)$$

By substituting the correlation function into Eq. (3-12), one obtains

$$\frac{1}{N_I} = \frac{1}{\sqrt{2\pi}\,\sigma_f T} \, \text{erf}(\sqrt{2}\,\pi\sigma_f T) - \frac{1}{2\pi^2\sigma_f^2 T^2} [\exp(-2\pi^2\sigma_f^2 T^2) - 1] \qquad (3\text{-}16)$$

where erf is the error function.

Lhermitte [228] has shown that for correlated samples, the difference between continuous integration [Eq. (3-16)] and the sum of N_I discrete samples equally spaced in time T [Eq. (3-10)] is small. Thus, in Eq. (3-16) we make the substitutions

$$T = \frac{N}{f_r} \text{ (for dependent sampling)}$$

and

$$\sigma_f = \frac{2\sigma_v}{\lambda} \text{ (the doppler equation)}$$

where N is the actual number of samples, f_r is the sampling frequency, σ_v is the standard deviation of the velocity distribution of the clutter scatterers, and λ is the radar wavelength.

As the total observation time is increased, the second term of Eq. (3-16) becomes insignificant compared with the first. At the same time $\text{erf}(\sqrt{2\pi}\,\sigma_f T) \cong 1$. Thus,

$$N_I \cong 2\sqrt{2\pi}\, \frac{\sigma_v T}{\lambda} \quad \text{(for } T \text{ large)} \qquad (3\text{-}17)$$

This result leads to the definition of the equivalent time for independence

$$T_I = \frac{T}{N_I} = \frac{\lambda}{2\sqrt{2\pi}\,\sigma_v}$$

(3-18)

Equation (3-18) is illustrated in Fig. 3-12. We may interpret T_I as the time required between samples for independence; sampling at a more rapid rate gives practically no new information.

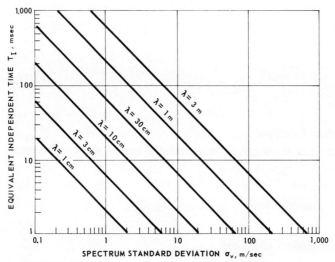

figure 3-12 *Equivalent independent sampling time for radar clutter.*

Assume that a given sampling frequency gives independent samples. As that frequency is increased, a point is reached where the samples become dependent. The transition between independent and dependent sampling is quite abrupt, as shown in Fig. 3-13, occurring approximately at $(\sigma_v/\lambda f_r)$ = 0.2. This figure has been derived from Eq. (3-10).[1] Thus, we may state, with only a small error, for

---

[1] Adapted from Lhermitte [228]. The spectrum width parameter σ_v used in this section refers to the width after envelope detection. If the predetection spectral width is available it should be multiplied by $\sqrt{2}$.

$$\frac{\sigma_v}{\lambda f_r} < 0.2$$

N_I may be determined by using Eq. (3-16) or Fig. 3-13, for

$$\frac{\sigma_v}{\lambda f_r} > 0.2$$

N_I may be determined by $N_I = N$. Alternately N_I may be exactly determined from Eq. (3-10).

The concept of independent sampling may be applied to the work of Marcum and Swerling where the number of pulses integrated is understood to equal N_I.

One word of caution is due. Marcum and Swerling's results were obtained for noise with a chi-squared distribution of degree $2N$, which is the distribution one obtains (after low-pass filtering) when adding N independent squared gaussian variates. To be completely rigorous, one must also show that the mean value over the period T of a correlated variable follows the chi-squared distribution of degree $2N_I$. This seems

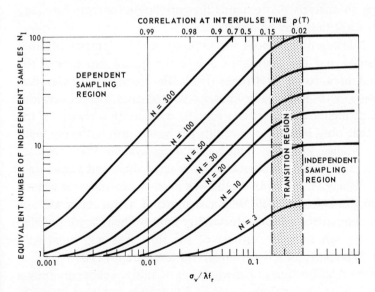

figure 3-13 *Independent samples for partially correlated waveforms (gaussian correlation function assumed).*

like a reasonable assumption since N_I was defined in such a way that the second moment of the distribution of the mean correlated variates corresponds to that of the $2N_I$ degree chi-squared distribution.

3.7   LIMITATIONS ON INDEPENDENT SAMPLING

Two important phenomena limit signal-to-clutter ratio improvements by pulse integration. One is the nonstationary and nonergodic nature of the clutter statistics. The average radar cross section of land and weather is nonuniform in space. Therefore, although pulse integration can reduce the time fluctuation of the clutter signals, this technique has no effect on the spatial nonuniformities. Thus, one must set the threshold detector's level according to the statistics of the spatial fluctuation as well as the properties of the integration.

The second limitation is due to possible long-term correlation of the clutter echoes. Although the clutter spectra have been represented by the gaussian function, some investigations [49] have shown that the temporal correlation function of sea clutter contains long tails that decay slowly.[1] For most correlation analyses, these tails would be regarded as a dc offset or a system bias and consequently would be subtracted. It is only through an analysis that extends the correlation function for many seconds that these tails may be revealed.

A pictorial description of both the improvement in detectability that can be obtained and the limitations due to the spatial structure of the seas is found in a paper by Croney [82]. He compared a radar that rotated slowly and had many closely spaced hits per beamwidth (integration of highly correlated sea echoes) with another version of the same radar that only had one or two hits per beamwidth but many rapid independent scans. For the rapidly scanning system there was an improvement in recognizing objects such as small ships and buoys on a cathode-ray display. The improvement was much more apparent when a storage tube display was used, which is more of a true integrator. A quantitative comparison of the detectability improvement is not possible because of the variations in contrast of the slow scan and rapid scan photographs. The need for a linear integrator is apparent [160].

Since the correlation times of many types of clutter echoes are only a few milliseconds at carrier frequencies of 3,000 MHz or higher, it would

[1] These tails might be due to long-term waves or swell structures that may have periods of several seconds.

seem that considerable improvement in detectability can be obtained by linear integration of signals in partially correlated clutter echoes; however, the problem of where to set the threshold level is much more complicated when clutter is present than when only receiver noise is present. For a given false alarm rate, the threshold level in an automatic detection system must be set as a function of the noise power and the number of pulses integrated. In a clutter-free system the distribution of the noise and the number of pulses are known quantities; thus, the threshold detector requires only a measured value of the noise power to fix the threshold level. The number of independent samples of clutter and the average echo power, however, vary in time and space. Thus, in a clutter environment, not only is a measure of clutter fluctuation power required; but additionally an estimate of the number of independent samples N_I is required. One way to achieve this is to measure at each range interval the ratio of the mean-squared fluctuation to the square of the average signal. Since the probability density function of the clutter was assumed to have the form of the chi-squared distribution of degree $2N_I$, this ratio would uniquely determine the value of N_I. An alternate method is to set the threshold on the basis of a minimum expected value of N_I, which is based on the minimum expected spectrum width. As shown in later chapters, a minimum value of the spectrum width for precipitation clutter or chaff can be calculated by using $\sigma_v \geq 1.0$ m/sec. The value of σ_v for sea clutter must be chosen as a function of sea state or wind velocity.

3.8 CUMULATIVE DETECTION OF A RADAR TARGET

In the previous chapter the radar equations were written to determine the range at which a target can be detected with a certain probability P_D on a single scan. In many radar situations the primary purpose is to detect approaching targets *before* they reach a given range. In general, the typical surveillance radar has been searching for targets for a considerable period of time before the S/N ratio has increased to a level that yields the desired detection probability. The probability that the target has been detected on at least one of j scans (or *dwells*)[1] can be

[1] The term dwell is introduced here since a surveillance radar often transmits several pulses in a single direction during one scan but has no integration device. Thus, the time between *dwells* is generally a few milliseconds while the time between scans is a few seconds.

written

$$P_c = 1 - (1 - P_D)^j \qquad (3\text{-}19)$$

where P_c is the cumulative probability of at least one detection in j scans and R_D is the single scan (or *dwell*) detection probability.

The first question to be addressed in this section is whether to build a linear integrator or simply to wait for a detection to occur in one of j scans. The false alarm number n' for the entire process is 10^6 in the examples.

The efficiency of looking more often and setting a higher threshold per dwell versus using a linear integrator is a function of the fluctuation characteristics of the targets and the desired certainty of detecting the target.

TABLE 3-1   Cumulative Detection and Incoherent Integration for Steady Targets*

Number of dwells, j	Cumulative detection statistics			Incoherent integration (S/N) per dwell, dB	Integration improvement factor, dB
	P_{Di}	n'	(S/N) per dwell, dB		
1	0.90	10^6	13.1	13.1	
2	0.684	2×10^6	12.2	10.8	1.4
3	0.535	3×10^6	11.8	9.4	2.4
4	0.44	4×10^6	11.4	8.5	2.9
6	0.32	6×10^6	11.1	7.0	4.1
10	0.206	10^7	10.6	5.5	5.1
30	0.074	3×10^7	9.7	2.0	7.7
100	0.023	10^8	9.2	-1.2	10.4
1,000	0.002	10^9	8.0	-6.8	14.8

*Steady target detection probability = 0.9, and overall false alarm number $n' = 10^6$.

The required value of (S/N) per dwell has been calculated using Fehlner's [113] curves for two extreme cases[1] of target fluctuation. The

[1] Of the original Swerling models. As in previous discussions, (S/N) is the predetection signal-to-noise ratio.

overall detection probability P_D is 0.9. Table 3-1 shows the (S/N) per dwell required for 1 through 1,000 dwells. In one case, the threshold is slightly raised, and 90 percent of the time the target signal plus noise (at the target location) will add and exceed the threshold on one or more of the dwells. The required S/N per dwell is shown in the fourth column. The second case assumes the use of a linear integrator as discussed in Sec. 3.2. The last column in the table shows the relative improvement obtained by the use of the linear integrator. For the steady target, the *relative* improvement of the linear integrator is quite close to the square root of the number of dwells, and the technique of simply waiting for a detection is not very efficient.

TABLE 3-2   Cumulative Detection and Incoherent Integration for Fluctuating Targets*

Number of dwells, j	Cumulative detection statistics			Incoherent integration (S/N) per dwell, dB	Integration improvement factor, dB
	P_{Di}	n'	(S/N) per dwell, dB		
1	0.90	10^6	21.1	21.1	
2	0.684	2×10^6	15.6	15.0	0.6
3	0.535	3×10^6	13.5	12.4	1.1
6	0.32	6×10^6	11.1	8.9	2.1
10	0.206	10^7	9.7	6.8	2.9
30	0.074	3×10^7	7.5	2.8	4.7
100	0.023	10^8	6.0	-0.5	6.5
1,000	0.002	10^9	3.9	-6.0	9.9

*Fluctuating target detection probability = 0.9, and Swerling Case 2 overall $n' = 10^6$.

The steady or nonfluctuating target is not very common in radar. Table 3-2 shows the relative performance of waiting for a detection versus linear integration for the Swerling Case 2 fluctuating target (independent fluctuation from pulse to pulse). The last column shows that there is relatively less improvement for linear integrators with fluctuating targets. If the fluctuation is slow compared to the interpulse period, the computation becomes more difficult; but Johnson [195] has shown that cumulative detection only allows 2 dB reduction in the required signal-to-noise ratio for $j = 4$, $P_D = 0.9$, $P_N = 10^{-6}$ and a Swerling Case 3 target. If the target had the same amplitude distribution but fluctuated rapidly (Case 4), the required signal-to-noise ratio is reduced by about 5.8 dB. This emphasizes the need for obtaining several *independent* looks on targets that fluctuate slowly if high detection probabilities are desired. This can be accomplished by mechanically (or electronically) scanning the antenna more rapidly so that there are several scans (and probably independent observations of the target) in the desired detection time. Alternate techniques are pulse-to-pulse frequency or polarization shifting.

Barton [20, pp. 142-149] discusses the apportionment of the number of pulses per beamwidth per scan for a given total number of pulses, and he concludes that about four scans should be used for $P_D = 0.9$, $P_N = 10^{-8}$. This is based on the rapidly fluctuating target and also considers beam shape losses. It would seem that $j = 4$ to 6 is also a good value for the slowly fluctuating target since the fluctuation loss vanishes[1] at $P_D \approx 0.33$. This assumes that the target return is uncorrelated from scan to scan. If there is clutter or interference, the fluctuation rate of the clutter would also have to be accounted for in the optimization of the number of scans. With clutter, the linear integrator has the additional advantage of reducing the *relative* fluctuations of the clutter (discussed in Secs. 3.5 and 3.6).

3.9 DETECTION RANGE FOR AN APPROACHING TARGET

Another implication of the use of cumulative detection is that the basic volume search radar equation must be modified for the case of an approaching target with uniform radial velocity. The volume radar equation from Chap. 2 was written as

$$R^4 = \frac{\bar{P}_T L_T A_e L_R L_P L_a L_c \sigma_t T_{SC}}{4\pi K T_s \psi (S/N)} \tag{3-20}$$

where T_{SC} is the scan time of the surveillance radar and ψ is the total solid angle to be searched.

In the approaching target case, the time that the target is within the instrumented maximum range may well be many times T_{SC}. Mallett and

---

[1] For values of P_D of less than 0.33 at $P_N = 10^{-6}$, less *(S/N)* is needed for fluctuating than for steady targets.

Brennan [242] showed that there is an optimum choice of T_{SC} (the frame time in their paper) for a constant velocity target and with the other radar parameters fixed. They make the substitution

$$\frac{\Delta}{V_c} = T_{SC}$$

where Δ is the radial distance traversed by the target in the scan time T_{SC} and V_c is the radial closing velocity of the target. They then define (in the notation of the previous sections)

$$R_0{}^4 = \Delta R_1{}^3 \tag{3-21}$$

$$R_1{}^3 = \frac{\bar{P}_T L_T A_e L_R L_P L_a L_c \sigma_t}{4\pi K T_s \psi V_c} Q_i{}^3 \tag{3-22}$$

where R_0 = the detection range for $(S/N) = 1$

R_1 = an artificial range used in their optimization of Δ including the parameters of the radar

$Q_i{}^3$ = a correction factor less than unity relating to the amount of incoherent integration performed during one scan and for false alarm numbers other than 10^6

The interesting feature derived from this equation is that for coherent integration on each beam position, the detection range varies as the cube root of the power aperture product for the optimum scan time.

The optimization parameter δ is then defined

$$\delta = \frac{\Delta}{R_1} \tag{3-23}$$

and varies with the desired P_c and the fluctuation model but is independent of the beamwidth, transmit frequency, waveform, and receiver bandwidth. The value of Δ for optimum performance is thus proportional to R_1, and longer range radars should have longer frame times for a given target velocity. The frame time expression is

$$T_{SC} = \frac{\delta R_1}{V_c} \tag{3-24}$$

The optimum values for δ versus P_c (from Mallett and Brennan [242])
are shown in Fig. 3-14 for the steady target and Swerling's Cases 2 and 3
for $n' = 10^8$. Coherent integration is assumed during each dwell ($Q_i = 1$).
The optimum values of δ are not critical for P_c less than or equal to 90
percent, and the detection range will decrease by less than 7 percent for
a range of 0.5δ to 2δ for this cumulative probability.

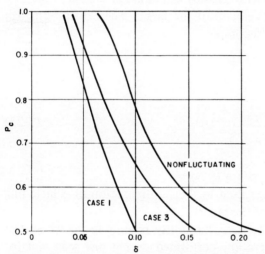

figure 3-14 *Optimum value of δ vs. cumulative
probability of detection. (After Mallett and
Brennan [242].)*

The resulting detection ranges with respect to R_1 for the optimum
values of δ are shown in Fig. 3-15 for values of P_c from 0.5 to almost
unity. The detection range varies from 0.155 to 0.196 for $P_c = 90$
percent for the three cases.

The detection range for unity (S/N) and coherent integration during
each frame is expressed by

$$R_0^{\;4} \;=\; \Delta R_1^{\;3} \;=\; \delta R_1^{\;4} \tag{3-25}$$

and the detection range is then

$$R^4 \;=\; \frac{\delta R_1^{\;4}}{(S/N)} \tag{3-26}$$

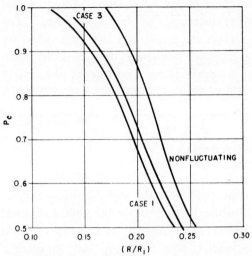

figure 3-15 *Normalized range (R/R_1) at optimum δ vs. cumulative probability of detection. (After Mallett and Brennan [242].)*

$$\left(\frac{S}{N}\right) = \frac{\delta}{(R/R_1)^4} \tag{3-26}$$

where (S/N) is the required signal-to-noise ratio per scan.

For $P_c = 0.9$, $n' = 10^6$, and Swerling Case 1, the optimum value of $\delta \doteq 0.042$, and $R/R_1 \approx 0.155$. Then from Eq. (3-24)

$$T_{SC} = \frac{(0.042)\,R}{V_c\,(0.155)} \approx \frac{0.27\,R}{V_c}$$

For a 200,000-ft detection range (~ 33 nmi) and $V_c = 2,000$ fps, $T_{SC} = 27$ sec. For very high velocity targets, such as ballistic missiles and satellites with $V_c = 20,000$ fps, the optimum scan time for 2×10^6 ft detection (330 nmi) is also about 27 sec, showing that coherent integration should be carried out as long as possible even for the fluctuating target with $P_c = 0.9$ and $n' = 10^8$. The signal-to-noise ratio required is about 18.7 dB as compared to 21.1 dB for the single look.[1] For the same criteria and a steady target, the optimum scan time would be about 41 sec. The use of the term *optimum* is not intended to imply

---

[1] The assumption in this case is that the target is uncorrelated after time T_{SC}.

an optimum detection process since it will be shown in Chap. 4 that *sequential detection* can further increase the detection range.[1] The detrimental case of a slowly fluctuating target can also be avoided for many targets by changing the transmit frequency from pulse to pulse. This has the effect of changing a Swerling Case 1 or 3 target to a rapidly fluctuating target. This effect will be expanded upon in Chap. 5 on radar targets.

3.10 DETECTION STATISTICS WITH RADAR OPERATORS AND DISPLAYS

While most of this report is concerned with the statistics of automatic detection processes, it is worthwhile to summarize the ability of radar operators to detect targets on a radar display and some of their limitations in an attempt to resolve two common, but conflicting, statements.

1. "A well-trained attentive operator can detect targets quicker than an automatic system especially in a cluttered or jamming environment."

2. "An operator loss factor D_0 must be used in the radar equations to make experience conform with theory."

Later chapters will show that a simple fixed threshold detector as used in the early automatic systems is far from optimum with respect to environmental problems. A *saturation effect* occurs unless *both* the transmitted waveform and the receiver system are designed to handle these problems. Thus, the first statement may be true with the proper operator and display system if the target is between regions of high clutter.

The second statement is also true in most current systems unless a highly motivated trained operator is given an *optimized* display and an extended attention span is not required of him. In the past, many subtle receiver losses were also attributed to the operator.

The factors that contribute to operator losses will be expanded upon for three types of displays: (1) the PPI (plan position indicator) display, which is essentially a map of the area with the operator's location usually at the center; (2) the B scope, which is a rectangular display of range

---

For general references see Lawson and Uhlenbeck [223], Barton [20], and Blake [44].

[1] This subject is also discussed in detail in Brennan and Hill [51].

versus azimuth; and (3) the A scope, which is signal amplitude (or power) versus range.

These loss factors encompass two areas representing the losses when the display is not matched to the radar parameters and when the ideal radar operator is not present or when he is confronted with too much information. Some of these display losses in addition to nonmatched receivers[1] are enumerated below:

1. The ratio of the limit or saturation level of a video receiver or display with respect to the average noise is shown in Fig. 3-16a for a PPI display. This does not apply to wideband IF limiting discussed in Secs. 4.4 to 4.8.

2. Inadequate brightness of the display causes losses as shown in Fig. 3-16b for A scopes. It has also been shown that for PPI displays an excessive brightness can also cause several dB of loss.

3. Defocusing of the CRT parallel to the sweep direction for PPI or A scopes gives losses as shown in Fig. 3-16c. This is a form of collapsing loss.

4. There is a loss for PPI or A-scope displays for displayed pulse widths that are either very short or very long, but the optimum width is not critical from 0.3 to about 5 mm.

5. Insufficient average noise deflection on an A scope causes considerable loss as shown in Fig. 3-16d.

6. A large number, 50 to 100, of range elements requiring observation on an A scope causes a significant loss. (A 1-μsec pulse on a 50-nmi display has about 600 elements.) A decrease in detectability of about 1.5 dB when going from 2 to 60 possible signal positions is shown in Lawson and Uhlenbeck [223, p. 233].

7. In a similar manner to (6) above, there is a whole class of losses that occur when the operator is presented too much information. This is partially due to the dimensions of the spot in both range and azimuth (for a PPI) either being too small with respect to the size of the display or being away from the fixation point of the operator. Baker [14, chap. 7] discusses several studies of this effect. He describes experiments that show that these effects can cause a loss of several decibels in detectability for factors of 2 to 4 in target "pip" size versus screen areas. In a similar manner he cites experiments where the operator was only required to observe one-fourth to one-half the PPI area and the visibility

---

[1] In earlier studies these receiver losses were often attributed to the operators.

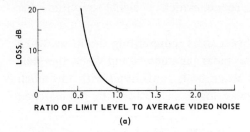

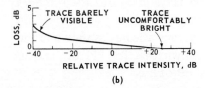

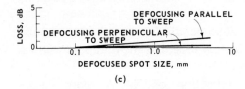

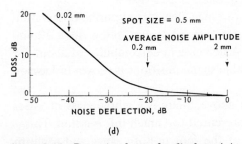

figure 3-16 *Detection losses for displays. (a)*
Ratio of limit level to average video noise; (b)
relative trace intensity; (c) defocused CRT
spot size; (d) noise deflection on A scopes.
(From Barton [20].)

increased by about 2.5 dB. In a similar study Barton [20] reproduces
data from Ashby et al. [11] that show similar results. This effect is often
alleviated by having multiple displays and assigning operators to smaller
sectors.

8. Another effect that is described in Baker [14] is that the operator of a PPI pays the most attention to the region of 0.4 to 0.8 of the radius of the PPI display, and his *threshold* may be over 5 dB higher at 0.1 to 0.3 of the radius. This effect is much less apparent on the B scope, which does not bunch up the sweep lines near the center, or where sensitivity time control STC is used. In cases where it is essential to locate close-in targets, the inverted PPI, or IPPI, has been suggested. In this display the range sweep starts at the perimeter and ends at the center. An experiment by Hickson et al. (reproduced in Baker [14, p. 83]) showed the detectability of targets as a function of target range for both types of display. The IPPI was clearly superior at close range; but when the curve was reversed to show position on the scope, the IPPI had about the same values as the PPI.[1] These curves are reproduced in Fig. 3-17. Other experiments show the B scope to be slightly superior to either type of display at all ranges. When radar clutter is displayed, it generally has greater brightness because of the inverse R^n law dependencies at close range in addition to the greater brightness from the bunching of the sweep lines. The overall effect is alleviated by sensitivity time control[2] STC and by the IPPI.

9. Finally, there is loss due to the operator's loss of interest or motivation after a period of time (about an hour). Baker [14] quotes other sources that indicate operators detecting 30 to 50 percent fewer targets after an hour's watch.

It should not be inferred that all of the above losses are additive, as a radar operator can compensate for some of the above factors with the adjustments available to him. On the other hand, these losses should be kept in mind in the subsequent discussions on automatic processes where the losses can be calculated and invariably result in several decibels of penalty from the *optimum*.

The discussion in Chap. 2 dealt with the basic range equation and the losses involved in detecting targets. The S/N ratio graphs (Figs. 2-2 and 3-3) considered the *single* target pulse detection cases. In most radar displays the multiplicity of return pulses per beam position reduces the required (S/N) considerably for given detection criteria. The operator or the cathode-ray tube CRT display acts as an integrator in a similar

---

[1] This author believes the tests were with constant brightness blips rather than an R^4 law.

[2] See Sec. 4.2. Typical values for n can be determined from Chaps. 6 and 7.

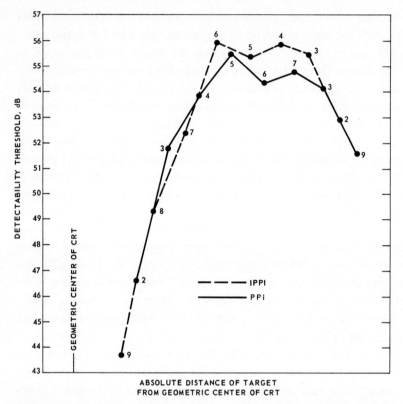

figure 3-17 *Equivalence of detectability on PPI and inverted PPI displays. (From Baker [14].)*

manner to integration in the mathematical sense as described by Marcum [244]. This is illustrated in Fig. 3-18, which is derived from several sources. The solid line is from the M.I.T. Radiation Laboratory studies and later replotted by Blake [44] to show the visibility factor in decibels (V_0 (50) in Blake's notation)[1] for a PPI display versus the number of pulses adjusted to 0.5 probability of detection.[2] Plotted on the same curve are some later data by Erdmann and Myers [106] from experiments on an FPS-3 radar rotating at 10 rpm. It appears that these data are for a cumulative detection probability of close to 100 percent

---

[1] Roughly equivalent to $P_D = 0.5$.
[2] The pulse length was 1 μsec, the PRF was 800 pps, and the scan period was 8 sec in one set of data.

and that they have more data points between 2 and 12 pulse integrations. Since the curve is to show the ability of the trained operator under close to ideal conditions, only the more optimum noise voltage settings of Erdmann and Myers' data[1] are shown. The data are not generally applicable since the meaning of a *detection* was not completely described and the spectrum of the noise was narrower than that of the signal. The most interesting feature of their data is that the improvement in detectability (reduction in S/N) is almost proportional to the number of pulses up to 32 pulses. For comparison purposes, the square-law detection curve taken from Marcum [244] for a $P_D = 0.5$ and a false alarm number of 10^6 is also shown.

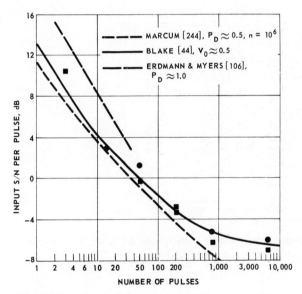

figure 3-18 *Required input signal-to-noise vs. number of pulses for PPI displays.*

An interesting experiment described in Lawson and Uhlenbeck [211, p. 243] shows that the azimuth beamwidth itself has negligible effect for beamwidths of 1 to 45° but that the number of pulses integrated determines the visibility.

[1] The pulse length was 2 μsec, the PRF was 400 pps, the scan period was 6 sec, and range display was 50 nmi.

A similar visibility factor curve (Fig. 3-19) is taken from Blake [44] for A-scope CRT displays and an equivalent detection probability of 0.5. The number of pulses per second ranges from 10 to 3,000. This type of curve is more appropriate for the acquisition mode of a tracking radar rather than for the surveillance mode. In this case, the low values of signal-to-noise ratio that are required would seem to follow from the relatively small number of target cells that the operator must observe as compared to the PPI display.

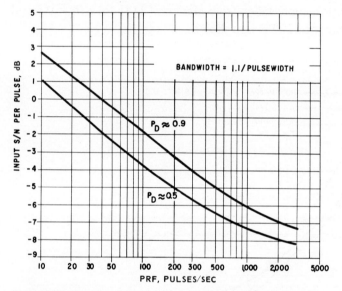

figure 3-19 *Required signal-to-noise ratio (per pulse) on an A-scope display. (After Blake [44] and Lawson and Uhlenbeck [223].)*

Detectability on displays is degraded by correlated clutter in a similar fashion to threshold devices, but the statistics are not known at this time.

This discussion of operator performance has been mostly based on a steady target signal and detection during a single scan. In practice the operator usually can obtain the results attributed to cumulative detection probabilities. If his attention is focused on a *peak* of a scintillating target, he can usually track the target through the nulls since his observations are localized on the display.

3.11 SUMMARY

The basic statistics of the detection problem have been summarized without regard to the type or complexity of the receiver, the processing technique, or the transmitted waveform. It has also been shown that the amplitude distribution and fluctuating characteristics of the target have a strong effect on the required (S/N) and thus the detection range. These characteristics will be discussed in Chap. 5 on the available experimental measurements and some of the theories.

The statistics of environmental noise with a significant correlation time have been covered; specific values will be given in the individual chapters on clutter. The chapters on sea and land clutter also indicate where the amplitude distributions deviate from the Rayleigh law.

Various losses from coherent processing have been shown for incoherent linear integration and for displays, but it must be emphasized that detection was the criterion. In the cases where parameter estimation is the goal, the input S/N is usually high with respect to unity, and the detector losses may become trivial. Numerical values can be computed from the graphs of this chapter of Fehlner [113], Marcum [244], and Barton [22].

The basic shortcoming of all the information in this chapter is the assumption that there is a prior knowledge of the rms value of the noise. With clutter or jamming and in some cases when looking near the sun, galaxies, or even the surface of the earth, this information is not available; and a measurement of the total noise in any portion of space is required when setting a threshold for detection. This uncertainty also occurs with radar displays that generally have 12- to 15-dB dynamic range above receiver noise. While the target return may be 50 dB above the receiver noise level, the display may have been completely saturated by rain, backscatter, or jamming. Several of the techniques for automatically or adaptively setting a threshold will be discussed in the next chapter.

4 AUTOMATIC DETECTION BY NONLINEAR SEQUENTIAL AND ADAPTIVE PROCESSES

4.1 INTRODUCTION

The radar equations and the statistics of detection described in the previous chapters were based on the return signal from a target exceeding a fixed threshold. It was assumed that this threshold was established with respect to either the rms value of the system noise or the environmental noise prior to searching for the target.[1] In the typical surface radar situation environmental effects or electronic countermeasures make that assumption of limited value. Figure 4-1 shows, in an overly simplified way, the effects of the range dependence of the mean radar backscatter from various types of clutter at a transmit frequency of 3,000 MHz (S band). In this figure backscatter is normalized to the radar cross section

---

[1] The block diagram used with the Marcum and Swerling analysis (Chap. 3) shows that the mean-square noise is measured at the time of arrival of the target echo.

of a target at a given range (left-hand ordinate); the right-hand ordinate is an estimate of the radar cross section of a target that can be detected in the presence of the clutter at the corresponding range. The required target signal-to-clutter ratio is assumed to be 20:1 (13 dB). This is a typical value for a narrow-beam surveillance radar. The purpose of this plot is to illustrate that:

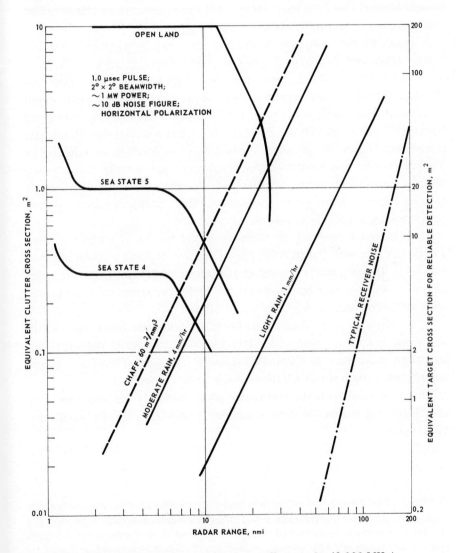

figure 4-1 *Environment for typical S-band surveillance radar (3,000 MHz).*

1. The various types of radar clutter have different range dependencies and considerable time variations of intensity (rainfall rate, chaff density, sea state).

2. For the chosen parameters of the radar, the clutter effects exceed the receiver noise by several orders of magnitude.

3. The radar cross section of the clutter is comparable to that of typical aircraft (1-10 m^2) and far exceeds the cross sections attributed to small missiles or reentry bodies.

A graph of this nature (see also Fig. 1-2) can be prepared from the clutter equations of Chap. 2 and the clutter cross sections of Chaps. 6 and 7. It illustrates the dynamic range requirements of the receiver and the amount of clutter rejection that should be obtained with signal processing. This graph was drawn for a surface radar that is scanning the horizon[1] from an elevation of about 150 ft, but a similar situation exists in airborne radars. While radars for detection or tracking of satellites or planets receive less clutter backscatter, they are usually one to two orders of magnitude more sensitive and are subject to variable solar and galactic noises, atmospheric anomalies, and varying antenna system temperatures, especially at low elevation angles.

The electronic jamming problem can easily be illustrated by assuming a noise jammer with 1-watt/MHz power density (P_J) and a 10-dB antenna gain (G_J). If the jammer is located at a range of 50 nmi from the radar and the radar receiver bandwidth is 1.0 MHz, the jamming noise power will exceed the receiver noise by about 50 dB.

It is obvious then that *in addition* to being designed for clutter rejection, the radar must ultimately include a threshold detector that adapts in some way to the interference at the time when (or location where) the statistical test for detection is to be carried out.

The following sections summarize some methods for accomplishing this, including the use of various nonlinear amplifiers and truly *adaptive* techniques.

[1]The rapidly sloping plot of sea- and land-clutter backscatter beyond the horizon is arbitrary. The land horizon from a ship effectively occurs at greater range than the sea-clutter horizon because of hills, etc., which extend above the geometric horizon. Quantitative values of reflectivity are given in Sec. 5 on ship targets and in Chap. 6.

4.2 DYNAMIC RANGE PROBLEMS—STC AND IAGC

The dynamic range limitation of radar receivers was one of the first problems encountered in radar design. The dynamic range of a system will be defined here as the larger of two ratios: the ratio of the return signal power that will saturate the receiver to the minimum detectable signal; or the ratio of the saturation signal power to the mean-square noise level.

It was shown in the sections on displays (Chap. 3) that the best operator performance is obtained with the receiver noise visible on a PPI display or with several tenths of a millimeter deflection on an A scope. The dynamic range under these conditions is only 15-20 dB for the PPI and perhaps 20-30 dB for the A scope. Since the hypothetical radar environment shown in Fig. 4-1 has a ratio of signals in excess of 10^7 (70 dB) at a range of 5 nmi, some form of compression of the amplitude of the signals must precede the display.

One of the earliest and most widespread techniques used to cope with this problem is the sensitivity time control, or STC, circuit. This circuit simply reduces the gain of the amplifiers (usually at the intermediate frequency stage) at close range where maximum sensitivity is not generally required. The gain may be varied as R^n, where n may be chosen between a value of 2 for weather effects and as much as 4 for land and sea clutter. The maximum gain is reached at the range where the clutter barely exceeds the noise. On a PPI, the STC circuit also tends to reduce the excessive brightness at the center of the display that is due to the *bunching* of the radial sweep lines.

A second technique for eliminating saturation is the instantaneous automatic gain control (IAGC). IAGC acts in a similar manner to the automatic volume control (AVC) on a radio receiver. The rectified output of the detector is integrated for several pulse widths in a low-pass filter, and this output is used as negative feedback to vary the gain of the amplifier that precedes the detector. This reduces the overall gain for extensive clutter with a slight loss in the detectability of the target signal. An alternate version of this technique can be used to counteract rapid variations in the noise level due to broadband electronic jamming. If the bandwidth of the system, including the IF amplifiers, is wider than the matched filter, the noise level can be sampled prior to the matched filter. If the total input noise is sufficiently broadband, the number of independent samples of the noise can be increased by the ratio of the IF to matched-filter bandwidths. This will allow a better estimate of the mean-square noise. A study of such a system with a bandwidth ratio of five has been reported by Manske [249]. With Monte Carlo methods, he showed that the degradation in the required signal-to-noise ratio was less than 1 dB compared to having a complete knowledge of the mean noise power for signal-to-noise ratios of less than 10 dB. More sophisticated variations of this approach are discussed in Sec. 4.10.

A third common technique is based on the knowledge that the target return signal for a pulse radar will have approximately the same width as the transmit pulse. In contrast, clutter returns or interference may be many pulse widths in duration. The optimization of the receiver for a given pulse width may take the form of a high-pass filter (fast time constant (FTC) circuit) or a pulse-width discriminator. The pulse-width discriminator inhibits return signals which are either shorter or longer than the expected target return signal. Since the spatial distribution of most clutter returns includes many signals with an extent comparable to the pulse width, the clutter rejection by these techniques alone is quite limited.

The fourth of the basic techniques is the use of a logarithmic amplifier to reduce the dynamic range of the input signals. Since the logarithmic receiver when used by itself (prior to the detector) has little effect on detection[1] or on performance in clutter, it will only be discussed when it is used in conjunction with the pulse-width discriminator or FTC circuits.

4.3 LOGARITHMIC AMPLIFIERS FOLLOWED BY DIFFERENTIATING CIRCUITS OR PULSE-LENGTH DISCRIMINATORS

The logarithmic amplifier followed by a fast time constant (FTC) circuit or by a pulse-width discriminator is another technique for reducing the effective dynamic range of clutter or jamming signals at the threshold detector or display. Its operation can be described as a form of adaptive detection where the goal is to maintain the *noise* level at the output of

---

[1] When the outputs of a logarithmic amplifier are added incoherently, it has been shown [147] that there is 0.5-dB loss in detectability for 10-pulse integration and a 1.0-dB loss for 100-pulse integration.

the receiver roughly constant and independent of the *mean* value of the detected input noise or clutter. The main purpose of such schemes is generally to aid the radar operator by reducing the amount of interference on the display. There is actually no improvement in the signal-to-noise ratio (S/N) or signal-to-clutter ratio (S/C). On the other hand, there is no other simple technique for maintaining the noise level constant in an environment similar to that of Fig. 4-1 where the interfering signals are changing temporally (sweep-to-sweep) as well as spatially (along the sweep).

The log-FTC technique is based on similarities between the probability distribution functions of noise and clutter. It has been shown that, if the scattering surface defined by the pulse length and antenna azimuthal beamwidth contains many scatterers, the amplitude distribution[1] of the return signals often approximates the Rayleigh distribution for most cell sizes as does the envelope of broadband noise after narrowband filtering and detection. The Rayleigh density function can be written:

$$p(x)\, dx \; = \; \frac{x}{\alpha^2} \; \exp\!\left(\frac{-x^2}{2\alpha^2}\right) dx \qquad x \geq 0$$

$$= \; 0 \qquad\qquad\qquad x < 0$$

(4-1)

and

$$\bar{x} \; = \; \alpha\!\left(\frac{\pi}{2}\right)^{\!\frac{1}{2}} \; = \; \text{the mean value}$$

$$\sigma_x \; = \; 0.655\,\alpha$$

where x is the instantaneous amplitude and σ_x is the rms fluctuation about the mean.

It can be seen that an important property of this distribution is that the fluctuation about the mean is proportional to the mean. Following the derivation by Croney [81], the transfer characteristic of an idealized logarithmic amplifier can be written

$$y \; = \; a \ln(bx)$$

(4-2)

---

[1] The spatial distribution of land clutter from a surface radar conforms more closely to the log-normal distribution, while the amplitude distribution from vegetation, etc., is closer to the Rayleigh distribution or Rice distribution at a given point. See also Chaps. 6 and 7 of this book and Lawson and Uhlenbeck [223, chap. 6].

where a and b are constants of the amplifier and y is the amplitude of the output. Croney showed that the variance of the output is

$$\text{Variance} = E(y^2) - [E(y)]^2 = \left(\frac{a^2}{4}\right)\left(\frac{\pi^2}{6}\right) \tag{4-3}$$

The rms fluctuation has a constant value of $\approx 0.64a$, which is independent of the input level.

Unfortunately a logarithmic characteristic cannot be maintained down to zero input (the output would then be $-\infty$), but it has also been shown [81, 119] that, if the logarithmic function is maintained to 15-20 dB below the rms level of the receiver noise, the output rms value would only vary by about 0.25 dB from the ideal. This has the effect of increasing the false alarm rate for a fixed threshold system.[1] The increase in the false alarm probability for a practical logarithmic amplifier has been calculated [119] where the amplifier has the characteristic of $y = a \ln(1 + bx)$ and the threshold has been set for $P_N = 10^{-8}$ on receiver thermal noise. Table 4-1 shows the change in the false alarm probability for different logarithmic amplifier constants. If $b\sigma$ is only 10, there will be an order of magnitude increase in the false alarm rate for heavy clutter. Since the clutter signals are often 60-80 dB above the rms receiver noise, the dynamic range requirement of the log-amp is substantial.

The purpose of the differentiator circuit following the log-amp is to remove the mean value of the clutter or noise return, and it is only effective if the clutter is uniform (in range) for a period that is long compared to the time constant of the FTC circuit. The nonuniformity in space of all types of clutter limits the usefulness of this device. In practice, the time constant for the pulse buildup is set at about 1.5 times the pulse length, and the discharge time constant is kept to about 0.2 pulse lengths to prevent *shadows* from occurring immediately after targets or discrete clutter. The shorter discharge time constant tends to accentuate the negative going peaks of the clutter at the output of the log-amp, and the resulting flyback adds to the noise level. This causes about a 3-dB S/N loss [81] for short discharge time constants. Another

---

[1]It is assumed that the mean value of the output of the log-amp has been removed by a high-pass filter.

**TABLE 4-1   Change in P_N for Various Noise Levels and
Logarithmic Amplifier Constants**

Product of standard deviation of thermal noise σ and amplifier constant b	False alarm probability for clutter- or jamming-to-receiver noise equal to:	
	10 dB	> 40 dB
1	$10^{-4.2}$	$10^{-2.6}$
2	$10^{-5.2}$	10^{-4}
5	$10^{-6.5}$	$10^{-5.7}$
10	$10^{-7.1}$	$10^{-6.7}$
40	$10^{-7.8}$	$10^{-7.6}$
200	$10^{-7.9}$	$10^{-7.9}$
∞	10^{-8}	10^{-8}

limit on the effectiveness of this technique is due to the correlation (in time) of the discrete clutter spikes from sweep-to-sweep on a radar display with many hits per beamwidth. For sea clutter these spikes may be correlated for tens of milliseconds (Chap. 7), and the successive returns will incoherently integrate and appear as targets.[1] The improvement in (S/C) for uncorrelated clutter is a result of the incoherent addition of target signals in a noiselike background and is not a result of the log-FTC circuits themselves. For single-pulse detection the effectiveness of the log-FTC circuit also depends on how closely the clutter conforms to the Rayleigh distribution. For short-pulse ($<$ 1-μsec) surface systems, the amplitude and spatial distributions of sea or land clutter echoes are not Rayleigh, and the variance often increases while the mean or median value may decrease, as compared to longer pulse systems. Finn [119] has calculated the effect on the log-FTC receiver of clutter echoes that contain a coherent and incoherent component (Rice distribution). If the threshold has been set to yield $P_N = 10^{-8}$ on the basis of a Rayleigh distribution and the actual distribution has a coherent component equal to twice the fluctuating component, the threshold level will be increased and the false alarm probability will drop to 1.6×10^{-18}. This is a considerably higher threshold than is desired, and the minimum detectable target signal must be several decibels higher than if the correct distribution had been assumed in establishing the threshold.

[1] Destruction of the correlation of the clutter by changing polarization or frequency from pulse-to-pulse will be discussed in later chapters.

The alternate case of incorrectly assuming a Rice distribution is even more costly. If it is assumed that the coherent/incoherent ratio is two and the actual signal is Rayleigh distributed, the false alarm probability will increase from 10^{-8} to 0.33; and the automatic detection system will saturate.

An alternative to the use of the differentiating circuit is to follow the logarithmic amplifier by a pulse length discriminator (PLD) circuit. This technique, which is discussed both theoretically and experimentally in Hansen [158], has the same general effect as a log-FTC circuit on clutter and noise. It also rejects pulsed interference and forms of land clutter that differ substantially from the transmitted pulse length.

A typical block diagram of such a receiver is shown in Fig. 4-2. The log-FTC circuit would have the same block diagram with the exception that a high-pass filter and clipper to remove the negative going peaks would follow the logarithmic amplifier.

In a conventional linear receiver the band-pass filter shown in Fig. 4-2 is usually the matched filter. For a log-amp, even for a simple pulse transmission, this band-pass filter can only be made to approximate a matched filter since the noise-and-signal bandwidth of the logarithmic amplifier varies with the level of the input signals. The pulse length discriminator passes only pulses within fixed limits of the desired pulse length. These limits are established by the delay lines of length T_{DL} and τ, the bandwidth of the band-pass filter, and the tolerances of the components. Design data and typical circuits for PLDs are given in Millman and Taub [250, p. 310] and Hansen [158]. The influence of the band-pass filter can best be illustrated by considering extreme conditions. In addition to the usual losses for departure from a matched filter, a narrow bandwidth ($B \ll 1/\tau$) will stretch the pulses and prevent the rejection of impulsive noise. A wide-band filter ($B \gg 1/\tau$) will have at the output some clutter residue due to the difference between the clutter and noise power spectral densities [158] and the variable bandwidth of a logarithmic amplifier. In addition there is the added noise from the broader bandwidth. The log-PLD receiver causes losses in the detectability of targets in noise and clutter as does the log-FTC. The experiments by Hansen [158] show losses of 4-8 dB for single-hit detection when compared with a linear receiver and 2–4-dB losses for systems with postdetection integration. These results for two sets of circuit parameters are summarized in Figs. 4-3 and 4-4 for a steady-target signal and a false alarm probability P_N of 10^{-2}, 10^{-3}, or 10^{-6}. The

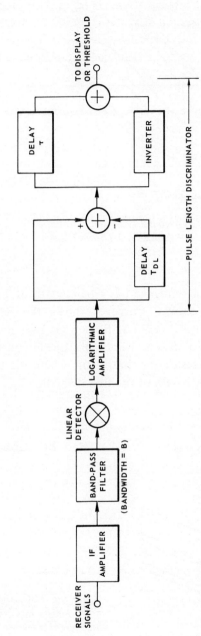

figure 4-2 *Log-amp pulse-length discriminator receiver.*

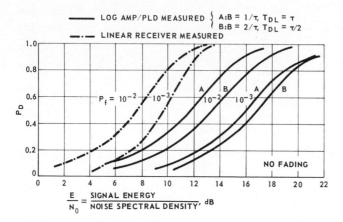

figure 4-3 *Measured detection probability vs. signal-to-noise ratio for a linear receiver and a log-amp/PLD. (From Hansen [158].)*

linear receiver used as a reference has a bandwidth of $1/\tau$. Two sets of circuit parameters are shown. Curve A shows the result when the band-pass filter has a width of $1/\tau$. Somewhat better detectability is achieved and only those pulses greater than $\approx 1.8\tau$ are rejected for strong signals. Curve B is the result for the circuit where the band-pass filter bandwidth is $2/\tau$ and the length of the first delay line T_{DL} is one-half the

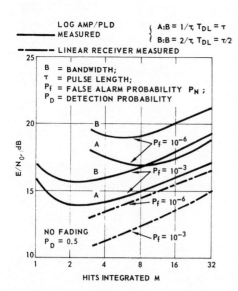

figure 4-4 *Measured total energy-to-noise density ratio vs. number of incoherent integrations. (From Hansen [158].)*

pulsewidth. While the detectability in this case is poorer, Hansen has shown that receive pulses that have greater duration than 1.4τ are rejected. Impulse noise less than $\tau/2$ in duration is also rejected for signal-to-noise ratios of 30 dB or less. Theoretical analyses of the losses in detectability are also available in the literature [158, 119, 147].

Other implementations of this same technique include the use of a logarithmic amplifier at IF (rather than at video) followed by a detector and a pulse length discriminator or FTC circuit.

The log-FTC and log-PLD have similar advantages in reducing clutter on displays. They both introduce a significant loss in detection characteristics if the effective time constants of the high-pass filter and PLD are short. It would seem that the log-PLD would be somewhat more effective in *lumpy* clutter (land, clouds, and small chaff bundles) since the mean value of the backscatter must remain constant for many time constants for the log-FTC to be effective, whereas the PLD can reject signals that are only 1.5 to 2 times the pulse length. A general discussion of this type of circuitry is also found in Lawson and Uhlenbeck [223, chap. 11] and Cope [73].

4.4 EFFECTS OF LIMITERS ON TARGET DETECTION

The use of a band-pass limiter[1] in the IF amplifiers of a receiver in order to compress the dynamic range of radar target signals, noise, and clutter appears to be one of the simplest means available to achieve automatic or adaptive detection of targets. On the other hand, the choice of waveform and receiver parameters (bandwidths, location, and characteristics of the matched filters, limiting level, etc.) and the calculation of the losses incurred in detection are quite complex for pulsed systems. This section will describe some of the general properties of limiters and the classes of radar waveforms that are amenable to various forms of limiting. In the absence of a detailed nonlinear analysis, it will only be possible to place upper and lower bounds on some of the effects. The emphasis will be on the target detection properties as a function of the various parameters rather than the maintenance of the fidelity of the input signal. The

---

[1] The terminology is standard but awkward. The amplitude of the waveform is *limited* by a symmetrical clipper, and, simultaneously, the band-pass of the waveform is *limited* in the frequency domain by a linear filter.

ability to maintain a constant false alarm rate (CFAR) for a wide dynamic range of input signals will also be emphasized along with some of the design compromises that are necessary.

The basic block diagram of a band-pass limited receiver is shown in Fig. 4-5.

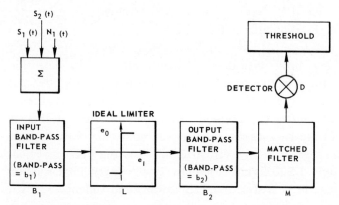

figure 4-5 *Block diagram of band-pass limited receiver.*

Not all of the blocks are necessary for all waveforms,[1] and in some cases the functions are combined:

$S_1(t)$ = target signal of interest at a carrier frequency f_1.

$S_2(t)$ = interfering signal, such as clutter, jamming, or interference from other radars.

$N_1(t)$ = system noise (gaussian) with a bandwidth $\gg S_1(t)$ prior to filter B_1.

B_1 = linear symmetrical band-pass filter with bandwidth b_1 and center frequency f_1. This may be the band-pass characteristic of the limiter itself.

L = limiter, which generally will be a *hard* limiter by definition having the property that the output signal e_0 = +1 if $e_i > 0$, = 0 if $e_i = 0$, and = −1 if $e_i < 0$, where e_i = the total input signal. (The limiter bandwidth is assumed to be greater than b_1.)

B_2 = linear symmetrical band-pass filter with bandwidth b_2 (less than b_1).

---

[1] The band-pass filter B_2 may be the matched filter, etc.

M = filter that is matched to the transmit waveform.

D = linear envelope or square-law detector.

T = threshold device.

Davenport [86][1] analyzed the effects of hard limiting on sinusoidal signals and gaussian noise. The spectrum of the noise is assumed to be narrow and to be centered on the carrier frequency f_1 of the CW signal. The band-pass filter B_2 has an ideal rectangular passband and is also centered on f_1. The ratio of the output signal-to-noise power ratio $(S/N)_0$, prior to the detector, to the input signal-to-noise $(S/N)_i$ was shown to asymptotically approach the following values:

$$\left(\frac{S}{N}\right)_0 \approx \left(\frac{\pi}{4}\right)\left(\frac{S}{N}\right)_i \quad \text{for} \quad \left(\frac{S}{N}\right)_i \to 0$$

$$\left(\frac{S}{N}\right)_0 \approx 2\left(\frac{S}{N}\right)_i \quad \text{for} \quad \left(\frac{S}{N}\right)_i \to \infty$$

(4-4)

The output $(S/N)_0$ is almost a linear function of the input $(S/N)_i$, and the maximum loss compared with a linear receiver is $\pi/4$ or 1.0 dB.[2] Davenport also showed that if the hard limiter were replaced by a "softer" square-root circuit, the maximum loss would be considerably smaller than 1.0 dB.

The analyses of Davenport were extended by Manasse, Price, and Lerner [225], who showed that the 1-dB loss occurred only if the noise out of B_1 was narrow-band. They showed that if the input signal-to-noise ratio was small and the bandwidth of the noise (at the output of B_1) approached infinity the loss in S/N approaches 0. When the signal and noise bandwidths are about equal, the maximum loss is only 0.6 dB.[3] They also indicated that the limiter degradation could be reduced by proper choice of the spectrum of the noise or even by adding noise whose spectrum fell outside of the filter B_2.

---

[1] Based on analysis techniques of Rice [311] and Middleton (see [386]). See also Sevy [341] for a discussion of modulated signals.

[2] The curve illustrating this effect can be seen as the uppermost one in Fig. 4-6.

[3] If the interference had a distribution function other than gaussian, the maximum loss remains quite small.

4.5 EFFECTS OF INTERFERING SIGNALS
IN SYSTEMS WITH LIMITERS

Cahn [58] showed that there is an additional degradation in the signal-to-noise ratio if there is an interfering signal $S_2(t)$ in addition to the desired signal $S_1(t)$ and receiver noise. His analysis applies when the desired signal $S_1(t)$ is much smaller than $S_2(t)$ at the limiter output. When $S_2(t)$ is also small compared to the noise, the loss approaches the 1.0-dB value reported by Davenport. On the other hand, when $S_2(t)$ is much larger than the noise [and also the desired signal $S_1(t)$] the loss in the ratio of S_1/S_2 approaches 6 dB. This is the *small signal suppression* effect that is common in FM radio systems. In radar problems $S_2(t)$ may represent clutter returns, the echoes from an undesired target, or interference from other radars.

Jones [196] also examined the case of two signals $S_1(t)$ and $S_2(t)$ and noise in a hard-limited receiver; two of his graphs are reproduced as Figs. 4-6 and 4-7. Figure 4-6 shows the ratio of the output signal-to-noise $(S/N)_0$ to the input signal-to-noise $(S/N)_i$ as a function of $(S/N)_i$ with the ratio of the interference S_2/S_1 as a parameter.[1]

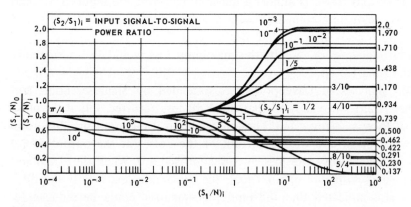

figure 4-6 *Ratio of the output SNR to the input SNR as a function of the latter for the ideal band-pass limiter. (From Jones [196].)*

The uppermost curve $S_2/S_1 = 10^{-3}$ is the same as that derived by Davenport and shows a 1-dB maximum loss. When the interference ratio

---

[1] Rubin and Kamen [328] have analyzed the case where $S_1(t)$ and $S_2(t)$ are separately filtered after limiting.

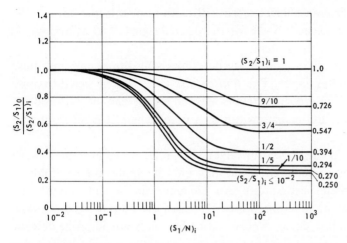

figure 4-7 *Ratio of output signal-to-signal power ratio to the input signal-to-signal ratio as a function of the larger input SNR. (From Jones, IRE [196].)*

reaches 0.2 and both signals are above the noise, $S_2(t)$ affects the signal-to-noise ratio of the desired signal. When $S_2/S_1 \geq 10$ and both are much greater than the noise, the signal-to-noise ratio of the desired signal is reduced by 3 dB. The intermediate cases are best described by the graph itself.

In Fig. 4-7 the ratios of the output signals are shown versus the signal-to-noise ratio of the larger signal. If both signals are less than the noise, the limiter is essentially a linear device. For two large signals, the smaller signal is suppressed by 6 dB. Some tentative conclusions would include:

1. If $(S/N)_i$ is very small, the limiting loss approaches $\pi/4$ until $S_2(t)$ dominates the limiter.

2. If the interference ratio S_2/S_1 is large and $(S/N)_i$ is also large, the loss in the $(S/N)_0$ of the desired signal is 3 dB.

3. If the interference ratio is much less than unity, the loss of the desired signal in all cases is small; there is actually an improvement in $(S/N)_0$ for $(S/N)_i$ greater than unity (this does not imply an improvement in detectability).

4. If $S_1(t)$ is within about 20 percent of $S_2(t)$ and they are both *large* with respect to the rms noise, the loss in signal strength of *either* the

signal or the interference may exceed 6 dB with respect to the noise.[1] The loss in S/N never exceeds 6 dB for the desired signal $S_1(t)$ assuming that $S_2(t)$ is the *noise* of the system.

5. The effects of the limiter are determined by the signals and noise levels at the limiter input. The *limiter loss* (i.e., loss in signal-to-noise ratio) and the suppression of the desired signal can be reduced by increasing the prelimiter bandwidth to a point where it is much greater than b_2.

The increase in the $(S/N)_0$ for large $(S/N)_i$ referred to in item 3 above does not improve detectability and may only be applicable to problems involving parameter estimation. An analogy can be drawn with the square-law envelope detector, which increases the output signal-to-noise ratio for large input signal-to-noise ratios but does not improve detectability[2] [225].

4.6 DETECTION STATISTICS OF PULSED SYSTEMS WITH LIMITING

The previous discussion was based on CW systems, and the results are not easily related to the general pulse radar detection problem. It is difficult to interpret the output of a *threshold detector* following a limiter and bandpass filter B_2 where the input is continuous, and its bandwidth and the noise bandwidth are identical. If the signal and noise spectra were identical and at the same carrier frequency, the amplitude of the limiter output would be almost constant with or without the signal being present. The implementation of a threshold type of detection system can be more easily visualized if the input noise bandwidth is much larger than the spectrum of the input signal and the output passband b_2 is very small compared to b_1 (B_1 will be assumed to limit the noise bandwidth). In this case, the output of B_2 will be small[3] compared to the *maximum possible* output of the limiter.[4] It is then easy to picture a threshold

---

[1] Both $(S/N)_1$ and $(S/N)_2$ at the output will remain greater than unity.

[2] The *detectability* of signals in noise is determined by the amplitude distribution of the detected output.

[3] The output noise power of the filter B_2 will be approximately b_2/b_1 times the maximum power output of the limiter for hard limiting in the absence of signal.

[4] By definition, the power output of a hard limiter has a constant value for any nonzero level of input signal, noise, or combination thereof. Alternatively, since noise, at least, is always present, a hard limiter is always saturated.

device that is set well above the detected output of B_2 for noise alone but below the maximum output for limiter-filter combination for $S_1(t)$ alone. As the input signal $S_1(t)$ is increased, the output of B_2 will increase almost linearly until the input signal is about equal to the total noise power into the limiter. The linear behavior of the output as $S_1(t)$ is increased is caused by the *softening* of the limiter by the wide-band noise that is passed by B_1. On the other hand, the output at the detector will not be increased by an increase in noise alone if the noise spectrum is flat over the entire passband of B_1 since the added noise is mostly outside of B_2. An example of this effect is illustrated in Fig. 4-8 where b_1/b_2 is a factor of four and there is postdetection integration of 32 pulses. Even though the mean-square value of the input noise was ten times that needed to limit, the output from each pulse was only about one-fourth of the maximum signal output. The noise level on the display remained constant, and a signal equal to the noise could easily be seen. This represents one case of a constant false alarm rate (CFAR) system where the number of threshold crossings due to noise (or jamming) is independent of the total power of the input noise as long as the noise spectrum is flat. The outputs of a 32-pulse coherent system are shown in Chap. 13.

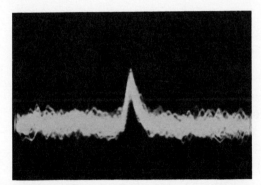

figure 4-8 *Postdetection integration of 32 hard-limited pulses. 10 dB limiting, 10-μsec trace; $b_1/b_2 = 4$; $(S/N)_i = -3$ dB in b_1.*

The following discussions will emphasize the effects of the bandwidths of filters B_1 and B_2 and the effects of interfering signals on the detection statistics. A study to determine the detection statistics for a pulse radar is described by Bello and Higgins [36], who analyzed the detection statistics of a pulsed signal in wide-band IF noise in which the total signal is hard limited and passed through a single-tuned high Q filter B_2. The bandwidth ratio b_1/b_2 is of the order of 150, which is the case where the noise output power of B_2 (and the detector) is small compared to the maximum signal output. The threshold detector can thus be set well above the noise but still below the maximum signal output. The result of their study indicates that when a limiter is introduced the input signal-to-noise ratio must be increased by 2.7, 2.1, 1.6, and 1.3 dB to achieve typical values of P_D for false alarm probability P_N settings of about 10^{-11}, 10^{-8}, 10^{-4}, and 0.017, respectively. These losses assume that the threshold is set with respect to the noise output without the limiter. With the limiter the noise output decreases; therefore, setting the threshold on the noise at the limiter output would tend to reduce these losses. If the threshold is set at the proper level above the noise output of the limiter, the 1.6-dB increase in $(S/N)_i$ for $P_N \approx 10^{-4}$ would drop a few tenths of a decibel. It is difficult to generalize Bello and Higgins' results because a matched filter was not used for the rectangular pulse and their total processing loss L_s, therefore, may include both the mismatch loss and the limiter losses. The noise output of a limited system is also studied in Silber [349]. The effect on S/N of the type of filter used for B_2 is discussed by Sevy [341].

4.7 LIMITING IN PULSE COMPRESSION AND PULSE DOPPLER SYSTEMS

Experimental results were reported by Bogotch and Cook [47] who used a linear FM pulse compression signal (chirp) having a time-bandwidth product (compression ratio) of 35:1. In their limiter tests $b_1 = b_2 =$ the frequency deviation of the FM pulse. With a dispersed signal, the *instantaneous* signal input to the limiter is smaller than the equivalent simple pulse by the compression ratio (for the same pulse energy). Their detection tests were concerned with the case of input signal-to-noise ratios less than unity, which is common in pulse compression (PC) systems. They showed that to achieve a given probability of detection, an increase in signal-to-noise ratio of 1.0-2.0 dB above a linear system was required for various values of limiting level above the rms noise.[1] In

---

[1] The 1.0-dB value was for limiting at the rms noise level, and the 2.0-dB loss was for about 40 dB of limiting. (The rms noise was 40 dB greater than a signal that would barely limit.) The false alarm probability was 2×10^{-4}.

other tests, they report that the (S/N) for detection, with the PC system, was 1.0 dB poorer than a linear pulsed CW mode followed by a $1.2/\tau$ band-pass filter. It appears from their experiments that for a typical pulse compression system the sum of the filter mismatch and limiting losses can be held below 1.5 dB if the compression ratio is large enough and if there is no interference other than broadband noise.

In multiple-pulse coherent radar systems, such as pulse doppler systems, the limiting loss can be minimized by a small *increase* in predetection bandwidth above a matched filter. It has been reported by Silber [349] that for a 30-pulse system with single pulse $(S/N)_i = 0.5$ and $b_1/b_2 = 2$, the detection probability dropped from 0.9 (linear system) to 0.7 (limiting). The noise was limited at its rms value, and the false alarm probability was 10^{-9}. For higher numbers of pulses the loss in detectability drops rapidly even with small bandwidth ratios.

When the number of pulses integrated in a hard-limited coherent pulse radar system is small, the prelimiter bandwidth must be widened to permit a threshold to satisfy two conditions:

1. The threshold level must be well above the sum of the detected rms noise samples out of the filter B_2 to achieve the desired false alarm probability.

2. The threshold level must be set below the maximum possible output of the combination of the limiter, postlimiter filter B_2, and the detector.

If there were only 10 pulses to be integrated, and $b_2/b_1 = 1$, the average sum of the noise samples would only be about 10 dB below the sum of the ten coherently integrated signal pulses if there were no noise. The two conditions stated above could not be met for low false alarm probabilities. The same difficulty exists for pulse compression systems with low compression ratios.

The effect of increasing the prelimiter bandwidth can be seen in Table 4-2, derived from Silber's analysis. The false alarm probability is 10^{-9}, the limiting is *hard*, and the desired detection probability is 0.5 for 20 pulses coherently integrated.

The limiting loss in this case is the deviation of the required input signal-to-noise from the Marcum computations for a linear system. The results are pessimistic because of the low false alarm probability ($P_N = 10^{-9}$) and the assumption that the rms noise is at least 13 dB above the limit level (very hard limiting). It can be seen that as the prelimiter bandwidth is increased and the input signal-to-noise ratio approaches zero, the 1-dB loss figure reported by Davenport is reached.

TABLE 4-2   Limiting Loss for Multiple Pulse Coherent Radar

b_1/b_2	Nb_1/b_2	$(S/N)_1$	Limiting loss, dB
1	20	$\gg 10$	$\rightarrow \infty$
2	40	1.1	≈ 3.2
3	60	0.57	2.1
4	80	0.38	1.6
5	100	0.29	1.4
∞	∞	0	1.0

Notes: N = 20 pulses
rms noise = 20 times the limit level
    $P_D = 0.5, P_N = 10^{-9}$
  $(S/N)_1$ = ratio of the signal power to total noise
        power in filter B_1 (per pulse)

A more general curve from Silber [349] (Fig. 4-9) shows the limiting loss as a function of the ratio of rms noise-to-limit level for several values of input signal-to-noise ratio. As the input noise-to-limit level ratio is decreased, the limiting loss is decreased; however, at the same time the CFAR action of the limiter is degraded, and an increase in the input noise level due to narrow-band jamming or clutter will increase the false

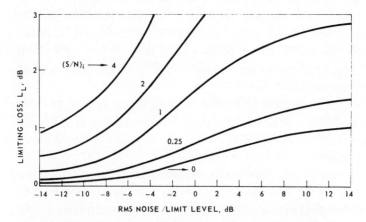

figure 4-9 *Signal loss due to limiting as a function of hardness of limiting with input signal to noise as a parameter. (From Silber, IEEE [349].)*

alarm rate P_N by several orders of magnitude. The 1.0-dB increase in the limiting loss in going from 0 to 40 dB limiting, reported by Bogotch and Cook [47], is in agreement with Fig. 4-9. Their input signal-to-noise ratio varied from $-$ 15 dB to unity.

4.8 SUMMARY OF LIMITER EFFECTS

The broadening of the spectrum of the *noise* input to the limiter has been shown to be effective in minimizing the limiter loss while maintaining a constant false alarm probability P_N. The retention of a low value of P_N in the presence of clutter $S_2(t)$ is a much more difficult problem. The spectrum of the clutter echoes has approximately the same width as the transmit signal $S_1(t)$ in pulse radar systems, and a widening of the input filter will only have a second-order effect unless both $S_1(t)$ and $S_2(t)$ are small. Since the clutter looks like a signal, the output of the postlimiter filter B_2 can also approach the maximum signal output for simple pulse radar systems with only clutter signals.

In multipulse or pulse compression systems, radar clutter often appears noiselike at the limiter. If the clutter is similar to noise and the number of pulses *or* the compression ratio is very large, the loss would seem to revert to the 1.0-dB maximum of Davenport's continuous signal analysis with *narrow-band noise*. The only widely published experimental work on the influence of strong interference is the previously cited paper by Bogotch and Cook [47]. They added a second chirp signal $S_2(t)$ to their limiter, displaced in time from the desired chirp signal $S_1(t)$, and measured the detection probabilities of $S_1(t)$ as a function of the overlap of the signals and the ratio $S_2(t)/S_1(t)$. The probability of detection of the desired signal dropped rapidly as $S_2(t)/S_1(t)$ exceeded unity for the case of 75 percent overlap. The limiting loss in these cases was quite high when the input signal-to-noise ratio was near unity[1] and was substantial even when $(S/N)_i$ was $-$ 5.2 dB. These severe losses can be attributed to two design requirements that are of great importance in the use of limiters for pulse radars as contrasted with communication systems:

1. The compression ratio or time-bandwidth product in the Bogotch and Cook experiment [47] was only 35:1 and leaves little margin between the threshold and the maximum signal output. This can be seen

[1]$(S/N)_i = + 0.8$ dB, $- 2.2$ dB (the desired signal).

from Table 4-2 (from Silber's analysis) by noting the loss for values of Nb_1/b_2 between 20 and 40. High-pulse compression ratios are needed for CFAR action in the presence of clutter echoes. The required ratio depends on P_N.

2. In a pulse radar detection system, the important criterion is the false alarm rate in the absence of a target signal. The suppression of the noise by interference allows a reduced threshold *only* if the interference exists continuously. The cited studies of continuous interference can only lead to quantitative clutter calculations if the results can be modified for intermittent interference. If the system designer is to take advantage of the noise reduction by intermittent interference, the threshold level must be adaptive. This is discussed in Sec. 4.10.

Some special cases of limiting and digital quantization are expanded in connection with the digital processors described in Chap. 14.

4.9 SEQUENTIAL DETECTION

In the previous discussions of detection by means of setting a threshold after the detector, no justification was given as to what constitutes an optimal decision procedure. A technique, which has enjoyed wide acceptance in hypothesis testing, is known as the method of *maximum likelihood* or *probability ratio test* [87]. The likelihood function is the ratio of the probability density of signal-plus-noise to noise alone. From this function a statistical test for radar detection can be established that minimizes two errors:

1. Declaring that a target is present when in fact there is only noise (or clutter, jamming, etc.). The probability of this occurrence is the false alarm probability P_N.

2. Declaring that a target is not present when in fact there is a target signal. The probability of this *false dismissal* is simply one minus the probability of detection P_D.

From the likelihood function (or its logarithm) it is possible to compute the signal-to-noise ratio out of a matched filter that is required for a given P_D and P_N. This procedure will yield graphs similar to Figs. 3-3 through 3-6. If the returns from N incoherent pulses are collected and the signal-to-noise ratio on each pulse is small, a square-law detector followed by a linear integrator is a good approximation to a maximum likelihood detection criterion [167].

In the analysis of Chap. 3 it was assumed that N was fixed for a given beam position. In sequential testing the same procedures may be followed except that the sample size N is not fixed and two thresholds[1] are used [394, 294]. One pulse is transmitted, and the echo is observed for a given range cell. A decision as to whether a target is or is not present can often be made on this information, and there is no need for further transmissions in that beam position. If, on the other hand, the signal has an intermediate value and a decision cannot be made, a second pulse is transmitted in the same direction as the first; and its echo is added to the first. This is illustrated by Fig. 4-10 [294]. The upper and lower thresholds are represented by the dashed lines, and the ordinate is the integrated video voltage. A *decision* is made whenever the integrated video either exceeds the upper threshold (a detection) or falls below the lower threshold, indicating that a target echo of a given strength is not likely (a dismissal). Pulses are transmitted in the same direction until one or the other decision is made in the range cell of interest. It was shown that on the average fewer are required (less transmission time) in a given beam position than with a fixed sample size [167]. For a single range bin Preston [289] has shown that for $P_D = 0.9$ and $P_N = 10^{-8}$, less than one-tenth the time of that of a classical detector is needed to determine

---

[1] It is also possible to use three or more threshold levels with some additional advantages [117].

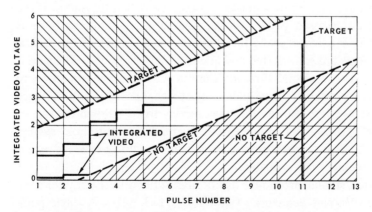

figure 4-10 *Pulse sequence illustrating probability ratio sequential detector for single range bin. (After Preston, IRE [294].)*

the absence of a threshold signal. In the presence of a threshold signal about one-half of the time is required. Alternately, the time saving can be traded for increased sensitivity.

In a search radar numerous range cells must be observed in a given beam position, and the absence of a signal crossing the lower threshold in any one of these bins becomes less likely. For a P_D = 0.86 and $P_N \approx$ 3×10^{-11} with 30-, 100-, and 300-range cells per sweep, the probability ratio sequential detector yielded 4.4 dB, 3.6 dB, and 3.2 dB improved sensitivity. These values were for binary quantized video (see Chap. 14) and would be somewhat higher for analog integration. Similar improvements are reported by Helstrom [167].

With a rotating radar the practical improvement is less dramatic since the time saved cannot be used to transmit pulses in another beam position. Alternately, if both thresholds are lowered to increase the sensitivity, the number of samples may increase to where the test has not been completed before the antenna has moved to a new beam position because of near threshold signals or clutter. For these reasons the primary interest in sequential detection is in array radars or in cases where average transmitter power needs to be reduced part of the time.

The problems of the time varying threshold of the radar *hanging up* in a given beam direction have precluded the recommendation of *basic* sequential detection even for array radars. Several simpler procedures have been proposed to limit the number of decisions and the maximum time per beam position. Helstrom [168] proposed a two-step procedure where the integrated sum of *n* pulses is compared with a threshold. If the threshold is not crossed in any of the range cells, the antenna beam is moved to the next sector. If the threshold is crossed in one of the cells, *m* additional pulses are transmitted and added to the previous *n*. The sum is then compared to a second threshold, and a firm decision is made as to whether there is a target. Essentially "most signal detection takes place in the first stage, with the second stage serving to keep down the rate of false alarms" [168]. The improvement over a single threshold system varies with the input signal-to-noise ratio and the choice of *n* and *m*, but 2-3 dB improvement is reasonable.

A somewhat similar procedure has been described by Finn, which he calls "Energy Variant Sequential Detection" [117, 118]. A single pulse (or pulse train if more energy is required) is transmitted, and the return compared with a threshold. If the threshold is crossed, a second higher energy pulse is transmitted and the signals compared with a second threshold. If both thresholds are crossed in the same range bin, a detection is declared. For 100- or 300-range bins per sweep 3-4 dB power savings are obtained. An additional 0.5 dB is gained by adding a third step to the process. Brennan and Hill [51] have studied the cumulative detection probability using this technique for the four Swerling target models. Similar improvements are obtained, and they show that for the slowly fluctuating targets (cases 1 and 3), it is desirable to transmit the second pulse immediately after the first since first-pulse detection implies a favorable target cross section σ_t at that time. Unfortunately, if a clutter echo has crossed the first threshold, it will probably cross the second one as a result of the relatively long correlation times of clutter echoes. Finn and Johnson [118] have studied the situation where strong clutter signals are present, but these clutter signals are uncorrelated from pulse to pulse. Significant improvement is still obtained over the simple threshold if the clutter can be decorrelated by leaving sufficient time between transmissions or by frequency agility; however, both types of detection still suffer from the existence of the strong clutter signals.

A final limitation on the use of sequential detection is the current tendency toward high-resolution radar. For most of the sequential techniques, the improvement drops to about 2.5 dB for 1,000 range or doppler cells per sweep and even less for 10,000 cells (with energy variant sequential detection). With 100 or more cells per sweep the second transmission should have considerably more energy than the first. When high resolution (many cells) is a requirement for target sorting, it is desirable to use a lower resolution waveform for the first pulse followed by a second or third waveform of higher resolution.

4.10 ADAPTIVE THRESHOLD TECHNIQUES

The use of hard limiters, sensitivity time control, and logarithmic amplifiers followed by the various forms of pulsewidth discrimination constitutes an attempt to adapt the radar to its environment by acting on the total signal prior to the threshold detector. An alternative is to measure the total signal input in a reference channel close to the time when a possible target signal arrives at the threshold. Some parameter of the total signal that is measured over a period of time[1] is then used to vary the threshold level by some statistical law (*maximum likelihood*, etc.).

---

[1] In one or more doppler channels for coherent systems.

This is not a new technique; it is implied in the analysis of Marcum [244] where he assumed that the noise level is measured prior to the determination of the threshold. The same procedure can be used for a general environment composed of noise, jamming, and clutter if the probability density function and power spectral density of the interference are known. The threshold is established as a function of the reference channel, which must contain a sufficient number of samples of the *noise* so that its output is an accurate measure of what will occur in the signal channel.

For a pulsed radar, the simplest implementation of an adaptive threshold is the measurement of the mean-square output of the matched filter or detector prior to the reception of the return signal. To derive an accurate estimate of the rms noise level, the noise samples must be averaged. (This may consist of detection of the noise, integration over time, and normalization with respect to the integration time.) When there is sufficient averaging time, this procedure can accommodate changes in the average noise level of the receiver. If, in addition, the antenna has already been pointed in the direction of the subsequent transmission, this procedure will adapt to slow changes in the environmental noise (sky noise, etc.) and to continuous noise jamming. The obvious shortcoming of this procedure is that there are no radar clutter signals until after the transmission, and this major contributor to the total *noise* of the system will not be accounted for.

The use of the word *time* in this discussion is an oversimplification, as it has been shown that the determination of the rms value of the noise to a given accuracy can also be obtained by integration over a broader frequency band prior to the matched filter [9, 119]. Such techniques can lead to an accurate threshold determination if the shape of the spectral density of the noise versus frequency is known along with the transfer characteristic of the matched filter. The accuracy of the determination of the threshold is thus related to the number of independent samples of the noise, whether the samples are taken in the time or frequency domains. The statistics of this determination are included in the next part of this section where the adaptive process is expanded to include measurements of the clutter return.

Figure 4-1 showed a typical total environment for an air-search radar where the clutter returns are extensive and span a dynamic range of over 40 dB. This broad range of signals often appears at a preset threshold where a 3-dB change in the *noise* level will raise the false alarm probability from 10^{-6} to 10^{-3}. In an automatic detection system this is an intolerable effect. Adaptive detection in this type of environment can be accomplished under certain restrictions. In order to keep the results general, the transmitted waveform will be arbitrarily assumed to have considerable extent[1] in both time and frequency, and the receiver implementation will be assumed to consist of multiple matched filters, each of which is centered at a possible target doppler frequency. A resolution cell will then be defined as being bounded by the smallest resolvable time and the smallest resolvable doppler frequency. It will also be assumed that both the receiver noise and clutter samples in all the cells in the region of the target are independent.

It has been shown [119, 154] that by sampling the outputs of a number of the cells surrounding the cell under test, a maximum likelihood estimate of the clutter distribution function can be made, and the threshold can be set with respect to the clutter and receiver noise. Finn [119] has shown that if the first-order probability density function of the clutter and noise is Rayleigh, then the desired threshold for a given value of P_N is

$$ T = K \left[\frac{1}{m} \sum_{i=1}^{m} \frac{x_i^2}{w_i} \right]^{1/2} \tag{4-5} $$

where K = constant dependent on P_N

   m = number of resolution cells used to establish the threshold

   x_i = linearly detected output of ith cell

   w_i = weighting factor to compensate for the range and velocity responses of the receiver (e.g., receive power may vary by $1/R^4$, $1/R^3$, etc.)

The significance of this equation is that, if the weighting factors w_i can be predetermined, the threshold setting is merely a constant times the measured power in the reference cells surrounding the target. In this case the threshold is adaptive to the average power of the interference.

As the number of cells in the threshold determination increases, the loss from the ideal case, where the noise statistics are completely known,

---

[1] The time extent is great enough to require multiple doppler filters. The results are also applicable to incoherent pulse radars.

decreases. The increase in the required signal-to-noise ratio for a given probability of detection P_D and false alarm probability P_N versus the number of independent samples has been calculated by Hall[1] [154], Finn [119], and Ares [9]. Examples are shown in Figs. 4-11 and 4-12. This *loss* can be interpreted as the necessary increase in the threshold level above the values calculated by Marcum, Swerling, Hall, etc. This increased threshold level is needed to allow for the fluctuation in the reference measurement due to a finite number of samples.

In an incoherent pulse radar with a single matched filter, the samples must be taken in the time (range) domain to adapt to the clutter. If the clutter is "patchy," a poor estimate of the statistics at any one location will result. A poor estimate of the clutter density also occurs at the leading and trailing edges of the clutter region. Finn [119] has also calculated a number of cases where there is a step discontinuity in the clutter level. If a target is in the clutter region, but the clutter occupies only one-half of the total reference region (40 cells), the false alarm probability will, in this case, only increase from 10^{-8} to 10^{-4} for step discontinuities of 20 dB or more.

In the previous analyses, it was assumed that the target return was not included in the determination of the threshold. In addition, the assumed range extent of the samples was small enough compared to the range of the cell of interest so that there was no range dependence of the clutter backscatter power. If the threshold is to adapt to distributed clutter, such as extensive rain, and the samples cover a 3-nmi interval, the weighting factors w_i must accommodate the inverse R^2 dependence of rain backscatter. This might apply to threshold determination in the first 20 to 30 nmi from the radar. An alternative would be to include an STC circuit prior to the threshold determination and video channels. Another drawback to sampling many range bins is that the sampled interval may contain several desired targets. The existence of these targets will raise the threshold and possibly prevent detection of some desired targets.

Finn has also compared detection statistics of clutter echoes having the Rice distribution with those having the Rayleigh distribution. The Rice distribution is characterized by a coherent or steady component in addition to an incoherent, fluctuating component. This distribution is often attributed to the temporal fluctuation of ground clutter returns (but not to the spatial distribution). For the case where the threshold has

---

[1]In a slightly different context.

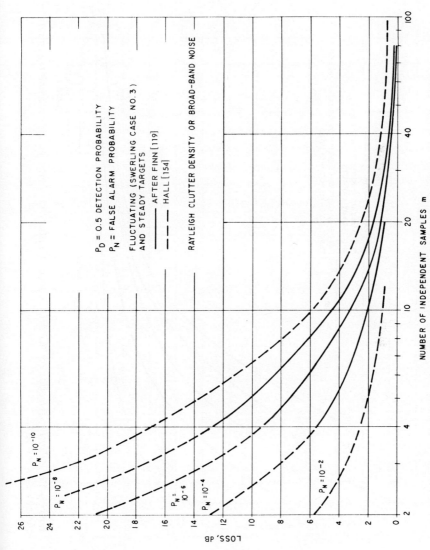

figure 4-11 *Adaptive detection or "CFAR." Loss vs. number of independent noise samples m to set threshold* $P_D = 0.5$.

137

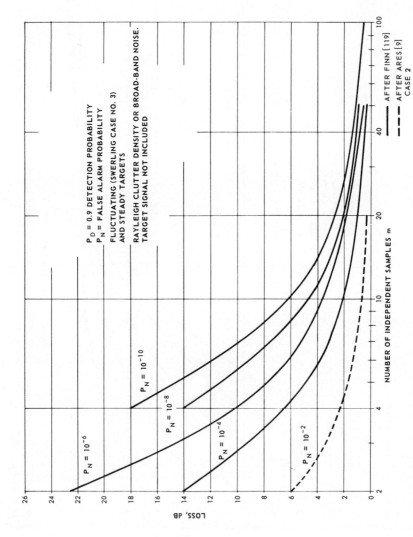

figure 4-12 *Adaptive detection or "CFAR." Loss vs. number of independent noise samples m to set threshold $P_D = 0.9$.*

been mistakenly set on the assumption of a Rayleigh distribution, the threshold is set higher than need be; and the target detection and false alarm probabilities are considerably reduced. For example, if the threshold is set for $P_N = 10^{-8}$ on the basis of a Rayleigh distributed sample of 30 cells and the actual signals have a ratio of the coherent to incoherent component of 2:1, the false alarm probability will drop to 4×10^{-15}. This causes an additional loss of several decibels in target detectability above the value obtained if the correct distribution had been assumed. The opposite assumption can also be made, and an expected 2:1 ratio of coherent to incoherent component used to set the threshold. In this case there is a relatively minor increase in P_N from 10^{-8} to 3×10^{-5}. This form of adaptive threshold is much less sensitive

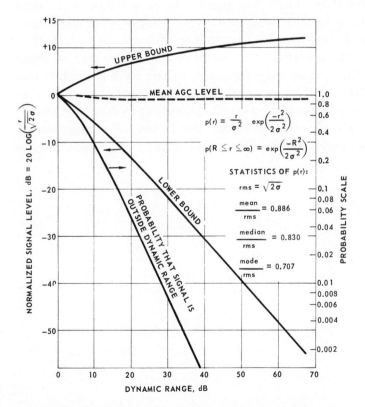

figure 4-13 *Dynamic range bounds for minimum probability of excluding a Rayleigh-distributed signal. (From Ward [396].)*

to the clutter distribution than the log-FTC circuit [119] described in Sec. 4.3.

4.11 DYNAMIC RANGE OF RAYLEIGH SIGNALS

The effectiveness of an adaptive technique using a *reference* channel is to some extent dependent on the dynamic range of the reference channel. A related design problem, which is especially important in digital receivers, is that of choosing where to set the available dynamic range with respect to the mean or median levels of the input signals. A graph to aid in this determination is shown in Fig. 4-13 from Ward [396]. The left ordinate is the input signal voltage expressed in decibels, and the abscissa is the dynamic range of the channel of interest. The probability that a signal will lie outside the dynamic range is shown as the right ordinate. The line marked "Mean AGC Level" is equal to the mean signal level for large dynamic ranges. The criterion of minimum probability of exceeding the dynamic range may not be optimum from the standpoint of detection, but it would seem reasonable to have a low probability that the peaks of the interference exceed the available dynamic range.

Rappaport [303] generalized on Ward's analysis to show the dynamic range centering for signal fluctuations that follow a Rician probability law. He shows curves that include several ratios of signal to clutter to noise.

5 RADAR TARGETS

5.1 GENERAL SCATTERING PROPERTIES—SIMPLE SHAPES

The radar range equation expresses the range at which a target may be detected with a given probability by a radar having a given set of parameters. This equation includes the target's *radar cross section σ* (abbreviated RCS) which is a measure of the proportion of the incident energy reflected back to the radar.[1] This returned energy varies with a multitude of parameters such as transmitted wavelength, target geometry, orientation, and reflectivity. Thus the question arises as to what is the proper description for radar evaluation purposes of a target's RCS.[2]

---

[1] A monostatic radar is assumed except as noted.
[2] A number of excellent bibliographies are available [75, 74].

The radar cross section of an object is proportional to the far-field ratio of reflected to incident power density, that is

$$\sigma = \left[\frac{\text{Power reflected back to receiver/unit solid angle}}{\text{Incident power density/}4\pi}\right]$$

Using this definition, consider the RCS of a perfectly conducting isotropic scatterer. The power intercepted by the radiator is the product of the incident power density P_I and its geometric projected area A_I. By the definition of isotropic scattering, this power is uniformly distributed over 4π steradians. For this isotropic scatterer then

$$\sigma_i = 4\pi\left[\frac{P_I A_I/4\pi}{P_I}\right] = A_I \qquad (5\text{-}1)$$

Thus, the RCS of such an isotropic reflector is its geometric projected area. The RCS of any reflector may be thought of as the projected area of an equivalent isotropic reflector which would return the same power per unit solid angle. A reflector which concentrates its reflected energy over a limited angular direction may have a RCS *for that direction* which exceeds its projected area. This indicates that, when specifying cross sections, one must also specify the aspect of the target. The RCS is also dependent on other parameters besides aspect angle.

The scattering properties of a sphere serve to illustrate the nature of the RCS wavelength dependence [354, p. 41]. There are three distinct regions of behavior for the RCS of a sphere of radius r. In the region for which $2\pi r/\lambda < 1$ (the *Rayleigh* region) the RCS is proportional to λ^{-4}. As the wavelength is decreased beyond the point where $2\pi r/\lambda = 1$, the *Mie* or *resonance* region is entered where the RCS oscillates between ever diminishing values approaching the optical cross section. Finally, for very small λ the optical region is entered where the fluctuations converge to the optical cross section πr^2. The following paragraphs give a basic description of scattering behavior for each region for perfectly conducting objects.

The Rayleigh Region

Although radar targets do not usually fall within the Rayleigh region (target dimensions smaller than the transmitted wavelength) other scatterers such as precipitation particles do. The wavelength dependence of the RCS for a sphere in this region is proportional to λ^{-4}, which also applies to most objects which have overall dimensions small compared to a wavelength. Siegel [346] shows that the critical parameter in determining RCS is the volume, slightly modified to account for the gross features of the shape. He finds that, for most bodies of revolution, the RCS at an aspect along the axis of symmetry is closely approximated by

$$\sigma = \frac{4}{\pi} k^4 V^2 F^2 \tag{5-2}$$

where $k = 2\pi/\lambda$, V is the volume of the scatterers in m^3, and F is a dimensionless factor which depends on the gross shape of the body. This shape factor may be neglected for spheroids except those of the very flat, oblate variety. Shape factors for other geometric shapes are also derived in Siegel's paper. This approach was found to lose accuracy as the scatterer became more flat; a flat surface has zero volume, but a nonzero cross section.

The RCS of any object in the Rayleigh region will be sensitive to polarization and aspect angle if one major dimension is very small compared with another.

Optical Region [80, 340]

The optical region in which most radar targets of practical interest reside is so named since the ray techniques of geometric optics may be applied to the problem of RCS estimation. Any smooth curved surface nearly normal to the incident field will give a specular return. From a consideration of the power reduction due to the divergence of the scattered beam, the RCS is found to be

$$\sigma \approx \pi R_1 R_2 \tag{5-3}$$

where R_1 and R_2 are the principal radii of curvature at the surface normal. In the optical region the RCS behavior with wavelength is monotonic although the RCS does not necessarily converge to a constant value. The RCS behavior with wavelength may be classified for many simple objects in terms of the principal radii of curvature at the point where the normal to the surface is parallel to the direction of incidence:

1. λ^{-2} dependence: two infinite radii of curvature (e.g., flat plates)
2. λ^{-1} dependence: one infinite, one nonzero (e.g., a cylinder)
3. λ^0 dependence: (*a*) one infinite, one zero (e.g., a wedge)
                                          (*b*) two nonzero finite (e.g., a spheroid)
4. λ^1 dependence: one nonzero finite, one zero (e.g., a curved edge)
5. λ^2 dependence: two zero (e.g., the apex of a cone)

Numerous examples of these classes are available in the literature [20, 38, 79, 184, 346].

The RCS of large complex reflectors may often be approximated by breaking the body into individual reflectors and assuming that the parts do not interact. In this case the total RCS is just the vector sum of the individual cross sections

$$\sigma = \left| \sum_k \sqrt{\sigma_k} \, \exp\left(\frac{j4\pi d_k}{\lambda}\right) \right|^2 \tag{5-4}$$

where σ_k is the RCS of the kth scatterer, d_k is the distance between the kth scatterer and the receiver. This formula has an important application in estimating the RCS of an array of scatterers.

The greatest use of the optical approximation is the calculation of specular returns or their sidelobes. This method may fail when there is a surface singularity such as an edge or a shadow; in this case the wavelength dependence is generally of a lower order than for a specular return alone. Surface singularities cause *second-order* effects, including *creeping* or *traveling* waves, which may actually be the dominant source of reflections when specular returns are weak [340].

Resonance Region

Approximate methods can provide a good estimate of the RCS when the characteristic size of the body is much smaller than the wavelength (Rayleigh region) or when it is much greater than the wavelength (optical region). In between lies a region where the geometry of the body is a critical factor, and neither the optical nor the Rayleigh methods can be easily applied to the problem of RCS estimation. Although a few techniques for RCS estimation in this region are available, no simple generalizations are possible. This subject is treated in several papers [184].

5.2 POLARIZATION SCATTERING MATRIX

Previous paragraphs have discussed the wavelength dependence of scattering. The polarization dependence is often of importance also. Consider a linearly polarized plane wave, propagating along the line of sight of the antenna, that is incident on a target. The incident field at the target can be resolved into components of electric field $E_H{}^T$ and $E_V{}^T$ along the H and V axes, respectively.[1] In general the component of scattered electric field E^S resulting from the incident H component *only* will have both H and V components. Expressed in terms of scattering coefficients, the H component of the scattered field is

$$E_H{}^S = a_{HH} E_H{}^T$$

and the V (orthogonal) component is

$$E_V{}^S = a_{HV} E_H{}^T$$

The first subscript in a_{HV} is for the transmitted component and the second for receive. The total scattered energy is the vector sum of that due to $E_H{}^S$ and $E_V{}^S$. When *both* components of incident field are present, the scattered field can be concisely stated in terms of the scattering matrix. (See [38, 133, 146, 235].)

$$\begin{pmatrix} E_H{}^S \\ E_V{}^S \end{pmatrix} = \begin{pmatrix} a_{HH} & a_{HV} \\ a_{VH} & a_{VV} \end{pmatrix} \begin{pmatrix} E_H{}^T \\ E_V{}^T \end{pmatrix}$$

A similar polarization matrix exists for circular or eliptical polarization. Expressed in terms of right and left circularly polarized waves E_R and E_L,

$$\begin{pmatrix} E_R{}^S \\ E_L{}^S \end{pmatrix} = \begin{pmatrix} a_{RR} & a_{RL} \\ a_{LR} & a_{LL} \end{pmatrix} \begin{pmatrix} E_R{}^T \\ E_L{}^T \end{pmatrix}$$

---

[1]While the axes are arbitrary, this text will only use the more conventional horizontal H and vertical V axes (H or V polarization).

If the reflector has circular symmetry about the line of sight axis, then the matrices have the properties

$$a_{HV} = a_{VH} = 0 \quad \text{and} \quad a_{HH}, a_{VV} \neq 0$$

$$a_{RR} = a_{LL} = 0 \quad \text{and} \quad a_{RL}, a_{LR} \neq 0$$

(5-5)

and, for most cases, regardless of symmetry

$$a_{HV} = a_{VH}, a_{RL} = a_{LR}$$

One important consequence of these properties is that a radar using the same circularly polarized antenna for both transmission and reception will receive little or no power from approximately spherical targets, such as raindrops, whereas the same transmit and receive linearly polarized antenna will receive most of the energy. This property has been used to ascertain the ellipticity of raindrops[1] and to reject weather clutter on both military and air traffic control radars.

Numerous experiments have been performed on the reduction of RCS of raindrop echoes with the use of *same-sense* circular polarization (σ_{RR} or σ_{LL}). In most of these radar experiments the reduction has been 15-30 dB where there are no terrain reflections. For surface radars and low-elevation angle beams (0-10°) the reduction is less [31]. Some experimental results are shown in Table 5-1 for different frequency bands and elevation angles. The reduction in rain echoes by circular polarization appears to be almost constant through X band with poor results for heavier rains.

Unfortunately the RCS of complex aircraft also reduces with same-sense circular polarization. Some typical values of the reduction in RCS with circular polarization below the values for linear polarization (σ_{HH} or σ_{VV}) are also shown in the table. One series of measurements of aircraft showed that $\sigma_{RR} \approx \sigma_{RL}$ and both were about 3 dB below σ_{HH} or σ_{VV} [133]. Most other experiments have shown a reduction in target RCS of about 5 dB for aircraft and even greater reductions for missiles and satellites.

A similar technique for increasing the target-to-precipitation echo ratio is to use crossed linear polarizations (RCS is then σ_{HV} or σ_{VH}).

---

[1]Both backscatter and attenuation vary with polarization for elliptical raindrops.

TABLE 5-1 Reduction in RCS Using Polarization Properties—Summary of Experimental Data

Frequency band	Precipitation Type	Elevation, deg	σ_{RL}/σ_{RR}, dB	Targets Type	Aspect, deg	$\dfrac{\sigma_{HH}\text{ or }\sigma_{VV}}{\sigma_{RR}\text{ or }\sigma_{LL}}$, dB	$\dfrac{\sigma_{HH}\text{ or }\sigma_{VV}}{\sigma_{HV}\text{ or }\sigma_{VH}}$, dB
L, 24 cm	Rain		22–30	Piston aircraft land	−8 el	6–8	7–8
	Wet snow		15				
S, 10 cm	Rain	<10	18				
	Rain over sea	0–10 (av)	20*				
	Rain over marsh	0–10 (av)	24*				
	Rain over land	0–10 (av)	27*				
	Rain over desert	0–10 (av)	34*				
C, 5.6 cm	Rain	0–10	17	Land	−8 el		6–10
				Vehicles	−3 el		5–16
				Trees	−3 el		0–6
				Ships	−3 el		−(2–8)
				Ships	−6 el		3–15
X, 3.2 cm	Thunderstorm	high	15	Aircraft	Nose and tail	Av 2.5	2.5–12
	Rain		26–28	Jet aircraft	Nose	4.5–6	10
	Bright band		13–20	Ships	0 el	6	
	Fine snow		26	Ships	−(3–6) el		7–15
Ku, 1.2 cm	Rain	high	26				
Ka, 0.86 cm	Rain	30	17–18	Jet and piston aircraft	Tail	3–5	7–11
					Nose	2–4	8–10
	Bright band	30	5–11	Trees	−3 el		4
	Dry snow	30	12–16	Vehicles	−3 el		8–14

*Theoretical.

147

While rain echoes are again reduced by 15-25 dB, the reduction in target RCS is generally greater than 7 dB [272].

5.3 COMPLEX TARGETS–BACKSCATTER DISTRIBUTIONS

Although it is possible to calculate the RCS of many geometrically simple shapes, most radar targets such as aircraft, ships, etc. are complex. The RCS of complex targets can be computed in many cases by using the RCS of simple shapes. The most common procedure is to break the target into component parts, each of which is assumed to lie within the optical region, and to combine them according to Eq. (5-4) [79, 80, 400]. This approach is usually used to determine the average RCS over an aspect angle change of several degrees. Realistic complex structures such as large aircraft may exhibit 10-15 dB RCS fluctuations for aspect changes of a fraction of a degree [211]. Since the precise target aspect is generally unknown and time variant, the RCS is best described in statistical terms.

The statistical distribution of the RCS is of great importance in predicting detectability. For fluctuating targets, the detection probability of a single echo is roughly the probability that the target echo alone will cross a detection threshold; receiver noise primarily affects the false alarm rate. This means that if the occurrence of a small RCS has a large probability an unacceptably low detection probability can result for a given mean signal-to-noise ratio. A variety of RCS distribution models have been used; their choice depends on certain assumptions about the nature of the target. Table 5-2 summarizes the most widely used statistical RCS models (see Chap. 3), describes their applicability, and cites pertinent references.

Steady-target or Marcum Model [244, 113]

The detection of a perfectly steady-target echo in receiver noise was originally analyzed by Marcum. This *nonfluctuating* model is not realistic for real radar targets except in limited circumstances, such as when the target is a sphere or is motionless over the observation time.

Chi-square of Degree 2m [368]

Specializations of the chi-square distribution encompass a large class of targets (see Table 5-2). The form of this distribution is given as item

TABLE 5-2 Statistical Models of Target Radar Cross Sections

Model	Density function of RCS, x	Applicability
(1) Marcum (*steady sinusoid* or *non-fluctuating target*)	One possible value	Perfectly steady reflector, e.g., spherical object or objects which are not rotating during the radar observation time [244, 113].
(2) Chi-square of degree $2m$ Also Weinstock cases $(0.6 \leq 2m \leq 4.0)$	$\dfrac{m}{(m-1)!\,\bar{x}} \left[\dfrac{mx^{m-1}}{\bar{x}} \right] \exp\left[-\dfrac{mx}{\bar{x}} \right]$	The distribution encompasses a class of target distributions (see items 3 and 4 below). Degree becomes higher as coherent component of target becomes greater [368, 400].
(3) Swerling cases 1 and 2 (*chi-square of degree 2, Rayleigh-Power* and *Exponential*)	$\dfrac{1}{\bar{x}} \exp\left(-\dfrac{x}{\bar{x}} \right)$ for $x \geq 0$	Random assembly of scatters, no single one dominant. Applies to complex targets having numerous reflectors. Case 2 statistics apply when fluctuations are independent on a pulse-to-pulse basis. Case 1 applies to pulse-to-pulse dependence [367, 113].

TABLE 5-2 (Continued)

Model	Density function of RCS, x	Applicability
(4) Swerling cases 3 and 4 (*chi-square of degree 4*)	$\dfrac{4x}{\bar{x}^2} \exp\left(-\dfrac{2x}{\bar{x}}\right)$ for $x \geq 0$	Has detection statistics nearly the same as when one large, steady reflection is combined with assembly of many small reflectors— RCS of large reflector same as sum of RCS of all the others. (See Rice distribution with $a = 1$.) Case 4 statistics apply when fluctuations are on pulse-to-pulse basis. Case 3 applies to pulse-to-pulse dependence [399, 113].
(5) Rice (power)	$\dfrac{1}{\bar{x}}(1 + a^2) \exp\left[-a^2 - \dfrac{x}{\bar{x}}(1 + a^2)\right] \cdot J_0\left[2\,ja\sqrt{1 + a^2(x/\bar{x})}\,\right]$ for $x \geq 0$	One large reflector and an assembly of small reflectors. a^2 is the ratio of RCS of the large to the sum of all the rest. More exact for this condition than chi-square members, but fewer detection statistics have been published [211, 313, 336];

TABLE 5-2 (Continued)

Model	Density function of RCS, x	Applicability
(6) Log-normal	$\dfrac{1}{xs\sqrt{2\pi}} \exp\left[-\dfrac{(\ln x - \overline{\ln x})^2}{2s^2} \right]$ for $x \geq 0$	Found to apply to many targets which have large mean-to-median ratios ρ – e.g., battleships, missiles, satellites, and aircraft near broadside aspects [164, 368].

NOTE: Detection statistics in varying amounts of completeness have been calculated for these target models in the literature cited. In all cases the interfering noise power is assumed to have an exponential probability density function (see item 3).

Definitions: x = RCS
$2m$ = degree of freedom
$\overline{x}$ = mean RCS
a^2 = ratio of RCS of large reflection to the sum of the small
J_0 = modified Bessel function
$j = \sqrt{-1}$
ρ = mean/median ratio
s = standard deviation of $\ln x$
$\ln x$ = mean of distribution with $\ln\ x$ as the variable

151

2 in Table 5-2 in which $2m$ is the *degree of freedom*. As the degree becomes higher, fluctuations about the mean become more constrained, that is, the *steady* component becomes stronger. The ratio of the variance to the mean is equal to $m^{-\frac{1}{2}}$. In the limit, as m becomes infinite, one has the steady-target case. Short term statistics of aircraft RCS have been found to fall into various members of the chi-square family.

Swerling Cases 1 and 2 (Chi-square of Degree 2)
[367, 113]

The statistics of Swerling cases 1 and 2 are most often referred to as the *Rayleigh-power* or *exponential* distributions. Swerling case 2 statistics apply when the RCS pulse samples are independent on a pulse-to-pulse basis (rapidly fluctuating). Case 1 statistics apply when the samples are correlated within a pulse group but are independent on a scan-to-scan basis (slowly fading). The pulse-to-pulse independence case is not applicable to most radar targets because of their narrow doppler spectra (see Sec. 5.8). This assumption may be valid when pulse-to-pulse frequency diversity is used (see Sec. 5.9). The Rayleigh model applies to a random assembly of scatterers, no one of which is dominant. It is the most widely used of all distribution functions for modeling large complex targets [38, 91, 255, 265]. Experimental evidence suggests this model for complex targets such as large aircraft, whose aspects undergo large changes. However, its use has prompted vigorous controversy when applied to a target which presents a limited range of aspects, such as a single target viewed over a short period of time [281, 256].[1] This point is considered further in Sec. 5.4 where specific aircraft RCS data are discussed.

Swerling Cases 3 and 4
(Chi-square of Degree 4)

Case 4 statistics apply when there is pulse-to-pulse independence and case 3 statistics apply to scan-to-scan independence. Swerling assumed this model represented the class of targets having a large steady reflector and a number of randomly oriented small reflectors although he did not attempt to justify the assumption on a physical basis. The exact distribution which fits this assumption was given by Rice.

---

[1] See also correspondence in *IRE Trans.*, vol. AP-9, pp. 227-229, March, 1961.

Weinstock Cases (Specializations of
Chi-square Members)

Weinstock [401] studied the RCS of simple structures to see whether
they could be incorporated into chi-square classes. He observed that the
RCS of certain simple shapes, e.g., cylinders, conformed to chi-square
distributions with degrees of freedom between 0.6 and 4, depending on
the range of aspects. Partially in response to these findings, Swerling
extended his work to include chi-square functions of arbitrary degree
[368].

Although the Rayleigh model is the only chi-square member which has
been measured in a number of instances, it is often possible to
approximate actual density functions by using the other chi-square
models [368]. This possibility is of great significance since very little is
available concerning the detection statistics for non-chi-square distributions. Weinstock considered how the requirements imposed on a radar
system will vary with the assumed density function if the only parameter
known is the mean or the median. He found that the best procedure was
to assume the Rayleigh function having the same *median* value as that
for the actual RCS. Even if actual density function were chi-square
having a degree between 0.6 and 4.0, the error in estimating system
performance would be far less than if the mean RCS were chosen
instead.[1] The median has the further advantage that it is easier to
measure than the mean. However, if system performance is to be
calculated for detection probabilities exceeding 90 percent, a statistical
model should not be assumed if it is possible to make detailed
measurements on the target of interest or a reasonable scale model.

Rice Distribution

When an assembly of scatterers includes a steady reflector whose
return is significant compared with the sum of all the others, the density
function has the form of the Rice distribution (see [211, p. 560]) given
by item 5 of Table 5-2, where a^2 is the ratio of the power from the large
scatterers to the power from all the rest. When $a^2 \to 0$, this distribution
converges to the Rayleigh density function, and when $a^2 \to \infty$, it converges
to the steady-target case. Scholefield [336] compared the four-degree

---

[1]At very low detection probabilities (e.g., < 30 percent) the mean value is a better
indicator of system performance.

chi-square case (Swerling 3 and 4) with the Rice distribution with $a^2 = 1$ (power from steady reflector equal to the power from the assembly). The detection statistics for the two agreed except for small differences at the extremes of detection probability. The Rice distribution is an exact representation of the steady-plus-assembly reflector case.

Log-normal Density Function

The RCS distributions of some targets do not fit into many independent scatterer classes and do not conform to any chi-square members. Some of these targets exhibit large values of RCS far more frequently than would any member of the chi-square class. Examples of this class of targets are satellites, missiles, and ships, which often have large mean-to-median RCS ratios.[1] The RCS of these targets can frequently be closely represented by the log-normal density function (item 6 of Table 5-2), which arises when the logarithm of a variate is normally distributed. Swerling derived procedures for approximating this function with chi-square members [368]. The log-normal function was also studied by Heidbreder and Mitchell [164] who derived the appropriate detection statistics. They postulated that the high tails of this density function arose in practice from large specular returns at a relatively few aspect angles. For conservative estimates of system performance, one should neglect these tails and resort to the Rayleigh assumption.

Summary of Statistical Modeling

1. Modeling is basically a coarse procedure. In no case should an assumed statistical model be used for very precise calculations of system performance (i.e., performance at very high- or at very low-detection probabilities).

2. If only one parameter is to be used to describe a complex target, it should be the median RCS. Detection statistics should then be calculated by using the Rayleigh model (Swerling cases 1 and 2) having the same median value. The choice between the pulse-to-pulse or the scan-to-scan independence assumption depends on the doppler spectrum width of the echoes and whether frequency diversity is used.

3. If the tails of the actual RCS distribution function are known to be much higher than those for the chi-square members, an upper bound on

---

[1]The mean-to-median ratio for the Rayleigh function is 1.44, and for the 4-degree chi-square is 1.18.

system performance may be established by using the statistics for the log-normal distribution function. These statistics have been expressed in terms of the mean RCS and the mean-to-median ratio.

4. If the target is approaching the radar and a high-detection probability is required before the target reaches a given range, cumulative detection curves should be used as described in Sec. 3.8.

5.4 AIRCRAFT RCS DISTRIBUTIONS

While it is possible to measure aircraft RCS at a given aspect, the precise aspect in flight is unknown and time variant. A plot of the aircraft return versus time appears as a noiselike function, even when the *nominal* aspect of the aircraft is *constant*. Because of the uncertainties about the aspect, the RCS is usually best described statistically. Usually, one assumes a Rayleigh distribution; this model has been suggested by many experiments on large aircraft undergoing several angular degrees of aspect change. Unfortunately, this model does not apply to all aircraft and to all radar detection and tracking problems. A good portion of the uncertainty lies in the observation time used to obtain the RCS distribution.

In one series of experiments by NRL, RCS distributions were obtained for a variety of aircraft [265]. Flight paths were straight and level, and various aspect changes were observed by the radar. Data was recorded over 10-sec intervals at P, L, S, and X bands. The Rayleigh model provided a good approximation to the distribution in many cases, although there were a large number of exceptions. The most consistent exceptions were for smaller aircraft and for all aircraft at broadside. Large multiengine aircraft gave good Rayleigh approximations for aspect changes of only 1 or 2°. Small aircraft were often fit very poorly by this model even when the aspect change was 8 to 9°. In most cases, deviations from the Rayleigh distribution tended towards higher order chi-square functions. In these cases the Rayleigh assumption would tend to provide conservative estimates of detection ranges for detection probabilities greater than 50 percent.

Small propeller aircraft had echoes from the blades at favorable parts of the rotation cycle that were often stronger than the residual echo from the remainder of the aircraft. For an F-51 aircraft, this condition occurred at nearly all aspects outside of the near-broadside regions of 70-110°. These propeller dominant echoes, which occurred about 20 percent of the time, would improve the F-51 detectability at low signal-to-noise ratios. Since the median RCS was unaffected by their presence, the 50 percent detectability would most likely be unaffected.

In tests at the Applied Physics Laboratory, RCS distributions at C band were plotted for a variety of conditions and were compared with chi-square functions. For a small two-engine jet, chi-square approximations were obtained with degrees ranging from 1 to 37. Widely different distributions were obtained from one data segment to the next even when the apparent azimuth aspects differed by only a degree or two. [1] Figure 5-1 contrasts pairs of measured density functions. In *a* and *b* the nominal aspects were similar although the precise aspect was unknown since the *crab* angle of the aircraft was unknown. Part *a*, perhaps, may be represented by the Rayleigh function and part *b* by the chi-square function of degree 22. In both cases the run length was 2.0 sec, although the results were similar when the run was extended to 8 sec. Parts *c* and *d* show density functions measured over 1 sec. Part *c* has the appearance of the Rayleigh function, whereas part *d* does not resemble any known function. These distributions point out that there is no simple solution to the RCS modeling problem for all aircraft.

It was pointed out in Sec. 5.3 that, lacking sufficient information about the actual target RCS, the best course is to assume the Rayleigh model having a median value equal to that of the aircraft in question. Even this is not a simple matter, since the median RCS observed over a measurement interval is highly dependent on the average aspects during that interval. For this reason, an average over all aspects, although it is often given, is not always useful unless perhaps one is concerned with the single-pulse detection probability for an aircraft where all aspects are equally likely.

Table 5-3 lists some representative median RCS measurements for selected classes of targets and aspects [265, 272, 388]. For most aircraft, the median RCS at nose aspect is smaller than that for other aspects, and this value can be used for conservative performance estimates. No distinction is made in Table 5-3 between horizontal and vertical polarizations since this makes little difference in the median RCS for most aspect regions. In measurements on C-54 aircraft, vertical polarization values were greater at nose aspects by 4 dB at L band, 6 dB at S band, and 1 dB at X band [272]. At other aspects, smaller differences were generally observed.

---

[1] About one-half of the analyzed distributions fell between degrees 1.4 and 2.7.

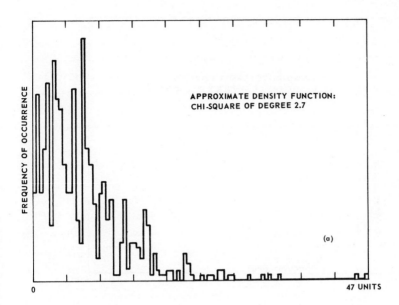

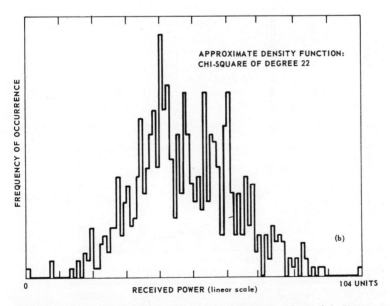

figure 5-1 *Probability density functions for small jet echoes. (a) observation time 2 sec, number of samples 624, aircraft aspect az. 0°, el. 12°, range 17.5 nmi; (b) observation time 2 sec, number of samples 624, aircraft aspect az. 0°, el. 14°, range 17.0 nmi.*

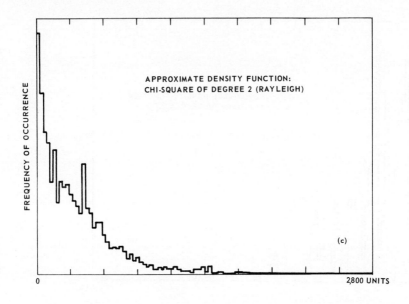

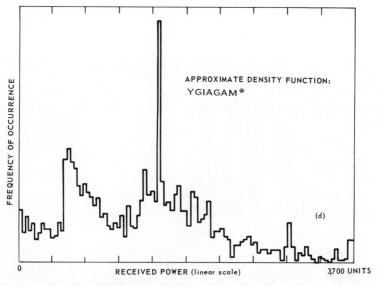

figure 5-1 *(c) Observation time 1 sec, number of samples 2,500, aircraft aspect az. 37°, el. 19°, range 8.5 nmi; (d) observation time 1 sec, number of samples 2,500, aircraft aspect az. 72°, el. 59°, range 4.9 nmi.*

*Your guess is as good as mine.

TABLE 5-3   Median Aircraft RCS for Various Classes and Aspect Regions, m^2

Class	Aspect	VHF 0.03–0.3	UHF 0.3–1.0	L 1–2	S 2–4	C 4–8	X 8–12
Large, heavy bomber or 707, DC-8, size jet	Nose			32	40	10–40	28
	Tail			15/63	250		25/500
	Broad			550/500	500		300/300
	Av		11	24/40	23/40		60
Medium, attack bomber, 727, DC-9, size jet	Nose	60/100	1.2/20/30	6/10/20			4
	Tail	100	6	20	20		
	Broad	300	200	300	280		800
	Av			12	12		11
Small, fighter or four-passenger jet	Nose	10	6	0.3–7.5	0.2–9.5	0.7–2	1.2
	Tail	10	3	2	0.6–15		
	Broad			5–90	35–300		20/30/65
	Av			0.7/1.3	1.3/2		1.3/5

5.5 MISSILE AND SATELLITE CROSS SECTIONS

The RCS of missile and satellite targets generally falls into a class that is between a simple geometric shape and a complex target such as a large aircraft. The resulting distribution of RCS is neither that of a point reflector nor a random array of scatterers. This is pointed out by full-size X band model measurements of a 54-in. satellite[1] having spherical symmetry reported by Kennedy [210]. An RCS density distribution is shown (Fig. 5-2) where the cross section is plotted in decibels. Two sets of measurements are shown: a single frequency set which shows a distribution that is more highly skewed than any of the Swerling models and a frequency stepping set with each point being the average RCS from eight frequencies spread over 500 MHz. In both cases the mean-to-median ratio of RCS is about 10. Hence, the majority of the values of RCS's are well below the *average* $\bar{\sigma}$. However, the median and the decibel average dBsm (the absolute geometric mean) are similar. This similarity and the appearance of Fig. 5-2 suggest the log-normal distribution.

With frequency agility the probability of both large and small RCS values decreases. This effect will be discussed in Sec. 5.9. Gaheen,

---

[1] Plus short stub antennas.

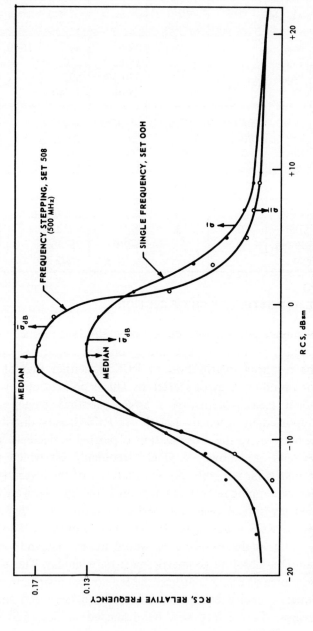

figure 5-2 *Probability density function plotted vs. dBsm. 54-in. satellite target (X band). (From Kennedy [210].)*

McDonough, and Tice [129] did further analysis on the same data and showed a comparison of the RCS distribution with RCA measurements on the larger Tiros[1] and Nimbus satellites. These data are shown (Fig. 5-3) along with the unusual RCS distribution on a conical-shaped missile model. This model had a cone angle of 23.5°, a length of 150 in., and a diameter of 53 in. at the tail. Note that frequency agility substantially reduces the number of RCS values below 0.01 m² for the missile target. On a single frequency these data show that the median cross section of the cone-shaped missile is about 1 m², while the 5 percent probability is

[1]The geometric cross-sectional area of the Tiros is about 3.4 m² or twice that of the 54-in. satellite.

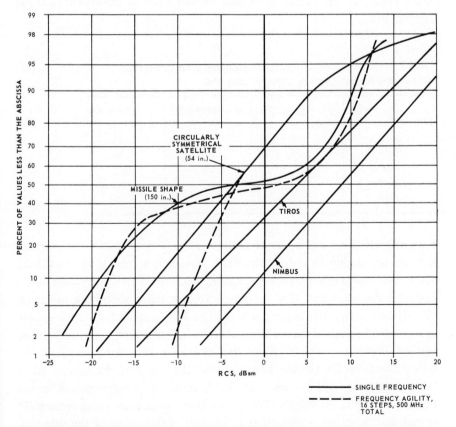

figure 5-3 *Cumulative distribution of satellite RCS. Average for all aspects. (From the data of [129].)*

about 20 dB less. Figure 5-3 also suggests the log-normal distribution which would appear as a straight line on this graph.

This satellite data should not be construed as being valid over a wide range of carrier frequencies. The median Tiros RCS value reported here (~ 2 m^2) is much higher than the estimated value at 30 MHz carrier frequency (0.08 m^2) [292]. An indication of the frequency sensitivity of RCS of satellites is illustrated by a series of model measurements of the roughly 105-in. Mercury space capsule reported by Mack and Gorr [241]. They show RCS values versus angle for simulated frequencies of 440 through 5,600 MHz. A summary of that data for vertical polarization is given in Table 5-4. The reference gives aspect angles that RCS's of 1, 10, and 100 m^2 are exceeded. In Table 5-4 this data is converted to a percentage exceeding a specified RCS. It can be seen that RCS values for the Mercury capsule do not have an obvious frequency dependence.

TABLE 5-4   Percentage of Aspect Angles That a Specified RCS is Exceeded for the Mercury Capsule, Vert. Pol.

Frequency, MHz	$\sigma \geq 100$m^2, %	$\sigma \geq 10$m^2, %	$\sigma \geq 1$m^2, %
440	0	28	70
933	35	69	98
1,184	8	28	62
2,800	7	47	96
5,600	2	13	60

In the previous examples of satellite targets there was no attempt in the satellite design to reduce the RCS. However, in missile systems there is a considerable advantage in presenting a low RCS to a defending radar. For this reason there is a tendency to use radar absorbing materials (RAM) or a shape like a cone-sphere. The RCS is small for a cone-sphere over a broad range of carrier frequencies when observed near the pointed end of the cone. The RCS of this object is the vector sum of the reflections from the tip, the joint between the cone and the sphere, and a "creeping wave" [344, 347, 399]. A series of measurements and a theoretical curve are shown (Fig. 5-4) where the nose-on cross section (in square wavelengths) is plotted as a function of the radius of the spherical portion (in wavelengths). The important points to note about this

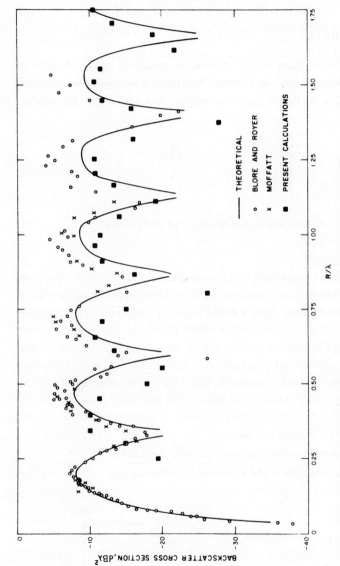

figure 5-4 *Experimental and theoretical nose-on cross section of a 15° half-angle pointed cone sphere. (From Weiner [399].) Sources of data are given in the reference listed at the back of the book.*

idealized target are that, neglecting the oscillations, RCS is proportional to λ^2 and the size of the object itself has little effect if the pulse length is greater than the target length. Descriptions of the RCS for high-resolution radar measurements are also given in [399].

5.6 MARINE TARGETS

The radar backscatter from a ship cannot be uniquely predicted due to interference resulting from the forward scattered electromagnetic waves. As an approximation, the received power at the radar can be written [211, 406]

$$P_r = \left[\frac{P_T G_T G_R \lambda^2 \sigma_t}{(4\pi)^3 R_t^{\,4}} \right] \cdot \left[16 \sin^4 \left(\frac{2\pi h_a h_t}{\lambda R_t} \right) \right] \tag{5-6}$$

where σ_t = radar cross section in the absence of reflections
   h_a = antenna height
   h_t = target height
The first term is the conventional free-space received power equation, and the second term is a modifier to the free-space equation that is periodic in elevation angle with a peak value sixteen. This is the familiar lobing effect of surface radars that doubles the detection range at some elevations but puts a null on targets which are situated on the surface of the water. There are several assumptions which must be explored before trying to evaluate the $\sin^4$ term in Eq. (5-6). The geometry can be illustrated by Fig. 5-5. The modification of the field strength at the target is then

$$\eta = \frac{\text{Field strength in the presence of the reflected wave}}{\text{Field strength if the target were in free space}}$$

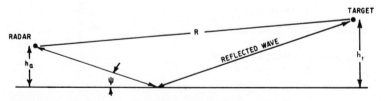

figure 5-5 *Geometry for propagation above a flat surface.*

Any difference in the amplitude of the reflected wave compared to the direct wave is assumed to be due to the surface reflection coefficient being less than unity. Since the two path lengths are not exactly the same, the relative phase of the reflected wave differs. The total reflection coefficient R_s is thus a complex quantity and can be expressed as $R_s = \rho \exp(j\phi)$, where ρ is the real part of the reflected amplitude and ϕ is the phase shift upon reflection.

It has been shown [406, 354] that for typical surface radars and target range $R_t \gg h_a$ or h_t and a flat surface with reflection coefficient R_s equal to minus one (180° phase shift) the relative phase of the reflected wave with respect to the direct wave is

$$D \doteq \frac{4\pi h_a h_t}{\lambda R_t} + \phi$$

where D = difference in phase between direct and reflected paths (radians)

$\phi = \pi$ for a flat smooth sea

Section 1.10 on forward scatter shows that the assumption that $R_s = -1$ is valid for low microwave frequencies ($\lambda > 30$ cm) and smooth seas. The relationship applies for grazing angles ψ of up to about 1° for horizontal or vertical polarization. Assuming that the ship is a point reflector, making the small angle substitution in Eq. (5-6) yields

$$P_r = \frac{4\pi P_T G_T G_R \sigma_t (h_a h_t)^4}{\lambda^2 R_t^8} \tag{5-7}$$

Thus, received power from a ship falls off as R^{-8} even without including the horizon effect.[1] Experimental verification of this is shown in Fig. 5-6 which illustrates relative power received as a function of range. The location on the ordinate is arbitrary. The R^{-8} dependence near the horizon is obvious on curve A for the 150-cm radar. The transition from the R^{-4} dependence (free space) to the R^{-8} dependence for the higher frequency radars occurs when $R = 4 h_a h_t/\lambda$ and is quite

[1] In Chap. 7 it is noted that the normalized backscatter from the sea itself σ_0 also falls off rapidly at low grazing angles.

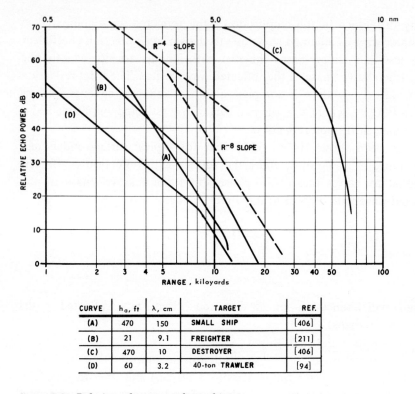

CURVE	h_a, ft	λ, cm	TARGET	REF.
(A)	470	150	SMALL SHIP	[406]
(B)	21	9.1	FREIGHTER	[211]
(c)	470	10	DESTROYER	[406]
(D)	60	3.2	40-ton TRAWLER	[94]

figure 5-6 *Relative echo power from ships.*

abrupt. The higher elevation measurement in Curve C shows a more gradual transition at more distant ranges.

As a result of the almost inseparable effects of sea reflection and the cross section of the ship itself, only the measurements of ship cross section in the R^{-4} region are of general interest. In 1965 the Naval Research Laboratory made a thorough study of the cross section of a 144-ft minesweeper primarily of wooden construction [85]. The measurements were made from an aircraft at P, L, C, and X bands. At broadside most of the median cross-section values were between 9.5 and 25 dBsm. These values were for both polarizations at C and X bands and radar depression angles of 1.9 to 3.5°. The median cross section σ_m for various depression angles and frequencies is shown for aspects 25° off the stern or bow in Table 5-5. The bow readings are somewhat higher which may be due to large metal equipment racks at that end of the

TABLE 5-5   Median Radar Cross Section of a Minesweeper at Various Frequencies and Depression Angles, dBsm*

Depression Angle, degrees	25° off bow				25° off stern			
	V_L	V_P	H_L	H_P	V_X	V_C	H_X	H_C
1.3	+19.5	+18.0	+17.5		+19.5	+17.5	+18.0	+19.5
1.35	17.5	17.5	23.5		19.5	18.0	22.0	18.5
1.4	18.5		21.5	+17.0	19.0	15.5	25.0	18.5
1.5	21.5	20.5	24.0	22.5	18.5	16.5	21.0	20.5
1.55	17.5	19.5	16.5	24.5	20.0	12.5	19.0	13.5
1.6	17.0	18.0	19.5	24.0	10.0	7.5	17.5	12.5
1.7	19.0	16.0	27.5		2.5	12.0	13.5	15.5
1.9	18.0	19.0	19.0		19.0	14.5	24.0	16.0
2.0	19.0	17.5	19.5		23.0	11.0	26.0	12.0
2.1	10.5	21.5	20.5	26.0	10.0	12.5	16.5	16.0
2.4	19.0	22.5	18.5	27.0	12.0	5.0	20.5	12.0
2.55	24.0	21.5	22.0	26.5	11.0	13.5	17.5	20.0
2.7	20.5	20.5	22.5	22.0	7.5	3.5	16.0	9.5
3.7	7.5	10.5			2.0	1.5	10.5	4.0
4.4	9.0	13.0			5.0	−2.5	17.0	9.0
5.7					0.5	+1.5	7.5	7.0
5.8	15.5	18.0		21.0	5.0	3.0	12.5	7.5
6.9	12.5	18.5			−3.5	−2.0	8.5	6.5
8.7	14.0	16.5	−1.5		−3.0	−10.5	5.5	−2.0
σ max	24	23	28	28	23	24	27	27

*From Daley [85].

ship.[1] It is also suspected that the lower cross-section values for low depression angles at P band are due to forward scattered interference. V_P in the table corresponds to vertical polarization at P band, etc. Maximum values of RCS are given which show virtually no dependence on transmit frequency.

In many cases the ship returns deviated from a Rayleigh distribution. The data has an 11- to 17-dB spread between the 10 and 90 percentiles of the cumulative distribution. There is some evidence that the RCS

[1] Bow measurements at P and L bands were as high as + 40 dBsm.

distribution of ships may be represented by the log-normal distribution [164].

It can be seen from the table that there is a tendency for the radar cross section of ships to decrease at higher depression angles. It will be shown in Chap. 7 that the backscatter from the sea increases at higher depression angles. The combination of these two factors makes it more difficult to detect ships from high-altitude aircraft or satellites. Even at low depression angles, such as with a radar on one ship looking for another ship, there is a sea-clutter problem at high sea states. Some of the practical aspects of detecting ships are described by Harrison [161] and Croney [82].

5.7 MISCELLANEOUS AIRBORNE REFLECTIONS

A modern high-power, sensitive radar will observe numerous reflections from various regions of the atmosphere. In addition to echoes from aircraft, missiles, and precipitation, there are a variety of reflections which are not as easily explained. This section will summarize a few of the backscatter properties of some of the *phenomena* which appear on radar displays. The use of the vague term *phenomena* is to emphasize that while many of the reflections are from discrete objects such as birds or insects there are also numerous echoes from atmospheric anomalies including clear air layers at the tropopause [13], breaking gravitational waves [171], and *ring echoes* [354, p. 553]. It should be recalled that a radar which can detect a 1 m^2 RCS target at 200 nmi will also detect a 10^{-4} m^2 object at 20 nmi unless specific techniques are used to inhibit detection of slowly moving low RCS objects at near range.

The term *angel* has been used for many years to describe the radar reflections from a location in the atmosphere that does not contain a *known* discrete target. Since it has been shown that many of the angels are undoubtedly due to the presence of small birds and insects, it is useful to present the measured cross section of these *targets*. Table 5-6 provides a summary of the RCS of birds and insects at several carrier frequencies.

Since birds typically have dimensions comparable to a wavelength (in the *Mie* region), the RCS is a strong and oscillatory function of wavelength. This can be seen from the first four entries in the table where the reported RCS at S band (10 cm) is higher than P band (71 cm) or X band (3.2 cm). It was also reported that the RCS distribution

TABLE 5-6 Typical Mean RCS of Single Birds and Insects, dBsm*

Type, length	Aspects observed	Frequency Band (vert. polarization)		
		UHF, 0.45 GHz	S, 3.0 GHz	X, 0.9 GHz
(1) Sparrow	Head			-46
	Broadside			-32
	Tail			-47
	Average	-56	-28	-38
(2) Pigeon	Head			-40
	Broadside			-20
	Tail			-40
	Average	-30	-21	-28
(3) Duck	Head	-12		
(4) Grackle		-43	-26	-28
(5) Hawkmoth, 5.0 cm		-54	-30	-18
(6) Worker bee, 1.5 cm		-52	-37	-28
(7) Dragonfly		-52	-44	-30

*After Glover et al. [137] and Konrad, Hicks, and Dobson [217].
Averaged over all aspects unless specified otherwise.

appears to follow a log-normal distribution with the mean-to-median ratio of about 2.5 for birds whose dimensions are greater than about four wavelengths and a smaller ratio for smaller birds.

The RCS of a number of insects was also measured with three radars at P, S, and C bands [137]. Since these objects generally fall in the Rayleigh region, their RCS increases rapidly with increasing transmit frequency. The RCS of some of the insects (especially the hawkmoth) are quite large at X band.

While many radar reflections can be attributed to birds, insects, precipitation, etc., there are a number of other echoes which cannot be easily identified. Many of these can be attributed to fluctuations in the refractive index of the atmosphere. Radar reflections have been reported from the region of the tropopause [13, 170]. This is a layer of air at an altitude of 8 to 16 km that delineates the separation between the troposphere and the stratosphere and is the region near which clear air turbulence (CAT) is often found to occur. Since CAT has a considerable adverse effect on jet aircraft, the possibility of detecting it by radar is currently receiving considerable attention. Clear air echoes were reported virtually every day of one six-weeks operation with a set of three large high-powered radars at UHF, S and X bands. Some of the time they appear as *braided structures* that are attributed to breaking gravitational waves [171]. In another set of experiments, regions which showed clear air echoes were simultaneously probed with an aircraft and were found to be turbulent [170]. An example of clear air echoes is shown on Fig. 5-7. An RHI display is shown with echoes that appear like braided structures (S-band radar). The sky was optically clear in these regions at the time of the photos.

It is of interest to note how multifrequency radars are used to distinguish between material particles, such as insects or clouds and discontinuities in the atmosphere. It has been shown (see Chap. 6) that the reflectivity per unit volume η for particle diameters less than about 0.06λ is given by[1]

---

[1]In Chap. 6 $\Sigma\sigma$ is used instead of η for the normalized volume reflectivity. The units of $\Sigma\sigma$ or η in this text are m^2/m^3 or m^{-1}.

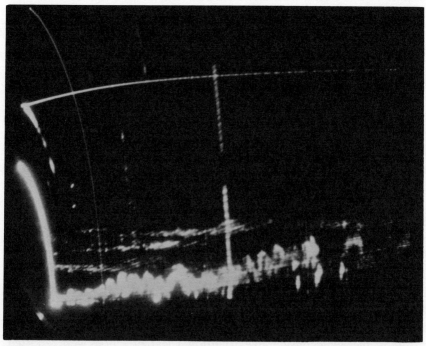

figure 5-7 *The range-height display showing clear-air convective cells. The abscissa is range out to 20 mi, the ordinate is altitude with the nearly horizontal line the 20,000-ft marker. Note also some layered echoes above the convective regions. (From Katz [207].)*

$$\eta = \pi^5 \lambda^{-4} |K|^2 \Sigma D^6 \tag{5-8}$$

where λ is the wavelength and D is the droplet diameter. The value of

$$K = \left| \frac{m^2 - 1}{m^2 + 2} \right|$$

where m is the complex refractive index. The equation for the reflectivity of a refractively turbulent medium can be approximated by [207]

$$\eta \sim \overline{(\Delta n)^2} L_e^{-2/3} \lambda^{-1/3} \tag{5-9}$$

where $\overline{(\Delta n)^2}$ is the mean-square refractive index fluctuation and L_e is related to the size of the turbulent eddies. The important difference between relations (5-8) and (5-9) is that particles like raindrops have a strong wavelength dependence while changes in refractivity are almost independent of wavelength. Thus, particulate matter or insects are seen primarily at the higher frequencies. Although the reflectivity η of the clear air echoes is only about 10^{-15} m^{-1}, they can be seen with high-powered radars at 10- to 20-km ranges.

There are many other reports of reflections from unidentified objects many of which are caused by reflections resulting from anomalous propagation (ducting) in certain areas of the world.

5.8 SPECTRA OF RADAR CROSS-SECTION FLUCTUATIONS

The spectral width of the echoes from a complex target can have a considerable effect on the choice of radar processing technique and the specific parameters. In a doppler tracking system (CW or pulse doppler) it is desirable to reduce the doppler filter bandwidth to the limits imposed by the fluctuations of the target echoes, the stability of the transmitter, and the doppler shifts due to acceleration. Using this minimum doppler bandwidth will maximize the signal-to-noise ratio and generally the signal-to-clutter ratio. To reduce the effect of target fading there should be additional postdetection (incoherent) integration of the outputs of the doppler filters for a time duration greater than the correlation time of the target echoes.

Unfortunately the relatively long correlation times of airborne targets means that in a tracking radar, the duration of a target fade may cause excessive range and angle errors. In a surveillance radar, the target cross section may remain in a null for the entire time the beam illuminates the target. This means that excessively long processing times may be required to benefit by incoherent integration. One technique which offsets the need for long processing times is frequency agility, which is discussed in Sec. 5.9.

The benefit derived from incoherent integration is illustrated in Fig. 5-8, which shows the probability density function at the output of an incoherent integrator when the input signal was an aircraft echo [102]. Four integration times were examined: 10, 40, 160, and 640 msec. As the integration time increased, the probability of occurrence of a small RCS value was reduced. Figure 5-9 shows how the ratio of the standard deviation of the RCS to its mean decreased with increasing integration

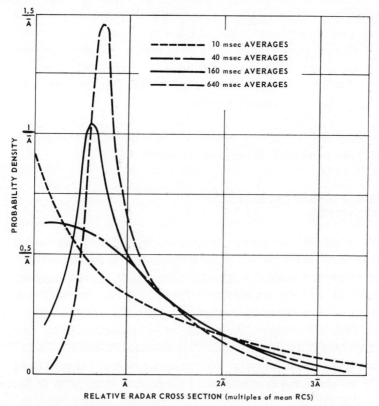

figure 5-8 *Probability density functions of X-band aircraft RCS (nose aspect). (After Edrington [102].)*

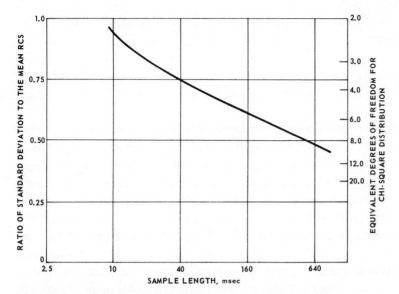

figure 5-9 *Relative standard deviation of aircraft RCS (nose aspect, X band). (From Edrington [102].)*

time. The rate of reduction depends on the power spectral density of the echoes, as described in Secs. 3.5 and 3.6.

The spectra of RCS fluctuations of aircraft can be conveniently described in terms of three effects. First, the airframe spectrum is due to the relative motion between various scattering points. This relative motion occurs as the aircraft aspect changes. The resulting spectral width is proportional to the transmitted frequency. In a propeller aircraft a second effect is caused by the amplitude modulation from the propellers. Returns from jet aircraft may also contain modulation components at aspects which allow reflections from the interior of the engines. The spectral width from this component is not proportional to the transmit frequency but will appear as modulation components about the airframe spectrum at multiples of the propeller rotation rate. The third effect is due to reflections from the rotating propeller or turbine blades themselves. At a given aspect the apparent center of reflection of a set of blades shifts back and forth in a periodic manner as the blades rotate. The radial component of this motion will generate a spectrum typical of phase modulation spectral energy spread out with no one line dominant. The width of this component and its center frequency will depend on the transmit frequency, the target aspect, and the blade rotation rate [132].

The width of the airframe spectrum is expected to follow the relationship

$$\Delta f = K\left(\frac{L_0}{\lambda}\right)\left(\frac{\Delta\theta}{\Delta t}\right)$$ (5-10)

The constant factor K is a proportionality factor. The factor L_0/λ is a *characteristic* length of the target, such as wingspan or body length, in wavelengths. The factor $\Delta\theta/\Delta t$ is a measure of the rate of change in target aspect. This could be a random fluctuation in aspect or a systematic change. The aspect change is due to random motion such as vibration, yaw, pitch and roll, and to systematic motion such as turning, maneuvering, or in a crossing flight. There is some indication that the length factor L_0 and the random component of $\Delta\theta/\Delta t$ are somewhat compensating. Larger aircraft have a smaller random motion due to their greater inertia. The systematic component of $\Delta\theta/\Delta t$ depends on the particular aircraft flight pattern. The work of Vogel [388] implies that, for a given $\Delta\theta/\Delta t$, the rate of signal fluctuation is proportional to L_0/λ. He used a model aircraft and simulated wavelengths of 26, 44, 96, and 160 cm. At all four wavelengths the envelope of the RCS versus azimuth plot was similar; however, the lobe structure detail depended on the wavelength. The number of lobes in the azimuth plot was very nearly $4L_0/\lambda$, where L_0 was taken to be the wingspan.

The three spectrum components are illustrated in Fig. 5-10 from Gardner [132]. In this example, the greatest density of energy is at the doppler frequency of the airframe spectrum although a greater percentage of the total energy resides at the propeller doppler spectrum. The propeller lines appear about 8 dB below the airframe line, but their frequency extent is considerably greater. In other tests by Gardner, the propeller spectrum of a receding two-engine aircraft was most clearly defined between the aspects of 15 and 37° from the tail. In this aspect range the propeller lines were 8 to 10 dB below the airframe line. The propeller spectrum of an approaching two-engine aircraft was most clearly defined for aspects less than 18° from the nose. In these tests, the width, center frequency, and relative power of the propeller spectrum varied with the aspect. For the aspects in which the propeller spectrum was strong its 3-dB width varied from 200 to 1,000 Hz. In other tests, strong propeller components have been observed at all aspects outside the near broadside range (70-110° from the nose) [265].

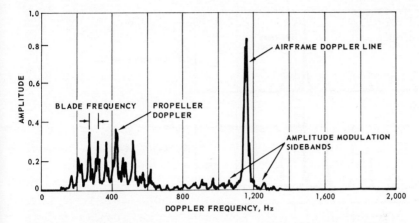

figure 5-10 *Doppler frequency spectrum of a propeller-driven aircraft (DC-7). (From Gardner [132].)*

The percentage of energy contained in the propeller spectrum is also time variant. Figure 5-11 shows the spectrum of a small T-28 single-engine aircraft at nose aspect. These spectra are taken from unpublished experiments performed at the Applied Physics Laboratory. Part *a* shows the spectrum as measured over an 0.8-sec interval. Parts *b* to *e* show how the energy was partitioned over successive 0.2-sec segments. (Notice that the vertical scale for each plot is different.) In this series of spectra, the total power in the propeller spectrum is relatively constant, whereas the energy contained in the airframe spectrum varies from one 0.2-sec segment to the next.[1]

The turbine spectrum present in jet aircraft echoes is illustrated in Fig. 5-12 from Gardner [132]. He detected the turbine spectrum at aspects up to 60° relative to the aircraft nose for a number of different aircraft. A lower sideband was observed at an average of about 5 dB below the aircraft line. An upper sideband was also observed although it was very much weaker. Gardner reasoned that the engine ducts were acting as wave guides which allowed ample amounts of RF energy to be propagated down and returned from the turbine blades. Since the turbine echo energy is often considerable, it can easily be interpreted as a separate target at a different velocity. Hynes and Gardner observed

---

[1] These illustrations have been normalized so that the maximum vertical deflection is constant.

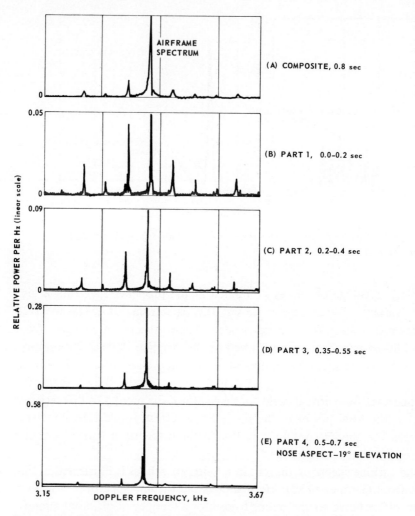

figure 5-11   *C-band spectra of T-28 aircraft showing propeller modulation.*

turbine modulation components which were offset from the airframe line by the frequencies $\Delta f = nc$, where n is the compressor rps and c is the number of compressor blades [182]. A general observation made was that the upper sidebands were again much lower in amplitude than the lower sidebands. The turbine spectrum has also been observed after envelope detection,[1] which indicates that amplitude modulation

---

[1]Unpublished experiments performed at the Applied Physics Laboratory.

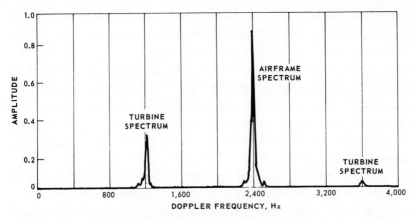

figure 5-12 *Doppler spectrum of an approaching single-engine turbojet fighter aircraft. (From Gardner [132].)*

components are also present. Although Gardner was not able to observe the turbine spectrum at S band for receding aircraft, it has been detected by others at C band[1] and at X band [132].

The percentage of total energy outside the airframe spectrum was measured in a series of experiments using coherent C-band radar.[1] Table 5-7 summarizes some preliminary findings. The percentages cited in this table refer not only to the propeller and turbine lines, but also to any other components outside the airframe spectrum (e.g., vibration or amplitude modulation components).

TABLE 5-7 Percentage of Total Spectral Energy outside of Main Airframe Spectral Line—Coherent Measurements

Aircraft	% energy outside of airframe spectrum			Aspect angle	No runs in av
	Minimum	Average	Maximum		
T-28, small prop. aircraft	12	29	75	Nose	8
Lear jet, small two-engine	0	22	59	0–30°	6
Medium multi-engine jet	4	11	17	Nose	7
Large four-engine jet, DC-8 or 707 size	17	28	38	Nose	3

[1]Unpublished experiments performed at the Applied Physics Laboratory.

A few tests have been conducted from which the width of the airframe spectrum can be deduced. Table 5-8 summarizes measurements for which the aircraft has undergone small aspect rate changes (0 to 0.5 deg/sec). Average spectrum widths are given in frequency units and in velocity units to remove the effects of the transmit frequency. The spectrum width is seen to be relatively insensitive to the particular type of aircraft. This tends to support an assertion made earlier that the length of target and its random component of motion are somewhat compensating.

One example of the airframe spectrum for a small T-28 propeller aircraft is illustrated in Fig. 5-13. In this example the target flew a constant altitude radial path. Its radial velocity decreased in 0.8 sec because it was at a close range (3.5 nmi). The spectrum resolution is 7 Hz in a and 3.5 Hz in b through e. In the lower four plots there is a doppler shift of the dominant spectral line of about 50 Hz. The relative strengths of these four are shown at the upper left of each. The complex composite spectrum a is now easily explained. The spectral line on the left of a is the result of the response during interval b. The noisy response during period c is of relatively low amplitude and does not contribute a single line. The strong clean line in d is the central line of a, and the moderate amplitude line of e appears on the right of a. Thus, what at first appeared to be a wide fluctuation spectrum is probably due to target acceleration. This points out one of the difficulties in trying to relate coherent and incoherent spectrum measurements. Another series of coherent spectrum measurements is shown in Fig. 5-14 for two small jets which were attempting to steadily maintain a wingtip separation of 300 ft. Differences in their instantaneous velocities are evident. Because coherent spectrum measurements include the effects of acceleration, the spectrum widths so measured should be considered upper limits on the RCS fluctuation spectrum.

When the spectrum analysis is performed after envelope detection, it is no longer easy to separate the airframe spectrum from the other components. Part B of Table 5-8 summarizes some incoherent measurements made on jet aircraft. The spectrum width increases with the transmitted frequency and is roughly comparable to that for the airframe component of coherent measurements. In part C the spectrum width is tabulated for propeller aircraft. Notice that the width seems to be independent of the transmit frequency although higher frequency tails were present in the spectra which did increase with transmit frequency.

TABLE 5-8 Airframe Spectrum Width for Nearly Constant Aspects

(A). Coherent measurements	Av aspect angle, degrees	Xmit. band	Record time, sec	$\Delta f_{0.5}$ Av 3-dB width Hz	$\Delta f_{0.5}$ Av 3-dB width m/sec	No runs in av	Refs.
Small prop. plane	Nose	C	0.8	14	0.38	8	APL
Small two-engine jet	25–35	C	0.8	10	0.26	3	APL
Two-engine propeller	140–175 6–43	S	1.2	17.5	0.92	13	[132]
Four-engine propeller	135	S	5.2	7	0.37	1	[132]
Multiengine jet		S	5.2	15	0.8	1	[132]
		X	5.0	27	0.45	4	[182]
Small helicopter		C	0.8	69	1.8	4	APL
(B). Incoherent measurements* — jet aircraft							
Jet fighter	All	L	5	2.4	0.3	4	[265]
	All	S	5	3.7	0.2	6	
	All	X	5	6.2	0.1	3	
Jet	Nose	X	5	12.4	0.2	3	[103]
Three- and four-engine jets	Nose and Tail	Ka	24	28	0.13	9	
	Broad	Ka	24	17	0.08	6	
(C). Incoherent measurements* — propeller aircraft							
Four-engine bomber	Nose, 127	L	5	29	0.35	4	[265]
	Broad	L	5	4.2	0.05	2	
	Nose, 127	S	5	20	0.10	3	
	Broad	S	5	7	0.04	2	

*Spectrum width for incoherent measurements is defined as twice the distance along the frequency axis from the spectrum peak to the −3-dB point.

TABLE 5-8 (Continued)

Av aspect angle, degree	Xmit. band	Record time, sec	$\Delta f_{0.5}$ Av 3-dB width		No runs in av	Refs.
			Hz	m/sec		

(C). Incoherent measurements* — propeller aircraft *(Continued)*

	Av aspect angle, degree	Xmit. band	Record time, sec	Hz	m/sec	No runs in av	Refs.
Four-engine bomber	Nose, 127	X	5	23	0.37	2	
	Broad	X	5	4.5	0.07	1	
Small prop. aircraft		L	5	18	0.22	1	
		S	5	18	0.09	1	
		X	5	22	0.35	2	
Four-engine aircraft	Nose	K_a	28	22	0.09	2	
	Broad	K_a	11	30	0.13	1	

*Spectrum width for incoherent measurements is defined as twice the distance along the frequency axis from the spectrum peak to the –3-dB point.

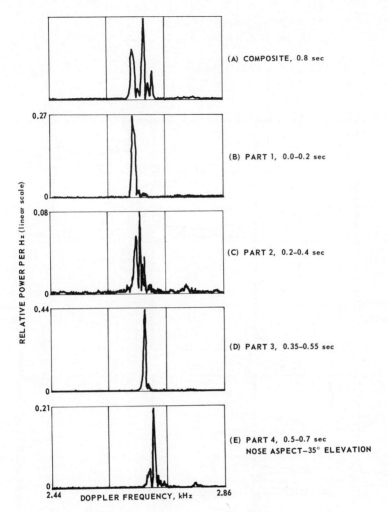

figure 5-13 *C-band power spectra of T-28 (propeller) aircraft.*

It is possible that the incoherent spectrum for propeller aircraft is dominated by amplitude modulation components from the propellers. The major energy in the propeller doppler spectrum most likely consists of frequency modulation components which do not appear after envelope detection. By comparing parts B and C, it appears that the incoherent spectrum for jet aircraft is dominated by the airframe spectrum. The time required for independent sampling may be estimated by using the 3-dB width of the spectrum. For the gaussian spectrum, the

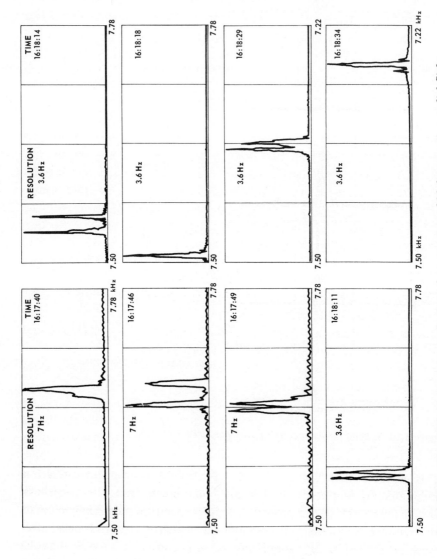

figure 5-14 *C-band doppler spectrum of echoes from two jets—300-ft separation, radial flight.*

standard deviation and the 3-dB width are related by $\sigma_f = 0.42\,\Delta f_{0.5}$. Using this relationship along with the results of Sec. 3.6, the independence time is estimated by $T_I \simeq 1/\Delta f_{0.5}$. This results in independence intervals ranging from 30 to 300 msec. Thus, unless polarization or frequency agility is used, the pulse-to-pulse independence assumption is not satisfied for most aircraft detection radar problems. These conclusions regarding spectral width and independence time should be regarded as tentative answers. Much remains to be understood concerning target RCS spectra.

5.9 FREQUENCY AGILITY EFFECTS ON TARGET DETECTION AND TRACKING

A frequency agility or diversity radar transmits several pulses on different carrier frequencies during a single look at the target. The received signals are detected and either linearly integrated or otherwise combined with some logic technique. While diversity techniques have been accepted as a valuable tool in the communication field for reducing fading losses, the benefits in radar are not always obvious. In detection and tracking radars frequency agility can

1. Deny an enemy the possibility of concentrating all of his jamming power at a single carrier frequency (spot jamming)

2. Reduce the probability that a target will have an aspect angle which gives a null in its RCS

3. Reduce range and angle tracking errors caused by finite target extent (glint) and multipath effects

4. Allow improvement in the target signal-to-clutter fluctuations (S/C) for incoherent pulse radars

The first item has been mentioned in Chap. 1 and in Gustafson and As [151]. The second and third benefits result from the modification of the RCS distribution and will be discussed in this section. The fourth benefit will become obvious after a discussion of the effects of multiple frequency transmissions on clutter echoes. Chapter 3 showed that the most difficult targets to detect with high probability were those whose echoes fluctuated slowly. (Swerling cases 1 and 3.) If such a target presents a null in cross section on a given pulse, its aspect may not change significantly to increase the RCS for successive transmit pulses during a single scan. The echo from a complex target is the vector summation of the reflections from every scatterer on the target, and the phase of each scatterer's contribution is determined by the relative spacing of each reflecting point. If the maximum radial separation between scatterers is a large number of wavelengths, a fractional change in carrier frequency will drastically change the vector summation of the echo. Hence, the frequency-shifted echo may be decorrelated from that using the original wavelength. This will occur even if there is no change in target aspect. This is illustrated in Fig. 5-15 for the 54-in. satellite target described in Sec. 5.5. The RCS is shown as a function of carrier frequency for a series of aspect angles near a null. The deep null at 85.7° aspect occurs only over a frequency span of about 40 MHz. A radar that transmits sufficiently different carrier frequencies in one scan can thus

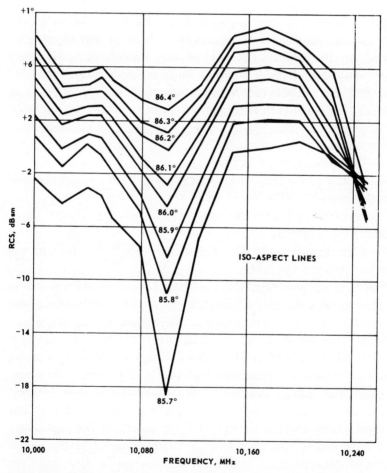

figure 5-15 *RCS vs. frequency near a null for 54-in. satellite. (After Kennedy [210].)*

receive decorrelated echoes from a stationary target.[1] A target that is described as a Swerling case 1 for a single frequency transmission becomes a case 2 target for a frequency agile radar, provided the frequency separation is sufficient. Similarly, a case 3 target becomes a case 4. If a high detection probability ($P_d \geq 0.9$) is desired, the case 2 or 4 target requires considerably less energy than the case 1 or 3 target for multiple pulse transmissions per scan. This is because the probability of making a threshold decision during a signal null is greatly reduced. The next paragraphs give some preliminary answers to the questions of how much frequency shift is needed to decorrelate the target echoes and how greatly the detectability is improved.

Since there is no unique description of a complex target, the decorrelation of target echoes with frequency shifting can only be described for some idealized cases. Examples are a volume filled with a large number of scatterers (rain, chaff, etc.) or a number of reflection points distributed along a line target. Theoretical and experimental verification of the correlation coefficient for volume scatterers is given in Chaps. 6 and 7. For uniformly distributed reflecting points along a line target of length L, Birkemeier and Wallace [42] have shown that the correlation coefficient ρ of the square-law echoes is

$$\rho(\Delta f) = \frac{\sin^2[2\pi\Delta f(L\sin\theta)/c]}{[2\pi\Delta f(L\sin\theta)/c]^2} \tag{5-11}$$

where Δf = carrier frequency shift
 $L\sin\theta$ = projection of target length onto radial dimension
  c = velocity of light
This equation assumes that the pulse length $c\tau/2$ is greater than $L\sin\theta$. They also define a critical frequency shift Δf_c which yields the first zero of $\rho(\Delta f)$ and hence one definition of *decorrelation*

$$\Delta f_c = \frac{c}{2L\sin\theta} = \frac{c}{2L_0} = \frac{150 \text{ MHz}}{\text{length in meters}} \tag{5-12}$$

L_0 is defined as the effective radial extent of the scatterers. The term $c/2L_0$ can be seen to be the inverse of the radial target extent in radar

---

[1] The frequencies may be contiguous or separated by the interpulse period.

time. Since $c \approx 3 \times 10^8$ m/sec, the critical frequency shift is 150 MHz divided by the effective target length. Ray [304], using different criteria, suggests that 45 MHz/L_0 is sufficient for decorrelation with $\rho(\Delta f) = 0.5$. Gaheen, McDonough, and Tice [129] suggest that 75 MHz/L_0 is adequate (for a correlation coefficient of 0.4). Since most targets are not a large collection of linearly distributed scatterers, these values may not completely result in decorrelated returns from a target at all aspects. However, they should be sufficient to fill in the deep nulls. The correlation function versus frequency shift for the satellite target is shown for three $10°$ aspect sectors in Fig. 5-16 from Kennedy [210]. In sector 1 where the cross section is quite small (0.5 m^2), there is considerable decorrelation of the received power with a frequency shift of $c/4L_0$. In sector 2 where the cross section is quite large (21 m^2) there is little decorrelation. In this region decorrelation is not essential or even desired since the reflections will be strong at all frequencies. By averaging eight frequencies spaced over a 250-MHz band, the nulls that were below -25 dB m^2 were enhanced by 17 dB. Those that were between -15 and -25 dB m^2 were enhanced by an average of 10 dB.

If the single-frequency target fluctuation model is known, the improvement in detectability for N pulses can be determined from the appropriate Swerling curves [244, 113].[1] Without frequency agility and with integration times that are short compared to the correlation time of the target echoes, case 1 or case 3 should be used. If the frequency shift of the agility radar is sufficient to decorrelate the target echoes, case 2 or 4 for N pulses integrated should be used. The improvement in detectability is the difference in (S/N) between the two sets. For partially decorrelated target echoes, N should be replaced with the number of independent target echoes N_i. This will give pessimistic answers since the noise fluctuation is further reduced by $N - N_i$ integrations.[2] The improvement in detectability is illustrated in Fig. 5-17 [129] for the missile target studied by Kennedy and described in Sec. 5.5. The figure shows that the probability of the target echo exceeding any given small value x is improved drastically for 16 frequency increments even for a total spread of 250 MHz. In this case $\Delta f \approx 16$ MHz which is less than $\Delta f/4L_0$. Doubling the frequency spread gives only slight additional improvement because most of the gain comes from the first few independent samples.

[1] See Fig. 3-4 as an example.
[2] A more exact method has been suggested by Swerling [368, 369].

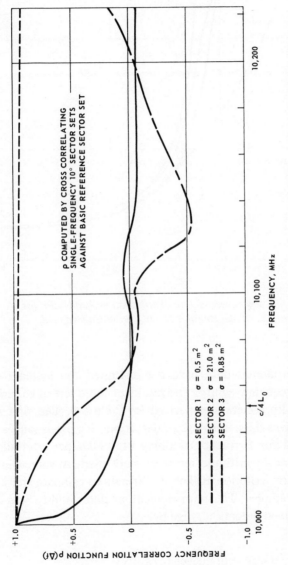

figure 5-16 *Normalized frequency correlation functions for 54-in. satellite target. (After Kennedy [210].)*

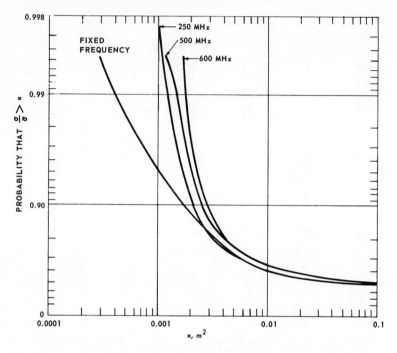

figure 5-17 *Missile target cross-section distribution vs. bandwidth for 16 frequency increments. Missile length = 150 in. (After Gaheen et al. [129].)*

Some experimental data have demonstrated the increase in detectability with frequency agility. The results of three sets of experiments on unidentified aircraft are summarized in Table 5-9. The first set at K_u Band [197] gives the percentage of detections at various ranges for single frequency and for frequency stepping of 1 MHz per interpulse period. Both linear and logarithmic receivers were used. A somewhat unusual digital detection criterion makes it difficult to estimate the number of independent samples. The improvement in detectability was estimated from the observed detection range by

$$\Delta\left(\frac{S}{N}\right) \approx \left(\frac{\text{Range with frequency agility}}{\text{Range with single frequency}}\right)^4$$

Each detection included about 30 carrier frequencies spaced by 1 MHz.

TABLE 5-9 Experimental Results Showing Improved Target Detectability in dB with Frequency Agility Transmission [151, 197, 398]

Percentage detections	Boeing aircraft tests, 16.5 GHz increased sensitivity, ∼30 pulses $\Delta f = 1$ MHz 1 pulse		Swedish tests		RRE tests, 2.8 GHz improvement with sum of two frequencies $\Delta f = 215$ MHz
	Linear detector	Log-amp	Jet fighter	Small prop.	Improvement in sum channel, dB
$P_d \approx 95\text{-}97$	7.8 dB	8.1 dB	----	----	----
95	6.4	7.2	----	----	(7)*
90	4.0	4.5	6.4	----	(4)*
80	2.7	2.7	4.7	7.8	7.0
70	2.1	2.4	3.8	7.0	2.5
50	1.6	2.0	3.0	3.3	1.3
30	1.3	1.6	----	1.8	1.2
10	1.2	1.4	----	0.7	----

*Experimental frequency agility compared to theoretical single frequency for Rayleigh fluctuating target.

189

The unexpected detectability improvement at low detection probabilities is not explained.

The second set of measurements summarized on Table 5-9 is from a frequency-jumping search radar test in Sweden [151]. The targets flew a controlled inbound course with a fixed-frequency or frequency-jumping track radar. Outbound targets had a 20 percent greater tracking range to the first *dropout* in the frequency-jumping mode.

The third set of columns of Table 5-9 represents similar data for an unidentified aircraft reported by Waters [398]. The detected echoes on two carrier frequencies separated by 215 MHz were summed and compared to one of the channels. There were not enough tests to determine (S/N) improvement for the single-frequency transmission above $P_D \sim 80$ percent. Therefore the numbers in parentheses were estimated by extrapolating the single-frequency data at the lower values of P_D. A portion of the experimental improvement (1.5-2 dB) is accounted for by the postdetection of summation of two channels that would be obtained even if they had been on the same frequency.

Since in all tests the targets observed were not identified, it can only be conjectured from the data that their reflectivity distribution functions were between Swerling case 1 and case 3. In these studies the improvements for $P_D \geq 70$ percent are quite dramatic. Similar improvements are expected for satellite and missile targets. Improvements from frequency agility will be considerably less if the targets fluctuate during a single scan or if only the cumulative detection probability need be high (P_D per scan is low).

Angle and range *glint* errors will also be reduced with frequency agility, but it is more difficult to quantify the expected improvement. If the target can be represented by a line of scatterers perpendicular to the line of sight, there will be a *time-correlated* angle error signal for a single-frequency radar with the spectral energy of the error signal concentrated near zero frequency.[1] Since the time constant of angle tracking loops is generally quite long, the angle error signals will follow the slowly varying phase front of the complex target echo. With several assumptions as reported in [42, 255], the spectrum of the angle error signals for a target of linearly distributed scatterers can be written

---

[1] Receiver noise and fading errors are neglected in this discussion.

$$W_u(f) = \frac{\overline{U^2}}{2f_m}\left(2 - \frac{3f}{f_m} + \frac{3f^2}{f_m{}^2} - \frac{f^3}{f_m{}^3}\right) \quad \text{for} \quad f \le 2f_m$$

$$= 0 \text{ elsewhere}$$

where $\overline{U^2}$ = the mean-square value of the perturbation signal $U(t)$
$f_m = f_0\,\Omega\,(L\,\cos\theta)/c$ and
f_0 = the carrier frequency, Hz
Ω = the rotation rate of the line target about its center (small angles assumed)

$L\,\cos\theta = L_x$ = the projection of the target perpendicular to the line of sight
The variance of $U(t)$ can be shown to be independent of Ω while the spectral width is proportional to Ω [42]. Frequency agility essentially spreads the *glint* spectral energy outside the angle-servo bandwidth. Another way of viewing this effect is that frequency agility allows a number of independent samples of the angle of arrival to be averaged within the passband of the angle servos [129]. Methods of analysis of the reduction in angle error variance are given in [42, 304]. Under certain conditions the maximum reduction in the angle error variance due to glint can equal the number of frequencies transmitted. This assumes that the separation between adjacent frequencies is greater than $c/2L_x$, where L_x is taken as the target projection perpendicular to the line of sight. Substantial glint reductions on both aircraft and ship targets have also been measured experimentally [151]. For the aircraft targets, the glint error signals up to 2-4 Hz were considerably reduced. With large ship targets the bearing errors were one-fourth to one-half as large with frequency-jumping as with single-frequency transmission. Range error fluctuations can also be reduced by frequency agility which reduces the wander of the apparent *centroid of reflection*. For aircraft targets where the length is comparable to the wingspan, the same amount of frequency shifting per pulse ($\sim c/2L_0$) improves detectability and reduces angle and range glint significantly.

This section has shown that in many cases frequency agility is desirable to modify target echoes and eliminate nulls in the RCS. It is shown in Secs. 6.5 and 7.6 that frequency agility can also decorrelate distributed clutter echoes. By integrating these echoes, the RCS is made to approach its mean value while the variance of the clutter is reduced. This improves the signal-to-clutter ratio just as incoherent integration improves the signal-to-noise ratio.

6 ATMOSPHERIC EFFECTS, WEATHER, AND CHAFF

Meteorological phenomena have two major effects on radar: the signals are attenuated by clouds, rain, snow, and the atmosphere itself; a relatively large signal is reflected from raindrops, hail, and snowflakes. Attenuation effects become quite significant above X band (9,300 MHz), but backscatter from snow and rainfall generally dominates the detection and tracking problem at frequencies down to L band (1,300 MHz). To compound the problem, the backscatter spectrum from precipitation and chaff is broadened because of wind shear, vertical fall, and air turbulence all of which cause limitations in the ability of doppler processors to separate targets from clutter on the basis of their relative velocities. The backscatter coefficient of chaff dipoles will be shown to have much less dependence on the carrier frequency than precipitation echoes, but both

For general references see Atlas, "Advances in Radar Meteorology" [12], and Battan, "Radar Meteorology" [25].

have similar amplitude and spectral distributions. The equations and numerical values in this chapter apply to monostatic radars with linear polarization.

6.1 ATMOSPHERIC ATTENUATION

Atmospheric attenuation was thoroughly described by L. V. Blake [44] in 1962, with subsequent modifications in August, 1963, and more recently by Rice et al. [310] in 1965.[1] Figure 6-1 from Blake is included

[1] Experimental work on the 35-, 50-, and 94-GHz regions can be found in [138, 78, 175], respectively.

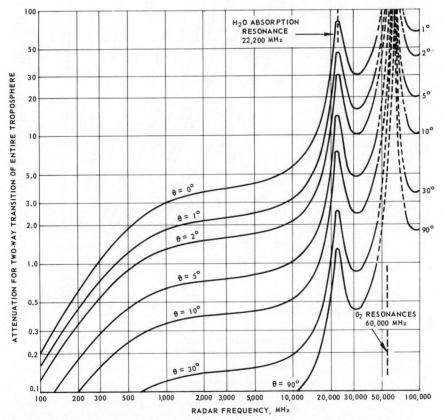

figure 6-1 *Radar attenuation for transversal of entire troposphere at various eleva-tion angles. Applicable for targets outside the troposphere. Does not include iono-spheric loss, which may be significant below 500 MHz in daytime. (After Blake [44].)*

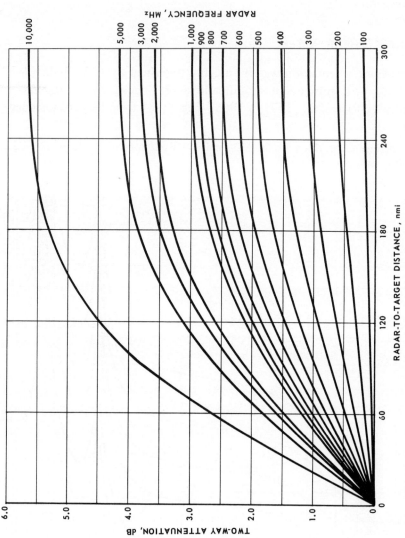

figure 6-2 Radar atmospheric attenuation, 0° ray elevation angle, 100-10,000 MHz. (After Blake [44].)

here to show the normal atmospheric attenuation throughout the entire troposphere. The attenuation for a horizontal beam at sea level as a function of range and frequency is shown in Fig. 6-2. Numerous other curves are given in the references. The attenuation coefficient per mile at sea level can be obtained from the slope of the curves near zero range in Fig. 6-2.

6.2 ATTENUATION DURING PRECIPITATION

Common rainfall rates are defined in the following chart:

Type	Rainfall rate r, mm/hr
Drizzle	0.25
Light rain	1.0
Moderate rain	4.0
Heavy rain	16.0
Excessive rain	40.0

Many theoretical calculations have been made on the attenuation caused by rainfall, but they have not been confirmed very closely by experiment. An excellent summary of experimental data was given by Medhurst [248] in 1965 for the region of 8 to 60 GHz. Measurements have been made at 94 GHz by Hoffman et al. [175], but in this case it is difficult to separate the effects of the attenuation by the normal atmosphere from those by the rainfall. A summary *mean* attenuation curve from the Medhurst data is given in Fig. 6-3, which also includes additional data points subsequent to Medhurst (see [198, 101, 138, 162]). The mean-value curve has been extrapolated to 6,000 MHz from theoretical considerations. The wide range between the minimum and maximum attenuation is due to various effects, a considerable part undoubtedly being due to the attenuation (in decibels/nautical mile) not really being linear with rainfall rate. This is especially true at X band, where attenuation theoretically varies as $r^{1.3}$. A second factor that causes a wide variation in attenuation for a given precipitation rate is change in drop size distribution for different types of rain [12, 248]. A third factor that has an effect on the total attenuation is the increase in absolute humidity that often accompanies a rainstorm.

With a few exceptions [257] the reported measurements of attenuation versus rainfall rate are considerably higher than the earlier theoretical relationships especially at rainfall rates of 0.5 to 4 mm/hr. The attenuation (in decibels) at 9 to 15 GHz for light rains is apparently twice the value reported in such earlier texts as Gunn and East [150], Skolnik [354], and Battan [25]. For systems analysis it would seem that a linear relationship of attenuation (in decibels) versus rainfall rate (in millimeters/hour) is adequate. Since the Medhurst mean-value curve drawn on Fig. 6-3 seems somewhat severe for rains of 1-4 mm/hr, a dotted line is shown that reflects some later measurements. The slope of this line indicates that attenuation (in decibels) varies as (carrier frequency)$^{2.8}$ between 6 and 40 GHz.

Snow has a similar attenuation effect, but there is less experimental data for snow than for rain. An equation from Gunn and East [150] can be used.[1]

One-way attenuation α at $0°$C (decibels/kilometer)

$$\alpha = \frac{0.00349\,r^{1.6}}{\lambda^4} + \frac{0.0022\,r}{\lambda} \tag{6-1}$$

where r is the snowfall rate (millimeters of water content/hour) and λ is wavelength (centimeters).

The attenuation due to precipitation is a highly statistical phenomenon related to both time and location. The percentage of time that rain exceeds a given rate is shown for eastern England, Haiphong, North Vietnam, and three typical locations in the United States in Fig. 6-4; in [55, 384] similar data are given for other locations throughout the world.

A second essential statistic relates to the fact that heavier storms occur at lower altitudes and have a smaller extent (diameter) than less severe storms. This tends to reduce the overall effect of the increase in attenuation with rainfall rate since the total attenuation is the product of the attenuation coefficient and the extent. For high-elevation beams the mean intensity of rainfall versus altitude (from [310]) can be roughly

---

[1] Bell [35] has reported measured peak attenuations at 3.6 and 11 GHz that are far in excess of values predicted by this formula. At 11 GHz he recorded as much as 25-dB attenuation (one way) over a 55-km path when there was snow of only 2 mm water content in one hour.

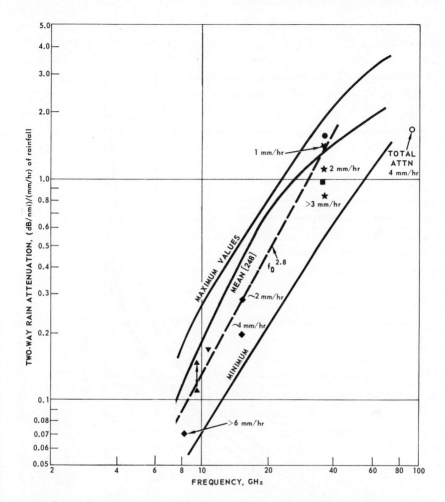

figure 6-3 *Measured values of rain attenuation vs. frequency (two-way path).*
——— *Medhurst [248] ; ★ Godard [138] ; ◆ Blevis, et al. [46] ; ▼ Easterbrook
and Turner [101] ; ● Emerson [198] ; ○ Hoffman [175] ; ■ Harrold [162] ;
▲ USSR av.; — — — Modified mean attenuation.*

expressed as $r/r_s = \exp(-0.2\,h^2)$, where r_s is the surface rainfall rate and
h is the height above ground (kilometers). For surface radars and nearly
horizontal antenna beams the extent can be estimated from the work of
James [189], who suggested the following relationship for rainstorms.
(See also [310, pp. 3-5].)

$$\text{Diameter (nautical miles)} = 25.9 - 14.7 \log r \qquad (6\text{-}2)$$

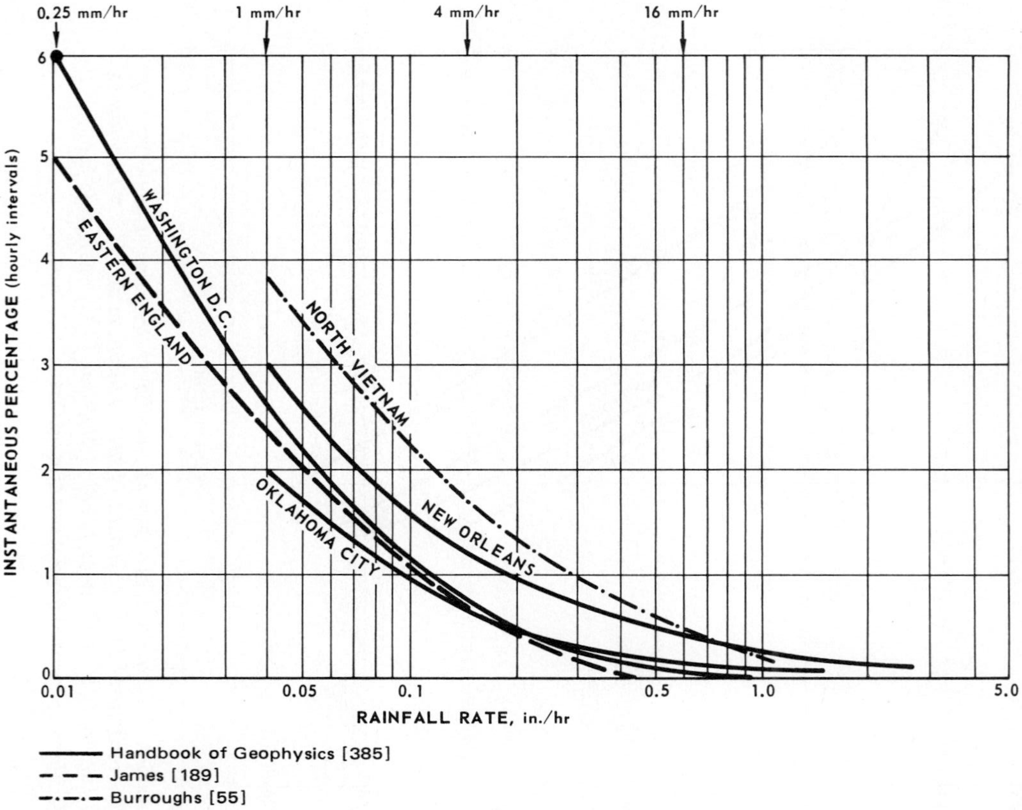

figure 6-4 *Percentage of time rainfall exceeds a given rate.*

Legend (bottom of figure):
——— Handbook of Geophysics [385]
– – – James [189]
–·–·– Burroughs [55]

X-axis: RAINFALL RATE, in./hr
Y-axis: INSTANTANEOUS PERCENTAGE (hourly intervals)

Curves labeled: NORTH VIETNAM, NEW ORLEANS, OKLAHOMA CITY, WASHINGTON D.C., EASTERN ENGLAND

Top axis marks: 0.25 mm/hr, 1 mm/hr, 4 mm/hr, 16 mm/hr

At X band (9.3 GHz) the maximum attenuation (two ways) reaches about 10 dB at 20 mm/hr for the theoretical attenuation relationships or about 15 dB using the dotted curve of Fig. 6-3.

Attenuation from heavy clouds can also be substantial at the higher microwave frequencies. *Natl. Bur. Std. Tech. Note* 101 [310] contains values of the one-way attenuation coefficient for several frequencies and cloud densities. These are summarized in Table 6-1. These attenuation data roughly follow a λ^{-4} variation, which contrasts somewhat with the lower order law for rainfall above X band determined by the experiments summarized by Medhurst [248]. A convenient approximation for the attenuation when the temperature is 20°C is to multiply the total attenuation (decibels) given in Table 6-1 by 0.6 [310]. A cloud density of 1.4 gm/m^3 corresponds to a heavy cumulus cloud and that of 4 gm/m^3 corresponds to very heavy nimbostratus clouds. The attenuation for ice clouds with the same water content is about two orders of magnitude below those of the table.

TABLE 6-1 α = Attenuation in Typical Clouds (One Way)*

Frequency GHz	Coefficient of attenuation at 0°C, (dB/km)/(gm/m^3)	Attenuation	
		Cloud density of 1.4 gm/m^3, dB/nmi	Cloud density of 4 gm/m^3, dB/nmi
9.4	0.086	0.22	0.63
17	0.267	0.69	2.0
24	0.532	1.38	3.9
33	0.99	2.6	7.4

*After Gunn and East [150, p. 539]. See reference for other temperatures.

6.3 BACKSCATTER COEFFICIENT OF RAIN, CLOUDS, AND SNOW

Radar backscatter from raindrops limits the detection performance of long-range radars at almost all microwave frequencies. The Rayleigh scattering theory generally applies to all microwave frequencies up to and including X band (except for excessive rain at X band). The backscatter is proportional to the sixth power of the drop size in the

Rayleigh region.[1] Unfortunately, the variations in drop-size distribution for various rainfall rates tend to limit the exactness of theoretical calculations to about 3 dB. The most representative relationship used by radar and radar meteorology texts [354, 12, 25] is

$$Z = 200 \, r^{1.6} \text{ (temperate latitudes)} \tag{6-3}$$

where[2] r = rainfall rate, mm/hr

Z = reflectivity, mm^6/m^3

An alternate relationship, proposed by Atlas [12], is $Z = 214r^{1.39}$. Measurements of Florida rains have fitted the equation $Z \approx 286r^{1.43}$ and in Indonesia $Z \approx 311r^{1.44}$ [257]. The lower values of the exponent in other experiments [12] generally correspond to microwave frequencies above X band with drop size large with respect to 0.1 λ. Measurements at 35 GHz and higher place the exponent between 0.95 and 1.35; values for the constant term increase to almost 1,000. Godard [138] suggests an exponent of 1.2 at 35 GHz.

The Z relationship is not in a useful form for radar system evaluation. A conversion factor to pulse radar received power, attributed to J. R. Probert-Jones [296], is

$$P_r = \frac{1.1 \times 10^{-23} P_t G_t G_r \theta \phi \tau Z L_a}{\lambda^2 R^2} \tag{6-4}$$

including the beam shape losses where in the units of the reference

P_t = transmitted pulse power, watts

P_r = received pulse power, watts

G_t, G_r = antenna gains, dimensionless

θ, ϕ = one-way, half-power antenna beamwidths, degrees

R = range to clutter cell, nmi

λ = wavelength, cm

τ = pulse length, μsec

Z = reflectivity, mm^6/m^3 (Z is not wavelength dependent at low frequencies in the Rayleigh region)

L_a = atmospheric and rain attenuation, two way

---

[1] See Sec. 5.7 for general formula.
[2] In units commonly used in meteorology texts.

The ratio of returned power from a target to the returned power from the clutter can then be derived from the classicial radar equation (see Chap. 2)

$$\frac{P_{\text{tar}}}{P_{\text{rain}}} = \frac{K\lambda^4 \sigma_t}{\theta \phi \tau Z R^2} \tag{6-5}$$

where K = constant (primarily to keep units consistent, ≈ 200 depending on antenna pattern and assuming 3 dB of beam shape loss on both target and clutter); it also includes the refractive index for rain

  σ_t = target cross section, m^2

Since $[(\pi/4)\,\theta \phi \tau R^2]$ is the volume of the range cell (not in consistent units) and Z is a parameter generally only used by radar meteorologists, it has been found that a more useful form for the radar target detection criterion, where receiver noise power is small when compared with clutter power, is

$$R^2 = \frac{\sigma_t L}{(\pi/4)\,\phi\theta\,(c\tau/2)\,(S/C)\,\Sigma\sigma} = \frac{10^{10}\,\sigma_t L\lambda^4}{284\,(\pi/4)\,\phi\theta\,(S/C)\,Z\tau} \tag{6-6}$$

where beam shape losses for the target and the rain are assumed equal, and

where (S/C) = the signal-to clutter ratio for a given probability of detection (S/C is ≈ 13 dB for a typical single-pulse surveillance radar with detection probability = 0.5 and false alarm probability $\approx 10^{-8}$)

  θ, ϕ = one-way, half-power antenna beamwidths, radians

  $c\tau/2$ = pulse length, m

  σ_t = target cross section, m^2

then $\Sigma\sigma$ = backscatter coefficient, m^2/m^3, including wavelength effects (Skolnik [354]), also denoted as η by some authors

  L = processing losses appropriate to the rain backscatter case

  τ = pulse duration, sec

  λ = wavelength, cm

R = target range, m, assuming that the rain fills the volume of the range cell occupied by the target

Table 6-2, derived from both theoretical and experimental data, gives $\Sigma\sigma$ versus frequency and type of precipitation. Interpolation between radar bands on the table should be as λ^4 except for the highest frequencies and rainfall rates. Interpolation between the rainfall rates under the same restrictions should follow an $r^{1.6}$ law.

The Z-r relationships reported for snow vary from $500r^{1.6}$ for ice crystals to $2,000r^{2.0}$ for wet snowflakes, where r is the rate of fall but now refers to the water content of the melted snow in millimeters/hour. In converting the Z-r relationship to backscatter power, the substantially lower refractive index (about one-fifth) of ice and snow reduces the backscatter power by a factor of five for a given value of Z. Battan [25] indicates that snow causes about twice the backscatter of rain of the same water content at 1 mm/hr and three times the backscatter at 3 mm/hr. A single set of measurements by the author of January 26, 1966 gave the same (± 2 dB) backscatter at C band for a uniform snow as would be calculated for a uniform 1 mm/hr rain from Eqs. (6-3) and (6-4); interpolating weather bureau data for January 26 gives an equivalent water content fall rate of about 1 mm/hr.

The values of $\Sigma\sigma$ for clouds in Table 6-2 are useful only as models since there is no uniform description of clouds. The identical values of $\Sigma\sigma$ on the table for the P- and L-frequency bands are experimental points determined by the Tradex system investigators [83]. A series of measurements of the reflectivity for clouds has been reported in the USSR and is abstracted on Table 6-3 [282]. The mean value of reflectivity, the variation, and the mode are given for several cloud types along with several vertical profiles. It should be noted that most of the values are higher than earlier estimates [354, p. 543]. Reflectivity from rain clouds (Ns) is only about 3 dB below that of 1 mm/hr rain at the freezing level.

The backscatter from precipitation particles can be represented by a random assembly of scatterers, no one of which has a dominant effect on the sum. The RCS of the random assembly may be derived by assuming that the relative phase of each scatterer's return is statistically independent of the others and is uniformly distributed between 0 and 2π. This is an example of the two-dimensional *random walk* problem where each step is $\sqrt{\sigma_k}$. This, in turn, leads to the well-known *Rayleigh power*

TABLE 6-2 Reflectivity of Uniform Rain, m²/m³

Type	Probability of occurrence, Washington, D. C., %	$\Sigma\sigma$, dB m⁻¹ transmit frequency, GHz						
		P 0.45	L 1.25	S 3.0	C 5.6	X 9.3	K$_u$ 24	K$_a$ 35
Heavy stratocumulus clouds, 4gm/m³		-140[1]	-140[1]	-118	-108	- 98		
Fog, 100-ft visibility				-120	-110	-100		
Drizzle, 0.25 mm/hr	6			-102	- 91	- 82	-64[2]	-57[2]
Light rain, 1 mm/hr	3		-107	- 92	- 81.5	- 72	-54	-47
Moderate rain, 4 mm/hr	0.7		- 97	- 83	- 72	- 62	-46	-39
Heavy rain, 16 mm/hr	0.1			- 73	- 62	- 53		-32[2]

(1) Tradex data [83].
(2) May be in error by 3 dB (rainfall only).
(3) To convert to m²/nm³ add 98 dB.
(4) Tropospheric attenuation not included.
(5) m⁻¹ also represents m²/m³.

TABLE 6-3 Reflectivity Statistics for Clouds

	Cirrus, Ci		Altocumulus, Ac or Altostratus, As		Stratocumulus, Sc		Nimbostratus, Ns	
$\overline{Z}\left(\dfrac{mm^6}{m^3}\right)$	0.20		0.53		0.50		106.7	
σ_z (variation)	0.54		1.24		0.60		262.0	
Z mode	0.05		0.05		0.50		50.0	
Z, dB below 1 mm/hr rain	−30		−26		−26		−3	
Specific examples of vertical profiles	Height, m	$\overline{Z}$	Height, m	$\overline{Z}$	Height, m	$\overline{Z}$	Height, m	$\overline{Z}$*
	7,460	0.11	4,880	0.35	900	0.5	1,250	3.4
	7,820	0.09	5,210	0.80	1,150	0.43	1,750	5.9
	8,180	0.07	5,540	1.16	1,370	0.31	2,250	9.0
	8,540	0.07	5,880	1.14	1,600	0.31	2,750	36.0
	8,900	0.07	6,250	0.58	1,800	0.31	3,200	272.5†
							3,400	199.0
							3,600	84.5
							3,800	28.5
							4,000	15.6
							4,200	3.1
Peak cloud height, m	9,000		6,500		2,000		4,000	
Min. cloud height, m	6,900		4,400		700		750	

*$\overline{Z}$ = 200 for 1 mm/hr rain.
† Bright band, 0°C.
Ignatova, Petruskevskii, and Sal'man [282].

or *exponential* probability density function. When each scatterer is of equal cross section, the mean is simply the sum of the RCS of each element. It is sometimes erroneously assumed that individual areas of a large assembly of reflectors, for instance a rain cloud, have approximately the same RCS. In reality the amplitude fluctuation is quite large.

6.4 RADAR PRECIPITATION DOPPLER SPECTRA

It is convenient to describe the doppler spectra of radar signals returned from clouds or precipitation by four separate mechanisms (Atlas [12, p. 405]).

1. *Wind shear*: The change in wind speed with altitude results in a distribution of radial velocities over the vertical extent of the beam.

2. *Beam broadening*: The finite width of the radar beam causes a spread of radial velocity components of the wind when the radar is looking crosswind.

3. *Turbulence*: Fluctuating currents of the wind cause a radial velocity distribution centered at the mean wind velocity.

4. *Fall velocity distribution*: A spread in fall velocities of the reflecting particles results in a spread of velocity components along the beam.

A fifth mechanism may be considered in a system using a rotating or moving antenna (see Chap. 9 and Barrick [19]). This mechanism is independent of the characteristics of the precipitation itself and is not discussed in this chapter. If we assume that these mechanisms are independent, then the variance of the doppler velocity spectrum σ_v^2 can be represented as the sum of the variances of each contributing factor [1]

---

[1] The power spectral density function of sea, land, or weather clutter is also described [18] by a gaussian spectrum having a stability or spectrum parameter a defined by

$$G(f) = G_0 e^{-af^2/f_0^2} \qquad (6\text{-}7)$$

where $G(f)$ is the spectral density function of the clutter, f_0 is the transmitted frequency, a is the stability parameter (dimensionless), and G_0 is a constant that depends on the received average power.

A more convenient expression for the power spectrum is given in terms of the frequency spectrum variance σ_f^2

$$G(f) = G_0 e^{-f^2/2\sigma_f^2} \qquad \textit{(Footnote continued on next page)}$$

$$\sigma_v^{\ 2} = \sigma_{shear}^2 + \sigma_{beam}^2 + \sigma_{turb}^2 + \sigma_{fall}^2 \qquad (6\text{-}8)$$

Although independence is not strictly satisfied, Eq. (6-8) nevertheless provides a useful approximation. The spectrum width will be given in velocity units, m/sec; for most radar frequencies the spectra can be converted to frequency units through the doppler equation.

Wind Shear

Although wind shear (the change in wind velocity with altitude) has long been known to meteorologists, most current radar literature has failed to report its effect on precipitation spectra.[1] Wind shear can be the greatest factor contributing to the doppler spectrum spread of precipitation echoes in a long-range, ground-based radar. The change in wind velocity with height is a worldwide and year-round condition. An altitude of maximum wind velocity is always present although it experiences seasonal changes in height and intensity. The change in wind velocity with height can often be approximated by a constant gradient. A representation of the wind shear effect is shown in Fig. 6-5.

For elevation angles of a few degrees, the difference in radial components at the half-power points of the beam is

$$\Delta V_r \simeq |V_1 - V_2|$$

If we assume that the wind velocity gradient within the beam is constant, then for a gaussian antenna pattern this velocity distribution will have a standard deviation that is related to the half-power width by

$$\sigma_{shear} = 0.42 \, (\Delta V_r)$$

[1] G. R. Hilst [173] was one of the earliest radar meteorologists to analyze the effects of wind shear on precipitation spectra.

(Continued from footnote on preceding page) Alternately, this equation can be expressed in terms of the velocity spectrum

$$G(v) = G_0 \, e^{-v^2/2\sigma_v^2}$$

The relationship between the spectrum variance and the parameter a is then

$$a = \frac{f_0^{\ 2}}{2\sigma_f^2} = \frac{c^2}{8\sigma_v^2}$$

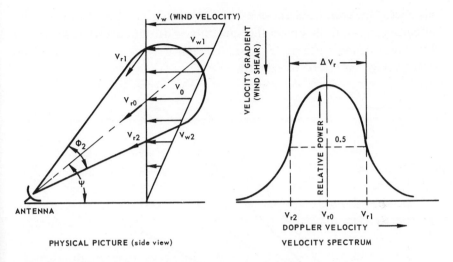

figure 6-5 *Effects of wind shear on the doppler spectrum.*

In terms of the wind speed gradient k this spectrum width is

$$\sigma_{shear} = 0.42 \, kR\phi_2 \qquad (6\text{-}9)$$

where k is the shear constant, i.e., velocity gradient (m/(sec)(km)), along the beam direction, R is the slant range to clutter (kilometers), ϕ_2 = two-way, half-power antenna elevation beamwidth (radians). For radar calculations the value $k = 5.7$ m/(sec)(km) is suggested for radars pointing in the direction consistent with that of the primary high-altitude winds.[1] Although this value may give excessive estimates of the average wind velocity, it is consistent with average values of shear observed for 0.5-2 km altitude thickness layers [262, 339, 362, 10]. Figure 6-6 illustrates the shear component as a function of range by using the value of $k = 5.7$ m/(sec)(km) for the maximum gradient in Eq. (6-9). For an arbitrary radar azimuth, the author has found that a shear constant $k = 4.0$ m/(sec)(km) is more appropriate. There will be some direction in which the wind shear component is a minimum although it may not be the crosswind at the beam center. Not only the wind speed but also the wind direction may change with height. As a result, in some

---

[1]While this value is based on average wind conditions in the Northeastern U. S., it is also representative of many parts of the world.

instances the crosswind shear effect may be as great as the downwind effect. Figure 6-6 illustrates the spectrum width due to wind shear as a function of range. The shear constant assumed is 4.0 m/(sec)(km), which is an average over 360° of azimuth.

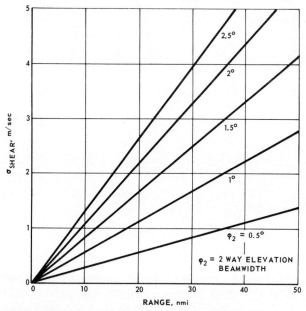

figure 6-6 *Spectrum width due to wind shear (average over 360° of azimuth).*

Although Eq. (6-9) seems to indicate that the spectrum width can increase indefinitely, in practice the value of σ_{shear} is limited to about 6 m/sec for beamwidths of 2.5° or less. This is partly because precipitation is limited at some altitude ceiling, such as 15,000 or 20,000 ft [262].

Beside the spectrum reduction due to the effects of a weather ceiling, the value of σ_{shear} at longer ranges will also be limited when a strong *bright band* exists. The *bright band* is a layer of small elevation extent that gives 5-10 dB greater intensity than the rain echoes that are above or below it (see Sec. 6-6). Since this layer dominates the total clutter power it will tend to place an upper limit on the width of the spectrum.

Spectra Caused by Wind Turbulence

Previous paragraphs have used the concept of a wind profile in terms of a mean velocity corresponding to a given altitude. Strictly speaking, because wind velocity is a nonstationary process, its mean cannot be defined without specifying the time over which the average is taken. Fluctuations about the mean are then called turbulence, an unpredictable variation that must be described statistically. (For purposes of this chapter, the averaging time shall be understood to be on the order of one second.)

A number of experiments using radar measurements have shown that the turbulence spectrum width is nearly independent of height, and an average value of the spectrum standard deviations for altitudes up to 3 km is $\sigma_v \simeq 1.0$ m/sec (although extreme values of 0.5 to 2.0 are also observed). For higher altitudes experiments indicate a slightly lower value of $\sigma_v \simeq 0.7$ m/sec [211, 324, 323, 356, 229, 262].

Beam Broadening

The spectrum component caused by beam broadening arises from the distribution of radial velocity components of a tangential wind blowing across a radar beam of nonzero width. The spectrum component will obviously have a zero mean and a standard deviation that may be derived in a manner analogous to the wind shear component.

$$\sigma_{\text{beam}} = 0.42 V_0 \theta_2 \sin\beta \qquad\qquad (6\text{-}10)$$

where θ_2 is the two-way, half-power antenna beamwidth in azimuth (radians), V_0 is the wind velocity at beam center, and β is the azimuth angle relative to wind direction at beam center. For most radar beamwidths (i.e., a few degrees) the beam-broadening component is quite small compared with the wind turbulence and wind shear components. For example, at a wind speed of 60 knots and a beamwidth of 2°, the maximum broadening component is 0.5 m/sec. To calculate the mean wind speed as a function of height, the gradient 4.0 m/(sec)(km) is recommended with a minimum wind speed of 5 m/sec at low altitudes (see Fig. 7-2).

Distribution of Fall Velocities

Because of the distribution of drop sizes in precipitation, a radar pointing vertically will observe a doppler spread resulting from a distribution of precipitation fall velocities.[1] Lhermitte [228, pp. 25-28]

---

[1] Fall velocities versus drop size are found in Medhurst [248] and are typically 4-6 m/sec for light rains.

states that the standard deviation of fall velocities for rain is approximately

$$\sigma_{fall} \simeq 1.0 \text{ m/sec (vertical velocity component)}$$

and is nearly independent of the rain intensity.[1] At elevation angle ψ the spread of radial fall velocities is

$$\sigma_{fall} = 1.0 \sin\psi \quad \text{m/sec} \tag{6-11}$$

The mean fall velocity for heavier rains is ~ 9 m/sec. For elevation angles of a few degrees, this component is insignificant when compared with the wind shear and turbulence components. For snow, $\sigma_{fall} \ll 1$ m/sec regardless of the elevation angle.

Summary of Precipitation Spectrum Components

From the previous discussion, the precipitation spectrum width may be computed by

$$\sigma_v = (\sigma_{shear}^2 + \sigma_{turb}^2 + \sigma_{beam}^2 + \sigma_{fall}^2)^{\frac{1}{2}} \quad \text{(m/sec)}$$

where $\sigma_{shear} = 0.42 \, kR\phi_2$ m/sec $(\sigma_{shear} \leq 6.0)$
$\sigma_{turb} = 1.0$, m/sec
$\sigma_{beam} = 0.42 \, V_0 \, \theta_2 \sin\beta$, m/sec
$\sigma_{fall} = 1.0 \sin\psi$, m/sec and
$k =$ wind shear gradient, m/(sec)(km) (5.7 for along-wind direction or 4.0 averaged over 360° azimuth)
$R =$ slant range, km
$\theta_2, \phi_2 =$ horizontal and vertical two-way beamwidths, radians
$\beta =$ azimuth relative to wind direction at beam center
$\psi =$ elevation angle
$V_0 =$ wind speed at beam center, m/sec

---

[1] There are often long *tails* on the low-frequency side of the doppler spectrum due to the slow fall rate of the smaller drops.

Experimental Data

The previous discussion indicated that wind shear and wind turbulence are the primary contributors to the clutter spectrum of precipitation. At close ranges the turbulence effect should dominate ($\sigma_v \simeq 0.7$ m/sec or 1.0 for altitudes less than 3 km); and, as range is increased, a point should be reached where the spectrum increases linearly with range because of the wind shear effect. This relationship has been verified in a series of experiments with a 5.2-cm radar having a two-way, half-power beam-width of 1.4° [262]. Figure 6-7 summarizes measurements of the range dependence of σ_v for this radar. The broadening effect with range is evident. The spectrum was found to be approximately centered at the radial wind speed component at the beam center. The measured spectral widths on some occasions were found to exceed 5 m/sec at short ranges. This condition was usually associated with very high wind shear gradients at low altitudes. The behavior of σ_v on one such day is shown in Fig. 6-8 (not included in Fig. 6-7); the wind velocity was 60 knots at 5,000 ft on this day.

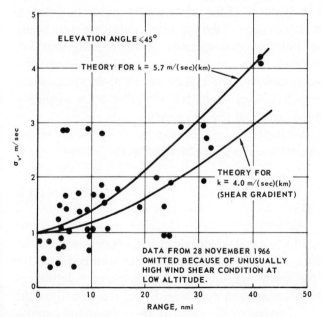

figure 6-7 *Spectrum standard deviation for rain echoes—composite graph from several tests. 1.4° elevation beam-width.*

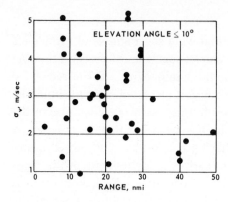

figure 6-8 *Spectrum standard deviation for rain on Nov. 28, 1966—unusually high shear at low altitudes, e.g., 60 knots at 5,000 ft [262].*

6.5 FREQUENCY CORRELATION OF PRECIPITATION ECHOES

In many radar applications it is desirable to get a good estimate of the mean value of rain or snow backscatter in as short a time as possible in order to establish a detection threshold. In addition, target detection is improved if the clutter echoes can be processed in such a way that the ratio of the variance to the mean of the processed return can be reduced. This goal can be met if the echoes from successive pulses are statistically independent and are detected and summed (incoherent integration). Unfortunately, the echoes from precipitation remain correlated for a number of milliseconds; however, the echoes can be made independent by shifting the carrier frequency from pulse-to-pulse by a frequency separation of at least the inverse of the pulse width.

This property of independence can be expressed by means of the *frequency correlation coefficient* which is defined as [245, 396]

$$\rho(\Delta f) = \frac{\overline{(I_0)(I)} - (\overline{I})^2}{\overline{I^2} - (\overline{I})^2} \tag{6-12}$$

where I_0 is the square of the signal amplitude at frequency f_0 and I is the square of the amplitude at $f_0 + \Delta f$. This function has been evaluated for the case in which the scattering volume contains many independent scatterers with more or less random positions. Under this condition the normalized frequency correlation function of the echoes from multi-frequency rectangular pulses can be written [395, 211]

$$\rho(\Delta f) = \left(\frac{\sin \pi\tau\Delta f}{\pi\tau\Delta f}\right)^2 \tag{6-13}$$

where τ = pulse length

Δf = the transmit frequency change

$\rho(\Delta f)$ = the normalized correlation coefficient (correlation of the echoes from two pulses transmitted closely in time but separated in frequency by the increment Δf)

The $(\sin^2 x/x^2)$ function falls to zero at $\tau\Delta f = 1$ and remains less than 0.05 for $\tau\Delta f > 1$, i.e., the clutter echoes are uncorrelated for frequency shifts greater than $1/\tau$.

The result of an experiment to verify this relationship for rain is shown in Fig. 6-9 [262]. The frequency shift Δf in all cases was 500 kHz, the pulses were approximately rectangular and of durations between 0.4 and 3.2 μsec, and the rainfall rate was high (20 mm/hr). The elevation extent of the illuminated area was about 400 meters. The experimental results conform closely to the theory.

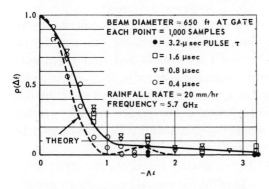

figure 6-9 *Correlation coefficient of rain echoes vs. frequency shift—pulse length product. (From Nathanson and Reilly, IEEE [262].)*

In conclusion, the signal-to-clutter ratio for spatially uniform clutter can be increased by frequency diversity and a postdetection integrator in the same way that the signal-to-receiver noise ratio is increased with postdetection integration.

6.6 SPATIAL UNIFORMITY OF RAIN BACKSCATTER

In Chaps. 3 and 4 it was noted that many techniques for detection of targets in clutter were dependent on the clutter not only having a Rayleigh amplitude distribution and a short correlation time but also being spatially uniform. While there does not appear to be any quantitative description of the spatial uniformity of the mean backscatter, series of experiments conducted at the Applied Physics Laboratory have shown the wide variations in spatial characteristics with a C-band radar [262].

Rain profiles were taken by transmitting a pulse train and fixing a range gate at a selected minimum range R_0. The magnitudes of the amplitude of the echoes from 100 to 200 pulses were added and the sum stored in the digital computer. The range gate was then moved one-half to one pulse length, and the procedure repeated. This process, which took from 10 to 60 sec, was continued until the desired profile was obtained. Examples of these profiles are shown in Figs. 6-10 and 6-11. The ordinate is the summation of the amplitudes of all the samples at a given range multiplied by the range to compensate for the increase in beam volume and reduction in received power with range. The abscissa is the radial or slant range.

The tests were primarily on a series of thundershowers that were extensive but obviously not uniform. The extent of the variation is indicated in Fig. 6-10*A* and *B*, which shows data that were taken only 2.25 min apart at an azimuth separation of only 2°. The extreme variations in reflectivity are quite apparent. Over 10 dB of variation was observed within 1 nmi (8 pulse lengths). The higher intensity regions gave reflectivity values that would be predicted by Eq. (6-4) using the higher rainfall rates observed on a rain gauge (10-40 mm/hr).

An example of what could be considered a uniform rain gave rather consistent recordings of 16 to 40 mm/hr for several hours. Radar profiles from this day are shown in Fig. 6-11*A* and *B*. These two were chosen to illustrate the relatively uniform backscatter at lower altitudes and the apparent *bright-band* effect at the melting level. At the melting level, snow acquires a coating of water that increases the dielectric constant; thus the reflectivity increases by about 7 dB. As melting occurs, the size of the crystals decreases and the fall velocity increases to perhaps five times that of the snowflakes. This results in a reduction in reflectivity and about 6 dB decrease in backscatter due to the lower density of the raindrops.

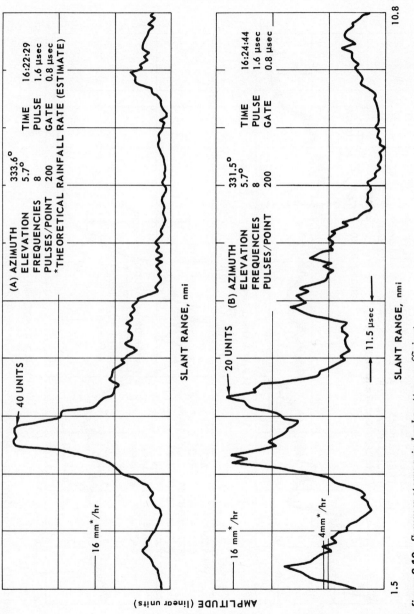

figure 6-10 *Summer storm rain-backscatter coefficient.*

215

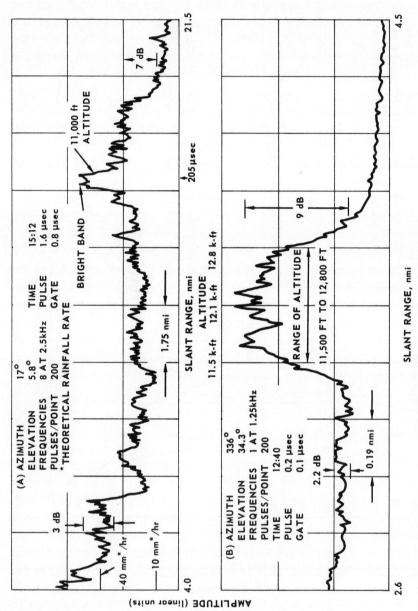

figure 6-11 *Heavy uniform rain–backscatter coefficient (with bright band).*

With only a few exceptions the *uniform* rain appeared to be uniform at the lower elevation angles with about a ± 2-dB peak-to-peak variation in backscatter coefficient over 1-2 nmi range intervals. The data on the figure and other data taken at about the same time indicate that spatial nonuniformities of less than 1 nmi in extent exceed any uncertainties in the measurements.

Unfortunately the wide disparity of measurements on different days precludes postulating any model for the gross uniformity of rain clutter. Changes in the mean backscatter may be larger than ± 10 dB in 10 nmi for showers or as small as ± 1 dB for the more uniform rains.

Similar runs were made on the uniformity of a snowstorm. The mean backscatter was more uniform than for any of the rain measurements.

An attempt was made to estimate the spatial extent of the major non-uniformities by computing the spatial correlation function of the *mean* backscatter. The sum of the magnitudes of the backscatter from the pulses at each range was corrected for the R^{-1} dependence and inserted into an autocorrelation program. The range displacement where the normalized autocorrelation drops to one-half its value of zero lag is thus a measure of the range extent of the variations of the mean backscatter. While the statistical sample was too small for a rigorous analysis, the *characteristic extent* of the nonuniformities (where $\rho = 0.5$) was 0.6 to 1.4 nmi in the showers. This value was 2-3 nmi for the uniform rain, including the bright-band effect, and in a snowstorm the extent was 1.4 to 1.9 nmi at 3° elevation.

6.7 TROPOSPHERIC REFRACTION EFFECTS[1]

Radio waves propagated through a medium with a varying index of refraction undergo both refractive bending and retardation of their velocity of propagation. These phenomena give rise to angle and range errors when a target is being tracked by a surface radar. The angular error is caused by the bending of the radar beam such that the wave front reflected from a target appears to be coming from a direction other than the target's true angular position in space; the range error is due to the increase in time necessary to travel over the curved path as compared to the straight-line path between target and receiver and the variation in propagation velocity.

[1] For general references see *Natl. Bur. Std. Tech. Note* 101 [310]; Berkowitz, "Modern Radar" [38]; Blake [44]; Durlach [99]; Ehling [104].

General practice has been to assume an atmospheric refractive index that decreases linearly with the height above the surface of the earth. The 4/3-earth-radius principle, based upon this linear assumption, leads to errors at long ranges and low elevation angles. For the results in the following graphs, a negative-exponential model of the refractivity-height function (refractive index versus height above the earth's surface) is used. The model is $n(h) = 1 + 0.000313 \exp(-0.04385h)$, where h is altitude in thousands of feed and $n(h)$ is the refractive index. This model is based upon a surface index of refractivity and is referred to as the CRPL (National Bureau of Standards, Central Radio Propagation Laboratory) exponential reference atmosphere.

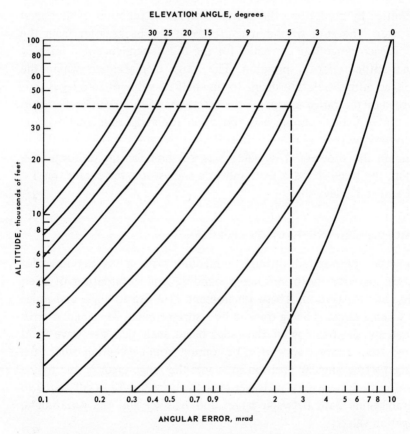

figure 6-12 *Target angular error due to refractive effect of earth's atmo-sphere (angular error is a function of target altitude and elevation angle). (From Shannon, Electronics [342].)*

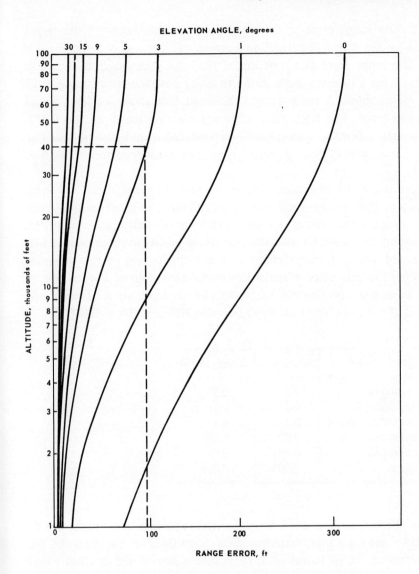

ELEVATION ANGLE, degrees

RANGE ERROR, ft

figure 6-13 *Target range error due to refractive effect of earth's atmosphere (range error is a function of target altitude and elevation angle). (From Shannon, Electronics [342].)*

The CRPL reference atmosphere was used to calculate the tropospheric angular and range errors presented graphically in Figs. 6-12 and 6-13. The angular error is the angle between the apparent ray path to the target from the radar and the straight-line path from the radar to the target. The range error is the apparent increased range of the target caused by the wavefront traveling a curved path instead of a straight-line one. The range error as a function of the total path length is shown in Fig. 6-13 for a one-way path. For a two-way path the errors in Fig. 6-13 must be doubled. A radar several thousand feet above sea level would have less error. The NRL radar coverage diagram based on the CRPL exponential reference atmosphere is reproduced from Blake [44] as Fig. 6-14. Experimental data on refractivity have been reported by Nichols [266].

These range and elevation angle errors should be considered as known *bias* errors since their magnitudes are calculable. The uncertainty in their exact magnitudes (about 10 percent) from simplifying assumptions concerning the state of the atmosphere is partly from unknown bias errors and partly from *precision* errors. Contributing to the precision errors are the effects of irregularities in the troposphere. Barton [64, p. 489] gives some *maximum*[1] rms values of the fluctuation of range and angle in the troposphere as follows (see also Millman [38, p. 317ff]):

Type of weather	Δ range, ft	Δ angle, mrad
Heavy cumulus	2.0	0.7
Scattered cumulus	0.5	0.3
Small scattered cumulus	0.1	0.15
Clear moist air	0.02	0.07
Clear normal air	0.005	0.03
Clear dry air	0.001	0.015

The above bias error models are limited to surface-to-air paths and do not take into account meteorological conditions at the time of the measurement. In meteorological radar investigations and in cases where true geometric target trajectories to the highest attainable degree of accuracy are desired, the reference atmosphere should not be used without modification. At the minimum it should be corrected on the basis of a directly or indirectly measured value of the refractive index at the radar location. The effort required to make refraction corrections on the basis of several point measurements of refractive index along an

[1]Presumably a low elevation angle two-way path traversing the entire atmosphere.

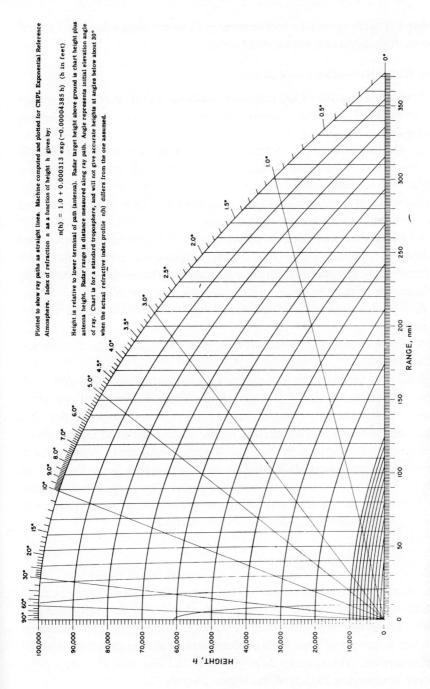

Plotted to show ray paths as straight lines. Machine computed and plotted for CRPL Exponential Reference Atmosphere. Index of refraction n as a function of height h given by:

$$n(h) = 1.0 + 0.000313 \exp(-0.00004385\,h) \quad (h \text{ in feet})$$

Height is relative to lower terminal of path (antenna). Radar target height above ground is chart height plus antenna height. Radar range is distance measured along ray path. Angle represents initial elevation angle of ray. Chart is for a standard troposphere, and will not give accurate heights at angles below about 30° when the actual refractive index profile n(h) differs from the one assumed.

figure 6-14 *Earth radar coverage diagram. (From Blake [44, 45].)*

221

entire ray path is usually too expensive to be justified primarily because the perturbations occur over small areas.

The Radio Refractive Index of Air

A recent summary of the state of knowledge of this parameter appears in Bean [26]. Bean's conclusions, with some paraphrasing, are as follows:

1. The differences in determined values of the constants in the expression for refractivity N (Bean's Eq. 10) are small compared with the errors from using the formula; i.e., the errors in using ordinary meteorological measurements (e.g., radiosondes) mask the errors caused by using published values of the constants.

2. The refractive index is effectively nondispersive for frequencies below, for example, 50 GHz. (N.B., ionospheric effects, multipath effects, and effects due to storm phenomena, which exhibit high coefficients of radiowave absorption, are not considered.)

3. The choice of atmospheric model for N depends on the application at hand.

4. The atmosphere, on the average, yields an exponential distribution of N with height.

Bean also indicates the divergence of a typical measured profile of N from the often assumed 4/3 earth profile. From his graphs it is evident that the 4/3 earth model has about the correct slope in the first kilometer above the earth's surface but decreases much too rapidly above that height. It is also seen that the observed refractivity distribution is more nearly an exponential function of height than a linear function as assumed by the effective earth's radius model. The exponential decrease of N with height is sufficiently regular to permit a first approximation of average N structure from surface conditions alone.

6.8 GENERAL PROPERTIES OF CHAFF

The use of chaff (or *window*) to confuse radar systems dates back to World War II; its popularity is primarily due to its simplicity and low cost relative to electronic countermeasures. Commonly chaff packages contain hundreds of thousands of aluminum or aluminum-coated glass or Mylar dipoles, which, when dispensed into the atmosphere, reflect radar energy comparable to that of the largest aircraft.

The chaff dipoles are designed to resonate at the frequencies of the radars that they are attempting to confuse. (Dipole length $\approx$ one-half radar wavelength.) Often the dipoles in a package are cut to different lengths to cover an entire radar band or several radar bands. The maximum backscatter cross section of a single dipole occurs when the dipole is oriented parallel to the E plane of the incident radiation and perpendicular to the direction of propagation. This gives a maximum radar cross section

$$\sigma_m = 0.857 \lambda^2 \qquad\qquad\qquad (6\text{-}14)$$

Schlesinger [335] has stated that if these dipoles are randomly oriented after dispersal into the atmosphere the overall cross section is approximately

$$\sigma = 0.18 \lambda^2 N$$

where N is the total number of dipoles and σ is the resultant radar cross section. He further calculates that if they are all cut from aluminum foil 0.001-in. thick, $\lambda/2$ long, and 0.01-in. wide the radar cross section is

$$\sigma \approx 3{,}000 \; W/f \quad m^2$$

where W is weight in pounds and f is transmit frequency in gigahertz. This would yield a 1,000-m^2 cross section for a single pound of narrow-band chaff at 3 GHz. (See also Cassedy and Fainberg [63] and Mack and Reiffen [240].)

The bandwidth of the resonant-length chaff is only about 10 to 15 percent of the center frequency; and if it is desired to cover the desired frequency range from 1 to 10 GHz with the 10 percent bandwidth assumption, a 1.0-lb package of broad-band chaff would give a 60-m^2 radar cross section anywhere within the frequency range.

The use of chaff is beyond the scope of this chapter, but there are two general tactical techniques used in military operations.

1. The dispersal of single packages from an airplane or missile or from a forward-fired, aircraft-launched rocket. The chaff simulates a false target or attempts to break the tracking loop of a radar in range or angle.

2. The dispersal in close sequence of large amounts of chaff to form a corridor or cloud which appears on the radar scope as clutter similar to rain. The clouds may stay at the desired altitude for hours.

Chaff dipoles have high aerodynamic drag; their velocity drops to that of the local wind a few seconds after they are dispensed.[1] Pilie' et al. [290] has shown that the horizontal velocity of chaff at X band (3 cm) is very close to the wind speed measured by radiosonde and smoke trail photography. The vertical component of the velocity (fall rate) depends to some extent on the dipole density and dimensions. This factor is discussed in Sec. 6.9 on the chaff spectrum. Experimental results are also reported by Totty et al. [376].

The use of chaff for frequencies below 500 MHz (60-cm wavelength) requires a change in packaging. The aluminum foil is wound rather than cut in flat dipoles, but it has basically the same function. The common description for this material is *rope*.

Chaff may be placed in orbit by a satellite for use as a reflector for long-range communication [376].

6.9 SPECTRA OF CHAFF ECHOES

The doppler spectrum of chaff return can be split into four components in the same manner as for precipitation:
1. Wind shear due to finite height of radar beams
2. Beam broadening due to finite width of the beams
3. Turbulence
4. Fall velocity distribution of the dipoles

In the case of precipitation it was shown that the wind shear and beam broadening effects were caused by the variations in the horizontal motion of the droplets within the radar beam. This motion of the rain was shown to follow the horizontal wind speed very closely. Experiments by Pilie' et al. [290] have also shown that there is very high correlation between horizontal wind and chaff velocities at altitudes of from 5,000 to 10,000 ft. Similar experiments showed the same high correlation at from 24,000- to 35,000-ft altitudes. It can be concluded that the calculation of the first two effects (due to wind shear) for chaff should be identical to that described for rain in Sec. 6.4.[2]

---

[1] Dispensing refers to ejection from an aircraft; dispersion is the process of forming a chaff cloud.

[2] The effects are identical if both the rain and chaff fill the radar beam.

Because of the difficulty of separating the turbulent component from the shear components of the chaff spectrum, there is little information available on the spectrum of the chaff after dispersion. One fairly extensive set of data was reported by Smirnova [356] in 1965 in the USSR. Unfortunately his radar parameters were not included, but it was implied that the antenna beamwidths were about 0.7° and that the measurements were made looking almost exactly downwind. The altitudes observed were from about 1,000 to 40,000 ft.

Several conclusions can be reached by interpreting the available information:

1. The turbulence component σ_{turb} above 12,000 ft varies between 0.35 and 1.2 m/sec with an average of about 0.7 m/sec or somewhat less than the assumed value of 1.0 m/sec for precipitation.

2. Below 12,000 ft, readings as high as 2 m/sec have been recorded, but an average value for from 1,000 to 12,000 ft is close to the 1.0 m/sec used for precipitation. The larger values may be partially due to the vertical fall velocity distribution [356].

3. An average of Smirnova's data on the turbulence velocity is included in Fig. 6-15. The lower altitude tests (1,500-13,000 ft) overlapped the high-altitude tests (8,000-32,000 ft). Since there is not a significant correlation between wind velocity and turbulence it will be assumed that the horizontal turbulence component is roughly constant at 1.0 m/sec for low altitudes and at about 0.7 m/sec above 12,000 ft. These points are shown as arrows on the figure.

These data are only partially consistent with Goldstein's data (at $\lambda = 10$ cm) from World War II [211]; he estimated a standard deviation of 0.4 to 1.0 m/sec[1] for low (< 10 knots) winds and 1.3 m/sec for 25-knot winds. Barlow's [18] general value of 3.5 fps (~ 1 m/sec) agrees more closely. It should be realized that Goldstein's report does not describe the wind direction, range, altitude, or the elevation of the beam. Since the 0.4 m/sec data point was for low winds and a different type of chaff, it can probably be disregarded. This leaves a range of σ_v between 0.6 and 1.3 m/sec.

The final component of the chaff spectra is the doppler spread due to the variation in fall velocity of the dipoles. Chaff dipoles fall more slowly than do the large raindrops that dominate the backscatter from

---

[1]The 0.4 m/sec curve is for chaff dispensed from a blimp; this chaff had different aerodynamic properties than the chaff used to obtain the other values.

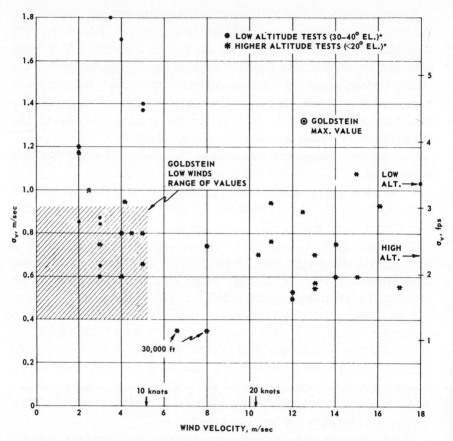

figure 6-15 *Turbulence σ_v of chaff vs. wind velocity. (After Smirnova [356] and Goldstein [211].)*

precipitation; dipoles have typical terminal velocities of 0.7 to 1 m/sec as compared to about 6 m/sec for 1 mm/hr rain. Theoretical and experimental studies of the fall velocity of two types of chaff were made by Jiusto and Eadie [191]. The first was aluminum foil 0.008 in. by 0.00045 in. and $\approx \lambda/2$ long at S and X band; the second was metallized nylon cylinders 0.0035 in. in diameter and 0.6 in. long. The experimental work was in a small altitude chamber without air turbulence. With both types of chaff, the measured median terminal fall rate was about 0.6 m/sec; this was only about 10 percent less than the theoretical value for the nylon chaff. At simulated altitudes of 10 km (33,000 ft) the median fall rate of the nylon was slightly less than 1 m/sec, and the S- and X-band aluminum foil had a fall rate of 1 m/sec.

The difference between the median and maximum terminal fall rate for the foil was about 1.4 m/sec. If this can be considered a 3σ value, then $\sigma_{fall} \approx 0.45$ m/sec. At 10-km altitude, the corresponding σ_{fall} would be ≈ 1 m/sec. The variation in the fall rate of the cylinders was about an order of magnitude less. Thus, the range of vertical differential velocities is apparently below 1 m/sec (≈ 0.5 m/sec) at low altitudes and, perhaps, close to the 1 m/sec attributed to rainfall only at higher altitudes ($> 20,000$ ft) where the shear components usually dominate anyway. Further experimental data are available in Totty et al. [377].

The measured spectra for chaff and rain are similar enough so that separate computations are not needed except for two factors:

1. If the vertical dispersion of chaff at high altitudes is small, there will be less shear effect.

2. Chaff has a higher fall velocity at high altitude (over 25,000 ft) and a low terminal velocity ~ 0.7 m/sec; rain, on the other hand, has a higher terminal velocity than its *initial* condition (clouds).

The general equation for the total variance of the velocity spectrum for chaff is then

$$\sigma_v{}^2 = \sigma_{shear}^2 + \sigma_{beam}^2 + \sigma_{turb}^2 + \sigma_{fall}^2$$

where σ_{turb} $\begin{cases} 3.4 \text{ fps} \approx 1.0 \text{ m/sec} & \text{below } 12,000 \text{ ft} \\ 2.3 \text{ fps} \approx 0.7 \text{ m/sec} & \text{above } 12,000 \text{ ft} \end{cases}$

$\quad\quad \sigma_{fall} = 0.45 \sin \psi \quad$ m/sec

The values for shear and beam broadening can be found in Sec. 6.4.

It will be emphasized in the chapters on processing that the mean velocity of chaff (and rain) is also important in many signal processing techniques, especially MTI, and that the spectra described here are centered at the mean radial velocity of the volume cell containing the chaff.

7 SEA AND LAND
BACKSCATTER

7.1 MONOSTATIC BACKSCATTER FROM THE SEA

Backscatter from the surface of the sea limits the performance of radar surveillance and missile guidance for airborne, shipboard, and coastal early-warning and defense systems. The following three sections will provide models of the normalized radar power return from the sea as a function of frequency, polarization, grazing angle, and sea condition. The statistics of these signals will be given in Secs. 7.5, 7.6, and 7.7.

The general relationships between the wind and the surface conditions of the sea are given in Fig. 7-1. This figure defines *sea state* as well as would seem appropriate to subsequent discussions. Two notes of caution must be given with respect to Fig. 7-1: combined conditions at any given time or place may vary considerably from these charts because of the wind history or geographical factors and *Beaufort sea state* is not the same as the *hydrographic sea state*. The latter will be used in the following discussions.

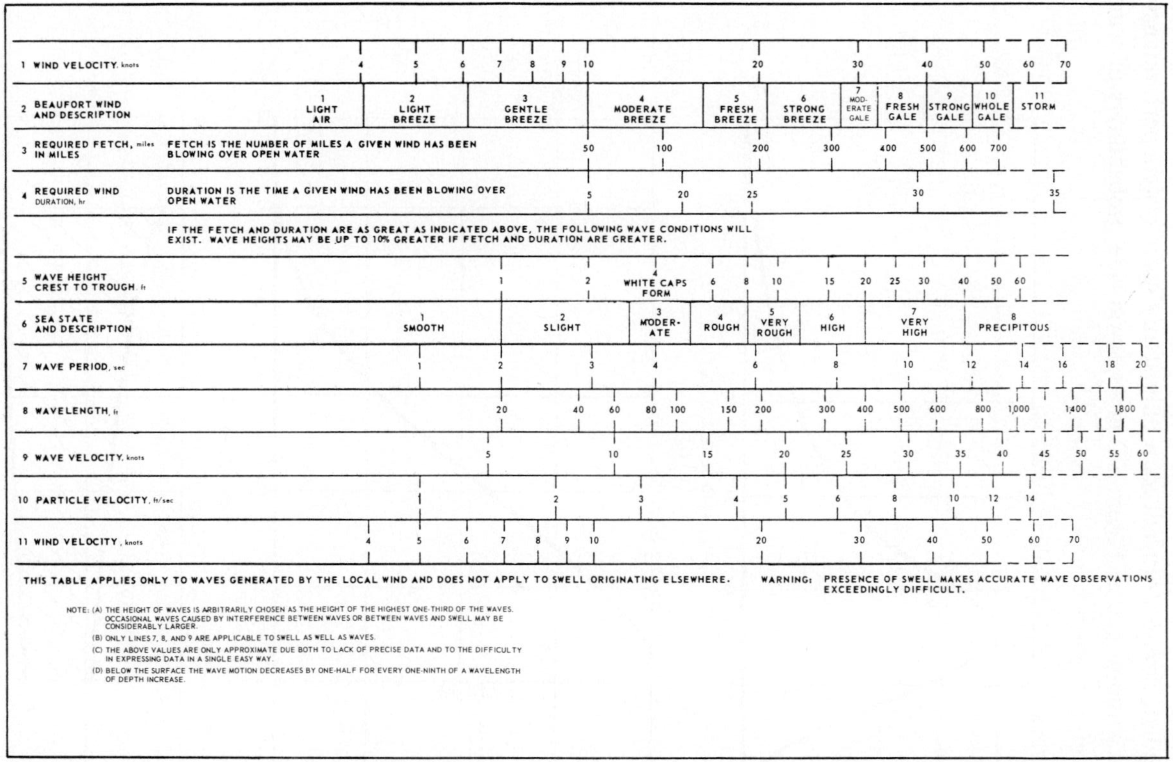
1 WIND VELOCITY, knots
4 5 6 7 8 9 10 20 30 40 50 60 70
2 BEAUFORT WIND AND DESCRIPTION
1 LIGHT AIR
2 LIGHT BREEZE
3 GENTLE BREEZE
4 MODERATE BREEZE
5 FRESH BREEZE
6 STRONG BREEZE
7 MOD-ERATE GALE
8 FRESH GALE
9 STRONG GALE
10 WHOLE GALE
11 STORM
3 REQUIRED FETCH, miles IN MILES
FETCH IS THE NUMBER OF MILES A GIVEN WIND HAS BEEN BLOWING OVER OPEN WATER
50 100 200 300 400 500 600 700
4 REQUIRED WIND DURATION, hr
DURATION IS THE TIME A GIVEN WIND HAS BEEN BLOWING OVER OPEN WATER
5 20 25 30 35
IF THE FETCH AND DURATION ARE AS GREAT AS INDICATED ABOVE, THE FOLLOWING WAVE CONDITIONS WILL EXIST. WAVE HEIGHTS MAY BE UP TO 10% GREATER IF FETCH AND DURATION ARE GREATER.
5 WAVE HEIGHT CREST TO TROUGH, ft
1 2 4 WHITE CAPS FORM 6 8 10 15 20 25 30 40 50 60
6 SEA STATE AND DESCRIPTION
1 SMOOTH
2 SLIGHT
3 MODER-ATE
4 ROUGH
5 VERY ROUGH
6 HIGH
7 VERY HIGH
8 PRECIPITOUS
7 WAVE PERIOD, sec
1 2 3 4 6 8 10 12 14 16 18 20
8 WAVELENGTH, ft
20 40 60 80 100 150 200 300 400 500 600 800 1,000 1400 1800
9 WAVE VELOCITY, knots
5 10 15 20 25 30 35 40 45 50 55 60
10 PARTICLE VELOCITY, ft/sec
1 2 3 4 5 6 8 10 12 14
11 WIND VELOCITY, knots
4 5 6 7 8 9 10 20 30 40 50 60 70
THIS TABLE APPLIES ONLY TO WAVES GENERATED BY THE LOCAL WIND AND DOES NOT APPLY TO SWELL ORIGINATING ELSEWHERE.
WARNING: PRESENCE OF SWELL MAKES ACCURATE WAVE OBSERVATIONS EXCEEDINGLY DIFFICULT.
NOTE: (A) THE HEIGHT OF WAVES IS ARBITRARILY CHOSEN AS THE HEIGHT OF THE HIGHEST ONE-THIRD OF THE WAVES. OCCASIONAL WAVES CAUSED BY INTERFERENCE BETWEEN WAVES OR BETWEEN WAVES AND SWELL MAY BE CONSIDERABLY LARGER.
(B) ONLY LINES 7, 8, AND 9 ARE APPLICABLE TO SWELL AS WELL AS WAVES.
(C) THE ABOVE VALUES ARE ONLY APPROXIMATE DUE BOTH TO LACK OF PRECISE DATA AND TO THE DIFFICULTY IN EXPRESSING DATA IN A SINGLE EASY WAY.
(D) BELOW THE SURFACE THE WAVE MOTION DECREASES BY ONE-HALF FOR EVERY ONE-NINTH OF A WAVELENGTH OF DEPTH INCREASE.

figure 7-1 *Wind waves at sea. (Reprinted from Undersea Technology, May, 1964, p. 34 by permission of copyright owners Compass Publications, Inc.)*

To select radar parameters for operating near or on the sea, the relative probabilities of specific conditions of wind speed and sea state are needed. Severe storms degrade almost any radar system performance, but they rarely occur. Figure 7-2 gives worldwide probabilities of wind

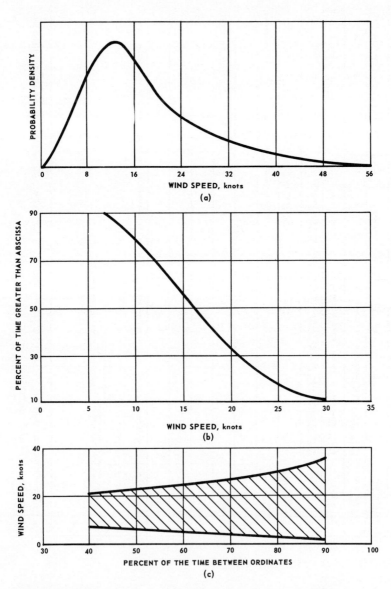

figure 7-2   *Surface wind speed statistics. (From Long et al. [234].)*

speed, derived by Long et al. [234] from the *U.S. Navy Marine Climatic World Atlases,* 1957. Table 7-1 gives the relative frequency of occurrence of wave heights in specific regions of the oceans.

Before proceeding with the discussion of the properties of the backscatter from the sea, it is useful to define some of the terms used to describe ocean waves.

1. *Sea state.* The sea state is a single number that grossly summarizes the degree of agitation of the sea and the characteristics of its surface. The relationship between wave height and wind velocity or sea state is not clearly defined; however, most tables and charts designate some relationship between sea state and rms wave height. (See Fig. 7-1.)

2. *Gravity and capillary waves.* Capillary waves are small wavelets that appear on the surface of the ocean. They are generally less than 1.7 cm in length and a few millimeters in height. Gravity waves are ocean waves and can be defined in two states: sea and swell. Sea is the state of the waves when they are being blown by the wind that raised them; swell is the state of the waves when they have escaped the influence of the generating wind. Gravity waves are virtually free from the effects of surface tension and are affected primarily by pressure, wind velocity, and gravity. Capillary waves are dependent upon surface tension.

3. *Wave height.* (*a*) Variance—Wave height variance σ_h^2 is the mean-square deviation from the mean wave height of the wave record. (*b*) Significant—the significant wave height is defined as the average of the highest third of the waves in the wave record and is usually denoted as $\bar{H}_{1/3}$. The relationship suggested by Burling is

$$\bar{H}_{1/3} = 4.0\,\sigma_h \qquad\qquad (7\text{-}1)$$

The significant waveheight is also related to the average wind velocity V by certain authors. Some of the relationships are

Pierson-Neumann : $\bar{H}_{1/3} = 7.06 \times 10^{-6} V^{2.5}$

Sverdrup-Munk : $\bar{H}_{1/3} = 2.66 \times 10^{-4} V^2$

Darbyshire : $\bar{H}_{1/3} = 6.18 \times 10^{-5} V^2$

7.2 SEA BACKSCATTER MODELS FOR LOW GRAZING ANGLES (0-20°)

In determining sea or land backscatter, the term σ_0 is used to represent the normalized mean (or median) backscatter from a surface area

TABLE 7-1   Relative Frequency of Occurrence of Wave Heights in Specific
Regions of the Oceans*

Region	Percentage of time significant wave height (ft) from					
	0–3	3–4	4–7	7–12	12–20	> 20
North Atlantic between Newfoundland and England	20	20	20	15	10	15
Mid-equatorial Atlantic	20	30	25	15	5	5
South Atlantic at latitude of southern Argentina	10	20	20	20	15	10
North Pacific between latitude of Oregon and southern coast of Alaska	25	20	20	15	10	10
East-equatorial Pacific	25	35	25	10	5	5
West wind belt of South Pacific at latitude of southern Chile	5	20	20	20	15	15
North Indian Ocean during northeast monsoon season	55	25	10	5	0	0
North Indian Ocean during southwest monsoon season	15	15	25	20	15	10
Southern Indian Ocean between Madagascar and northern Australia	35	25	20	15	5	5
West wind belt of southern Indian Ocean between Cape of Good Hope and southern Australia	10	20	20	20	15	15
Averages over all regions	22	23	20	15	9	9

*From Long et al. [234].

illuminated by a pulse radar.[1] To a first approximation this area A is $R\theta_2(c\tau/2)$ for small grazing angles, narrow azimuth beamwidths, and full antenna gain on the water, where R is the range from radar to center of cell, θ_2 azimuth beamwidth (3 dB) of radar (two way) and $(c\tau/2)$ is the pulse length in distance units for a two-way path (12.34 μsec = 1 nmi). Complete detection range equations for sea clutter limited radars are given in Chap. 2. The conventional backscatter parameter σ for radar can be approximated from $\sigma_0 = d\sigma/dA$. Then

$$\sigma \approx R\theta_2\left(\frac{c\tau}{2}\right)\sigma_0 \qquad (7\text{-}2)$$

for a small beamwidth, compared to a radian, and for low grazing angles.

Obviously, the above discussion did not take lobing and forward scatter into account. These items are virtually always included in the *models* for σ_0. The term σ_0 is also called the *normalized reflectivity* and is generally given as a mean value in decibels. If $\sigma_0 = -30$ dB, the *radar cross section* σ is 30 dB below a 1-m^2 target for every square meter of the sea that is illuminated. Since the power density on this surface is proportional to the sine of the incident angle, another term γ is often used for reflectivity, where $\sigma_0 = (\gamma)(\sin\psi)$, where ψ is the grazing angle.

The next sets of tables (Tables 7-2 to 7-6) are models for the backscatter coefficient σ_0. These tables were compiled using experimental data from numerous sources. They provide a complete and somewhat consistent set of numbers for the radar designer and system planners and evaluators. No attempt has been made to develop the theory of scattering from the sea surface or explain the anomalies in certain data. On the other hand, the points have been derived from a more extensive set of experiments than was the case heretofore.[2] In assembling such models it has been found that only a few extra data points make the tables converge rapidly. Separating by frequency, polarization, incidence angle, and see state in the tables seems to make the data more consistent. When further data are available, it will also be

---

[1] In a similar manner to the radar cross section of a target σ_t. The term σ_0 is only a constant for a given angle, carrier frequency, polarization, pulse length, and type (sea state in this case).

[2] The data were obtained primarily from [3, 4, 49, 59, 83, 77, 89, 94, 85, 84, 326, 122, 141, 145, 194, 193, 192, 211, 222, 234, 239, 259].

TABLE 7-2 Normalized Mean Sea Backscatter Coefficient σ_0 for Grazing Angle of 0.1°

Sea state	Pol.	Reflection coefficient, dB, below $1\ m^2/m^2$ at indicated carrier frequency, GHz						
		UHF 0.5	L 1.25	S 3.0	C 5.6	X 9.3	K_u 17	K_a 35
0	V							
	H			90*	87*			
1	V					65*		
	H			80	75*	71*		
2	V	90*	87*	72*	64	56		
	H	95*	90*	75*	67*	61*		
3	V				56*	51		
	H	90*	82*	68	60*	53*		
4	V				53*	48		
	H			58	55	48		
5	V					44		
	H		65*	53	48	42*		

*5-dB error not unlikely. Monostatic radar, 0.5- to 10-μsec pulse.

TABLE 7-3 Normalized Mean Sea Backscatter Coefficient σ_0 for Grazing Angle of 0.3°

Sea state	Pol	Reflection coefficient, dB, below $1\ m^2/m^2$ at indicated carrier frequency, GHz						
		UHF 0.5	L 1.25	S 3.0	C 5.6	X 9.3	K_u 17	K_a 35
0	V							
	H			83*	79	74*		
1	V			62*	60	58		
	H			74	71	66*		
2	V	80*		59*	55	52		
	H			66	60	56*		
3	V			55*	48	45		
	H			58*	50	46		
4	V			54*		43		
	H			50*		42	39*	
5	V	75*		50*		39		
	H			44	41	39	39*	

*5-dB error not unlikely. Monostatic radar, 0.5- to 10-μsec pulse.

TABLE 7-4 Normalized Mean Sea Backscatter Coefficient σ_0 for Grazing Angle of 1.0°

Sea state	Pol.	Reflection coefficient, dB, below 1 m^2/m^2 at indicated carrier frequency, GHz						
		UHF 0.5	L 1.25	S 3.0	C 5.6	X 9.3	K_u 17	K_a 35
0	V		68*			60*		
	H	86*	80*	73	70	66		
1	V	70*	65*	56	53	50	47*	
	H	84*	73*	65	56	51	45	40*
2	V	63*	58*	53	47	44	42	38*
	H	82*	65*	55	48	46	41	38*
3	V	58*	54*	48	43	39	37	34
	H	76*	60*	48	43	40	37	36
4	V	55*	45	42	39	37	34	32
	H		52*	42	39	36	34	
5	V		43	38*	35	33	32	31
	H	65*	50*	42	35	33	31*	

*5-dB error not unlikely. Monostatic radar, 0.5- to 10-μsec pulse.

TABLE 7-5 Normalized Mean Sea Backscatter Coefficient σ_0 for Grazing Angle of 3.0°

Sea state	Pol.	Reflection coefficient, dB, below 1 m^2/m^2 at indicated carrier frequency, GHz						
		UHF 0.5	L 1.25	S 3.0	C 5.6	X 9.3	K_u 17	K_a 35
0	V				60*	56*	52*	48*
	H	75*	72*	68*	63*	58*		53
1	V	60*	53*	52	49	45	43	41
	H	70*	62*	59	54	48	45*	43*
2	V	55*	53	49	45	41	39	37
	H	66*	59	53	48	42	41	40
3	V	43*	43	43	40	38	36	34
	H	61*	55*	46	42	39	37	37
4	V	38*	38	38	36	35	33	31
	H	54*	48*	41	38	35	34*	34
5	V		38	35	33	31	31*	30*
	H	53*	46	37	34	32	30*	

*5-dB error not unlikely. Monostatic radar, 0.5- to 10-μsec pulse.

TABLE 7-6   Normalized Mean Sea Backscatter Coefficient σ_0 for Grazing Angle of 10°

Sea state	Pol.	Reflection coefficient, dB, below 1 m²/m² at indicated carrier frequency, GHz						
		UHF 0.5	L 1.25	S 3.0	C 5.6	X 9.3	K_u 17	K_a 35
0	V		45*			49*	45*	44*
	H		60*			56*		
1	V	38			44	42	40	38
	H		56*		53	51		
2	V	35*	37	38	39	38*	34	33
	H	54*	53	51	48*	46	37*	
3	V	34*	34	34	34	32	32	31
	H	50	48	46	40	37	33	31
4	V	32*	31	31*	32	31	29	29
	H	48*	45		36	34	31	29
5	V	30*	30	28	28	28	26	26
	H	46	43	38	36	33	29	27

*5-dB error not unlikely. Monostatic radar, 0.5- to 10-μsec pulse.

useful to separate out data by wind and wave direction and by pulse lengths. Until that time, the models refer to

1. A time average over several tens of milliseconds.

2. An average of the upwind, crosswind, and downwind values where available (see Sec. 7.4).

3. Pulse lengths in the 0.5- to 10-μsec region with echoes having approximately Rayleigh distributions (see Sec. 7.7). Thus there is a spatial average with resolution cells large enough so that individual wave structure is not resolved.

Data points not conforming to these assumptions have been crudely adjusted to conform. An asterisk is shown where data are questionable or where there is a severe conflict, leading to an expected error of 5 dB or more.

The following ground rules have been observed and used for extrapolation and interpolation.

1. For a given entry on the table, the return from vertical polarization will equal or exceed that from horizontal; and the deviation will increase at lower sea states, lower grazing angles (below 1°), and lower transmit frequencies.

2. The backscatter increases with depression angle from 0 to $20°$ as θ^n, where n may be as high as three for low angles, low sea states, and low frequencies. The value of n decreases in the tables towards the lower right-hand corner (high frequencies and sea states) where it approaches zero.

3. The backscatter coefficient at low grazing angles always increases with transmit frequency as f^m for horizontal polarization, where m may be as high as three below 2 GHz for very low grazing angles (less than $1°$) and seas below state three. As the angle, sea state, or transmit frequency exceeds these values, the exponent drops toward zero.

4. The backscatter increases with sea state by as much as 10 dB/sea state for low seas and low frequencies but reduces to a negligible change at higher sea states and frequencies.

5. Sea state zero arbitrarily corresponds to a significant wave height less than 0.25 ft and winds less than 4 knots (see Table 7-1).

6. At small antenna depression angles the true grazing angle on the ocean is smaller because of the curvature of the earth. (Section 1.9 gives the true angles that should be used in entering the charts at 0.1 and $0.3°$.)

These generalizations were made to complete the tables; they should not be used for depression angles of greater than $20°$. At frequencies below 400 MHz there are little data, and with the relatively little backscatter power available, it is best to refer to [234, 153, 3, 66, 361]. Similarly above 50 GHz there are little data [53, 193].

7.3 SEA BACKSCATTER AT HIGH GRAZING ANGLES (30-90°)

The magnitude of σ_0 at $\psi = 90°$ (normal incidence) varies inversely with the sea state number, that is, σ_0 has its greatest magnitude, reaching about $+ 10$ dB for most frequencies, for sea state zero (smooth sea). As the depression angle decreases, σ_0 decreases. At depression angles around $60°$ this sea state dependence reverses and for values of ψ less than $60°$ the magnitude of σ_0 increases as the sea state number increases. The actual value of the crossover of the curves for σ_0 for different sea states varies with different parameters. σ_0 is modeled from experimental data in Tables 7-7 and 7-8 in the same manner as the low-angle reflection coefficients.

The magnitude of σ_0 also depends on the direction of viewing with respect to the direction of the mean wind velocity. About normal

TABLE 7-7 Normalized Mean Sea Backscatter Coefficient σ_0 for Grazing Angle of 30°

Sea state	Pol.	Reflection coefficient, dB, below 1 m²/m² at indicated carrier frequency, GHz						
		UHF 0.5	L 1.25	S 3.0	C 5.6	X 9.3	K_u 17	K_a 35
0	V		42*					
	H		50*					
1	V	38*	38*	40	40	39	38*	37*
	H		46*		48			
2	V	30	31	32*	34	34	31*	30
	H	42	41	40	42	44*	34*	
3	V	28	30	29	28	28	23*	23*
	H	40*	39	38	37	34	27*	
4	V		28	27	25	24	24	22
	H		37	37	35	33		
5	V	28	24	23	23	22*	21	20
	H	35	34	32	30	26*	22*	20*

*5-dB error not unlikely. Monostatic radar, 0.5- to 10-μsec pulse.

TABLE 7-8 Normalized Mean Sea Backscatter Coefficient σ_0 for Grazing Angle of 60°

Sea state	Pol.	Reflection coefficient, dB, below 1 m²/m² at indicated carrier frequency, GHz						
		UHF 0.5	L 1.25	S 3.0	C 5.6	X 9.3	K_u 17	K_a 35
0	V	32	33	34	35*	36*	28*	
	H	32	32	32		34*		26*
1	V	23*	22	24	26	28		26*
	H	22	24	25	26	26		
2	V	20*	21	21	23	20	18*	19*
	H	22	21		22	23		
3	V	18*	18*	19	18*	17	14*	14*
	H	21	20		20	21		
4	V	14*	15*		15*	14*	11	10
	H	21*	18*					
5	V	18*	15*	15	15	13*	6	4
	H	21*	18*	17	17	14*	8*	

*5-dB error not unlikely. Monostatic radar, 0.5- to 10-μsec pulse.

incidence σ_0 is almost independent of the direction of viewing, but for less than $60°$ the differences become larger as ψ decreases. σ_0 has a greater magnitude in the upwind direction than in the downwind direction, and it is smallest in the crosswind direction.

For angles near normal incidence, the magnitude of σ_0 is almost independent of polarization for all sea states. As ψ decreases, the magnitude of σ_0 for vertical polarization is greater than that for horizontal polarization for calm and medium sea states. For very rough seas, σ_0 for horizontal polarization may become equal to, or in some cases greater than, σ_0 for vertical polarization.

In scattering from the sea, a considerable amount of depolarization occurs for ψ less than $60°$, increasing as ψ decreases [85, fig. 9].

7.4 POLARIZATION AND WIND DIRECTION EFFECTS

The preceding models considered horizontal and vertical polarizations that were the same for transmit and receive. If the polarization is crossed on receive, the relative backscatter (σ_{HV} or σ_{VH}, where the first subscript refers to the transmit polarization) will be reduced. Researchers at Georgia Institute of Technology have summarized this result in [234, 235]. Figure 7-3 is reproduced from [235] to give an idea of the magnitude of the effect at C band and low grazing angles. Campbell [59] and Daley [85] have presented some additional data. Since the target cross sections for aircraft and ships are reduced by about the same amount as the relative backscatter (see Sec. 5.2), there seems to be little to gain in signal-to-sea-clutter ratio by this technique for an airborne or surface radar attempting to detect low-flying aircraft or surface targets.

Similarly, using the same sense of circular polarization reduces the relative sea backscatter from that of the larger of the linear polarizations σ_{VV}, but Long [234, 236] indicates that reduction of σ_0 by circular polarization should only be 6 dB.

It will be shown in Sec. 7.5 that the location of the center of the doppler spectrum is different for the horizontal and vertical components. Thus, circular polarization may negate much of the gain achievable with MTI systems.

Because of the limited amount of data, the previous charts gave an average value of σ_0 over all wind directions. Researchers at NRL in 1956, 1964, and 1965 made numerous measurements that generally showed

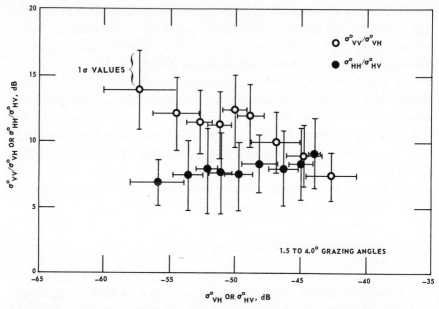

figure 7-3 *A comparison of σ_0 at 6.3 GHz for various polarizations and a wide variety of sea states. (From Long [235].)*

the largest backscatter occurring upwind at low angles. Table 7-9 shows some typical downwind-to-crosswind backscatter ratios at L band, where the effect is pronounced only for low angles and horizontal polarization. Upwind/downwind ratios at P band (74 cm) and L band (25 cm) from

TABLE 7-9   Backscatter from the Sea versus Wind Direction*

Measurements of downwind-to-crosswind ratio, dB, for various incident angles at L band				
Aspect angle, deg	Downwind/crosswind ratio ($10 \log_{10} (\sigma_d/\sigma_c)$)			
	HH	*VV*	*VH*	*HV*
1	−6.8	+2.8	+1.2	+2.0
3	−3.9	+2.0	+2.1	+0.3
8	−1.5	+3.2	+2.1	+2.2
15	−1.0	+3.2	+0.9	−0.4
35	+0.6		+2.5	+1.8

*From Macdonald *IRE* [239].

the 1965 NRL tests are shown in Fig. 7-4 [85]. The upwind component dominates for virtually all depression angles.

7.5 SPECTRUM OF SEA CLUTTER ECHOES[1]

Relationship between Sea Clutter Bandwidth and Sea State for Low Grazing Angles

One explanation for the frequency spread of signals returned from sea clutter is that the distribution of radial velocities of the scatterers[2] causes a distribution of doppler frequencies. When the spread of the velocity distribution increases, such as when the surface of the sea becomes more agitated, the clutter spectrum also broadens.

The width of the velocity spectrum[3] σ_v may be related to the width of the doppler spectrum σ_f by the familiar expression

$$\sigma_v = \frac{\lambda}{2}\sigma_f \tag{7-3}$$

where λ is the transmitted wavelength. If the scattering mechanism is the same for all wavelengths, then multiplication of the doppler spread by $\lambda/2$ should make spectrum measurements (expressed in velocity units) at different frequencies independent of frequency. Experimental observations from a number of different investigators are illustrated in Figs. 7-5 and 7-6. The carrier frequencies used ranged from 220 MHz to 9.5 GHz. (Most measurements were made at the higher frequencies.) Figure 7-5 presents the data of experimenters who analyzed the clutter spectrum prior to envelope detection, whereas Fig. 7-6 presents data of those who analyzed the spectrum after envelope detection. These figures show the spectrum half-power width as a function of wind speed. Sea agitation is shown in terms of hydrographic sea state since this seemed to be a most simple and convenient expression. In some cases the investigators did not specify the wind speed, but gave some other indication such as sea state or wave height. In these cases a value for wind speed was deduced. Since

---

[1] A considerable part of this section was prepared by J. P. Reilly.
[2] These scatterers may be either individual wavelets or wind-blown spray and foam.
[3] Spectrum measurements are expressed in velocity rather than frequency units so that the results of investigations at various frequencies can be compared.

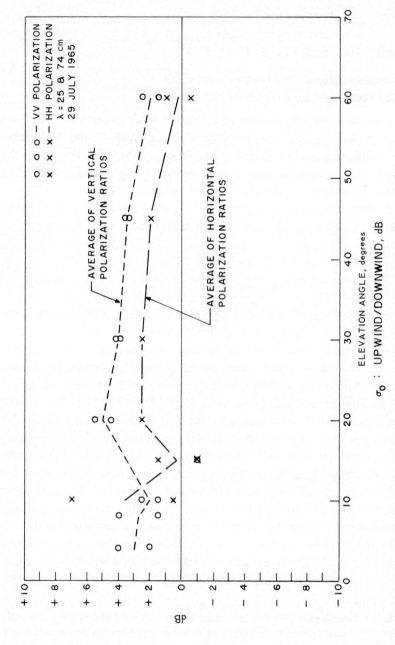

figure 7-4 *Upwind/downwind sea backscatter ratio at P and L bands. (From Daley [85].)*

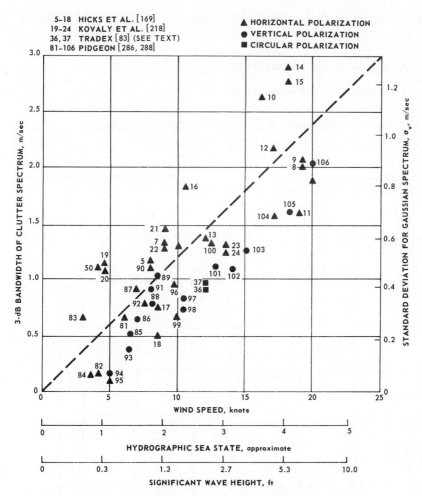

figure 7-5 *Variation of bandwidth for coherently detected sea clutter signals.*

in theory the spectra of coherently detected and envelope-detected signals differ, the two analyses must be treated separately.

For purposes of this section a coherently detected signal may be thought of as one whose spectral properties are examined before envelope detection. The spectrum bandwidth when predetection analysis is used is defined as the half-power width of the double-sided spectrum. Figure 7-5 shows an almost linear dependence of bandwidth on wind speed or sea state. Alternately, the spectrum width may be given in terms of its standard deviation σ_v. For a gaussian-shaped spectrum the relationship between the standard deviation and the half-power width ΔV is

$$\sigma_v = 0.42 \ \Delta V \quad \text{in velocity units}$$
$$\sigma_f = 0.42 \ \Delta f \quad \text{in frequency units, Hz}$$

(7-4)

In this section, bandwidth Δf for envelope-detected signals is defined as twice the width from the spectrum peak to the half-power point. A plot

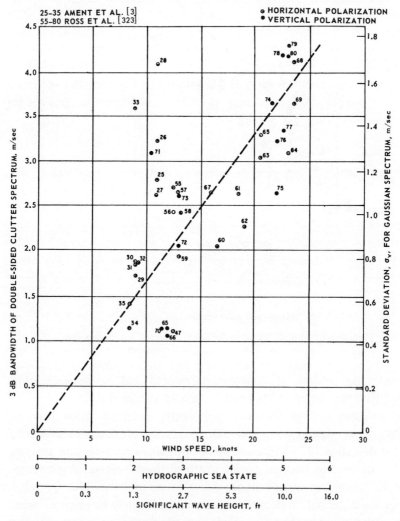

figure 7-6 *Variation of bandwidth for envelope-detected sea clutter signals.*

of postdetection spectrum bandwidth to corresponding values of wind speed is shown in Fig. 7-6. Although the scatter of points is larger than for the coherently detected signals, notice that for a given value of wind speed the envelope-detected bandwidth is larger than the coherently detected bandwidth. If we accept the interpretation of some authors that the doppler spectrum (predetection spectrum) is caused by the random motion of independent scatterers, then the video spectrum (postdetection spectrum) is caused by the velocity difference distribution of the scatterers. Thus, the second central moments of the two spectra can be related; the variance of the video spectrum being twice that of the doppler spectrum. This conclusion is a consequence of the fact that the variance of the difference of two independent random variables is the sum of the variances of the two variables. This result could also be arrived at by expressing the video spectrum as the convolution of the doppler spectrum with itself.[1] The variances of the two spectrum representations are related by

$$\sigma^2_{\text{video}} = 2\sigma^2_{\text{doppler}} \tag{7-5}$$

When the doppler spectrum is gaussian, the video spectrum is also gaussian. Therefore, the half-power width of the video spectrum should also be greater by the factor $\sqrt{2}$. Comparing Figs. 7-5 and 7-6, we see that this relationship is indicated.

Lhermitte suggests that the effect of observation time should also be considered [228]. Figure 7-7 shows the spectrum resulting from a group of scatterers with internal turbulent motion as an *instantaneous* spectrum that has a standard deviation σ_f. Since the group is assumed to have slow motion as a body, the instantaneous spectrum slowly shifts the position of its peak value. Measurements made after a period of observation long enough to include these slow variations would have a variance equal to the sum of the variances for the instantaneous spectrum and the slow variation, that is, the doppler spectrum has the variance[2]

$$\sigma_{dl}^2 = \sigma_{di}^2 + \sigma_g^2 \tag{7-6}$$

where σ_{dl}^2 is the variance of the doppler spectrum for a long observation

---

[1]This result is for a square-law detection process. See Davenport and Root [87, pp. 251-257].

[2]See also the discussion by Barrick [19].

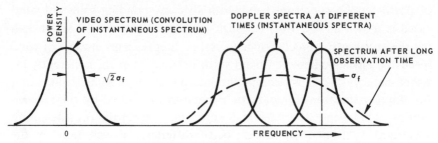

figure 7-7   *Video spectrum and doppler spectra showing the effects of observation time.*

time, σ_{di}^2 is the variance of the instantaneous doppler spectrum, and σ_g^2 is the variance due to the group velocity for the total observation time. The variance of the video spectrum, however, is independent of σ_g^2, that is, Eq. (7-5) still applies.

$$\sigma_{\text{video}}^2 = 2\sigma_{di}^2 \tag{7-7}$$

Equations (7-6) and (7-7) may be combined to relate the long-term doppler spectrum and the video spectrum

$$\sigma_{\text{video}} = \sqrt{2}\,(\sigma_{dl}^2 - \sigma_g^2)^{1\!/\!2} \tag{7-8}$$

Aspect with Respect to Wind Direction

Pidgeon [288] reported no observed dependence of spectrum width on wind direction. Curry [83] and NRL investigators [3] reported broader spectra when viewing crosswind than when viewing upwind or downwind. Kovaly [218] reported no difference for sea states one, two, and three, but observed an asymmetrical spectrum for sea state four when viewing downwind. Hicks et al. [169] observed a broadening of the downwind edge of the spectrum when viewing upwind or downwind. An attempt was made to plot the upwind and downwind spectrum width data from all investigators separately from the crosswind data, but there was no significant difference in the two plots. If there is a relationship between spectrum width and wind direction, it must be weak compared to other dependencies. Figure 7-8 illustrates that the mean velocity of the clutter spectrum is related to the wind direction by a cosine factor [288]. These measurements were made at 5.8 GHz with a surface radar.

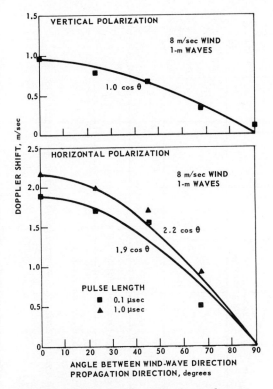

figure 7-8 *Mean doppler shift of sea echoes at various angles from the wind direction. (After Pidgeon [288].)*

Transmitted Polarization and Depression Angle Dependence

Most experimenters report little or no dependence of the spectrum width on the polarization (horizontal or vertical) [288].[1] This result may not hold for circular polarization, where the mean doppler is considerably different for the horizontal and vertical components (see Fig. 7-9). The data in Curry's paper [83, figs. 7 and 8] seem to verify this conjecture. The data points on Fig. 7-5 refer to only the lower velocity components. Similarly there is little dependence of the spectrum width on grazing angle in the range from 0-6° [169, 288, 83]. Since the

---

[1]The spectrum width from Pidgeon's data is somewhat lower on both polarizations than that reported by the other authors.

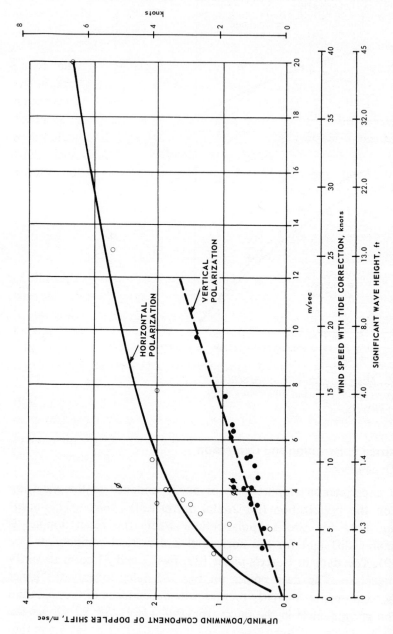

figure 7-9 *Mean doppler shift for sea clutter.* ○ *Horizontal polarization;* ● *vertical polarization;* φ *bistatic data.* *(From Pidgeon's data [288].)*

motion of the waves is primarily horizontal, the spectral width *may* decrease as the cosine of the grazing angle.

Pulsewidth Dependence

Goldstein [211, p. 579] and Pidgeon [288] report little or no dependence of the sea clutter spectrum width on pulse length in the range 0.1 to 10.0 μsec; however, Myers [258], using a narrow-beam radar and a pulsewidth of 0.008 μsec, reports much narrower spectra.[1] For wind speeds ranging from 8-15 knots, his X-band data indicate values of σ_v ranging from only 0.0001 to 0.0036 m/sec. With such short pulsewidths and narrow beamwidths the radar essentially observes single waves with a substantial mean doppler but a narrow spectral width. The effects of narrow pulses will be further discussed in Sec. 7.7.

Relationship between Wind Speed and the Mean Clutter Velocity

Few investigators have provided information from which the shift in the spectrum peak could be deduced. Figure 7-9 shows the data collected by Pidgeon [288, 286]. One interesting feature of this curve is that it shows an upper limit on the spectrum shift of about 7 knots. A second interesting feature is that the mean doppler shift for horizontal polarization is two to four times as great as the doppler shift for vertical polarization for similar wind and wave conditions. Although the abscissa in this graph is wind speed, Pidgeon [288] showed that the doppler shift for horizontal polarization is dependent upon both wind speed and wave height. The vertical polarization doppler shift is primarily dependent upon the wave height and is directly related to the orbital velocity of the gravity waves. The observations included depression angles between 0.1 and 10° and some bistatic geometries. In Pidgeon's paper he shows that smooth curves with a good fit can be drawn through his data points when the orbital velocity of the waves is taken into account.

7.6 SPATIAL AND FREQUENCY CORRELATION OF SEA CLUTTER

The spatial correlation of sea clutter is defined as the cross correlation between the signals returned from two separate patches of the sea in the

[1] The cell size studied was about 22 ft in azimuth and 5 ft in range, and the sampling rate was only five per second, which may have eliminated some high-frequency components.

radial dimension. The time interval separating the measurement of these two signals is assumed to be so small that there is negligible time decorrelation. At 5.7 GHz, Pidgeon [286] noted that the separation necessary to achieve independence was about the distance corresponding to a pulse length. Figure 7-10 illustrates the spatial correlation function versus radial displacement in units of pulse length. Included in Fig. 7-10

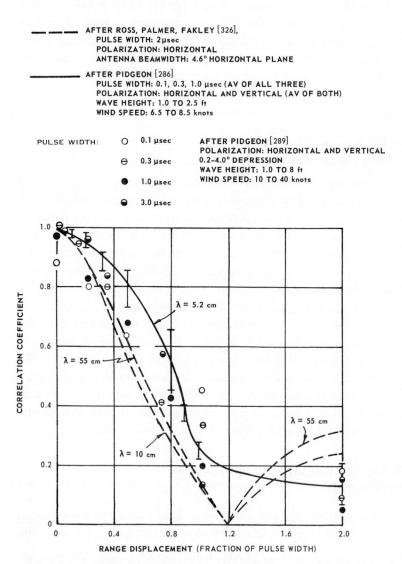

figure 7-10   *Spatial correlation of radar sea return.*

is another set of spatial correlation functions replotted from [326]. The pulse length dependence is evident at UHF, S band, and C band. Further measurements will be required to determine if the returns are uncorrelated in range for very short pulses (< 0.1 μsec) and narrow beamwidths, in which case the sea echo is resolved into individual waves (see Sec. 7.7).

In the measurement of the mean backscatter σ_0 from the sea, the number of independent clutter samples determines the accuracy of the estimated mean value of the backscatter. It has been shown in Sec. 7.5 that the sea return from a given cell is correlated for a period of many milliseconds and that to achieve accuracies of about 1 dB in the estimation of the mean requires measurement times in excess of 1 sec for a stationary radar. Airborne radar measurements of clutter acquire the necessary independent samples averaging over space. Frequency[1] is a third *dimension* for averaging clutter returns.

The frequency correlation of clutter is also of great importance for the detection of low-flying aircraft or surface targets by a stationary radar. It is shown in Chap. 3 that incoherent integration can improve the detectability of target returns in the presence of sea clutter if the sea clutter return is decorrelated from pulse to pulse.

Following the same approach as was described in Sec. 6.5 for precipitation echoes, it has been shown by Goldstein [141], Kerr [211], and Wallace [395] that for a large collection of independent scatterers the correlation coefficient of the intensity of the radar echoes from a rectangular transmit pulse can be expressed as

$$\rho = \left(\frac{\sin \pi \tau \Delta f}{\pi \tau \Delta f} \right)^2 \tag{7-9}$$

where τ is the pulse duration, Δf is the transmit frequency change, and ρ is the correlation coefficient. The correlation coefficient falls rapidly to zero at $\tau \Delta f = 1$ and remains near zero for $\tau \Delta f \geq 1$.

The return from the sea for small beamwidths and pulse lengths does not contain as large a number of scatterers as does an extensive rain; however, it would appear to contain a sufficient number to cause independence when the frequency separation exceeds the inverse of the pulse length. If the waves are resolved, the correlation function may take other forms, as described by Voles [389].

---

[1] Frequency in this section refers to the transmitter frequency.

An experimental program to determine the correlation coefficient of sea returns with frequency was performed by the Defense Research Laboratory of the University of Texas (Pidgeon [289, 286]). The experiments were performed primarily at C band (5.7 GHz) with both horizontal and vertical polarization and with a 2.5°, two-way beamwidth.

Figure 7-11 is a composite of the correlation coefficient versus $\tau \Delta f$ for the data. The upper curve is for the *infinite* collection of small scatterers.

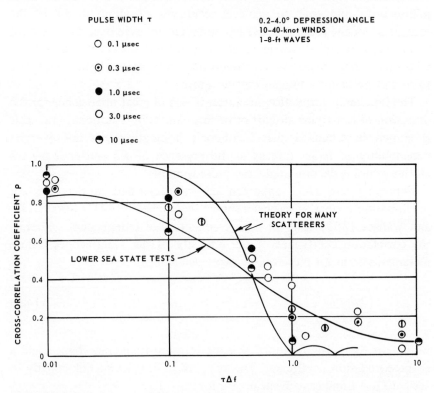

figure 7-11 *Frequency correlation of radar sea return at C band as a function of pulse length times frequency shift (vertical and horizontal polarization). (After Pidgeon [286, 289].)*

The experimental points for $\tau \Delta f \ll 1$ are less than unity due to the slight time decorrelation for the signals with small frequency separations. The solid-line data represent points taken at about 10° grazing angle, wind speeds of 3 to 9 knots, and wave heights of ½ to 2½ ft (Pidgeon

[286]). The correlation coefficients at $\tau\Delta f > 1$ for pulse lengths of 0.1, 0.3, and 1.0 μsec are all below 0.2, which indicates that the return is essentially decorrelated when considering the effects of receiver noise and finite sample lengths. The individual points are from higher sea state tests [289].

Figure 7-12 shows only the data points for the 0.1-μsec pulse transmissions at lower sea states; this was done in order to determine if the *spiky* clutter described in Sec. 7.7 is more correlated with frequency. Although the spread in the computed correlation coefficient is somewhat greater for horizontal polarization, the echoes seem decorrelated at $\tau\Delta f \geq 1$; however, the mean backscatter power that is common to all frequencies for the 5-sec computation period was subtracted before the correlation coefficient was computed. Thus an echo from an individual wave that persists for the entire measurement period would be taken out of the data.[1] As in any attempt to describe the correlation properties of a nonstationary process, care must be taken to ensure that the measurement is applicable to the type of radar processor that is under consideration.

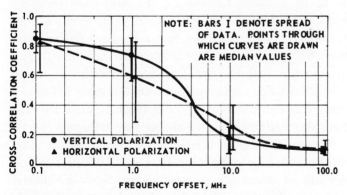

figure 7-12 *Frequency correlation of radar sea return (pulse length 0.1 μsec, C band).*

7.7 SHORT-PULSE SEA CLUTTER ECHOES

It is known that the amplitude statistics of sea and land clutter deviate from the Rayleigh distribution for short pulses, narrow beamwidths, and

---

[1]The correlation process describes only deviations from the echo in the region of interest.

low grazing angles. The *tails* of the distribution extend further from the mean value than would be expected from a *large* number of random scatterers. It is not difficult to visualize the resolution of ocean waves since the individual waves often have a spatial period of over 200 ft ($c\tau/2$ = 0.4 μsec). This is important in high-resolution or high-pulse compression ratio radar systems where a threshold detector is set at some arbitrary value above the mean value of the receiver noise or clutter. Several qualitative descriptions of *spiky* clutter have been reported, especially for horizontal polarization sea return. Partly because of the difficulty of recording sea clutter echoes from short-pulse radars, it is difficult to confirm a specific power distribution function for short sea-clutter echoes. The measurements with an 8-nsec, X-band, 0.9° beamwidth radar reported by Conlon [70] and Myers [258] of NRL suggest a log-normal distribution. This distribution, which was introduced for certain targets (satellites, birds, etc.), appears as a gaussian distribution when σ_0 *in decibels* is plotted on probability paper. The ratio of the standard deviation to the mean value of σ_0 is one parameter of interest since it is equal to unity for the Rayleigh distribution. Conlon's data show that value to be between 1.5 and 2.1 for short-pulse vertically polarized sea backscatter at ranges of 450 to 2,500 yards from a surface radar. The horizontally polarized echoes had a standard deviation-to-mean ratio of from 6.0 to in excess of 14.0 under the same conditions. In some data runs the echo that occurred 2 percent of the time exceeded the median echo by 20 dB. For vertical polarization this value was only 8 to 10 dB, and for a Rayleigh distribution 7 dB would be expected. The more highly skewed distributions occurred at about sea state two.

Recent measurements by Pidgeon [289] with a C-band surface radar cover the range of pulse lengths between 0.03 and 10.0 μsec with a horizontal beamwidth of about 3°. A-scope returns from the shorter pulse transmissions are shown in Fig. 7-13. The echoes are obviously spiky with the effect being more pronounced with the 30-nsec pulses.

An attempt to determine the change in distribution with pulse length is shown as Fig. 7-14. Since the complete distributions from Pidgeon's data are not available, only the ratio of the standard deviation to the mean is shown. An estimate of this parameter is also shown from the NRL data. This parameter is unity for a Rayleigh distribution. Pidgeon's tests were at C band with a 2.5° beamwidth, and the NRL tests were at X band with a 0.9° beamwidth. The lines drawn through the average

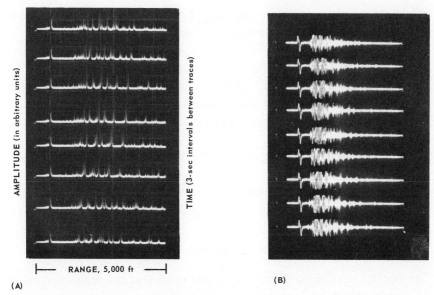

figure 7-13 *A-scope returns from sea clutter at C band, Sea state 5, trace length = 5,000 ft, about 3 sec between traces. (A) 30 nsec pulse-detected video; (B) 100 nsec pulse-bipolar video.*

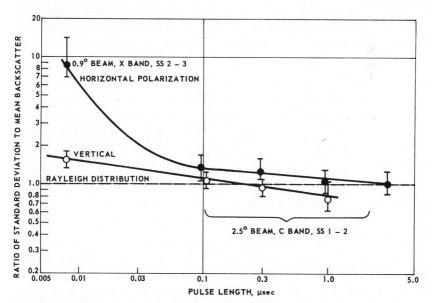

figure 7-14 *Distribution of short-pulse sea backscatter σ_0, standard deviation to mean ratio.*

values are merely to indicate the trend toward a highly skewed distribution at short pulse lengths especially on horizontal polarization.

Until further tests are made on short-pulse systems, only a few generalizations can be given for low grazing angles.

1. The clutter returns with pulse lengths of 1 μsec or greater, at S band (3 GHz) or above, *and* with beamwidths of 1° or greater conform to the Rayleigh law for most sea conditions. (Also see Lane and Robb [222].)

2. At 6-9 GHz the above relationship holds for pulse lengths greater than 0.25 μsec.

3. Short-pulse data for vertical polarization returns conform more closely to the Rayleigh law than those for horizontal polarization.

7.8   BISTATIC CROSS SECTION OF THE SEA

The bistatic microwave scattering characteristics of the sea surface must be known in order to predict the amount of power received at a receiver that is not at the same location as the transmitter. Pidgeon [284] performed bistatic backscattering experiments at C band with the transmitter and receiver separated but in the same vertical plane. Two types of polarizations were used: (1) vertical transmitter and receiver and (2) vertical transmitter and horizontal receiver. Results shown on Fig. 7-15 indicate that for the polarized component of the return signal VV the effective radar cross section of the sea σ_0 is independent of the receiver depression angles for transmitter depression angles between 0.2 and 3° below the horizontal and for receiver angles between 10 and 90°. The values of σ_0 for sea state three are approximately 10 dB larger than corresponding values for sea state one. There seems to be no direct relationship of σ_0 to wind velocity; a sea state three and a 10-knot wind gave larger returns than a sea state two and 20- to 30-knot winds. This does not, however, imply σ_0 is independent of wind velocity.

The cross-polarized component of the sea return (vertical transmitter and horizontal receiver polarization) gave 10 to 15 dB less return than did the polarized component (vertical-vertical) for transmitter depression angles ψ_t less than 1°. At depression angles near 3°, the cross-polarized component is only 5 to 8 dB less than the polarized component.

For sea states two and three, σ_0 values for the various receiver depression angles ψ_r are distributed equally about $K \sin \psi_t$, where K is a constant depending on sea state. This means that, within the angular

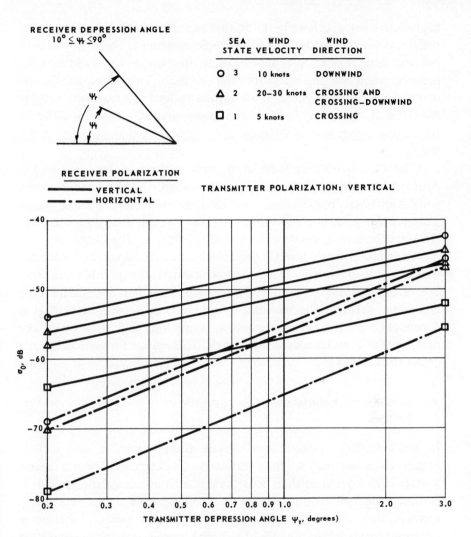

figure 7-15 *Bistatic cross section (C band). (After Pidgeon [284].)*

region measured, σ_0 is independent of ψ_r for moderate seas; however, for sea state one, the data for the smaller receiver angles appear mostly below $K \sin\psi_t$, whereas those for the larger angles appear mostly above that line, indicating an overall angular dependence greater than the sine law. Thus, while there may be a slight dependence on the receiver depression angle, it is not believed to be significant.

The data indicate that σ_0 is sine-law dependent on ψ_t; however, this dependence is due primarily to the change in power density on the water surface as the transmitter depression angle is changed. The power density per unit area is directly proportional to the sine of the angle that the incident radiation makes with the surface. Therefore, the actual scattering is σ_0 independent in the angular regions reported here, and the change in ψ_t is due to the change in power density caused by different transmitter angles not to a change in the scattering characteristics of the sea.

A set of X-band bistatic sea cross-section tests was also performed at Aruba, Netherlands Antilles [285]. A modified AN/APS-3 radar system with horizontal polarization was used to measure the radar signal scattered from the sea. The transmitter was located on a 100-ft cliff, and the bistatic receiver was flown at a 1,000-ft altitude. The illuminator was positioned in azimuth directly beneath the aircraft. Figure 7-16 gives the results of this operation. Monostatic backscatter data points taken from Sec. 7.2 for similar conditions are included on the figure. The close agreement between the monostatic and bistatic σ_0 verifies that σ_0 is dependent on transmitter depression angle and that it is essentially independent of receiver depression angle. This applies for vertical bistatic angles to at least $55°$.

7.9 BACKSCATTER FROM VARIOUS TERRAIN TYPES

In the detection of small, low-altitude aircraft, missiles, and surface targets, radar backscatter[1] from various land and cultural features creates a more severe problem than does the backscatter from the seas. This is because land and cultural features generally have a higher reflectivity factor σ_0 than does the sea surface at low depression angles. This effect is mitigated somewhat in the case of airborne targets since the doppler shift and spectrum width of land returns are small compared to those due to aircraft velocities. MTI techniques on stationary or slowly moving radars can have high clutter rejection ratios for most land objects (Chap. 9).

It is difficult to give an adequate statistical distribution of the backscatter characteristics of land for the following reasons:

---

[1]The discussion in the next sections applies primarily to monostatic target detection radars. Discussions of surface mapping radars or *scatterometers* can be found in Moore [253], Levine [237], and Floyd and Lund [122].

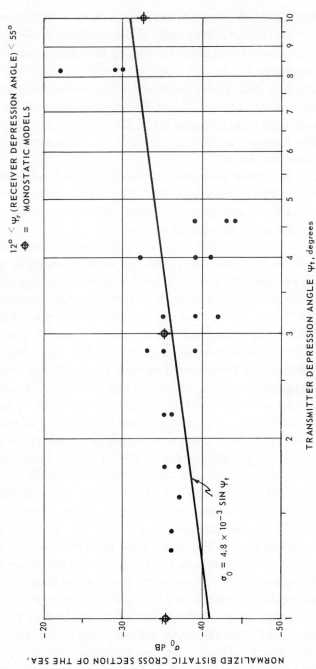

X BAND
0.5 μsec PULSEWIDTH
HORIZONTAL POLARIZATION
LOOKING INTO THE WIND
WIND VELOCITY, 15 – 23 knots

$12° < \Psi_r$ (RECEIVER DEPRESSION ANGLE) $< 55°$
⊕ = MONOSTATIC MODELS

$$\sigma_0 = 4.8 \times 10^{-3} \sin \Psi_t$$

TRANSMITTER DEPRESSION ANGLE Ψ_t, degrees

NORMALIZED BISTATIC CROSS SECTION OF THE SEA, σ_0 dB

figure 7-16 *Bistatic cross section of the sea (Beaufort 5). (After Pidgeon [285].)*

1. The statistical nature of the return from a given area cannot be related to the type of land as easily as the relatively convenient use of sea state descriptions. (Note that even the sea state description at any time is highly subjective.)

2. The land backscatter amplitude distribution at low grazing angles does not usually conform to the Rayleigh distribution because of the *shadowing* from hills, buildings, trees, etc.

3. The moisture content of the soil, or snow cover, can alter the backscatter coefficient.

4. The derivation of a mean or median value for σ_0 differs between land and airborne measurements. The fixed radar sites essentially perform a time average of a given clutter cell while an airborne measurement performs a spatial average.

Figure 7-17 gives several cumulative distribution functions of σ_0 from land-based radars. Two of the radars were operated at the Applied Physics Laboratory in Maryland for the detection of low-flying aircraft. Both of the profiles shown are for an azimuth angle for which the clutter return extended for several miles. The terrain consisted of rolling countryside with patches of 30-ft high trees and a number of small houses. The approximate peak values of the time fluctuation were plotted rather than the temporal average. The third distribution is from a Swedish forest area [232] with a radar of similar parameters to the X-band radar at APL. While the maximum values of σ_0 for these two areas are similar, it can be seen that the median value of the backscatter coefficient for the two APL radars differs by about 11 dB for the same terrain. This is undoubtedly due to the shadowing effect, which almost completely obscures close to 50 percent of the terrain. Essentially, the comparison of the median values of these two experiments would indicate a strong frequency dependence that other experimenters have not verified. The Swedish data in Fig. 7-17 from Linell [232] do not have as marked a shadowing effect as do the APL data, probably because the radar used by Linell was located atop a 100-ft waterworks tower. The APL radars are approximately 50 ft above the local terrain. A fourth cumulative distribution is shown for a mountainous area.

The low-angle backscatter measurements reported by Katz [204] and Erickson [108] and various studies at NRL, Ohio State University, Goodyear Aircraft Corp., and the University of Texas are crudely summarized in Tables 7-10 and 7-11.[1] Various general classifications of

---

[1] See list of references in Sec. 7.11.

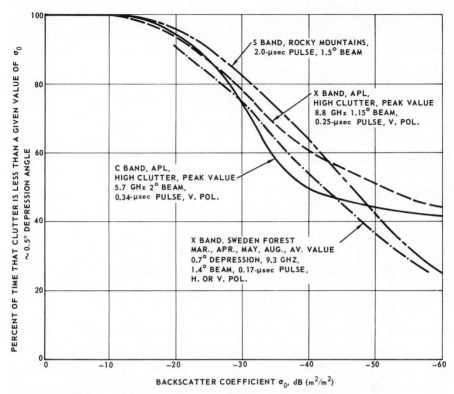

figure 7-17 *Land clutter backscatter distributions from surface radars.*

terrain are arranged on the chart in the order of increasing backscatter coefficient at low depression angles. The values of σ_0 (median) for each frequency are the average of horizontal and vertical polarization unless otherwise stated.[1] The results can be considered seasonal averages since the median return from vegetation and forests will vary by more than 9 dB, depending on the amount of foliage [232]. This same reference shows that the terrain backscatter was about 6 dB lower than the lowest seasonal average when there was a 4-in. snow cover at depression angles of about 1°.

The use of the term σ_{84} in Tables 7-10 and 7-11 is an initial attempt to define the statistics of the backscatter coefficient for a pulse radar; σ_m refers to the median value and σ_{84} refers to the value of the backscatter coefficient that will *not* be exceeded in 84 percent of the range cells. The difference between these values σ' has been found to be as high as 18 dB

---

[1] Some polarization ratios were included in Table 5-1.

TABLE 7-10 Land Clutter Reflectivity (0 to 1.0° Incident Angle)

Terrain type	Reflectivity, dB, below 1 m²/m² at indicated carrier frequency, GHz						
	UHF 0.5 σ_m, σ_{84}	L 1.2 σ_m, σ_{84}	S 3.0 σ_m, σ_{84}	C 5.6 σ_m, σ_{84}	X 9.3 σ_m, σ_{84}	K_u 17 σ_m, σ_{84}	K_a 35 σ_m, σ_{84}
Desert	30,- 28,-	45,35	-,31		38,30		
Cultivated land		32V,-		38,30	36,30		23H, 15H 18V, 10V
Open woods	24,18	34H,-	33,26		30,22		
Wooded hills	34,-	35,20 45,27	32,24 47,29	-,27	30,20 36,28		21H, 13H 13V, 8V
Small house Districts			35,26	35,26	30,24 36,-		
Cities	22,- 30,-	30,20			24,14		

Average of both polarizations except where noted.

σ_m = median backscatter coefficient in decibels below 1 m²/m²

σ_{84} = coefficient that 84 percent of the cells are below

τ = pulse length $\approx 1\ \mu$sec

θ_2 = beamwidth $\approx 2°$

TABLE 7-11 Land Clutter Reflectivity (10° Incident Angle)

Terrain type	Reflectivity, dB, below 1 m²/m² at indicated carrier frequency, GHz						
	UHF 0.5 σ_m, σ_{84}	L 1.2 σ_m, σ_{84}	S 3.0 σ_m, σ_{84}	C 5.6 σ_m, σ_{84}	X 9.3 σ_m, σ_{84}	Ku 17 σ_m, σ_{84}	Ka 35 σ_m, σ_{84}
Desert	40,34	40H,- 38V,-	25,-		26,20	26V,-	22V,-
Cultivated land	36,30	33,27	21,-	29,23	25,21	23,-	20,18
Open woods	20H,- 26V,20	23,-	26V,- 20H,-		23,18	22,-	19,-
Wooded hills	22,-		23,-		25,20	19,-	19,-
Small house Districts	23,17				23,13-17		
Cities	12V,- 6H,-	12H,- 18V,-	18V,- 10H,-		12-18V,6 14H,8		

Average of both polarizations except where noted.

σ_m = median backscatter coefficient in decibels below 1 m²/m²

σ_{84} = coefficient that 84 percent of the cells are below

τ = pulse length $\approx 1\ \mu$sec

θ_2 = beamwidth $\approx 2^{\circ}$

263

(unpublished Raytheon tests and [287]). In a surface pulse radar with narrow beams and short pulse lengths ($< 1 \mu\text{sec}$), the value of $\sigma_m + \sigma'$ should be used for low false alarm probability ($< 10^{-2}$) systems; but, depending on the false alarm criteria, $\sigma_m + 2\sigma'$ or $\sigma_m + 3\sigma'$ may be more appropriate. As the depression angle of the radar increases, the shadowing effect diminishes and the backscatter conforms more closely to the Rayleigh distribution (see Sec. 7.10). The following statements are very tentative, but their general trends are indicative of low depression angles and homogeneous terrain:

1. The median backscatter coefficient increases somewhat with frequency for most terrain types but usually not faster than linear with transmit frequency. The frequency effect on return from urban areas is quite small.

2. The median backscatter coefficient increases about linearly with depression angle from 0.5-10° below the horizontal. In some cases a reduced value is found at 3-5°.

3. There are polarization differences on individual measurements, but there is not a strong general effect.

7.10 COMPOSITE TERRAIN AT LOW GRAZING ANGLES

The description of land clutter return has been divided into types of terrain having backscatter coefficients that differ by a spread of about 15 dB between cities and cultivated land. If a surface radar is placed near a city a composite terrain return will be obtained. An example of this placement is shown in Fig. 7-18 [176]. The radar used to obtain these data was located near Huntsville, Alabama; the survey was made with an L-band radar using a 3-μsec pulse and a 1.8° azimuth beamwidth. The figure was drawn for 241 of the returns from the strongest *cells*; the Huntsville city area gave the group of higher values of σ_0, and the hilly countryside showed another distribution centered around $\sigma_0 = -40$ dB. It appears from the report that over 2,000 cells were observed but that about 90 percent of the cells were masked by the terrain and hence gave negligible return. The median value σ_m cannot be determined, and even the 84 percent value is probably indeterminate. The false alarm probability for a simple search radar, owing to clutter returns at this location, is determined by the relatively few large clutter

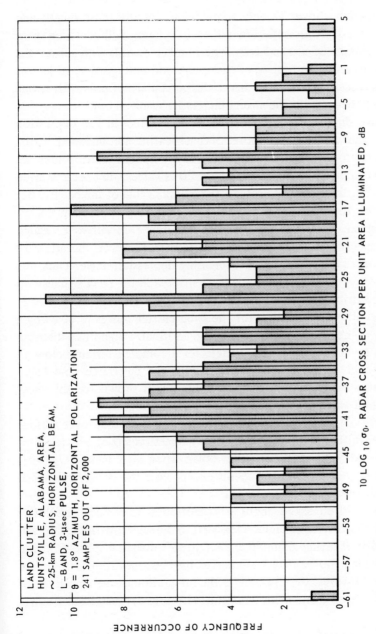

figure 7-18 *Histogram of occurrence of σ_0. (After Holliday [176].)*

265

cells unless a *crossgating*[1] system is used to inhibit the returns from the city area.

The histogram shown is only appropriate for the pulse length and beamwidth used. A decrease in either parameter will yield an even greater percentage of masked cells. The number of cells with substantial returns will increase, and the σ_0 of those cells containing strong reflectors will also increase. On the other hand, the percentage of cells with strong reflectors will decrease as the strong targets are resolved out.

Descriptive data on the increased shadowing effect at low depression angles and the seasonal variations of backscatter are included in the previously cited Swedish experiments [232]. Spatial distributions are available for depression angles of 0.4 to 5.0° (Table 7-12). At any given month, or for the all-season average, the median value of the backscatter coefficient σ_m decreases very rapidly for decreasing depression angles. This is a stronger effect than would normally be expected for *rough* surfaces (especially for the pine and fir forest); however, the arbitrary variable σ' increases rapidly with decreasing depression angle. Thus, the clutter power in a given direction is concentrated in the peak signals with negligible contribution from the shadowed areas. The use of σ_m and σ' (in decibels) allows calculation for the total clutter power received by a CW system if the spatial distribution is log-normal. Log-normal distribution seems to be the most appropriate for a complex terrain below 4° depression angle. The seasonal variation for the cultivated land backscatter is also interesting. A distinct minimum is shown for the snow cover in March and a peak return for the crops in August.

In a series of tests by the Naval Research Laboratory over various parts of the United States at from 5 to 9° depression angles, σ' was about 6 dB at 5° and 5 dB at 9° for either polarization at X band [97, 149]. The value of σ_m was -21 to -29 dB except near the city of Nashville, where σ_m increased to about -12 dB.

A question often asked about land clutter is how much subclutter visibility or improvement factor is needed to detect a low-flying aircraft. The difficulty of giving a simple answer to this question is illustrated in Fig. 7-19, which shows a trace similar to the A-scope trace of a surface radar. Here, however, the ordinate is the integration of the amplitude of the clutter echoes from many transmitted pulses. The abscissa is range

---

[1] An operator or automatic control system that prevents radar signals from a given location from being displayed or entered into a digital system as a possible target.

TABLE 7-12 Summary of Swedish Land-clutter Experiments

Depression angle, deg	Normalized Backscatter from cultivated flat land, dB											
	March (snow)		April		May		August		November		Average	
	σ_m	σ'	σ_m	σ'	σ_m	σ'	σ_m	σ'	σ_m	σ'	$\bar\sigma_m$	$\bar\sigma'$
5.0	-39	7	-30.2	5	-26.3	4.7	-23	4.6	-26.3	4.7	-26.4	4.8
2.5	-50.5	13	-38.2	10.7	-38.2	7.7	-30	6.5	-32.5	7.5	-35	8.1
1.25	-53.5	16.5	-48	14.2	-46.5	11.8	-38.5	10.5	-41	12.3	-43	13.7
0.4			<-50	>20								
Backscatter from pine and fir forest, dB												
0.7	-43.5	18	43.5	18	-43.5	18	-43.5	18	-36	15	-41.6	17.2
0.5					-49	~18					~49	~18
0.4					-56	~19					-56	~19

From Linell [232].

σ_m = median backscatter coefficient, dB

$\sigma' = \sigma_{16\%} - \sigma_m$ = standard deviation of coefficient, dB

$\bar\sigma_m$ = average σ_m, dB for April to November.

$\bar\sigma'$ = average standard deviation for April to November

Radar parameters: antenna height = 100 ft, wavelength = 3.2 cm, pulse length = 0.17 μsec, azimuth beamwidth = 1.4°. Horizontal polarization $\approx$ vertical polarization.

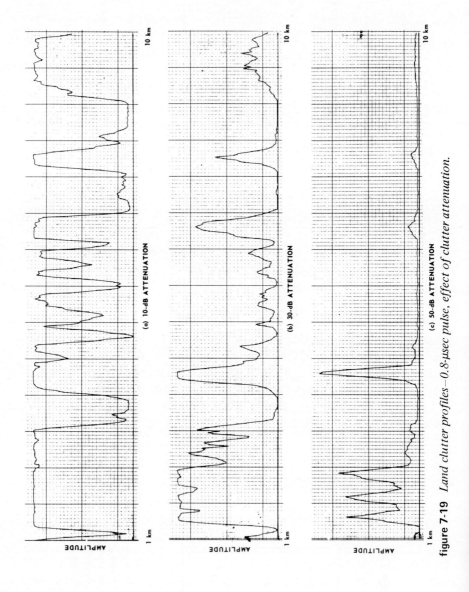

figure 7-19 *Land clutter profiles — 0.8-µsec pulse, effect of clutter attenuation.*

from the radar.[1] To obtain the traces, 10 dB, 30 dB, and 50 dB of attenuation were successively inserted ahead of the receiver. This allows determination of the effects of different amounts of clutter attenuation without having to construct a processor that actually discriminates between signal and clutter. In trace *a*, 10 dB of attenuation, the system was saturated for most of the trace. There is effectively 30 dB of *clutter attenuation* in trace *b* and 50 dB in trace *c*. Thus, even with 30-50 dB of *clutter attenuation*, some land echoes will be seen on the output display. The detectability of a target in such clutter will depend on the attentiveness of the operator and on the fluctuations of the target and land clutter (Sec. 7.12).

A second question of interest is how much is target detection improved if the pulsewidth is reduced or if pulse compression is used. The upper trace of Fig. 7-20 shows how the land clutter looks if the pulse length is 3.2 μsec. The same region as in Fig. 7-19 was observed with 10 dB of attenuation prior to the receiver. It is virtually impossible to detect a target at any time under these conditions. In *b* the pulse length was reduced to 0.8 μsec with the same peak power and in *c* the pulse length is 0.2 μsec. While the 16-to-1 reduction in the pulse length does not materially reduce the large reflectors, the chance of seeing a target between the high clutter regions is considerably improved. Although the reflectivity in the location where these traces were taken is high, there is little masking. The cumulative distributions of σ_0 for the three pulsewidths in this region are shown on Fig. 7-21. It can be seen on the lower right of the curves that the percentage masking is more pronounced for the shorter pulses and that strong reflectors (water towers, etc.) give higher peak values of σ_0 with shorter pulses.

It can be concluded that land-clutter echoes are rarely uniform or noiselike and that the relatively simple range equations for a clutter-limited radar in Chap. 2 must be used with care. A more reasonable detection criterion would be based on in just what percentage of the area of interest can a low-flying target be detected [149].

7.11 COMPOSITE TERRAIN AT HIGH INCIDENCE ANGLES (10-70°)

The backscatter from composite terrain at high incident angles does not have a strong shadowing effect and is less sensitive to polarization and

---

[1] The data were taken in Howard County, Maryland with a 1.4°, two-way beamwidth, 10-kW C-band radar with vertical polarization. No STC was used for Figs. 7-19 and 7-20.

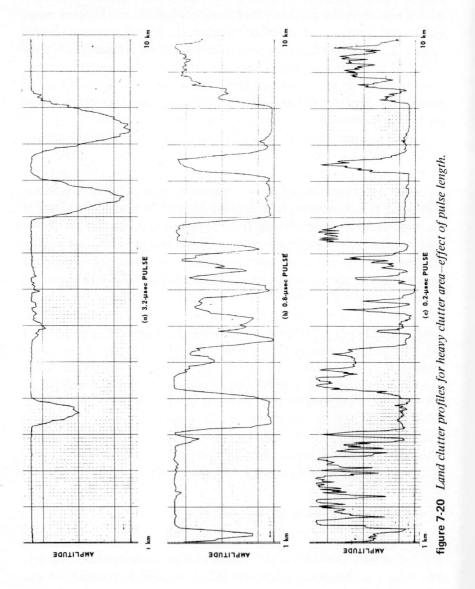

figure 7-20 *Land clutter profiles for heavy clutter area—effect of pulse length.*

(a) 3.2-μsec PULSE

(b) 0.8-μsec PULSE

(c) 0.2-μsec PULSE

AMPLITUDE

1 km

10 km

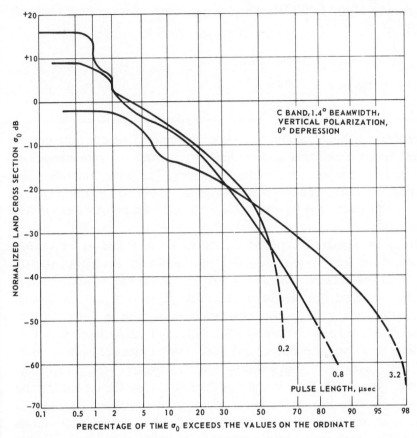

figure 7-21 *Cumulative distribution of land cross section σ_0 for high clutter azimuth.*

transmit frequency. To a first approximation the angle dependence can be removed by using the parameter γ which is defined as

$$\gamma = \frac{\sigma_0}{\sin \psi}$$

where ψ is the grazing angle and γ is usually expressed in dB (m²/m²). ITT Research Institute [185] (based to a great extent on the compilations of experiments by Katz and Fyler [204]) has shown that γ is almost independent of ψ, neglecting backscatter from essentially smooth surfaces such as deserts and roads. A convenient summary for all

frequencies (400-35,000 MHz) and all polarizations for grazing angles of
10-70° is

$$\gamma_{\max} \approx -3 \text{ dB}$$

$$\gamma_m = \gamma_{\text{median}} \approx -14 \text{ dB}$$

$$\gamma_{\min} \approx -29 \text{ dB}$$

In this oversimplification, $\gamma_{\max}$ is a maximum of various experiments
rather than a peak signal in time. Similarly γ_m and $\gamma_{\min}$ are *averages* of
the reported medians and minimum values. The maximum values are
dominated by the echoes from urban areas. Above 65° grazing angle γ_m
(and σ_0) increases rapidly to a value of 0 to 10 dB (m^2/m^2) at vertical
incidence.

There are numerous reports[1] on the values of γ and σ_0 for various
terrain types and frequencies, but the spread in the reported data
precludes any detailed summary as was given in the case of sea clutter. A
heavily averaged (in decibels) chart of γ_m and $\gamma_{\max}$ for several
frequencies and gross terrain classifications is given in Table 7-13. It
shows that for rough terrain (woods, hills, and cities) there is not an
obvious frequency dependence and that even for measurements of roads
and desert areas γ increases slightly less than linearly versus frequency.
On the other hand there is a strong dependence on the type of terrain
(roughness) especially at the lower microwave frequencies. The value of
$\gamma_{\max}$ is an indication of the average (in decibels) of the maximum values
reported by each experiment rather than an indication of the maximum
expected return.

The references indicate no obvious polarization dependence. The
spatial probability density functions seem to conform more closely to
the Rayleigh law than they do for low grazing angles. The temporal
distribution functions result from a combination of a fixed and a
fluctuating component, with the ratio depending on the terrain type.

7.12 SPECTRUM OF LAND CLUTTER

The fluctuation spectrum of echoes from vegetated terrain arises from
the relative motion of the scatterers (plants) as they move about in the

[1] See [109, 33, 76, 90, 97, 95, 108, 122, 131, 145, 149, 185, 190, 219, 204, 203,
202, 205, 253, 275, 276, 287, 307, 343, 371].

TABLE 7-13 Averaged Land-clutter Return $\gamma = \sigma_0/\sin \psi$, $\psi = 15\text{-}70°$ Grazing, in decibels below 1 m²/m²

Terrain type	Av. of median γ, dB	Carrier frequency, GHz											
		UHF 0.50		L 1.25		S 3.0		X 9.0		Ku 17		Ka 35	
		γ_m	γ_{max}	γ_m	γ_{max}	γ_m	γ_{max}	γ_m	γ_{max}	γ_m	γ_{max}	γ_m	γ_{max}
Deserts and roads	25	37	30	32	28	28	22	23	10	23	17	22	15
Cultivated land	22	32	18	--	--	--	10	18	10	19	10	16	5
Open woods	16	22	12	15	8	17	10	15	10	15	8	14	5
Wooded hills	14	16	--	--	--	--	--	13	6	15	8	13	--
Cities	11	6	-2	11	4	15	5	12	3	--	--	--	--

Data taken from incoherent, monostatic radars, with pulse width $\approx 3~\mu\text{sec}$.

wind. As the wind speed increases, the motion increases and the spectrum broadens. Experimental evidence shows that the spectrum width is almost directly proportional to the transmitted frequency, at least in the 1- to 9-cm range of wavelengths. Figure 7-22 displays the spectrum width as a function of wind speed as determined from the data of four different sources—the transmitted wavelength varied from 1.25 to 9.2 cm. An estimated fit to the data is shown by the broken line. The polarizations used are unknown except for the APL experiments (vertical polarization) and those of Fishbein et al. [120] (horizontal). The standard deviation of the clutter spectra σ_v in velocity units[1] was determined by estimating the best fit to the gaussian shape,[2] noting its

[1] See Secs. 6.4 and 7.5.
[2] The data of Fishbein et al. [120] showed large high-frequency "tails" in the power spectrum and approximated $P(f) = 1/[1 + (f/f_c)^3]$.

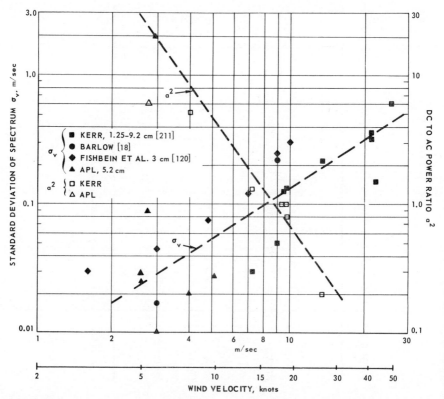

figure 7-22 *Spectrum width for land clutter, wooded terrain (fluctuating component).*

standard deviation, and converting it to velocity units through the doppler equation (each point of the APL data is an average of several measurements). In presenting these data, we make no distinction between those spectra resulting from coherent and those from square-law detection. There is a steady (dc) component superimposed on these fluctuations. The ratio of the power in the steady component (dc) to that in the fluctuating component (ac) is also shown on Fig. 7-22 for $\lambda =$ 5 cm [APL] and 9.2 cm [211]. In MTI systems the dc and ac terms must be considered separately.

8 SIGNAL PROCESSING CONCEPTS AND WAVEFORM DESIGN

The first seven chapters explored the concepts of detection, display, and target and clutter reflectivity that are common to many forms of radar processing. With proper interpretation these concepts are applicable to the various waveforms and signal processors to be discussed in the following seven chapters.

This chapter includes some of the general principles of waveform and processor design. The choice of transmit waveform and of the corresponding receiver configurations involves two separable design problems. The waveform must be chosen to optimize performance in some total environment as outlined in Chap. 1 [317]. The limitations are generally external to the radar though restrictions may be imposed by the type of transmitter or antenna. The currently accepted belief that there is no universal waveform is not surprising in light of the wide variety of waveforms used for electrical communication, a subject that has had over 100 years of intensive study. On the other hand, the inability to find a *best* waveform is not an excuse for failure to search for *locally optimum* waveforms for specific radar tasks and environments.

The design of the radar processors (the hardware) is somewhat separable since there are generally two or more ways to design a near optimum processor for a given waveform. Cost, complexity, and reliability are generally the bounds on processor design rather than physical realizability. Practical signal processors with less than a 2-dB deviation from maximum efficiency have already been constructed for random noise waveforms with bandwidths in excess of 20 MHz.

This chapter introduces the subject of waveform design with the now standard range-doppler ambiguity function description of Woodward [407] that was derived in the early 1950's. The ambiguity function discussion in Sec. 8.3 will only contain a summary of the results pertinent to the descriptions of the specific waveforms and processors of later chapters. Detailed descriptions of the general properties of various general waveform classes and their ambiguity functions can be found in many excellent references such as Siebert [345], Skolnik [354], Cook and Bernfeld [71, 39], Spafford [359, 358], Fowle [125], and Rihaczek [321]. These expand the pioneering work of Woodward.

Despite their wide study, a knowledge of ambiguity functions has not generally been a substitute for learning specific waveforms and processors. This is primarily due to several limitations in the *general* descriptions of the ambiguity function. While with modern computers it is not difficult to derive the ambiguity function for a specific waveform, it is not generally possible to derive a specific waveform by starting with an ambiguity function. There are exceptions to this for a specific class of waveforms (i.e., FM, phase-coded waveforms, pulse trains). Also the range dependence of various clutter and target echoes complicates the basic relationships of the above references if the target and clutter are not at the same distance from the radar. Finally a thorough and bounded description of the target and clutter environments is required for unique selection of appropriate waveforms. It is hoped that the descriptions of specific targets and clutter in Chaps. 5, 6, and 7 in conjunction with an analytical description of the environment (see Sec. 8.3) will remove some of this limitation.

8.1 MATCHED FILTERS

In the earlier discussions of signal detectability and the corresponding radar range equations in Chaps. 1 through 4, it was emphasized that the

signal-to-noise ratio should be maximized. The receiver transfer characteristic which achieves this end for white noise was derived by North
[269] and is called the "matched filter" for the transmit waveform (see
[378; 354, sec. 9; 71, chap. 2].

If the transmit waveform is represented by $f(t)$, its Fourier transform
is

$$F(\omega) = \int_{-\infty}^{\infty} \exp(-j\omega t) f(t) \, dt$$

If the receiver transfer function is $H(\omega)$, the output signal of the receiver
prior to envelope detection is

$$g(t) = \int_{-\infty}^{\infty} [\exp(j\omega t)] F(\omega) H(\omega) \, df \tag{8-1}$$

Let $g(t_0)$ be the maximum value of $g(t)$. The power spectrum of the
noise at the filter output is

$$G(\omega) = \frac{N_0}{2} |H(\omega)|^2 \tag{8-2}$$

where $N_0/2$ is the noise spectral density in watts/Hz at the filter input.
The factor one-half occurs because both negative and positive frequencies will be used in the analysis, and the usual definition of noise
density considers only positive frequencies. The average noise output
power is then

$$N = \frac{N_0}{2} \int_{-\infty}^{\infty} |H(\omega)|^2 \, df$$

The energy of the input signal can be written

$$E = \int_{-\infty}^{\infty} f^2(t) \, dt = \int_{-\infty}^{\infty} |F(\omega)|^2 \, df \tag{8-3}$$

The problem of receiver design is to maximize the ratio of the
magnitude of the filter peak power output for a given input signal energy
and noise power output.

$$\frac{|g(t_0)|^2}{EN} = \frac{\left| \displaystyle\int_{-\infty}^{\infty} F(\omega) H(\omega) \exp(j\omega t_0)\, df \right|^2}{\dfrac{N_0}{2} \displaystyle\int_{-\infty}^{\infty} |F(\omega)|^2\, df \displaystyle\int_{-\infty}^{\infty} |H(\omega)|^2 df} \tag{8-4}$$

This can be accomplished by use of Schwarz' inequality.

$$\left| \int_{-\infty}^{\infty} x(\omega)\, y(\omega)\, d\omega \right|^2 \leq \int_{-\infty}^{\infty} |x(\omega)|^2\, d\omega \int_{-\infty}^{\infty} |y(\omega)|^2\, d\omega \tag{8-5}$$

From this inequality, it follows immediately that

$$\frac{N_0\, |g(t_0)|^2}{2EN} \leq 1$$

and the ratio is equal to unity only if

$$H(\omega) = KF^*(\omega) \exp(-j\omega t_0) \tag{8-6}$$

where the asterisk denotes the complex conjugate, t_0 is the time delay to make the filter physically realizable, and K is a gain constant. As a result the peak signal-to-noise power ratio is

$$\frac{2E}{N_0} = \frac{2\,(\text{signal energy})}{\text{noise spectral energy density}}$$

Since there has been no specification of the waveform represented by $f(t)$, the output signal-to-noise ratio is independent of the shape or complex modulation of the waveform as long as the noise is white. In later sections, it will be shown that the *optimum* filter for detecting targets in strong clutter signals may not be the matched filter if the clutter location or spectrum is known. The impulse response function $h(t)$ of the filter described by Eq. (8-6) is

$$h(t) = K_2 f^*(t_0 - t) \tag{8-7}$$

In other words, the impulse response is a delayed time *image* or time inverse of the waveform multiplied by a simple gain constant. Since the output of a filter is the convolution of the input signal and the impulse response, the matched filter output $g_0(t)$ can be expressed as

$$g_0(t) = \frac{1}{T} \int_{-T/2}^{T/2} f(\tau) f(\tau + t_0 - t)\, d\tau$$

Thus in the absence of noise, the waveform is a time-shifted replica of the autocorrelation of the input signal.

If the interfering noise is gaussian and does not have a flat spectrum (is not white) but can be described by a power density spectrum $N(\omega)$, then the general optimum filter can be described [135, 71].

$$H_0(\omega) = \frac{KF^*(\omega) \exp(-j\omega t_0)}{N(\omega)} \tag{8-8}$$

Under these conditions the matched filter also maximizes the probability of detection for a given false alarm rate.

The matched-filter output being the autocorrelation function of the transmit waveform is a general result. This indicates that the matched filter may also be implemented as a cross-correlator between the echo from a point target and a time-delayed and doppler-shifted replica of the transmit signal.

It might seem that the matched filter might be difficult to realize. This is not the case for many common waveforms since it is general practice to construct a filter that yields within 1 dB of the sensitivity of the matched filter. Thus, the assumption that a matched filter is used is the basis of the range equations of Chap. 2. If compromises are made, the loss in sensitivity should be included in those equations.

As an example, consider the rectangular pulse of duration τ. The transfer function of the matched filter is the complex conjugate function of the spectrum of the signal [Eq. (8-6)]. Since the gain constant K and the time delay t_0 are of little importance, $H(\omega) = F^*(\omega)$. Thus for a rectangular video pulse, the magnitude of the filter gain is

$$|H(\omega)| = K\tau \left| \frac{\sin \omega\tau/2}{\omega\tau/2} \right| \tag{8-9}$$

Figure 8-1*A* shows a surprisingly simple matched filter with the response function of Eq. (8-9). When the rectangular video pulse is inserted into the filter, there is a linear rise at the output for time τ and then a linear fall to zero as shown on the right of Fig. 8-2. This is, of

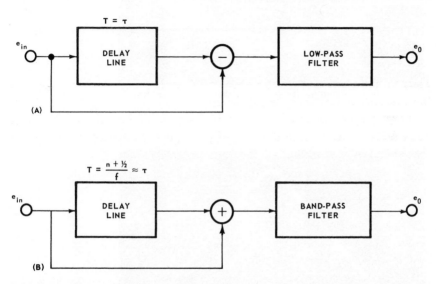

figure 8-1 *Matched filter for rectangular pulse of duration τ. (A) Bipolar video (two required); (B) IF pulse (carrier frequency = f).*

course, the autocorrelation function of the pulse neglecting time delays, losses, and gain terms.

Since pulsed radar signals are modulations ot an RF carrier, the single video pulse is not a sufficient representation of the signal unless a *homodyne* system with two quadrature channels and two matched filters is implemented when the receiver conversion is made from RF or IF to video signals. This subject will be discussed in Secs. 12.6 and 14.8.

Unless quadrature channels are instrumented, the maximum detection efficiency is obtained with band-pass filters at RF or IF. The matched filter for a rectangular pulsed sinusoid is shown as Fig. 8-1*B*. The filter must have a high Q and the length of delay line must be exactly an odd half cycle of the period of the carrier. If τ is large compared to a period of the carrier, the peak output of the filter will be within a few percent of the theoretical maximum [322]. It should

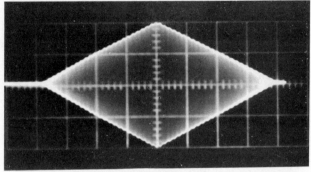

MATCHED-FILTER IF RESPONSE FOR STRONG SIGNAL
IF = 3.16 MHz

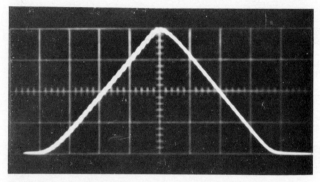

MATCHED-FILTER DETECTED VIDEO FOR STRONG SIGNAL

figure 8-2 *Filter responses for rectangular input pulse. Sweep speed = 1
μsec/cm, pulse length = 4 μsec. (Courtesy of S. A. Taylor.)*

also be noted that the impulse responses of both of the IF and video
filters are the time images of the waveforms to which they are
matched.

An experimental result with the IF matched filter is shown on the
top photo of Fig. 8-2. The S/N out of this filter was 0.9 dB (0.88 dB
theoretical) higher than with a single-tuned band-pass filter having the
optimum bandwidth of $0.4/\tau$.

It should be noted that maximizing predetection S/N maximizes detectability in noise only if the system is linear (a square-law or higher
order detector will often increase the output S/N but will not enhance

detectability [392, pp. 117, 167]. As stated previously, S/N in this book refers to a predetection power ratio unless stated otherwise.

The previous discussions showed that S/N in a matched filter receiver is independent of the transmitted waveform. Until about 1950 power amplifier transmitters were not generally available, and most transmitted waveforms were either CW or sine wave pulses. The advent of practical power amplifier chains led to an interest in the transmission of pulses with complex modulation and *pulse-compression* systems. The primary object of pulse compression was to transmit high energy with a long pulse and simultaneously to obtain resolution corresponding to a short pulse. One of the earlier attempts at pulse compression utilized an all-pass filter in the transmit chain to stretch a short pulse input. The all-pass filter had a nearly uniform amplitude response but a quadratic phase shift response versus frequency. A matched filter was then used to obtain the *pulse compression*. The properties of this waveform are best explained after an introduction to ambiguity functions.

8.2 THE RADAR AMBIGUITY FUNCTION

The study of radar waveforms would be quite simple if there were not a relative radial velocity between the radar, target, and the environmental interference (clutter). However, when there is a significant doppler shift, the reflections from even a point target are no longer replicas of the transmitted waveform. As a result the output of the *stationary target,* or zero-doppler, matched filter is not the autocorrelation function of the transmit waveform when there is relative motion. In addition the response from a second target or clutter at a slightly different range may appear at the matched-filter output when the desired target response is at its peak value. This overlap of signals will occur when the time extent of the waveform is greater than the differential time delay between the targets. As a result a special set of mathematical functions has evolved to allow interpretation of the output of a signal processor either when there is a target with a significant radial velocity or when multiple targets are present. These functions are called time-frequency autocorrelation functions or ambiguity functions and are based on the text by Woodward [407].

The time frequency autocorrelation function neglecting the time delay to the target can be written in symmetrical form [345]

$$\theta_u(\tau,\nu) = \int u\left(t - \frac{\tau}{2}\right) u^*\left(t + \frac{\tau}{2}\right) \exp(-j2\pi\nu t)\,dt \tag{8-10}$$

or in the more conventional nonsymmetrical form[1]

$$\chi_u(\tau,\nu) = \int u(t)\, u^*(t + \tau)\, \exp(-j2\pi\nu t)\,dt \tag{8-11}$$

where $u(t)$ is the transmit waveform, $u^*(t)$ is its complex conjugate, τ is the time displacement variable, and ν is the doppler shift. The subscript u denotes that the receiver filter is matched to the transmit waveform. The subscript uv will be used when the receiver is not matched to the transmit waveform.

Since the radar detector at the output of the matched filter usually removes the phase information, the function of interest is generally

$$\psi_u(\tau,\nu) = |\chi_u(\tau,\nu)|^2 = |\theta_u(\tau,\nu)|^2$$

$$= \left|\int u(t)\, u^*(t + \tau)\, \exp(-j2\pi\nu t)\,dt\right|^2 \tag{8-12}$$

This function is what is commonly referred to as the *ambiguity function*, and its plot is known as the *ambiguity diagram* (preferably a three-dimensional representation). In this notation $\tau = 0$ and $\nu = 0$ correspond to the time delay and doppler displacement of the target of interest; i.e., the ambiguity diagram origin is centered on the target location in the range-doppler plane or range-doppler space.

For the case of matched-filter reception, the origin of the ambiguity function may be thought of as the output of the matched filter which

---

[1]All integrals without limits are assumed to have either infinite limits or for at least the entire radar observation interval. Since there is no consistency in the notation for these functions, the symbols of many of the references have been altered to make this text more self-consistent. While the symbol ϕ is often used for doppler shift, it is used here for elevation angle, etc.

is tuned in time delay and frequency shift to the signal reflected from an idealized point source target. In this case τ becomes the time delay relative to the target position, and ν becomes the doppler relative to the target doppler [7]. The filter responses far from the desired target location are often called distal ambiguities. These are undesirable if other targets or clutter may appear at these locations. The response at $\psi_u(\tau, 0)$ is then the filter response to reflections at a different range but at the same doppler as the target. Also $\psi_u(0, \nu)$ is the response to reflections at the same range as the target but with other doppler shifts.

By the use of Parseval's theorem, the ambiguity function can also be written

$$|\chi_u(\tau, \nu)|^2 = \left| \int F(f + \nu) F^*(f) \exp[-j2\pi f\tau] \, df \right|^2 \tag{8-13}$$

where $F(f)$ is the Fourier transform of the wave form $u(t)$. It is assumed that the received echoes are mixed with the transmit carrier frequency, and the only frequency variable of interest is the doppler shift. An important inequality is

$$\psi_u(\tau, \nu) \leq \psi_u(0, 0) \tag{8-14}$$

that is, the ambiguity function is a maximum at the origin. There are many other relationships in the literature that are important for specific studies [295, 321]. The following are general assumptions implied in the development of the ambiguity function and its application in this text [359]:

1. Point targets are assumed. This allows the usual convention of normalizing the peak signal power P_s to unity at the origin of the ambiguity function

$$P_s = \psi_u(0, 0)$$

2. Target acceleration is assumed to be negligible.

$$a \ll \frac{\lambda}{T_d^2} \tag{8-15}$$

where a is target acceleration, λ is carrier wavelength, and T_d is the signal duration. See Kelly and Wishner [209] for extensions to accelerating targets.

3. Mismatch of the envelope of the target echo and the transmit waveform due to high relative velocities is negligible or

$$\frac{2v}{c} \ll \frac{1}{BT_d} \tag{8-16}$$

where v is the radial target velocity relative to the radar, c is velocity of light, and B is the signal bandwidth. (See Remley [308] for the effect when this constraint is violated.)

4. All signals are narrow band such that the term doppler shift is meaningful. This can be expressed as $B \ll f_0$, where f_0 is the carrier frequency. Rihaczek [321] shows that this is not a very severe restriction.

5. There is a small percentage difference in the range from the radar to the various targets in the region of interest of the ambiguity diagram. This widely used simplification is a poor approximation for CW radar or pulse-train transmissions when the target of interest is distant from the radar and clutter or undesired targets are close to the radar. Analytic expressions to account for the range dependence are found in Chaps. 9, 10, and 11.

There are several other properties which simplify the use of the ambiguity function. It is convenient to normalize the signal $u(t)$ to unit energy

$$\int |u(t)|^2 dt = 1 \tag{8-17}$$

It is also useful to define a cross-ambiguity function $\psi_{uv}(\tau, \nu)$ for use when the receiver filter is mismatched to the transmit waveform [364, 365, 359]

$$\psi_{uv}(\tau, \nu) = |\chi_{uv}(\tau, \nu)|^2 = \left| \int u(t) v^*(t + \tau) \exp[-j2\pi\nu t]dt \right|^2 \tag{8-18}$$

where the subscript u refers to the waveform properties and the subscript v refers to the mismatched filter. Then with point targets and the customary signal normalization

$$\iint \psi_u(\tau, \nu) \, d\tau d\nu \ = \ \iint \psi_{uv}(\tau, \nu) \, d\tau d\nu \ = \ 1 \qquad (8\text{-}19)$$

This important relation states that the total volume of the ambiguity function for all waveforms and filters is constant over the prescribed τ and ν space and must be less than unity over any finite region. Equation (8-19) is especially important in clutter and multiple target environments. The undesired portions of the ambiguity function (the range and doppler ambiguities away from the origin) can be rearranged and hopefully placed in a region that is of little importance (i.e., at doppler frequencies that are higher than expected from any target of interest). It will be shown that this can be accomplished if the approximate range and doppler separation of targets and clutter are known.

Before expanding on the further properties of ambiguity functions for various waveforms, it seems worthwhile to pause and illustrate the derivation of the ambiguity function for a linear FM pulse (chirp). This particular derivation is an abstract of an analysis by P. J. Luke. Similar derivations can be found in the literature [214, 71, 39, 125].

Ambiguity Function for Linear FM Pulse

Consider a signal represented in Woodward's notation as follows:

$$s(t) \ = \ \text{rect}\left(\frac{t}{\tau'}\right) \exp[j2\pi(f_0 t \ + \ \tfrac{1}{2} k t^2)] \qquad (8\text{-}20)$$

where $\text{rect}(Z) = 1 \quad \text{if} \quad |Z| < \tfrac{1}{2}$
$\qquad\qquad\quad\ = 0 \quad \text{if} \quad |Z| > \tfrac{1}{2}$

τ' is then the pulse length, and $f_0 + kt$ is the instantaneous frequency. This can be rewritten in the form $s(t) = u(t) \exp[j2\pi f_0 t]$, and $u(t) = \text{rect}(t/\tau') \exp[j\pi k t^2]$ is the complex envelope function.

The time-frequency autocorrelation function is computed from Eq. (8-11). As previously stated, it is customary to normalize the signal to unit energy. In the present case this is accomplished by dividing $u(t)$ by $\sqrt{\tau'}$. Substituting in Eq. (8-11), one obtains

$$\chi_u(\tau, \nu) \ = \ \frac{e^{-j\pi k\tau^2}}{\tau'} \int_{-\infty}^{\infty} \text{rect}\left(\frac{t}{\tau'}\right) \text{rect}\left(\frac{t + \tau}{\tau'}\right) \exp[-j2\pi(k\tau + \nu)t] \, dt$$
$$\qquad (8\text{-}21)$$

For $0 \leq \tau \leq \tau'$ the limits of integration are $-\tau'/2$ and $(\tau'/2) - \tau$, since the integrand is zero outside this range. Thus for $0 \leq \tau \leq \tau'$, one can obtain

$$\chi(\tau, \nu) = e^{j\pi\nu\tau} \frac{\sin[\pi(k\tau + \nu)(\tau' - \tau)]}{\pi(k\tau + \nu)\tau'} \tag{8-22}$$

For $-\tau' \leq \tau \leq 0$, the limits are $-\tau'/2 - \tau$ and $\tau'/2$, and

$$\chi(\tau, \nu) = e^{j\pi\nu\tau} \frac{\sin[\pi(k\tau + \nu)(\tau' - |\tau|)]}{\pi(k\tau + \nu)\tau'} \, \text{rect}\left(\frac{\tau}{2\tau'}\right) \tag{8-23}$$

which is valid for all τ.

The time-autocorrelation function may be obtained from Eq. (8-23) by setting $\nu = 0$; thus

$$\chi(\tau, 0) = \frac{\sin[\pi k\tau\tau'(1 - |\tau|/\tau')]}{\pi k\tau\tau'} \, \text{rect}\left(\frac{\tau}{2\tau'}\right) \tag{8-24}$$

The right side of Eq. (8-24) resembles the expression usually given for the compressed pulse $(\sin\pi k\tau\tau')/\pi k\tau\tau'$ which is actually the response of an ideal lossless delay equalizer. The differences are the factor rect $(\tau/2\tau')$ which makes the true matched-filter output zero for $|\tau| > \tau'$ and the argument of the sine function which in Eq. (8-24) contains an additional factor $(1 - |\tau|/\tau')$.

The zeros of Eq. (8-24) occur at those values of τ for which $k\tau(\tau' - |\tau|)$ is an integer. Since this is a quadratic expression in τ, the zeros are not uniformly spaced. The values of τ for which Eq. (8-24) has zeros are given by

$$|\tau| = \frac{1}{2}\tau'\left[1 \pm \sqrt{1 - \frac{4n}{D}}\right] \tag{8-25}$$

where $D = k(\tau')^2$ is the compression ratio. Note that the maximum integer n is less than or equal to one-fourth the compression ratio.

Simple Pulse Ambiguity Function

The ambiguity function for a simple rectangular pulse is

$$\psi(\tau, \nu) = \left[\frac{\sin^2 \pi\nu(\tau' - |\tau|)}{(\pi\nu\tau')^2}\right]\text{rect}\left(\frac{\tau}{2\tau'}\right) \tag{8-26}$$

where τ' is the pulse duration. (This notation is used where pulse length might be confused with the time delay variable τ.) The function rect is the pulse function defined by Woodward [Eq. (8-20)]. The ambiguity function is normalized so that $\psi(0,0) = 1$.

The linear FM chirp pulse in which the frequency varies at the rate k Hz/sec has the normalized ambiguity function expressed in Eq. (8-23).

Comparison of Eqs. (8-23) and (8-26) shows that the only difference is that ν in Eq. (8-26) is replaced by $k\tau + \nu$ in Eq. (8-23). The result is that the surface represented by Eq. (8-23) may be obtained from that represented by Eq. (8-26) by translating every point (τ, ν) to the point $(\tau, k\tau + \nu)$. Such a translation is a *shear* of the function parallel to the ν axis (see Figs. 8-3 and 8-4). A cross section of the surface at constant τ is unaltered in shape or size but is shifted in the ν direction by an amount proportional to τ. On the other hand, a cross section of the simple pulse ambiguity function at constant ν corresponds to a cross section of the FM ambiguity function along a line in the (τ, ν) plane having slope k with respect to the τ axis. These two cross sections have the same shape and for any given τ have the same amplitude, but the second is stretched relative to the first by the factor $\sqrt{1 + k^2}$.

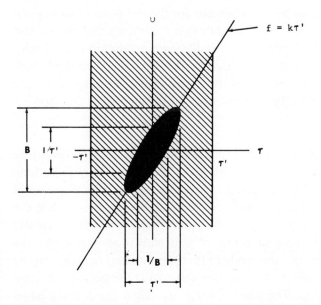

figure 8-3 *Ambiguity diagram for frequency-modulated pulse of sine wave.*

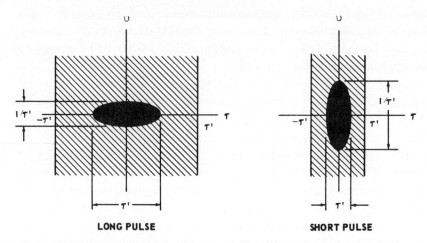

LONG PULSE   **SHORT PULSE**

figure 8-4 *Two-dimensional ambiguity diagram for a single pulse of sine wave. (After Siebert, IRE [345].)*

The direction of the shear is determined by the sign of k. For k positive (frequency increasing) the shear is toward negative ν for positive τ. For k negative (frequency decreasing) the shear is toward positive ν for positive τ.

A *top view* of the ambiguity diagrams for the FM pulse of bandwidth $B = k\tau'$ and for the single pulse of duration τ' is shown in Figs. 8-3 and 8-4. The dark areas generally represent the regions where the ambiguity function is above an arbitrary (say 3 dB) level. The hatched areas represent the regions where $\psi(\tau,\nu)$ has a smaller but nonzero value. The hatched areas are often called the sidelobes of the ambiguity function.

Other Waveforms

Ambiguity diagrams for many other common radar waveforms can be found in the literature [71; 358; 354, sec. 10]. The coherent pulse-train waveform is generated by sampling a sinusoidal carrier with N pulses of duration τ'. The interpulse period (constant in this case) is T, and the time duration of the train is T_d as shown on Fig. 8-5A. The frequently illustrated ambiguity diagram for $N = 5$ is shown as Fig. 8-5B. The majority of volume of the ambiguity diagram can be seen to be distributed into a number of small area spikes that are periodic in range delay τ and doppler f_d. The hatched areas are called the doppler ridges and are shown in more detail in Chap. 11. Note that there is a clear area of $T - 2\tau'$ between the ridges in the range dimension (as long as $T \gg 2\tau'$). If the undesired echoes can be placed in this region by choice of T and τ', they will not yield any response at the receiver output when the desired target is observed. The undesired echoes are then completely *resolved*

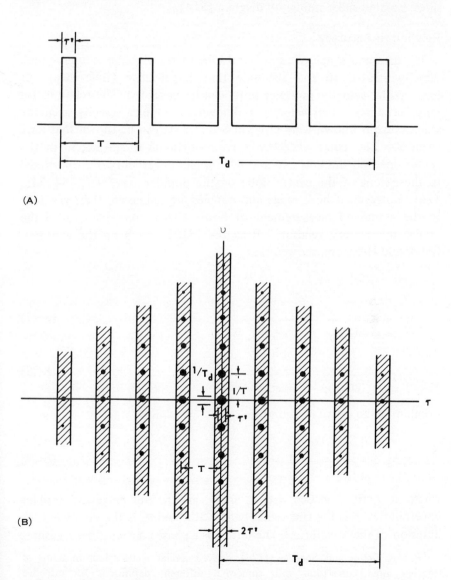

figure 8-5 *Ambiguity diagram for a pulse train consisting of five pulses. (A) Waveform; (B) ambiguity diagram for (A). (After Siebert [345].)*

from the target. The popularity of pulse-train or burst waveforms and pulse doppler processors for multiple target environments results from this interference-free region and the small area of the central lobe of the ambiguity function. It can be shown that there is no waveform that has a single isolated spike ambiguity diagram [314].

Relation to Accuracy

The area and shape of the central lobe of the ambiguity function are directly related to the accuracy that target-range (time delay) or target-radial velocity (doppler shift) can be measured. The relations for range accuracy when velocity is known or doppler velocity accuracy when range is known were given in Sec. 1.8. It can be shown that for a given S/N the range uncertainty is proportional to the extent of the central lobe on the time axis and the doppler uncertainty is proportional to the extent of the central lobe on the doppler axis [407, 354, 345, 124]. However, if both range and doppler are unknown, there is a limit to the combined measurement of both. This is sometimes called the "radar uncertainty relation." Rihaczek [317], based on the works of Gabor and Helstrom, showed that

$$\beta^2 t_e^2 - \alpha^2 \geq \pi^2$$

$$\sigma_\tau = \frac{1}{\beta (2E/N_0)^{\frac{1}{2}} (1 - \alpha^2/\beta^2 t_e^2)^{\frac{1}{2}}} \qquad (8\text{-}27)$$

$$\sigma_d = \frac{1}{t_e (2E/N_0)^{\frac{1}{2}} (1 - \alpha^2/\beta^2 t_e^2)^{\frac{1}{2}}} \qquad (8\text{-}28)$$

Combining the above relationships

$$\sigma_\tau \sigma_d \leq \frac{1}{\beta t_e (2E/N_0)} \left[1 + \frac{\alpha^2}{\pi^2} \right] \qquad (8\text{-}29)$$

where, as defined in Sec. 1.8, σ_τ and σ_d are the rms range and doppler uncertainties, β is the rms signal bandwidth, and t_e is the effective time duration of the waveform.[1] The term α is a phase constant that measures

---

[1] Fowle, Kelly, and Sheehan [124] derived similar relationships in terms of frequency-time correlation with somewhat different definitions for effective bandwidth and time duration. They also included a discussion of the effect of rotation of the axes of the ambiguity function on accuracy.

the linear FM content of the signal [317]. A high FM content can be shown to increase the uncertainty in simultaneous measurement of range and velocity. Unless $2E/N_0$ or S/N is extremely large, it is difficult to tell where the target is on the diagonal ridge of the FM ambiguity function. The resolution measurement properties of the linear FM waveform have been further clarified in a later paper by Rihaczek [318].

If there is no linear FM as in a uniformly spaced single frequency pulse train or other amplitude modulated waveforms ($\alpha \approx 0$), the uncertainty decreases as the time-bandwidth product $\beta\tau_e$ is increased. As in clutter rejection or countermeasures immunity, the time bandwidth product is a useful measurement of the *quality* of a waveform.

8.3 THE RADAR ENVIRONMENTAL DIAGRAM

The ambiguity diagram has been used as a tool for evaluating the choice of waveform in specific environments. It has proven quite successful for suggesting specific waveforms for resolving multiple targets in range or range rate and in certain highly specific clutter environments. With a surveillance radar system, which may simultaneously encounter several types of clutter with various spectral characteristics at unspecified locations, the ambiguity diagram alone is generally insufficient to make the choice of waveform. "In the extreme case, all signals (waveforms) are equally good (or bad) as long as they are not compared against a specific radar environment" [71, p. 70].

Chapters 6 and 7 have shown that different types of clutter have considerably different spectral, spatial, and amplitude distributions. Since the power reflected from land areas, sea surfaces, chaff, and precipitation often is far in excess of that from the target, it is necessary to choose the transmit waveform on the basis of these distributions. Fortunately, there are inevitably some bounds on the location, velocity, and intensity of clutter. The environmental diagram shown in Fig. 8-6 is a pictorial representation of a *clutter threat* or *clutter model* [264]. It can also be used to help suggest the appropriate waveform for a particular class of radars. The diagram in this form is useful for an air defense surveillance radar located at a coastal site.

The limits on the radial velocity of various types of clutter are given on the ordinate, while the range extent is indicated along the abscissa. In this example the scale is broken to allow all four major types of

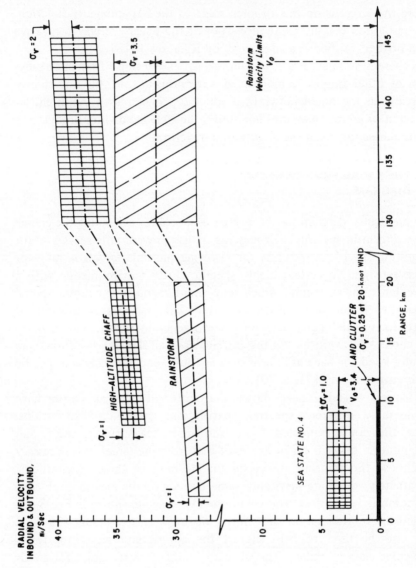

figure 8-6 *Environmental diagram for air-defense radar*

clutter to be presented simultaneously. Land clutter is shown extending from the origin to 10 km in range to represent the radar horizon for the particular location. The vertical extent of the bar is the standard deviation of the velocity spectrum for land-clutter echoes from a wooded area with a 20-knot surface wind.[1] Density of the cross-hatching or shading can represent the reflectivity of the clutter of interest. Land clutter is shown as a solid bar to indicate that it generally has the highest reflectivity of any form of clutter seen by a surface surveillance radar.

Sea clutter is modeled on the figure for a state four sea as a relatively intense type of clutter with a spectrum standard deviation of 1.0 m/sec. It is shown centered at 3.4 m/sec mean velocity. This is a reasonable value when looking directly into the wind or waves. On the other hand, the mean velocity may be zero when looking crosswind. The range of mean velocities $\pm V_0$ is given by the vertical dashed line to indicate that the cross-hatched region can be centered anywhere between ± 3.4 m/sec. The cross-hatching is more dense near the radar to show that the reflectivity near the radar is higher than at the lower grazing angles near the horizon. It becomes obvious that these models are derived on a statistical basis. Whoever specifies a clutter threat model essentially defines under what conditions the radar system must meet full specifications. While there is a finite probability of sea state seven occurring, it would be costly to expect full performance in that environment.

Figure 8-6 also models the clutter threat for a rainstorm which may occur 5-10 percent of the time. Mean velocities of 28 m/sec are not uncommon at moderate altitudes. As was the case for sea clutter, crosswind values of the mean velocity of precipitation clutter will drop to zero. The increase in the spectral width with range shows the combined effect of wind shear and the increasing vertical extent of the antenna beam at longer ranges. The standard deviation of the spectrum can be obtained from Chap. 6 when the beamwidth is known. Similarly, the mean radial velocity V_0 can be approximated by a linear increase with the height of the beam center or with slant range.

Finally, a high-altitude chaff corridor is illustrated in the same format. To be effective in confusing the radar, its reflectivity should be somewhat higher than a rainstorm. At all altitudes chaff moves at

---

[1] The mean velocity of the trees V_0 is obviously zero.

the velocity of the prevailing winds. Since the vertical extent of a chaff corridor is usually less than that of a rainstorm, the standard deviation of the spectrum will usually have a lower value at long ranges than for rainstorms.

In all of the examples given, the spectral width has been expressed in meters/second and must be multiplied by $2/\lambda$ to convert to doppler frequency. The relationship between the physical motion of the clutter scatterers and the spectral width has been found to conform to the doppler equation throughout most of the microwave region of interest.

The cross-hatching for rain and chaff is shown on the figure to have lower density in the shear regions, indicating that the power spectral density of the echoes is lower for a given reflectivity than for signals with a narrow spectral width.[1] It should be noted that normalized reflectivity per unit frequency rather than received power spectral density is indicated by the cross-hatching.

Target threat models can also be superimposed on this diagram, with the origin of the figure representing the radar location. This technique of presentation, sometimes called an *R-V* diagram, has been used to show the range and velocity bounds on a ballistic target with impact points near the radar site.

In its simplest form, the environmental diagram gives a pictorial description of the clutter and should aid the radar system designer by giving the numerical limits of the clutter characteristics for his design. If the carrier frequency is defined, the ordinate can be converted to hertz. If the radar performance in severe clutter is also desired (e.g., higher sea states, wind speeds, etc.) a second environmental diagram can be drawn.

Just as the ambiguity diagram is not a panacea for choosing waveforms in a multiple target environment, the environmental diagram will not suggest a unique waveform for maximum clutter rejection although it may suggest certain desirable characteristics of the waveform or the processing technique. For example, if land echoes are the only clutter threat, a waveform and receiver response function with a *notch* or null at zero velocity, such as one obtains with a moving target indicator (MTI), is obviously desirable. For other types of clutter the MTI (see Chap. 9) technique must be modified. It is well known that MTI systems designed for land use do not work very well on ships since the

---

[1] This is important in CW or pulse-doppler systems which may have a doppler filter bandwidth which is narrower than the clutter spectrum.

location of the MTI notch may be in error by the mean doppler of the sea clutter V_0 relative to ship's motion. *Clutter-locked* systems, in which the ship's (or aircraft's) velocity and V_0 are removed by compensating for the relative mean velocity of the clutter, have considerably improved the performance that can be obtained on moving platforms.

If rain and chaff are part of the clutter threat, the required width of the notch and the mean compensating frequency are often excessive especially at the higher carrier frequencies; and other signal processing techniques must be used.

Since the axes of both the ambiguity and environmental diagrams can be drawn to the same scale, they can be superimposed to reveal range-doppler regions where clutter energy will be received. Preferably, the ambiguity diagram (AD) should be a transparent overlay, as its origin represents the expected target range and velocity. This location may occur over a large part of the environmental diagram (ED). The clutter outputs of a matched-filter receiver will then appear with the range-doppler characteristics defined by the intersections of the nonzero portions of the two diagrams. High clutter outputs will correspond to superposition of the dense areas on both diagrams, while the occurrence of either one alone will not yield a clutter output. In evaluating a search system, the origin of the AD must be successively located on the ED at all target ranges and velocities within the specified threat model. An intersection of a high sidelobe on an AD with a low-density rainstorm is obviously less serious than an intersection with high-density (large σ_0) land clutter. While a quantitative measure of the clutter performance of a given waveform has not been worked out using the ED, it would seem reasonable that a computer program could be written to yield the performance obtained with a given waveform for various target ranges and velocities. Some analytical descriptions and references will be given in the next section.

The diagonal ridge on the AD for a linear FM search radar waveform will obviously intersect with the rainstorm or chaff returns at some locations. The clutter output is then proportional to the *width* of the ridge. Since the width of the ridge (say at the 3-dB points) is inversely proportional to the time-bandwidth product or pulse-compression ratio, the clutter output is minimized for high compression ratios.

With a stationary radar and a CW waveform (Chap. 10), there is no intersection of the narrow horizontal ambiguity ridge and the clutter regions on the ED for targets with radial velocities of greater than $\pm$ 40 m/sec. Thus, if target detection is not required in the doppler region corresponding to less than $\pm$ 40 m/sec, the theoretical clutter output of narrow-band doppler receiver filters is zero. As a result, the theoretical clutter rejection of a CW radar system is infinite. In practice, however, while the receiver output contains no high doppler-frequency clutter, the front end of the receiver often becomes saturated since it receives all of the clutter energy. CW system performance is generally limited by transmitter noise on the clutter echo, spillover signals, and the intermodulation products of all the clutter signals.

8.4 SPECIAL PROPERTIES OF AMBIGUITY FUNCTIONS

The previous section illustrated where different types of clutter fit onto the ambiguity plane. In a similar manner multiple-target models or ballistic missiles with wakes and tank fragments can be illustrated as an overlay for the ambiguity diagram. This section will outline some of the techniques for use of the ambiguity function for performance estimation in a complex environment.

Since the processors of interest are assumed to be linear to the output of the matched filter, the response from multiple targets or clutter can be obtained by superimposing the individual responses. The target responses must be shifted on the ambiguity plane according to their differential range and velocity relative to the target. When there are numerous scatterers as with most types of clutter, the individual responses will add powerwise. This is equivalent to adding the respective ambiguity functions.

It has been shown that for matched systems [401, 317, 314]

$$\int |\chi_u(\tau, \nu)|^2 d\tau \ = \ \int |\chi_u(\tau, 0)|^2 \exp(-j2\pi\nu\tau)\,d\tau \qquad (8\text{-}30)$$

Rihaczek [317] pointed out that this relation illustrates that the volume distribution of the ambiguity function in the doppler domain is completely determined by the value of the ambiguity function on the delay axis. Specifically the volume distribution in the doppler domain is given as the Fourier transform of the squared envelope of the ambiguity function on the τ axis or the squared envelope of the autocorrelation function of $u(t)$. Also the volume distribution is independent of the phase of the frequency spectrum.

Rihaczek also gives the dual relation to Eq. (8-30).

$$\int |\chi_u(\tau, \nu)|^2 d\nu = \int |\chi_u(0, \nu)|^2 \exp(-j2\pi\nu\tau) d\nu \qquad (8\text{-}31)$$

This relation states that the ambiguity volume distribution in the time-delay domain is determined by the value of the ambiguity function on the doppler axis. When the signal envelope is chosen, introducing phase modulation can merely redistribute the volume at each delay.

Thus it can be shown that the complex waveforms of later chapters which reduce the ambiguities in one domain will merely redistribute them into the other domain where they may or may not be detrimental.

A subject of considerable interest is whether waveforms can be selected where the significant portions of the ambiguity function are disjoint from the significant clutter regions on the environmental diagram for all values of target range and velocity. This subject will be discussed in connection with optimum waveforms in the next section. However, it is useful to point out some of the studies of waveforms that have ambiguity-free or *clear regions*.

Price and Hofstetter [295] have shown that there is an upper bound to the size of the *clear region* that can surround a central spike (at $\tau = \nu = 0$). The normalized area of this region is four. This can be seen most easily from Fig. 8-5B where there are four rectangles (T in width and $1/T$ in height) surrounding the spike (shaded black) at the origin. If the pulses of the train were impulses and the waveform had infinite duration, the vertical hatched area through the origin would have negligible value. This result implies that the maximum range-doppler area that can be unambiguously explored has unit area (i.e., the area between any four spikes). This result has been explored for other waveforms and rotations of the axes of the ambiguity diagram [71, 317]. Thus a pulse-train waveform may be desirable to search for or track targets if all undesired clutter or targets are known to be within $\pm T/2$ in range delay or $\pm 1/2T$ in doppler shift.

An extension of this result has been derived for a thumbtack ambiguity function which consists of an impulse of volume V_0 at the origin. It was shown [295] that this impulse is surrounded by a plateau of height V_0 and average sidelobe level $(V_0)^{1/2}$. The thumbtack ambiguity function results from the transmission of a noiselike waveform and can be approximated by a long duration random binary phase-coded waveform (Chap. 12) among others. This type of waveform is desirable when there are few undesired targets or clutter only over a portion of the ambiguity plane.

In the more general environment there is an average of the shaded areas of the ambiguity and environmental diagrams. If it is desired to quantify the *interference* due to the clutter, the clutter from a complex of n scatterers of average cross section $\bar{b}$ distributed (in the radar beam) can be described by a probability density function $P(t, f_d)$. Fowle, Kelly, and Sheehan [123, 124] derived several relationships for the signal-to-interference ratio S/C.

The interference energy can be written

$$E\,|e(\tau, \nu)|^2 \;=\; n\bar{b} \iint \psi(\tau - t, \nu - f_d)\, P(t, f_d)\, dt\, df_d \qquad (8\text{-}32)$$

where $\psi(\tau, \nu)$ is the single-target ambiguity function and (τ, ν) is the point in the time-frequency plane at which the output of the processor is observed (the target location). If a target of cross section σ_t is inserted at a point (τ, ν) the signal-to-clutter ratio is

$$\frac{S}{C} = \frac{\sigma_t}{n\bar{b} \iint \psi(\tau - t, \nu - f_d)\, P(t, f_d)\, dt\, df_d} \qquad (8\text{-}33)$$

where $C = E\,|e(\tau, \nu)|^2$ of Eq. (8-32). While this equation may be rewritten to solve for n and thus the maximum number of scatterers, it will suffice to state that $n\bar{b}$ is a measure of the clutter power within the beam and the product of the two terms in the integral represents the intersection of the ambiguity and environmental diagrams.

One special case of interest is an isolated target embedded in the clutter. Then

$$\frac{S}{C} = \frac{\sigma_t}{n\bar{b} \iint \psi(\tau, \nu)\, P(\tau, \nu)\, d\tau\, d\nu} \qquad (8\text{-}34)$$

If n is large, $\bar{b}$ is small and the clutter is uniformly distributed over time T_c and doppler shift W_c [123]

$$\frac{S}{C} = \frac{\sigma_t T_c W_c}{n\bar{b} \iint \psi(\tau, \nu)\, d\tau\, d\nu} \qquad (8\text{-}35)$$

where the region of integration includes the entire clutter space.

8.5 OPTIMUM WAVEFORMS FOR DETECTION IN CLUTTER

All of the waveforms discussed in Chaps. 9-14 are optimum for some particular clutter environment within the constraints of cost and complexity. The very short sinusoidal pulse waveform has a thin vertical ridge ambiguity diagram. This thin ridge will have a small common area with the distributed clutter regions of the environmental diagram. Similarly, a simple CW waveform may be the optimum for a surface radar if the targets of interest have a radial velocity in excess of 40 m/sec. The purpose of this section is to briefly refer to some of the numerous studies of optimization for clutter environments.

Optimization When the Relative Doppler Shift Is Zero or Unknown

The design of optimum processors for detecting signals in clutter has been approached from several directions [8, 1, 71, 277]. Urkowitz [380; 354, sec. 12.4] considered the form of the optimum receiver when the interference was entirely clutter and the target may have the same radial velocity as the clutter. For stationary clutter, the received power spectrum is identical to the power spectrum of the transmitted signal. That is, $N(\omega) = |F(\omega)|^2 = F(\omega) F^*(\omega)$. Then from Eq. (8-8) and neglecting time delay the *optimum* clutter filter transfer function can be written

$$H_{\text{opt}}(\omega) = \frac{K}{F(\omega)} \qquad (8\text{-}36)$$

where $F(\omega)$ is the Fourier transform of the received signal and K is a constant. Since this filter would have an infinitely wide total band pass, the analysis emphasized the more practical band-limited case. In this latter case, the improvement I in signal-to-clutter ratio S/C was found to be proportional to the receiver bandwidth in the absence of receiver noise. In the usual case when noise is present, it has been shown by Manasse [243], Rihaczek [319], Brookner [54], Urkowitz [383], and others that increasing the signal bandwidth and using a matched filter is a better solution than merely increasing bandwidth when there is no doppler separation of targets and clutter or when the separation is unknown.

In Manasse's study it was assumed that the clutter consisted of a large randomly distributed ensemble of very small independent point scatterers. He showed that the transfer function of the optimum filter when noise is present is given by

$$H(f) = \frac{F^*(f)}{(N_0/2) + k\,|\,F(f)\,|^2} \tag{8-37}$$

where $N_0/2$ is the (additive) receiver noise spectral density, k is a constant, and $|\,F(f)\,|^2$ is the energy spectrum of the received signal. Manasse also showed that the single-pulse optimum signal-to-total interference ratio can be expressed as

$$\left(\frac{S}{N+C}\right) = A^2 \int \frac{|\,F(f)\,|^2}{N_0/2 + k\,|\,F(f)\,|^2}\,df \tag{8-38}$$

The constants A and k are determined by the nature of the target and clutter echoes and the geometry. Two important conclusions were drawn for the single pulse case from the relationship of Eq. (8-37).

1. When no clutter is present $S/(N+C)$ depends only on signal energy-to-noise density as shown by matched-filter theory.
2. When noise is negligible $N_0 = 0$ or when clutter is dominant $C \gg N$, $(S/N + C)_{opt}$ depends only on effective system bandwidth and does not depend on transmitted pulse energy (i.e., the Urkowitz [380] result).

It was also shown that when the spectrum of the pulse is flat as with a rectangular envelope linear FM or chirp pulse, a further improvement is obtained.

Rihaczek [319] expanded the work of Urkowitz, Manasse, and others and showed that for negligible differential velocity between the target and clutter the matched filter is nearly the optimum filter whenever the clutter-to-noise ratio is less than five. He also studied the case when there was a differential doppler ν_0 between target and clutter. In that case the bracketed term in the denominator of Eq. (8-38) is replaced by $|F(f - \nu_0)|^2$. The optimum waveform is then obtained when $F(f)F(f - \nu_0) = 0$ for $F(f) \neq 0$. This is essentially stating that the ambiguity function of the signal should have no volume at those portions of the time-frequency plane where the doppler shift is ν_0.

Analytical and numerical results of this type were also obtained by Brookner [54] who emphasized that the existence of an optimum filter was dependent on the presence of nulls in the signal spectrum. A knowledge of the target-clutter doppler is extremely important in taking advantage of the nulls. Westerfield, Prager, and Stewart [401] presented a study of waveform optimization for the noise-free case but where the target has a doppler shift and the clutter has a spectral spread. Using the ambiguity function, they show that the spectrum of the transmit signal should either be much narrower or much wider than the clutter spectral spread.

Van Trees [385] using a *hypothesis testing* approach showed that waveform design is the most important step in combating reverberation (the sonar man's clutter). The second step is to consider how much would be gained by optimum filter design rather than a matched filter. Van Trees' work was expanded by Urkowitz [383] and some of the restrictions on the clutter distributions were removed.

This discussion has neglected the analyses of the multiple-target environment in the preceding references. The reduction of close-in (proximal) range or doppler *sidelobes* of the ambiguity diagram for a given waveform is reserved for later chapters. One special case of a nonlinear FM waveform with inherently low-range sidelobes is illustrated in Sec. 8.7.

Optimization Based on Relative Doppler Shift

The previous analyses emphasized the optimization for waveforms which were constrained to a single envelope and the target-to-clutter differential was not known. There has also been a considerable effort in optimizing performance in clutter when there is known to be a significant doppler shift. The best known example is the MTI processor (Chap. 9) where a null in the spectrum of the received echo is placed at the mean doppler velocity of the clutter. It will be shown that rejection of clutter signals having finite spectral width is improved as the number of cancellation stages is increased. Thus, the optimum *waveform* becomes the coherent pulse burst or pulse train (Chap. 11); and the MTI processor is a clutter-rejection filter rather than a matched filter. However, it will be shown that when target velocity is unknown, increasing the number of pulses (the total energy) does not improve the signal-to-noise ratio when the MTI processor alone is used. Thus, a further optimization is to cascade an MTI clutter-rejection filter with a matched filter for the pulse train. The best performance is obtained when the time delay to the target is less than the interpulse period (see Kroszezynski [221] and Kaiteris and Rubin [200]).

The optimization of pulse-train waveforms and processors has been studied in a number of excellent references [8, 92, 200, 309, 316, 330, 331, 359].

Many of the specific cases will be examined in Chap. 11. However, it is worthwhile at this time to point out a pitfall in using the basic ambiguity function to compute the performance of a long pulse-train waveform in an extended clutter environment (i.e., *uniform* rain). This results from the general assumption given in Sec. 8.2 that the echo from a given scatterer at any location in the time-frequency plane of the ambiguity diagram yields the same power at the receiver. With this assumption the range dependence $P_R \sim 1/R^n$ is neglected. It can be seen from the data in Chaps. 5 through 7 that the value of the exponent n is four for point targets, two for uniform volume scatterers and three to four for surface clutter. The range-law dependence can severely degrade the performance obtained from long duration waveforms when distant targets must compete with *ambiguous* close-in clutter. One case of a constant amplitude, constant frequency, uniformly spaced pulse train is illustrated on Fig. 8-7. The ordinate is the improvement factor I or the output signal-to-clutter ratio of the matched filter divided by the input signal-to-clutter ratio. The volume clutter in this example extends from the radar to 2 msec of time delay. The abscissa is the number of pulses transmitted times the interpulse period T. The target range delay for this figure is 1 msec and the interpulse period $T_0 = 77, 80,$ or 83 μsec. The number of pulses that can be transmitted in the round-trip time to the target is 13. The radar-target doppler is 18,750 Hz which is at the optimum doppler for the 80-μsec interpulse period (curve B). However, when the target echo from the first pulse enters

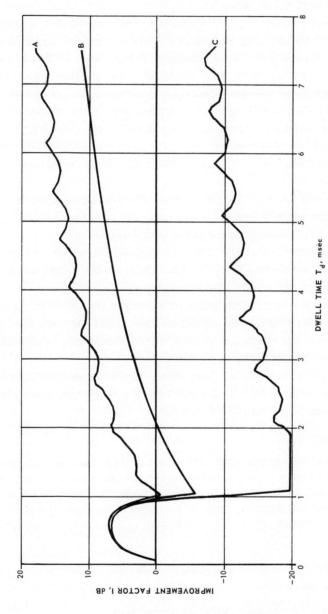

figure 8-7 *Improvement factor for uniform pulse train vs. dwell time for target at 1-msec time delay. Uniform volume clutter from 1 μsec to 2 msec. Target to clutter doppler = 18,750 Hz. (A) 77-μsec interpulse period; (B) 80-μsec interpulse period; (C) 83-μsec interpulse period.*

the receiver, the clutter power close to the radar from the thirteenth pulse is also entering the radar.

It can be seen from Fig. 8-7 that increasing the number of pulses improves the matched-filter output S/C until the pulse-train duration T_d is about six-tenths the target-range delay. At $cNT/2 \approx R_t$ the performance is poorer than that obtained from a single pulse even when the target is at the optimum doppler relative to the clutter. As more pulses are transmitted, the improvement increases since the clutter-doppler sidelobes are decreasing relative to the target. This general concept has also been explored for close and distant targets by Ares [8, 6] and Kaiteris and Rubin [200]. It will be further explored in Chap. 11.

While it would be nice to give a new set of guidelines for general waveform design when the environment is not well defined, I will suggest only two that have been inferred but not stated explicitly.

1. The time duration of the waveform should either be very much greater than, or much less than, the time delay to the target when the target is at a range where detection is mandatory.

2. The spectral spread of the transmit waveform should either be very much greater than that of the clutter or very much narrower than the clutter. However, the minimum spectral spread should not be much less than the spectral width of the target echo.

While much has been learned from Woodward's ambiguity diagram, it seems fitting to close with a timely quotation by Woodward made about 14 years after his book was published [408].

A Futility Theorem

There is continued speculation on the subject of ambiguity clearance. Like slums, ambiguity has a way of appearing on one place as fast as it is made to disappear in another. That it must be conserved is completely accepted but the thought remains that ambiguity might be segregated in some unwanted part of the *t-f* (time-frequency) plane where it will cease to be a practical embarrassment.

He then proceeded to dispose of "grandiose clearance schemes."

8.6 DESIRABILITY OF RANGE-DOPPLER AMBIGUITY

It can be seen from the discussion in this chapter and in Chaps. 9 through 14 that there are common waveforms that yield a wide variety of ambiguity function shapes. The contours of the main lobe (at 3 dB down from the peak) may be circular or elliptical with either large or small enclosed areas. The major axis of the elliptical shapes may be parallel to the range axis as with a CW transmission or parallel to the doppler axis as with a short-pulse transmission. As was shown in Sec. 8.2, it may be a diagonal ridge with the rotation angle controlled by the change of carrier frequency per unit time. The contours of the main lobe of the ambiguity function may also have relatively small area by displacing some of the total volume to other regions of the ambiguity plane. This may be achieved with pulse-train waveforms (Chap. 11) or *noiselike* phase-coded waveforms (Chap. 12). It will be shown in later chapters that the relative location of the target and the clutter or false target regions on the ambiguity plane may be the dominant factor in the choice of waveform. This section will point out some general hardware considerations that also affect the choice of waveform class and hence the ambiguity function shape.

The area of the ambiguity plane that is of interest for this discussion is a rectangle with dimensions determined by the maximum target range R_m and the maximum target-doppler frequency $\pm f_{dm}$ (assuming equal positive and negative maximum radial velocities). This may be called the range-doppler coverage area. Let the 3-dB contour of the central lobe of a thumbtack-type ambiguity function have a small area determined by $1/T_d$ in doppler where T_d is the effective coherent transmission time and by $c/2B$ in range where c is the velocity of light and B is the effective or rms transmission bandwidth. Then the number of range-doppler *cells* to be examined n for a target is approximately the ratio of the area of the target *rectangle* to the main lobe *rectangle* of the ambiguity diagram of the waveform. With a consistent definition of T_d and B this can be expressed as

$$n \approx \frac{4R_m f_{dm} BT_d}{c} \qquad (8\text{-}39)$$

This relation merely states that the use of thumbtack-type ambiguity functions with large time-bandwidth products BT_d requires that many range-doppler cells be examined for the existence of a target. In most waveform-receiver combinations the range cells appear sequentially at the receiver output. Thus, it is necessary to examine $2T_d f_{dm}$ doppler channels per range cell to determine the presence of a target. For example if the waveform has a duration of 0.1 msec and the maximum target doppler f_{dm} is $\pm$ 50 kHz, approximately 10 doppler channels must be instrumented. If the required transmission bandwidth B for clutter-reduction, target resolution, or accuracy is 2 MHz and the maximum range coverage R_m is 164 n.mi. $2R_M/c = 2.0$ msec), these doppler channels must be examined for 4,000 range cells.

In one sense this thumbtack ambiguity resulting from the use of a *noiselike* waveform is optimum if unambiguous target location in both range and velocity are required. However, the price of instrumenting the 10 doppler channels per range gate may in some cases be exorbitant compared to the total radar cost. It is possible to reduce the hardware complexity if some ambiguity in target location in range and doppler is permissible. For example, the transmission bandwidth requirement may be needed for clutter reduction only, and from a system standpoint exact determination of range or velocity is superfluous. In this case a waveform with a range-doppler ambiguity such as linear FM (chirp) may be adequate. The central lobe of the linear FM ambiguity function contains about $2T_dB$ resolution cells although many of these are generally outside the target range-doppler coverage. Then the output of the single channel FM matched filter can be said to contain simultaneous observations of a larger portion of the range-doppler coverage than with a noiselike transmission.

Various types of waveforms have different tolerances to doppler shifts or *range-doppler coupling factors* [71, chap. 9; 320, 305]. Some examples are shown on Fig. 8-8. The ordinate is the moving-target echo amplitude at the output of the receiver normalized to the stationary-target output amplitude. The abscissa is the normalized doppler shift $(\nu/\Delta f)$. With the parameters of the above example, the frequency deviation of the waveform Δf is 2 MHz and the maximum doppler shift is 50 kHz. Then the maximum normalized doppler shift is defined as $f_{dm}/\Delta f$ or 0.025 in this case. The assumption for this figure is that there is only a single doppler channel implemented to reduce hardware cost while accepting the loss in signal-to-noise ratio. If the reduction in amplitude is slight at f_{dm}, the hardware saving may be justified. The waveforms studied, the time-bandwidth product of the examples, and the references for the data in this text and elsewhere are

    A. Linear FM or chirp $T_dB = 100$, Chap. 13 ([71, chap. 9])
    B. Frank polyphase code $T_dB = 100$, Chap. 12 [71]

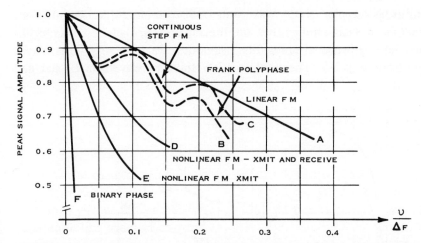

figure 8-8 *Signal amplitude vs. normalized doppler shift for waveforms with different degrees of range-doppler coupling. (After [71, 305].)*

C. Contiguous step FM $T_d B$ = 100, Chap. 13 [71]
D. Nonlinear FM with Taylor spectrum weighting on transmit and receive $T_d B$ = 50 [305]
E. Nonlinear FM with Taylor spectrum weighting on transmit $T_d B$ = 50 [305]
F. Binary phase code $T_d B$ = 63, Chap. 12
G. Linear FM with Taylor weighting on receive $T_d B$ = 50 [305]

It can be seen from Fig. 8-8 that the linear FM waveform has the most tolerance to doppler shift. At $\nu/\Delta f$ = 0.025 there is a negligible signal loss and the 10 doppler channels need not be implemented. However, there is considerable ambiguity in the simultaneous determination of range and velocity. This ambiguity is the price paid for a single-channel receiver. It can also be seen that the contiguous step FM (stepped chirp), the Frank polyphase codes, and the particular nonlinear FM on transmit and receive used in the example[1] also yield less than 1-dB signal loss at $f_d/\Delta f$ = 0.025. The binary phase-code waveform with TB = 63 yields very poor performance at a normalized doppler shift of 0.025 unless the appropriate doppler channels are implemented. The curves of Fig. 8-8 are only appropriate for the time-bandwidths specified.

There is another factor to be considered when multiple targets may be encountered. The time (range) sidelobes increase when there is a

[1] Another example of nonlinear FM is given in Sec. 8.7.

significant doppler shift. Waveform and receiver filter combinations which have small ambiguities on the range axis, $[|\chi(\tau,0)| \ll |\chi(0,0)|]$ often have substantial ambiguity (sidelobes) for $f_d \geq \tfrac{1}{2}T_d$. This can be seen from Fig. 8-9. The nonlinear FM transmit waveform E that had

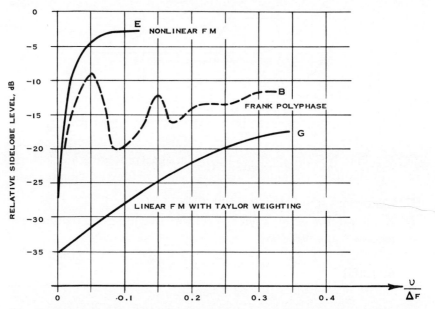

figure 8-9 *Ratio of mainlobe to sidelobe peak of ambiguity function vs. normalized doppler shift for waveforms with high FM content. (After [71, 305].)*

about a 2-dB signal loss at $\nu/\Delta f = 0.025$ has about -9-dB peak range sidelobes. On the other hand the linear FM waveform with Taylor weighting on receive maintains a -33-dB range sidelobe level.

In a similar manner the pulse-train waveforms with their *bed of nails* ambiguity functions may have more than one peak in the range-doppler coverage. The adverse effect of these ambiguous peaks with extended clutter is discussed in Chap. 11. When clutter is not a problem, the ambiguities in range or doppler may be tolerable and result in simpler hardware. The parameters of many current pulse-doppler systems deliberately have ambiguities in one coordinate of the ambiguity plane to achieve accuracy or resolution in the orthogonal coordinate.

The essential point of this section is that when ambiguities in the range-doppler coverage can be tolerated there may be considerable hardware savings. This is perhaps the primary reason for the continuing popularity of chirp systems in current radars. Alternately ambiguity in certain regions of the range-doppler coverage can be traded for accuracy or resolution in range or doppler as with pulse-train waveforms.

8.7 SPECIAL TECHNIQUES—OPTICAL SIGNAL PROCESSING

While later chapters are organized by the type of waveform that is transmitted, it is well to point out that there are at least two classes of *general* signal processors. The first class would be the special purpose digital processor. Specific configurations for digital signal processing are shown in Chaps. 11 through 14. These perform the same function as the electronic analog filter signal processors. The second class of generalized processors uses optical techniques to generate and decode the radar waveforms.

Both the digital and optical processors can be constructed at low cost, weight, and volume and with high reliability. Rather than duplicate discussions of the various types of optical processors found in the literature [38, pp. 245-273; 71, pp. 501-517], this section will describe a specific[1] optical pulse expansion/compression system that can generate and decode a complex pulse waveform.

Optical signal processing can currently provide pulse expansion/compression for radar signals with large compression ratios (to several thousand) and bandwidths to 25 MHz. It offers advantages not available with conventional electronic techniques: (1) multichannel capability in a unit essentially the same size as that required for a single channel system and (2) wide choice of signal frequency modulation for compressed-pulse sidelobe suppression, as well as control of the ambiguity function shape.

One form of an optical signal processor matches a moving waveform against a stationary stored waveform in an illuminated optical aperture.[2]

---

[1]The author is grateful to Dr. L. Slobodin of Lockheed Electronics for supplying the description and figures for this section.

[2]Patent No. 3,189,756 issued to L. Slobodin et al. and assigned to the Lockheed Company.

The moving waveform, produced in a column of transparent fluid or solid by an electromechanical transducer, is an acoustic equivalent of an applied electrical signal; the stationary waveform is a nonuniform, transmission-type diffraction grating which contains the reference signal frequency modulation. Fluctuations in the intensity of light diffracted by the accoustic waves and transmitted through the grating are sensed by a phototube which yields an output signal functionally related to the correlation of the two waveforms. The expanded transmit pulse is created when a very narrow beam of diffracted light produced by a small group of acoustic waves scans the entire grating. The compressed pulse is produced when the received signal waveform in the form of diffracted light is matched against the stored waveform of the grating. The optical signal processor operates at intermediate frequency; thus, the fine waveform (phase) structure of the signal is preserved.

Following is a description of the operation of this optical processor. The parameters referred to in the description are for the specific unit from whence the signals in Fig. 8-12 were obtained. The arrangement of the optical system is shown schematically in Fig. 8-10. The operation is as follows: Light from the point source is collimated by lens L_1 onto an ultrasonic light modulator (ULM). The point source is produced by focusing the light from a 100-watt mercury arc or equivalent on a pinhole. The ULM consists of a small light-transparent tank (several cubic inches) of water with two broad-banded quartz transducers mounted at opposite ends of the tank. Within the tank and centered in the coincident ultrasonic propagation paths of the transducers is located a transparent glass replica containing a series of lines of varying periodicity similar to a nonuniform diffraction grating. The replica is, in effect, a stored signal history, the distribution of the replica lines corresponding to the signal frequency modulation.

When the transmit transducer is excited by a 0.07-μsec pulse of 35 MHz, it sets up ultrasonic waves in the water which scan the replica and modulate the intensity of the collimated light along the wavefront of the light with a periodicity corresponding to the ultrasonic wavelength. The ultrasonic waves disturb the water in only a narrow region compared to the length of the replica. The replica is about 1 in. long and the effective length of the ultrasonic waves produced by the narrow pulse is about 1/200 in. Since the line spacing in the replica is the same as the ultrasonic wavelengths in the water, the moving ultrasonic *grating* in effect beats with the replica lines and causes fluctuations in the intensity

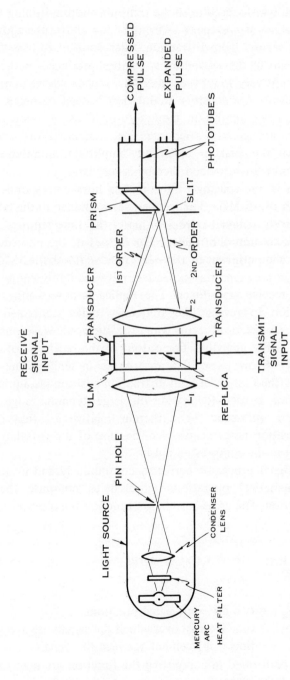

figure 8-10 *Optical signal processor. (Courtesy of L. Slobodin, Lockheed Electronics.)*

of the diffracted light focused onto the transmit photomultiplier tube by lens L_2. A slit-prism arrangement in front of the phototube allows only the light in the second light diffraction order position to pass through. The quiescent light in the second-order position produced by the replica in the ULM is kept very low by employing a replica whose lines have a sinusoidal amplitude (or density) modulation instead of being sharply defined (square-wave amplitude modulation). Thus, diffracted orders above the first are greatly attenuated or eliminated (as would be harmonics above the first in electronic amplitude modulation of a carrier). The output S/N provided thereby is very large.

The variation in the spacing of the replica lines corresponds to the signal bandwidth of 15 MHz. The frequencies contained in the 0.07-μsec pulse are selectively delayed over 16.7 μsec, the time required for the ultrasonic waves to scan the replica. The output of the phototube is a 16.7-μsec electronic pulse having a 15-MHz bandwidth about 35 MHz.

For pulse compression, the expanded pulse, suitably amplified, is applied to the receive transducer. The replica is now scanned in the opposite direction to reverse the frequency delay contained in the expanded pulse. Correlation occurs when the ultrasonic waves match the replica over the entire aperture. The output of the receive phototube at correlation is a narrow pulse, 0.1 μsec or so in length. The receive phototube views the first-order light diffraction position via a prism. The first-order position is used to provide the large dynamic range. At the somewhat larger quiescent light therein relative to that of the second-order position poses no problem because of the correlation gain (20 dB) inherent in the compressed pulse.

The optical signal processor performs continuously and in negligible time the mathematical operations required to compute the cross-correlation function. The cross-correlation function is expressed as

$$R_{1,2}(\tau) = \lim_{T \to \infty} \frac{1}{T} \int_{-T/2}^{T/2} f_1(t) f_2(t + \tau)\, dt$$

where $f_1(t)$ and $f_2(t)$ = two different signal functions

T = time interval over which the signals are averaged

τ = time displacement between the signals

The operations performed in computing the function are seen to be (1) multiplication of the signal function $f_1(t)$ and $f_2(t)$ after one function is incrementally displaced by a time τ with respect to the other and (2) the integration of the signal function product over a time interval T approaching infinity in the mathematically defined case. In actual practice we are restricted to a finite time such that the integration limits represent an interval large compared to that required for the envelope of the correlation function to go substantially to zero. In the optical system $f_1(t)$ is the stored signal history on the replica while $f_2(t)$ is represented by the propagating ultrasonic waves in the ULM. The displacement time τ varies as $f_2(t)$ sweeps past the stationary replica $f_1(t)$. The ULM aperture equivalent to 16.7 μsec of ultrasonic propagation time in water is the integration interval T. Actual integration is performed by having all the *signal-carrying* light emerging from the ULM focused onto the photodetector. Thus the entire mathematical operation to compute the correlation function is performed by the optical system. If $f_1(t)$ and $f_2(t)$ are duplicate signals, the process is an autocorrelation.

The optical signal processor can provide any desired signal frequency modulation, such as linear, gaussian, random, or other, tailored to specific requirements. With this wide choice of signal design, the high sidelobe structure of a linear FM signal can be avoided. When linear FM is used, sidelobe suppression requires an amplitude weighting filter that introduces a 1.5- to 2-dB loss in the signal-to-noise ratio of the compressed signal (Chap. 13). To make up for this loss, the transmit power must be increased up to 60 percent. The optical system achieves sidelobe suppression by waveform design rather than amplitude weighting. The 1.5 dB or more weighting loss is thus avoided, while the signal has a rectangular shape for efficient transmission. A further advantage is that a variety of different signal modulation *codes* can be simultaneously stored (as diffraction gratings) in the optical system and any one of them selected manually or automatically to accommodate specific radar tasks under changing environments.

The frequency modulation employed in the optical system to be described is gaussianlike. The modulation characteristic is shown in Fig. 8-11. In gaussian frequency modulation the use of frequencies within the signal bandwidth is weighted such that frequencies at the band edges are used least while frequencies near or at the band center are used most. The rate of change of frequency is gaussian, i.e., most of the time length of the signal history is devoted to the center-band frequencies. Since we are dealing with a finite bandwidth, the gaussian function is of necessity

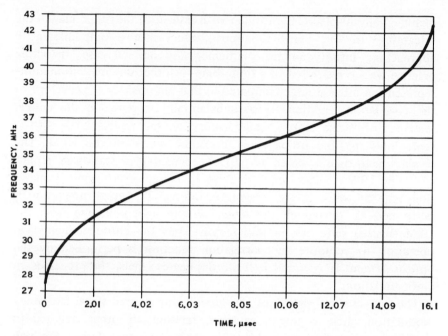

figure 8-11 *Gaussian frequency modulation characteristic. (Courtesy of L. Slobodin, Lockheed Electronics.)*

truncated. It is this truncation which controls the residual sidelobes. For the illustrated case, the truncation point was 1.76σ which yielded theoretical first sidelobes of -33 dB with reference to the main lobe. This was selected by design. Smaller sidelobes are obtainable by employing larger multiples of σ.

The compressed pulse obtained by correlating with the expanded pulse is shown as Fig. 8-12. The pulse is nominally 0.1-μsec wide with first sidelobes -32 dB from main lobe. The sidelobes are too small to be visible beyond the width of the oscilloscope trace.

Optical signal processing has two independent variables—time and space—while electronic processing is limited to time. Thus an optical system can handle a two-dimensional operation (i.e., range and doppler) without scanning. If one dimension is a filter channel, the second dimension can provide many independent filter channels, limited only by the number of resolvable positions across the optical aperture. When implemented in the multichannel arrangement, an optical processor can

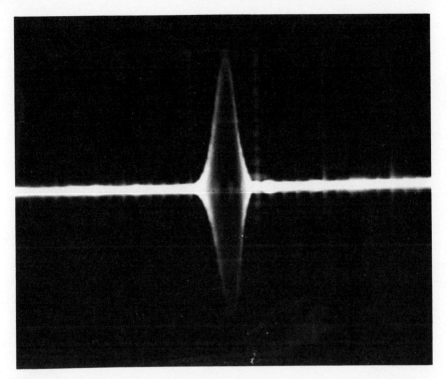

figure 8-12 *Nonlinear FM compressed pulse, 35-MHz carrier, 15-MHz base band-width, – 32 dB first sidelobes, 3-μsec trace. (Courtesy of L. Slobodin, Lockheed Electronics.)*

simultaneously process signals in, for example, 20 or more doppler channels *without* increasing the size of the optical aperture. Weight, volume, and cost savings over multichannel electronic systems are evident.

9 MOVING TARGET INDICATORS (MTI)[1]

MTI systems comprise the most widely used class of radar processors for detecting moving targets in a background of clutter. Clutter is distinguished from receiver noise by its relatively narrow, low-frequency spectrum, which implies that these echoes are correlated from one sample to the next. Because of this property it is possible to reduce the effects of clutter with filters that reject energy at the clutter frequencies but that pass the doppler-shifted echoes from targets having higher velocities than the clutter. A processor that distinguishes moving targets from clutter by virtue of the differences in their spectra is called a moving target indicator or simply MTI. The simplest MTI processor, the single delay-line canceler, subtracts two successive echoes from the same location; reflections from stationary objects cancel, while those from moving targets produce fluctuating signals.

[1]This chapter was prepared by J. P. Reilly.

MTI processors built around delay-line cancelers have been used since World War II. Early systems were primarily limited by such system instabilities as oscillator incoherence; however, subsequent equipment refinements have in many cases shifted the main source of limitation from equipment instabilities to the characteristics of the clutter itself.

The description and analysis that follow are somewhat different than what have appeared in the radar literature. They include a new approach to the prediction of MTI performance for detecting targets located in a background of precipitation, chaff, and sea clutter. The approaches in current texts to the problem of precipitation often lead to overly optimistic predictions, primarily because of three important omissions. First, precipitation (and chaff) clutter spectra depend on a wind shear effect, the change in wind velocity with altitude. Although this phenomenon has long been known to meteorologists, texts used by radar designers have failed to report it. Second, previous analyses have not considered the limitations on MTI performance imposed by the average clutter velocity, but have considered only the effects of fluctuations about this average ([354, chap. 4; 20 chap. 7]). Third, the effects of clutter lying beyond the unambiguous radar range are generally not considered. These three factors often account for unexpectedly poor MTI performance in precipitation.

As presented, the analysis is applicable primarily to stationary or slowly moving surface radar systems; MTI operation of a radar on a rapidly moving platform, such as an airplane, involves unique problems, the solutions of which are mentioned here only in passing. Descriptions and analyses of MTI operation from high-speed platforms can be found in [354, pp. 155-162; 95].

MTI canceler systems maximize signal-to-clutter ratios only for highly correlated interference.[1] Uncorrelated interference, such as receiver noise, is not affected by the MTI processor. It should be understood that to maximize the signal-to-noise ratio the radar system also contains a matched filter for the individual pulses. It is also assumed that the clutter-to-noise ratio is large. It is meaningless to speak of clutter power being reduced to levels that are below the basic thermal noise limitation.

Several basic types of MTI processors are possible. One distinction between them is the type of information processed in the returned

---

[1] The optimum two-pulse processor is identical to the single canceler, and the optimum three-pulse processor is nearly the same as the double canceler when the clutter is highly correlated from pulse to pulse [62; 221; 392, chap. 6; 7].

signals, that is, whether the phase, the amplitude, or both phase and amplitude are processed. As one would expect, systems that only use phase or amplitude do not match the performance of those systems that use both phase and amplitude.[1]

9.1 PHASE-PROCESSING MTI

The main elements of the phase-processing MTI system, referred to by some writers as a *coherent* MTI system, are illustrated in the block diagram of Fig. 9-1. Since the system must distinguish moving targets from stationary clutter by virtue of the doppler frequency produced by moving targets, the phase coherence within the system itself must be held within close tolerances. This coherence is provided by a stable local oscillator (STALO) and a coherent oscillator (COHO) which establishes the intermediate frequency. The STALO translates the signal from the transmitted RF to an intermediate frequency. The COHO provides a reference signal for coherent detection of the received echo. In the simplest processor this phase-detected signal is processed in a delay-line canceler that forms the difference between two signals separated in time by the interpulse period. One branch of the canceler circuit contains a

[1] Also see [392].

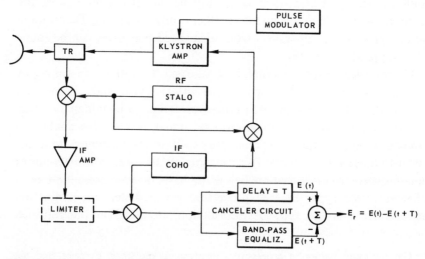

figure 9-1 *Phase-processing MTI system.*

band-pass equalizer, whose function is to match the frequency response of the delay-line branch. The operation of the MTI system when both signal and clutter are present is analyzed by considering their separate effects. This is a valid approach only if the system is linear. Undoubtedly the addition of a limiter (shown as an option in Fig. 9-1) affects the signal-to-clutter ratios although no attempt is made to calculate the effect in this chapter. The analysis of the effects of limiters in Chap. 4 will be extended in Chap. 14 to include MTI with limiters and digital quantization. Some experimental evidence indicates that considerable losses may result from limiting.[1]

Consider the operation of the canceler in response to an echo from a target at a particular range R_0. (See Skolnik [354, pp. 123 ff].) The signal presented to the canceler is

$$E_1 = E \sin(2\pi f_d t + \varphi_0)$$

where φ_0 is the phase shift due to range $= 4\pi R_0/\lambda$ (relative to that of the reference oscillator), E is the amplitude of uncanceled signal, and f_d is the doppler frequency. The signal at an interpulse period later is

$$E_2 = E \sin[2\pi f_d(t + 1/f_r) + \varphi_0]$$

where $1/f_r = T$ is the pulse repetition time. The signal output E_r from the subtractor is

$$E_r = E_1 - E_2 = -2E \sin\left(\frac{\pi f_d}{f_r}\right) \cos\left[2\pi f_d \left(t + \frac{1}{2f_r}\right) + \varphi_0\right] \qquad (9\text{-}1)$$

One assumption implicit in this formulation of the difference voltage is that each branch of the canceler has unity power gain. In later analyses this same assumption is made when considering clutter alone. No loss of generality is incurred when calculating signal-to-clutter ratios since the gain cancels in the ratio. As Eq. (9-1) indicates, the difference voltage is a sine wave at the doppler frequency whose amplitude depends on the

---

[1]See [344, chap. 17].

relationship between the doppler frequency and the pulse repetition frequency. The solid line of Fig. 9-2 describes the gain of the canceler circuit as a function of doppler frequency. In this figure S_0 and S_i represent the peak output and input signal power. Notice that certain doppler frequencies exist for which the output is zero. They occur whenever

$$f_{d \text{ blind}} = n f_r \qquad n \text{ an integer} \tag{9-2}$$

The corresponding *blind speeds* are

$$V_{d \text{ blind}} = \frac{\lambda}{2} n f_r \qquad n \text{ an integer} \tag{9-3}$$

As explained in Sec. 9.5, the signal power gain equals two when averaged uniformly over all values of f_d/f_r.

The MTI systems under consideration are pulsed systems. Therefore, a target's signal does not have the appearance of a continuous sinusoid at the doppler frequency; rather, the MTI output for a target echo can be considered to consist of a rectified sample of the voltage E_r [Eq. (9-1)]. The value of the difference signal depends on the phase φ_0. As illustrated by the dashed curves of Fig. 9-2, *blind phases* exist in

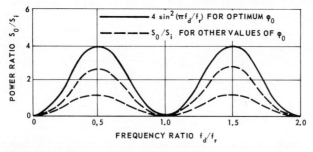

figure 9-2 *Power response of single-delay phase processor.*

addition to the blind speeds. These phase nulls may be eliminated by using two cancelers to process signals in quadrature phase and then taking their vector sum. The loss resulting from blind phases is not as serious a problem as the loss caused by blind speeds if there are many transmit pulses per beamwidth; however, there is an average loss in clutter rejection of about 3 dB for the phase-only system. (See Sec. 9.3.)

AMPLITUDE-PROCESSING OR ENVELOPE-PROCESSING MTI

The amplitude-processing, or more correctly *envelope-processing,* MTI, diagrammed in Fig. 9-3, is one example of what is termed a *noncoherent* system in other literature. An advantage of this system is that the local oscillator need not be as stable as in other systems. A disadvantage of this system is, paradoxically, that clutter must be present in relatively large amounts to detect moving targets. This can be understood by referring to the phasor diagram of Fig. 9-4. This figure shows that the amplitude of the voltage $E_s(t)$ presented to the canceler varies because of the doppler phase change of the target echo. The amplitude difference indicated in the figure is the output of a canceler circuit that takes successive amplitude differences.[1] Of course, if no clutter were present, the envelope of the signal would remain constant on a pulse-to-pulse basis. Thus, the signal out of the canceler would be zero for targets in the absence of clutter.

The amplitude-processing MTI also possesses blind speeds and blind phases that depend on target range and doppler frequency. Referring to

[1]In practice the detected video signals are generally mixed to a convenient IF since it is difficult to obtain wide-band video time delays.

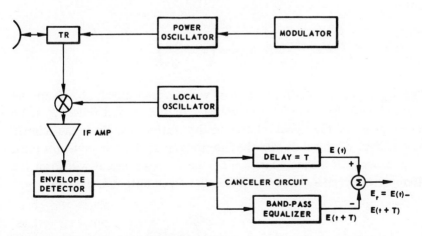

figure 9-3 *Envelope-processing MTI system.*

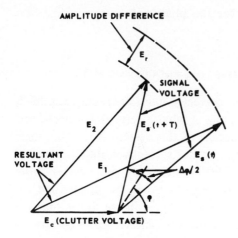

figure 9-4 *Phasor diagram of signals and clutter.*

the phasor diagram of Fig. 9-4, the resultant voltage amplitude may be expressed in terms of the signal and clutter amplitude by using the law of cosines

$$E_1^2 = E_c^2 + E_s^2 - 2E_c E_s \cos\left(\varphi - \frac{\Delta\varphi}{2}\right)$$

At a time later by the interpulse time, the target echo is assumed to have changed phase by the amount $\Delta\varphi$. For this time the amplitude can be written

$$E_2^2 = E_c^2 + E_s^2 - 2E_c E_s \cos\left(\varphi + \frac{\Delta\varphi}{2}\right)$$

where E_c, E_s are the clutter and signal voltage, respectively; E_1, E_2 are the magnitude of signal-plus-clutter voltage for the second pulse; φ is the average phase of the signal relative to the clutter; and $\Delta\varphi$ is the relative phase change of signal during the interpulse period. If the detector prior to the canceler is a square-law device, the canceler circuit will form the difference between E_1^2 and E_2^2. Thus, the signal output is given by

$$E_r = E_1^2 - E_2^2 = 4E_c E_s (\sin\varphi)\left(\sin\frac{\Delta\varphi}{2}\right)$$

and since $\Delta\varphi = 2\pi f_d/f_r$,

$$E_r = 4E_c E_s \left(\sin \frac{\pi f_d}{f_r}\right)(\sin\varphi) \qquad (9\text{-}4)$$

Equation (9-4) demonstrates that, just as in the phase processing MTI system, both blind speeds and blind phases exist and that the power output for a target's signal depends on the presence of clutter. If the envelope detector used in the processor were other than a square-law device, the shape of the response curves would change; but the blind speeds would remain the same.

9.3 VECTOR-PROCESSING MTI

The vector-processing MTI makes use of both the amplitude and phase information of the echo signals. One version performs the cancellation at IF rather than at video, as indicated by the block diagram of Fig. 9-5. Consider the response of this system to a target's echo. The IF signal at a particular time is

$$E_1 = E \sin[2\pi(f_{IF} + f_d)t + \varphi_0]$$

where E is the amplitude of the IF signal, f_{IF} is the intermediate frequency, f_d is the doppler frequency shift, and φ_0 is the initial phase of echo. At an interpulse time later the signal is described by

$$E_2 = E \sin[2\pi(f_{IF} + f_d)(t + T) + \varphi_0]$$

The output of the canceler is

$$E_r = E_1 - E_2$$

$$= 2E \sin\left[\frac{\pi(f_{IF} + f_d)}{f_r}\right] \cos\left[2\pi(f_{IF} + f_d)\left(t + \frac{T}{2}\right) + \varphi_0\right]$$

One requisite of the system is that, when the doppler frequency f_d is

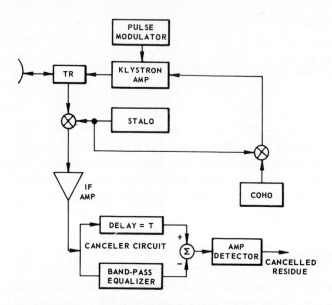

figure 9-5 *Vector-processing MTI.*

zero, the amplitude of the difference signal is zero. The above equation shows that this requires that the pulse repetition frequency and the intermediate frequency be related by

$$f_{IF} = n f_r$$

When f_{IF} and f_r are so related, the response of the system, as measured at the output of the amplitude detector, follows the solid curve of Fig. 9-2.

Another vector-processing MTI configuration is basically a phase-processing system (Fig. 9-1) that uses two channels in quadrature phase following the IF amplifier.[1] The squares of the outputs of the two canceler circuits would then be added. Such a processor would also have the response of the solid curve of Fig. 9-2. The canceler systems analyzed in this chapter are assumed to be vector-processing systems. If instead of combining two quadrature channels only one is used, as in a phase processor, roughly a 3-dB loss in performance is suffered. This arises

[1] Wainstein and Zubokav [392, p. 213] prove that such a processor optimizes the signal-to-clutter ratio in a two-pulse canceler.

because a single-channel processor requires roughly a 3-dB increase in signal power over the double-channel system to maintain a detection probability of 50 percent. (See Sec. 12.6. The loss reduces to ~ 2.4 dB for systems with 2–8 postdetection integrations.)

9.4 CLUTTER-LOCKING MTI SYSTEM

The presence of a mean velocity component of clutter can drastically affect MTI performance as demonstrated in a succeeding section. This mean velocity component can originate either from the average motion of the clutter itself or from the motion of the radar platform (as in a moving ship or aircraft). It is desirable to remove the average clutter frequency from the radar signal. One method for doing this known as the *clutter-locking* technique is illustrated by the block diagram of Fig. 9-6. In this technique the average doppler frequency of clutter signals from several range intervals is determined by averaging the phase change in each interval during the interpulse time. The mean doppler over range is compensated for by placing a phase shifter that adjusts the signal for the average phase change[1] in one branch of the canceler circuit. The response of this canceler would follow that of the one shown in Fig. 9-2 except that the frequency axis would be offset such that the null normally at zero velocity would occur at a frequency corresponding to the mean clutter velocity.

[1]The clutter-locking system is referred to as a noncoherent system in most literature because the phase of the signal is compared to the clutter rather than the local oscillator. This is a most unfortunate usage because the meaning is quite different from that as applied to other radar processors where *noncoherent* refers to envelope processing (for which phase information is discarded).

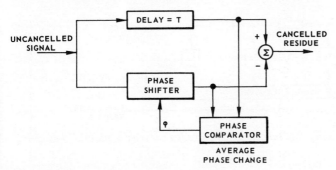

figure 9-6 *Example of clutter-locking circuit.*

Since the phase-averaging process involves a finite number of samples, the mean velocity at the range of interest can only be estimated. If a target were present in one of the range cells, its doppler frequency would have some effect on the average unless the integration time constant was long compared to a pulselength. Furthermore, if more than one type of clutter were present (for example, precipitation and land or sea clutter) and if there was a wide disparity in their mean velocities, the clutter-locking technique would not be effective. Another disadvantage of this system is that it would reject strong target signals in the absence of clutter since the phase compensation would be made for the target alone. Therefore, in such a system, it is desirable to have an adaptive technique whereby the clutter-locking circuits can be bypassed when the clutter power is small.

9.5 MULTIPLE-CANCELER SYSTEMS

Multiple delay-line cancelers provide greater clutter rejection than the single-canceler circuits described previously. One form of multiple canceler is composed of cascaded sections of single-canceler circuits (Fig. 9-7). The configuration using n cascaded sections is equivalent to the weighted sum of the $n + 1$ pulses.

It is easily verified that, for an n-stage canceler, the weighting that gives an operation equivalent to the cascaded canceler follows the binominal coefficients of $(1 - x)^n$. By forming successive differences as in

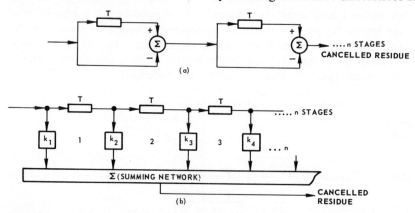

figure 9-7 *Multiple delay-line cancelers. (a) Cascaded canceler sections; (b) weighted summer.*

Eq. (9-1), the peak target power response for an n-stage canceler is obtained [306].

$$\left(\frac{S_0}{S_i}\right)_n = 2^{2n} \sin^{2n}\left(\frac{\pi f_d}{f_r}\right) \tag{9-5}$$

This equation shows that the blind speeds are independent of the number of canceler stages. The signal power gain when averaged over all possible target-doppler frequencies for an n-stage canceler is

$$\left(\frac{\overline{S_0}}{S_i}\right)_1 = 2 = 3 \text{ dB}$$

$$\left(\frac{\overline{S_0}}{S_i}\right)_2 = 6 = 7.8 \text{ dB}$$

$$\left(\frac{\overline{S_0}}{S_i}\right)_3 = 20 = 13 \text{ dB} \tag{9-6}$$

$$\vdots$$

$$\left(\frac{\overline{S_0}}{S_i}\right)_n = 1 + n^2 + \left[\frac{n(n-1)}{2!}\right]^2 + \left[\frac{n(n-1)(n-2)}{3!}\right]^2 + \cdots 1$$

If combined phase and amplitude processing (vector or IF processing) were not used, the multiple canceler would experience a 3-dB loss in performance because of the blind phases as explained for the single canceler.

One disadvantage of the cascaded-section canceler is that, because of the $\sin^{2n}$ response, the rejection of target signals near the blind frequencies becomes more severe as n is increased. It is possible to obtain other MTI response curves with multiple-canceler systems using various feedback and feed-forward circuits as illustrated in Fig. 9-8 ([354, pp. 132-135; 38, chap. 3; 381; 403]). The advantage of the feedback system is that the clutter notch may be shaped along with a very flat signal

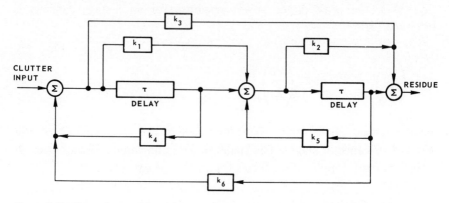

figure 9-8 *General two-delay filter using feedback and feedforward techniques.*

response between blind speeds. The disadvantage is that this response is only for a steady-state condition; this requires that a number of pulses be processed before the steady-state condition is met.

9.6 STAGGERED PRF SYSTEMS

The blind speeds inherent in the previously described MTI processors can pose serious limitations on target detection. The use of a varied pulse repetition frequency provides a technique for extending the first blind speed. (See [126, 279].) For a system that has two repetition intervals available T_1 and T_2 the canceler may form the successive differences

$$E_r = [E(t) - E(t + T_1)] - [E(t + T_1) - E(t + T_1 + T_2)]$$

This is equivalent to the weighted sum

$$E_r = E(t) - 2E(t + T_1) + E(t + T_1 + T_2) \tag{9-7}$$

Using similar techniques, it is possible to process combinations of more than two pulses by weighting them according to the coefficients of the binominal series.[1]

---

[1] When staggering is used, however, weightings other than binominal may be more desirable, as pointed out in Sec. 9.11.

Consider the two-period (three-pulse) system that has the ratio of interpulse spacing $T_1/T_2 = a/b$ (a and b are integers). The first true blind speed would occur at the frequency that satisfies the following equation:

$$\left(f_d\right)_{\text{blind}} = \frac{a}{T_1} = \frac{b}{T_2} \tag{9-8}$$

As the stagger ratio T_1/T_2 is increased, the depth of the nulls between the blind speeds in the response characteristics is increased as shown in Fig. 9-9.

9.7 LIMITATIONS ON MTI PERFORMANCE—CLUTTER STATISTICS

This section introduces two quantities that provide a measure of MTI performance,[1] clutter attenuation (CA) and system improvement factor I. Clutter attenuation is defined as the ratio of input clutter power C_i to output clutter power C_0.

$$\text{CA} = \frac{C_i}{C_0} \tag{9-9}$$

The system improvement factor (similar to Steinberg's definition [38, p. 494]) is defined as the signal-to-clutter ratio at the output of the MTI system compared with that at the input, where the signal is understood as that averaged uniformly over all radial velocities, that is,

$$I = \frac{\overline{S_0}/C_0}{S_i/C_i} \tag{9-10}$$

which can be expressed as

---

[1]MTI performance is sometimes measured in terms of *subclutter visibility*; however, there does not appear to be a standardized definition of subclutter visibility. Because the meaning of this term is quite ambiguous, it will not be invoked as a measure of MTI performance. On the other hand, the term *improvement factor* has not appeared with any contrary definition to the author's knowledge.

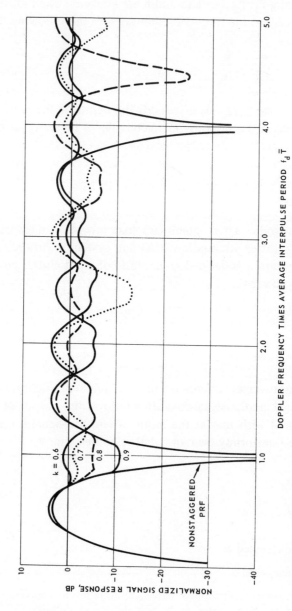

figure 9-9 *Response curves for three-pulse staggered canceler (k = stagger ratio).*

$$I = \frac{\overline{S}_0}{S_i} CA \qquad\qquad (9\text{-}11)$$

The following paragraphs show how the improvement factor is limited by the environment, the system configuration, and the radar parameters.[1] These limitations are discussed individually and then in Sec. 9.14 are related to the radar equations of Chap. 2.

Early MTI systems were limited primarily by the instabilities of the radar itself; however, subsequent developments in hardware have often shifted the main cause of limitation to the statistical properties of the clutter. The role of the clutter statistics can be appreciated from the following analysis.

Consider the residue signal from a single canceler

$$E_r(t) = E(t) - E(t + T)$$

where T is the interpulse time. Because of the noiselike appearance of clutter, it is not possible to predict the residue at any given time; instead, the residue power must be averaged over the ensemble of canceled signals.

$$\overline{E_r{}^2(t)} = \overline{[E(t) - E(t + T)]^2}$$

For stationary statistics this average is

$$\overline{E_r{}^2(t)} = 2E^2(t) - 2R(T) \qquad\qquad (9\text{-}12)$$

where $R(T)$ is the autocorrelation function of the clutter evaluated at $\tau = T$. The autocorrelation function may be expressed in terms of the normalized correlation function

$$\rho(\tau) = \frac{R(\tau)}{\overline{E^2(t)}}$$

By recognizing that $\overline{E^2(t)} / \overline{E_r{}^2(t)} = CA$, Eq. (9-12) can be expressed as

---

[1] Limitation due to quantization errors in a digital MTI are discussed in Chap. 14.

$$CA = \frac{1}{2[1 - \rho(T)]} \tag{9-13}$$

When the improvement factor is expressed, the factor 2 in the denominator is canceled because the average signal gain for the single canceler also equals 2 (see Eq. 9-6).

$$I_1 = \frac{1}{1 - \rho(T)} \tag{9-14}$$

This result shows the improvement factor depends only on the correlation function of the clutter signal evaluated at a single point—the interpulse time. A parallel development yields the expression for the double canceler

$$I_2 = \frac{1}{1 - \frac{4}{3}\rho(T) + \frac{1}{3}\rho(2T)} \tag{9-15}$$

The clutter improvement factor may be evaluated by assuming the gaussian function to describe the clutter spectrum; both theory and experiment suggest this function, and it is convenient to manipulate mathematically. At any rate, the results are not very sensitive to the precise spectrum shape, but rather to the width (second moment) of the spectrum. For the gaussian spectrum[1] [306]

$$I_1 = \frac{1}{1 - \exp[-2(2\pi\sigma_v/\lambda f_r)^2]\cos 4\pi V_0/\lambda f_r}$$

$$I_2 = \frac{3}{3 - 4\exp[-2(2\pi\sigma_v/\lambda f_r)^2]\cos 4\pi V_0/\lambda f_r + \exp[-2(4\pi\sigma_v/\lambda f_r)^2]\cos 8\pi V_0/\lambda f_r} \tag{9-16}$$

where σ_v is the standard deviation of clutter spectrum (meters/second), V_0 is the mean value of clutter spectrum (meters/second), λ is the

---

[1]When $V_0 = 0$, the resulting expression is identical with that normally encountered in the literature (Skolnik [354, p. 147] or Steinberg [38, pp. 494, 495]).

radar wavelength (meters), and f_r is the pulse repetition frequency (second^{-1}).

Expressions for higher order cancelers are derived in [306]. At this point, we may verify an earlier statement that the signal-to-noise ratio is not affected by the MTI processor. The improvement in the signal-to-noise ratio is easily evaluated for a white spectrum by evaluating the correlation functions of Eqs. (9-14) and (9-15) as impulse functions centered at $T = 0$, so that $\rho(T) = 0$. The result is that the improvement in signal-to-noise power is unity, that is, the average signal-to-noise ratio is not improved over that of a single pulse. For usable MTI operation (e.g., for $I_1 > 10$ dB) the arguments of the exponential and cosine functions in Eqs. (9-16) are much less than unity. Therefore, a good approximation may be obtained by taking the first two terms in the Taylor series to approximate e^{-x} and $\cos x$, which gives

$$I_1 \approx \frac{\lambda^2 f_r^2}{8\pi^2(\sigma_v^2 + V_0^2)} \tag{9-17}$$

The expression for the double canceler may be obtained by taking the first three terms of the Taylor series. For $V_0 = 0$, the result is

$$I_2 \approx \frac{\lambda^4 f_r^4}{128\pi^4 \sigma_v^4} \tag{9-18}$$

Figure 9-10 illustrates the improvement factor limitation for delay-line canceler systems when $V_0 = 0$. These curves apply to either clutter-locking systems or those that receive clutter of zero mean doppler. The reduction in improvement factor due to a nonzero mean clutter doppler is illustrated in Fig. 9-11, which shows the loss (in decibels) from a system where $V_0 = 0$. The loss is seen to depend on the ratio V_0/σ_v.[1]

Equation (9-16) can also be evaluated with values of σ_v and V_0 that correspond to the experimental values of clutter. For example, Fig. 9-12 shows the limitations on the single-canceler clutter improvement factor

[1]Figure 9-11 accurately describes the loss in the linear regions of Fig. 9-10. The nonlinear regions are, of course, most unsatisfactory for MTI performance because of the poor improvement factor.

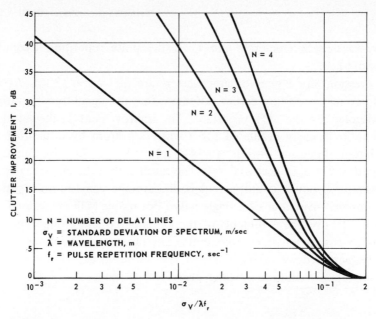

figure 9-10 *Improvement factor for delay-line canceler as limited by clutter spectrum ($V_0 = 0$).*

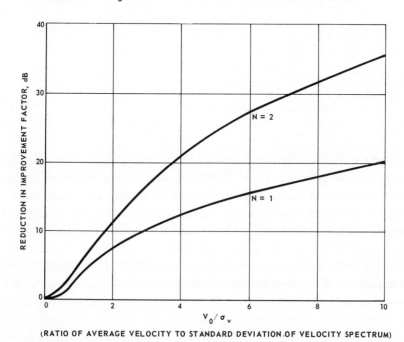

(RATIO OF AVERAGE VELOCITY TO STANDARD DEVIATION OF VELOCITY SPECTRUM)

figure 9-11 *Reduction in improvement factor due to an average velocity component of the clutter spectrum.*

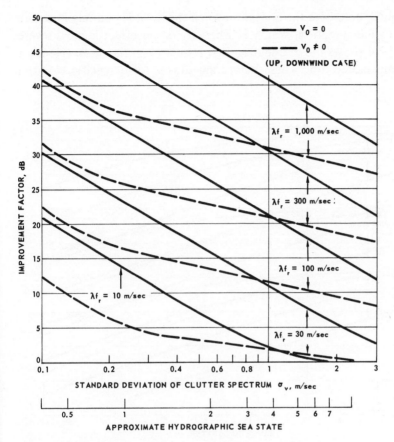

figure 9-12 *Limitation of single-canceler improvement factor by sea clutter (horizontal polarization).*

imposed by the spectrum of sea clutter as a function of hydrographic sea state. Particular values of σ_v and V_0 may be obtained from the models of Chap. 7.[1] It is evident that the presence of a mean doppler component of sea clutter has a considerable effect on the improvement factor.

For most radars the clutter improvement factor for precipitation is a function of range. This is so because the spectrum width of precipitation clutter is a function of range (because of the wind shear effect described in Chap. 6). Figure 9-13 illustrates an example of the improvement factor limitation for precipitation clutter when clutter locking is used.

[1] The worst case values of V_0 and σ_v for horizontal polarization are shown. (See Reilly [305] and Nathanson and Reilly [264].)

This figure represents the limitation when the radar is pointing in the up or downwind direction, which is when the shear effect is most severe. Figure 9-14 shows the degradation that results when the mean velocity is not compensated for. These curves are drawn for a specific elevation

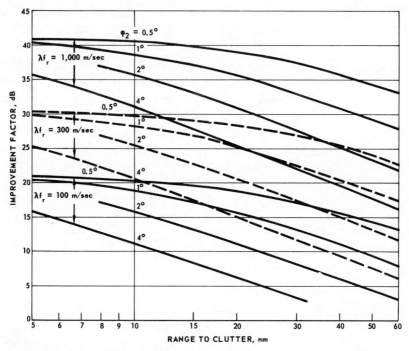

figure 9-13 *Single-canceler improvement factor as limited by precipitation clutter. Clutter locked, radar looking up/down-wind, wind shear (k) = 5.7 m/sec/km.*

angle since the shear-versus-range relationship is a function of angle and beamwidth.

9.8 ERRORS IN ESTIMATING MEAN CLUTTER VELOCITY

The preceding section demonstrated that the mean velocity component of clutter can seriously degrade MTI performance. One system that compensates for the mean velocity component is the clutter-locking system, described in Sec. 9.4. One type of clutter-locking system determines the mean clutter frequency by averaging the phase differences between the returns from two pulses over several range intervals.

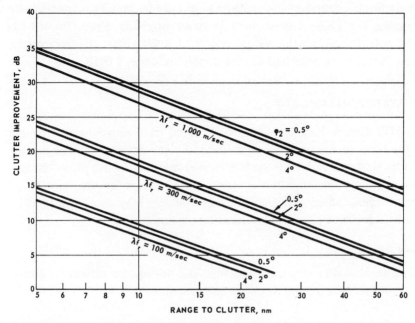

figure 9-14 *Single-canceler clutter improvement as limited by precipitation. No clutter locking, radar looking up/down-wind, elevation angle = 2°, two-way elevation beamwidth = ϕ_2.*

Since the averaging process involves a finite number of samples, the mean velocity at the range of interest can only be estimated, that is, the estimate of the mean will exhibit a statistical fluctuation. The variance of the mean velocity estimate is related to the variance of the spectrum itself by

$$\sigma_{v0}^{2} = \frac{\sigma_v^{2}}{n}$$

where n is the number of independent samples used to form the average, σ_{v0}^{2} is the variance of the estimate of the mean velocity, and σ_v^{2} is the variance of the clutter spectrum.

As shown in Sec. 9.7, the loss in clutter improvement depends on the ratio V_0/σ_v. Thus, the rms ratio for the clutter-locking system is

$$\frac{V_0}{\sigma_v} = \frac{1}{\sqrt{n}} \qquad (9\text{-}19)$$

For example, consider a clutter-locking system that estimates V_0 by averaging the phase change in two range intervals. From Eq. (9-19) $V_0/\sigma_v = 0.707$. Using this value, Fig. 9-11 reveals that the loss for the single canceler is 1.5 dB and for the double canceler, 3.0 dB.

9.9 SYSTEM INSTABILITIES

The MTI system is able to distinguish fixed from moving target echoes on the basis of the phase change caused by the doppler shift of moving targets. Unfortunately, the system is unable to distinguish phase changes caused by system instabilities from those caused by the motion of targets. Thus, phase and amplitude instabilities may cause the canceler circuit to produce a residue for fixed target echoes.

Phase Errors

The residue due to a phase change $\Delta\varphi$ during the interpulse period may be expressed for the IF processing system as

$$E_r = E_1 - E_2$$

$$= E\cos(2\pi f_{IF}t + \varphi_0) - E\cos[\omega_{IF}(t + T) + \varphi_0 + \Delta\varphi]$$

where E is the magnitude of the uncanceled signal, $\Delta\varphi$ is the phase error over the interpulse period, IF is the intermediate frequency, and T is the pulse repetition period. This equation may be reduced to[1]

$$\frac{E_r}{E} \simeq 2\sin\frac{\Delta\varphi}{2}\,\sin\left(2\pi f_{IF}\,t + \varphi_0 + \frac{\Delta\varphi}{2}\right)$$

The ensemble average of the magnitude of E_r^2/E^2 is recognized as the reciprocal of the clutter attenuation factor. For small values, $\sin\Delta\varphi \cong \Delta\varphi$, which results in

$$\frac{1}{CA} = \frac{\overline{|E_r|^2}}{E^2} = \overline{\Delta\varphi^2} \qquad \text{for the single canceler}$$

[1]Subject to the constraint that $f_{IF} = nf_r$; see Sec. 9.3.

Using Eq. (9-11), the improvement ratio as limited by phase instability is

$$I_1 = \frac{2}{\overline{\Delta\varphi^2}} \qquad \text{limitation due to phase instability} \qquad (9\text{-}20)$$

Equation (9-20) expresses the clutter improvement limitations due to phase instabilities that can arise from transmitter STALO or COHO variations in a single-canceler system.

A similar analysis of the double-canceler system leads to an interesting result. The form of the three pulses at the input of the double canceler is

$$E_1 = E \cos(\omega_{IF} t + \varphi_0)$$

$$E_2 = E \cos[\omega_{IF} (t + T) + \varphi_0 + \Delta\varphi_1]$$

$$E_3 = E \cos[\omega_{IF} (t + 2T) + \varphi_0 + \Delta\varphi_1 + \Delta\varphi_2]$$

where $\Delta\varphi_1$, $\Delta\varphi_2$ are respective phase errors in radians during the first and second interpulse periods. The normalized residue is

$$\frac{1}{CA_2} = \frac{|E_1 - 2E_2 + E_3|^2}{E^2} \qquad \frac{|E_{r2}|^2}{E^2}$$

Two special cases of E_{r2} are of interest. In the first case, the phase errors from one pulse to the next are assumed equal (as in an oscillator with constant or slow drift). In this case the peak residue for small $\Delta\varphi$ gives

$$\frac{1}{CA_2} = \frac{\overline{|E_{r2}|^2}}{E^2} \cong \overline{\Delta\varphi^4} \qquad (\text{for } \Delta\varphi_1 = \Delta\varphi_2)$$

In the second case, the phase errors from one pulse to the next are assumed statistically independent, that is, $\overline{\Delta\varphi_1 \Delta\varphi_2} = 0$. In this case the peak residue is

$$\frac{1}{CA_2} \cong 2\overline{\Delta\varphi^2} \qquad \text{for } \overline{\Delta\varphi_1 \Delta\varphi_2} \cong 0$$

The improvement ratio for the double canceler is

$$I_2 = \frac{6}{\Delta\varphi^4} \qquad \text{for } \Delta\varphi_1 = \Delta\varphi_2 \quad\left.\begin{array}{c}\\ \text{limitation due to}\\ \text{phase instability}\\ \\ \end{array}\right\} \qquad (9\text{-}21)$$

$$I_2 = \frac{3}{\Delta\varphi^2} \qquad \begin{array}{l}\text{for } \Delta\varphi_1 \text{ inde-}\\ \text{pendent of } \Delta\varphi_2\end{array} \qquad\qquad\qquad\qquad (9\text{-}22)$$

For phase errors that are only partially correlated, the solution for I_2 lies between the values corresponding to Eqs. (9-21) and (9-22). An analysis of systematic instabilities is given by Barton [20, pp. 199-210].

Pulse Jitter

Consider two echoes returned from a stationary target where the delay-line time between pulses jitters about the nominal interpulse time T. The output of the canceler would consist of two spikes of width equal to the difference between the timing errors for the two pulses ϵ_1 and ϵ_2 as illustrated in Fig. 9-15. Prior to threshold detection, the pulses are assumed to be passed through a low-pass filter whose bandwidth is approximately $1/\tau$ and rectified. The peak value of the residue out of such a filter is approximately

$$E_r = \frac{E(\epsilon_1 - \epsilon_2)}{\tau}$$

where ϵ_1, ϵ_2 are the timing errors for the first and second pulses.

The normalized residue power averaged over the ensemble for the two pulses is

$$\frac{\overline{E_r}^2}{E^2} = \frac{2\overline{(\epsilon_1 - \epsilon_2)}^2}{\tau^2}$$

which, for independent ϵ_1, ϵ_2, is

$$\frac{1}{CA_1} = \frac{\overline{E_r}^2}{E^2} = \frac{4\sigma_\epsilon^2}{\tau^2} \qquad \text{for the single canceler}$$

where σ_ϵ^2 is the variance of the leading time of the pulse.

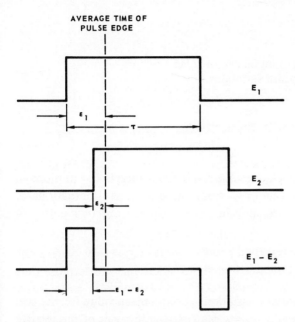

figure 9-15 *Illustration of pulse-to-pulse jitter.*

Using a similar procedure, the residue from the double canceler for the two impulses reduces to [306]

$$\frac{1}{CA_2} \leq \frac{12\sigma_\epsilon^{2}}{\tau^2}$$

Taking the equality for the more pessimistic value, the improvement ratio as limited by pulse jitter for either a single or double canceler can be expressed as

$$I_1 = I_2 = \frac{\tau^2}{2\sigma_\epsilon^{2}} \qquad \text{limitation due to pulse jitter} \qquad (9\text{-}23)$$

Another cause of cancellation residue that is quite similar to pulse position jitter results from variations in pulse width. Since a pulse width variation of δ produces a single spike of the same width, the improvement factor can be immediately written by comparison with that for pulse jitter

$$I_1 = I_2 = \frac{\tau^2}{\sigma_\delta^2} \quad \begin{array}{l}\text{limitation due to pulse} \\ \text{width variations}\end{array} \qquad (9\text{-}24)$$

where σ_δ^2 is the variance of the pulse width.

Amplitude Instabilities

Another source of canceler residue from fixed targets is variations in signal amplitude, which can arise from pulse-to-pulse fluctuations in transmitted power or in signal gain. The residue from two pulses of different amplitudes is

$$E_r = \Delta E_1 - \Delta E_2$$

where ΔE_1, ΔE_2 are the voltage variations about a mean value for the two pulses. The normalized residue power is expressed in terms of the relative amplitude fluctuation as

$$\frac{\overline{E_r^2}}{E^2} = \frac{\overline{(\Delta E_1 - \Delta E_2)^2}}{E^2}$$

which, for independent fluctuations, becomes

$$\frac{1}{CA_1} = \frac{\overline{E_r^2}}{E^2} = \frac{2\sigma_E^2}{E^2} \qquad \text{for the single canceler}$$

The normalized residue for the double canceler is

$$\frac{\overline{E_{r2}^2}}{E^2} = \frac{\overline{(\Delta E_1 - 2\Delta E_2 + \Delta E_3)^2}}{E^2}$$

Again assuming independent fluctuations,

$$\frac{1}{CA_2} = \frac{\overline{E_{r2}^2}}{E^2} = \frac{6\sigma_E^2}{E^2} \qquad \text{for the double canceler}$$

The clutter improvement ratios follow from Eq. (9-11).

$$I_1 = I_2 = \frac{E^2}{\sigma_E^2} \quad \text{limitation due to amplitude fluctuation} \qquad (9\text{-}24a)$$

9.10 ANTENNA MOTION LIMITATIONS

The motion of the radar antenna can degrade MTI performance because of the resulting spectrum broadening of the clutter echoes. The broadening can be broken into two effects. The first is due to the translational motion of the antenna; the second to the rotational motion.

The spectrum broadening due to translational motion is significant in a rapidly moving system such as would be found on an aircraft. The spectrum of ground or sea echo broadens because of two effects. One is the change in the radial velocity component of the ground return as a function of the changing incident angle over the area intercepted by the radar beam. The second effect is due to the distance the aircraft moves in an interpulse period. Because of the change in distance, the radar does not receive echoes from the identical patch of scatterers from pulse to pulse. The result is a fluctuating signal. Translational broadening depends on platform velocity, radar depression and azimuth angle, antenna pattern, pulse width, wavelength, and altitude [95, 112, 135, 166]. There are specialized techniques to compensate for these motions, but these will not be discussed here.

Another important cause of clutter spectrum broadening is the angular rotation of the antenna. As in the case of translational motion, the radar does not receive echoes from the identical patch of scatterers from pulse to pulse. Steinberg [38] analyzes this component by considering a gaussian two-way antenna beam pattern

$$G(\theta) = G_0 \exp\left(-\theta^2/2\sigma_{\theta 2}^2\right) \qquad (9\text{-}25)$$

where $\sigma_{\theta 2}$ is the equivalent two-way standard deviation of the antenna power pattern (radians) and G_0 is the on-axis antenna power gain factor. The voltage of the echo from an elemental reflector would also have a gaussian time function as the antenna scans the patch of reflectors. The voltage would then vary with time as

$$E(t) = K \exp(-t^2/2\sigma^2)$$

where $\sigma = \sqrt{2}\,\sigma_{\theta 2}/\alpha$ is the standard deviation of the time function, α is the rotational rate of the antenna (radians/second), $\sigma_{\theta 2}$ is the two-way standard deviation of the antenna pattern, and K is a scaling factor. When many independent scatterers are present, numerous spectra, each having the form of Eq. (9-25), are superimposed. The resulting spectrum retains the shape of the individual spectra, but the scaling factor K is increased.

The power spectral density function of the return is obtained by taking the squared magnitude of the Fourier transform of $E(t)$, which is

$$G(\omega) = G_0 \exp(-\omega^2/2\sigma_\omega^2)$$

where $\sigma_\omega = (\alpha/2\sigma_{\theta 2})$ (radians/second), ω is the radian frequency variable, and σ_ω is the standard deviation of the spectrum (radians/second). The spectrum standard deviation due to scanning is expressed in hertz by dividing σ_ω by 2π. Furthermore, the standard deviation of the antenna pattern $\sigma_{\theta 2}$ may be replaced by an equivalent expression for the two-way half-power azimuth beamwidth θ_2, which for a gaussian pattern is related by $\theta_2 \cong 2.36\,\sigma_{\theta 2}$. The result of these substitutions is

$$\sigma_s = \frac{\alpha}{5.35\,\theta_2} \text{ Hz} \tag{9-26}$$

where σ_s is the spectrum width induced by antenna scanning (hertz) and θ_2 is the two-way horizontal beamwidth (radians).

The effect of the scanning spectrum spreading σ_s can be more readily compared to that of clutter motion σ_v by converting σ_s in Eq. (9-26) to equivalent velocity units through the doppler equation

$$\sigma_s = \frac{\alpha\lambda}{10.7\,\theta_2} \text{ m/sec} \tag{9-27}$$

with λ in meters and α in radians/sec.

Figure 9-16 shows σ_s for various beamwidths and rates of scanning. To determine the MTI performance limitation caused by antenna scanning, the value of σ_s may be used along with Fig. 9-10 to obtain a value of the improvement factor. The result is nearly identical when the more

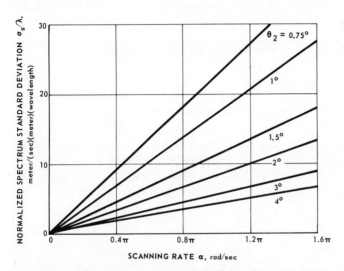

figure 9-16 *Normalized spectrum width due to antenna scanning (two-way half-power horizontal beamwidth = θ_2).*

realistic $\sin \theta/\theta$ pattern is assumed [148]. The limitation due to antenna scanning can be significant; however, this loss can often be eliminated by using an electronically scanned antenna.

9.11 LIMITATIONS ON STAGGERED PRF SYSTEMS

The blind speeds inherent in MTI operation pose a serious limitation to target detection. Section 9.6 described a staggered PRF system that increases the first blind speed.

One disadvantage of the staggered PRF system is that as the difference in interpulse times is increased, the clutter attenuation decreases (when the pulses are weighted binomially).

The improvement factor for the staggered system I_s is derived in a manner parallel to that used to derive Eq. (9-15), resulting in

$$I_s = \frac{1}{1 - \frac{2}{3}[\rho(T_1) + \rho(T_2)] + \frac{1}{3}\rho(T_1 + T_2)} \qquad (9\text{-}28)$$

The improvement factor for the staggered system can be compared with that for a double canceler system with interpulse time $T = (T_1 + T_2)/2$ by using Eq. (9-15) in the form

$$I_2 = \frac{1}{1 - \frac{4}{3}\rho[(T_1 + T_2)/2] + \frac{1}{3}\rho(T_1 + T_2)} \tag{9-29}$$

Equation (9-28) can be expressed in terms of the stagger ratio

$$I_s = \frac{1}{1 - \frac{2}{3}[\rho(T_1) + \rho(kT_1)] + \frac{1}{3}\rho[T_1(1 + k)]} \tag{9-30}$$

where $k = T_2/T_1$.

As was previously explained, the gaussian function is a reasonable choice for the spectrum of clutter; the autocorrelation function is also gaussian and may be expanded into a Taylor series.[1]

$$\rho(\tau) = \exp \frac{-\tau^2}{2\sigma_\tau^2} = 1 - \frac{\tau^2}{2\sigma_\tau^2} + \frac{\tau^4}{8\sigma_\tau^4} \cdots$$

Using the first three Taylor series terms in Eqs. (9-29) and (9-30) results in the ratio

$$\frac{I_2}{I_s} \cong \frac{4\sigma_\tau^2}{3\,T^2} \frac{(1-k)^2}{(1+k)^2} + \frac{16}{(1+k)^4}\left[k^2 - \frac{(1-k)^4}{12}\right] \tag{9-31}$$

where $T = (T_1 + T_2)/2$, $k = T_2/T_1$, I_2 is the improvement factor without staggering, I_s is the improvement factor with staggering. Figure 9-17 illustrates this ratio with the abscissa in terms of I_2 rather than σ_τ^2/T^2. This figure demonstrates that as the stagger ratio departs from unity the clutter attenuation is decreased toward the performance of the single canceler.

Ares has shown that by choosing nonbinomial weighting functions the velocity response and the clutter rejection notch may be shaped almost

---

[1]For a gaussian spectrum $\sigma_\tau = 1/2\pi\sigma_f = \lambda/4\pi\sigma_v$ (see Sec. 9.7). The first few terms of the Taylor series are adequate except for small cancellation ratios.

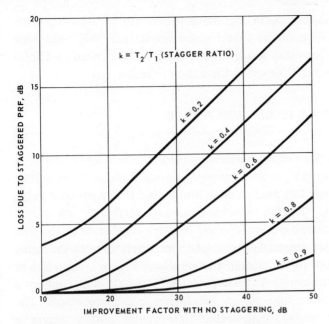

figure 9-17 *Improvement factor loss with staggered PRF system using binomial weights (the period of the unstaggered canceler is the same as the average period of the staggered system).*

independently [7]. He shows that if one wishes to maximize the signal-to-clutter ratio for a particular target velocity the pulses must be amplitude weighted and phase shifted, the amount depending on the clutter correlation. Ares presents an alternate weighting criterion, which he terms *maximally flat*. This criterion is useful if one is constrained to use only amplitude weighting. For the three-pulse canceler, the maximally flat weights are

$$C_1 = 1, \quad C_2 = -\frac{T_1 + T_2}{T_2}, \quad C_3 = \frac{T_1}{T_2}$$

where C_1 and C_2 are the weights and T_1 and T_2 are the interpulse periods. The response curves of Fig. 9-9 were determined using the maximally flat weights. This curve was found to be only slightly altered when the binomial weights were used instead. Although the maximally flat weights are less than optimum, the improvement factor using them was found to be as good as that for the nonstagger PRF case using binomial weights. Optimum weighting factors are also derived by Capon [62] using a somewhat different optimization criterion.

9.12 LIMITATION FROM SECOND-TIME-AROUND CLUTTER

In a two-pulse canceler, the second signal returned from volume clutter at a time corresponding to a particular range R_1 contains not only the clutter echo from that range but also an echo due to the first pulse at a range $R_1 + cT/2$, where $cT/2$ is the unambiguous range. Although this second echo, called *second-time-around clutter,* is much smaller than the first because of its greater range, it is completely uncorrelated with the signal in the first pulse. For this reason, it receives absolutely no attenuation in the delay-line processor.[1] A three-pulse MTI might also contain a *third-time-around* echo and so on for higher order processors. The following analysis shows that second-time-around clutter places a limitation on the maximum achievable improvement factor.

Consider the residue signal from a two-pulse canceler when second-time-around clutter is present.

$$\overline{E_r^{*2}(t)} = \overline{\{E_1(t) - [E_1(t + T) + E_2(t)]\}^2}$$

where $E_1(t)$, $E_1(t + T)$ are the clutter echoes from range R_1; $E_2(t)$ is the second-time-around clutter echo from range $(R_1 + cT/2)$; and $E_r^*(t)$ is used to denote a residue signal that contains second-time-around clutter. This may be expressed as

$$\overline{E_r^{*2}(t)} = \overline{[E_1(t) - E_1(t + T)]^2} - \overline{2E_2(t)[E_1(t) - E_1(t + T)]} + \overline{E_2^2(t)}$$

The first term is the residue signal when no second-time-around clutter is present. The middle term vanishes because $\overline{E_1 E_2} = 0$ (E_1 and E_2 are independent). Thus,

---

[1] An MTI canceler that processed many pulses in each range cell would have a steady-state condition in which second-time-around clutter is canceled if the interpulse period is fixed (except for magnetron transmitters).

$$\frac{\overline{E_r^{*2}(t)}}{\overline{E_1^2}} = \frac{1}{CA_1} + \frac{\overline{E_2^2(t)}}{\overline{E_1^2(t)}} \tag{9-32}$$

Chapter 6 demonstrated that for volume clutter the returned power is inversely proportional to the square of range, that is,

$$\overline{E_1^2} = \frac{K}{R_1^2} \quad \text{and} \quad \overline{E_2^2} = \frac{K}{(R_1 + cT/2)^2}$$

Therefore, Eq. (9-32) can be expressed as

$$\frac{1}{CA_1^*} = \frac{1}{CA_1} + \frac{R_1^2}{(R_1 + cT/2)^2} \tag{9-33}$$

where CA_1^* refers to the clutter attenuation when second-time-around clutter is present and CA_1 refers to the clutter attenuation when it is not. It is obvious from Eq. (9-33) that CA_1^* cannot be increased beyond the point where

$$\frac{1}{CA_1^*} = \frac{R_1^2}{(R_1 + cT/2)^2}$$

Using the definition of the clutter improvement factor [Eq. (9-11)],

$$I_1^* = \frac{2(R_1 + cT/2)^2}{R_1^2} \quad \begin{array}{l}\text{limitation due to second-}\\ \text{time-around clutter}\end{array} \tag{9-34}$$

It is interesting to note that beyond one-half the unambiguous range the maximum clutter improvement for the single canceler is about 13 dB.

This equation is pessimistic in that it assumed that the clutter completely filled the beam at the first ambiguous range. If a maximum altitude for rain or chaff is assumed, the second-time echoes will only partially fill the beam. Equation (9-34) can then be rewritten

$$I_1^* = \frac{2(R_1 + cT/2)^2}{R_1{}^2 \Delta F_2} \qquad \text{limitation due to second-} \atop \text{time-around clutter} \qquad (9\text{-}35)$$

where ΔF_2 is the percentage of the beam volume that is filled by clutter at a range of $R_1 + cT/2$.

The limitation for the double canceler can be expressed using a development parallel to the preceding, but where the third-time-around clutter is assumed to be negligible.

The residue signal that includes the second-time-around clutter for the double canceler is

$$
\begin{aligned}
E_{r2}^{*2} &= \overline{\{E_1(t) - 2[E_1(t + T) + E_2(t)] + [E_1(t + 2T) + E_2(t + T)]\}^2} \\
&= \overline{[E_1(t) - 2E_1(t + T) + E_1(t + 2T)]^2} - \overline{\{2[E_2(t) - E_2(t + T)]} \\
&\quad \overline{[E_1(t) - 2E_1(t + T) + E_1(t + 2T)]\}} + \overline{[2E_2(t) - E_2(t + T)]^2}
\end{aligned}
$$

As before, the middle term drops out because $\overline{E_1 E_2} = 0$. The first term is the residue for the double canceler in the absence of second-time-around clutter. This equation can be reduced in a manner similar to the preceding, resulting in

$$I_2^* \cong \frac{6(R_1 + cT/2)^2}{R_1{}^2 \Delta F_2} \qquad \text{limitation due to second-time-} \atop \text{around clutter if } I \gg 1 \qquad (9\text{-}36)$$

The effect of second-time-around clutter is illustrated in Fig. 9-18 for a surface radar with a 1.1° two-way beamwidth. The interpulse period corresponds to 63 nmi. The term ΔF_2 is estimated from the 4/3 earth profile. The rain is assumed to be uniform up to an altitude of 15,000 ft. The increase in I at longer ranges is the result of the majority of the beam being above the rain at the ambiguous range. Some clutter, however, may remain in the sidelobes of the antenna. This analysis does not treat the problem of clutter from sidelobes.

9.13 MTI SYSTEM IMPROVEMENT FACTOR

In the previous paragraphs we have discussed the factors that limit the clutter improvement ability of MTI radars. These individual limitations

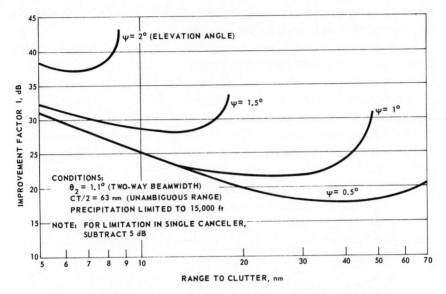

figure 9-18 *Limitation on improvement factor due to second-time-around clutter for double canceler MTI.*

may be combined to give the system clutter improvement. To obtain the system improvement, recall that the clutter improvement is defined in terms of the residue power at the output of the canceler. If each source of residue power is independent, then the total residue power $\overline{E_{rt}^2}$ is simply the sum of the separate causes $\overline{E_{ri}^2}$, that is

$$\overline{E_{rt}^2} = \overline{E_{ra}^2} + \overline{E_{rb}^2} + \overline{E_{rc}^2} + \cdots \overline{E_{ri}^2}$$

Then, the total system improvement ratio I_T may be calculated from

$$I_T = \frac{\overline{S_0}}{S_i} \, CA = \frac{\overline{S_0 E^2}}{S_i \overline{E_{rt}^2}}$$

It is evident from the above equation that the system improvement ratio can be determined from the individual ratios by

$$\frac{1}{I_T} = \frac{1}{(I)_a} + \frac{1}{(I)_b} + \frac{1}{(I)_c} + \cdots \qquad (9\text{-}37)$$

where I_T is the total system improvement factor (power ratio) and $(I)_i$ is the improvement as limited by the ith cause.

To illustrate the relative importance of the various limitations, consider a radar that has the following parameters:

> $\lambda = 50$ cm (transmitted wavelength)
> $f_r = 0.6$ kHz (pulse repetition frequency)
> $\tau = 3$ μsec (pulse width)
> $\varphi_2 = \theta_2 = 1.5°$ (two-way vertical and horizontal beamwidth, respectively)
> $\alpha = 0.4\pi$ rad/sec (antenna scanning rate—12 rpm)
> $n = 2$ (number of cascaded cancelers)

The type of cancellation scheme is a vector canceler. Assume the system instabilities are as follows:

> $\sigma_\epsilon = 20$ nsec (timing jitter)
> $\sigma_\delta = 15$ nsec (pulse-to-pulse variation)
> $\Delta\varphi = 10$ mrad (pulse-to-pulse independence) (rms phase instability at the canceler frequency)
> $\sigma_E/E = 0.005$ (amplitude instability pulse-to-pulse variation)

Let the source of clutter be precipitation at 50 nmi and the wind shear gradient be 4 m/(sec)(km) where $\psi = 2°$ (elevation angle). The precipitation is assumed to exist up to 15,000 ft altitude. Table 9-1 lists the clutter improvement limitation due to each individual cause. The additional loss due to the mean velocity component is given when no clutter locking is used or, alternately, when clutter locking is provided on the basis of two samples. The table shows that the overall system improvement is 13.8 dB when clutter locking is not used or 24.6 dB when it is used. In this example, the primary limitation is due to motion of the clutter.

9.14   RANGE EQUATIONS FOR MTI SYSTEMS

The clutter improvement factor describes the ability of the MTI system to improve the signal-to-clutter ratio; however, this factor may not always be directly used to compare the performance of two different systems since they could have the same clutter improvement even though one had a lower signal-to-clutter ratio to start with. The detection range

TABLE 9-1 System Limitations for Example MTI

Source of limitation	I_2, dB	Determined by
I. Clutter motion		
1. Spectrum width $\sigma_v = (\sigma_{shear}^2 + \sigma_{turb}^2)^{1/2} = 4.1$ m/sec	34	σ_{shear} and σ_{turb} from Chap. 6
2. Antenna scanning ($\sigma_s = 2.3$ m/sec)	44	I_2 from Fig. 9-10
		σ_s from Fig. 9-16
3. Total σ_v (Root sum square of 1 and 2) = 4.7 m/sec	31.5	I_2 from Fig. 9-10
4. Additional loss due to mean velocity component		V_0 from Chap. 6
		Loss from Fig. 9-11
(a) No clutter locking ($V_0/\sigma_v = 3.2$)		V_0 from Chap. 6
Loss = 17 dB		Loss from Fig. 9-11
(b) With clutter locking ($V_0'/\sigma_v = 0.707$)		V_0'/σ_v from Eq. (9-19)
Loss = 3 dB		Loss from Fig. 9-11
5. Total limitation due to motion		
(a) No clutter locking	14.5	Item 3 less item 4a
(b) With clutter locking	28.5	Item 3 less item 4b
II. Timing jitter	40	Eq. (9-23)
III. Pulse width variation	46	Eq. (9-24)
IV. Phase instability	45	Eq. (9-22)
V. Amplitude instability	46	Eq. (9-25)
VI. Second-time-around clutter	——*	Eq. (9-36)
VII. Total system improvement factor		
1. No clutter locking	13.8	Eq. (9-37) items I-VI
2. With clutter locking	24.6	Eq. (9-37) items I-VI

*In this example, the radar beam is completely out of the clutter at the first ambiguous range. Therefore, second-time-around clutter causes no limitation. (Altitude is determined with the aid of a 4/3 earth chart.)

in clutter, on the other hand, provides a basis for system evaluation and comparison. The following paragraphs develop expressions for the detection range for MTI radars.

MTI Detection Range for Targets in Volume Clutter

Equation (9-17) approximated the clutter improvement factor for a single canceler by

$$I_1 \cong \frac{\lambda^2 f_r^{\,2}}{8\pi^2 (\sigma_v^{\,2} + V_0^{\,2})} \tag{9-38}$$

The values of σ_v and V_0 for precipitation may be related to range as discussed in Chap. 6. The following presents some of the environmental factors described in that section in a form that will be useful for the present discussion.

Chapter 6 showed that the spectrum variance for precipitation clutter $\sigma_v^{\,2}$ is due primarily to the effects of wind shear and turbulence, that is, $\sigma_v^{\,2} = \sigma_{\text{shear}}^2 + \sigma_{\text{turb}}^2$. For a constant wind velocity gradient within the beam the factor σ_{shear} can be approximated

$$\sigma_{\text{shear}} \approx 0.42 \, k\Delta h = 0.42 \, kR\varphi_2 \tag{9-39}$$

where Δh is the altitude spread of the radar beam (meters), k is the component of wind velocity gradient in the direction of the radar beam [meters/(second)(meter)] (for the up or downwind case a typical value is $k \cong 5.7$ m/(sec)(km), φ_2 is the vertical half-power beamwidth—two-way path (radians), R is the range to clutter (meters), and σ_{shear} is in meters/second. The clutter improvement factor may be related to range by combining Eqs. (9-38) and (9-39). For $V_0 = 0$ this is

$$I_1 \cong \frac{\lambda^2 f_r^{\,2}}{8\pi^2 (\sigma_{\text{turb}}^2 + 0.18 \, k^2 R^2 \varphi_2^{\,2})} \tag{9-40}$$

Chapter 2 developed the expression for detection range in volume clutter. Assuming that the required signal-to-clutter ratio in the MTI system may be decreased by an amount equal to the clutter improvement factor,[1] the range equation becomes[2]

$$R^2 = \frac{4\,(\ln 2)\,\sigma_T\,L\,I_1}{\pi\theta_2\,\varphi_2\,(c\tau/2)\,(S/C)\,\Sigma\sigma_i}$$

(9-41)

where the value of I_1 is given by Eq. (9-40), which suggests two special cases. In one case the shear effect is dominant (as when the radar looks up or downwind at distant ranges); in the other case the turbulence effect is dominant (as when the radar looks crosswind or for short ranges).[3] For the shear dominant case, Eq. (9-41) becomes

$$R^4 = \frac{2.8\,(\ln 2)\,\lambda^2 f_r^2\,\sigma_T\,L}{\pi^3 k^2 \theta_2^{\,3}(c\tau/2)\,(S/C)\,\Sigma\sigma_i}$$

(9-42)

$$(\sigma_{shear}^2 \gg \sigma_{turb}^2)$$

This range equation is seen to depend on the product λf_r; (S/C) refers to the signal-to-clutter ratio at the input to the detector that is required for a given P_D.

In the turbulence limited case Eq. (9-41) becomes

$$R^2 = \frac{(\ln 2)\,\lambda^2 f_r^2\,\sigma_T\,L}{2\pi^3 \theta_2\,\varphi_2\,(c\tau/2)\,(S/C)\,\Sigma\sigma_i\,\sigma_{turb}^2}$$

(9-43)

$$(\sigma_{turb}^2 \gg \sigma_{shear}^2)$$

MTI Detection Range for Targets in Area Clutter

Chapter 2 developed the expression for the detection range of targets in area clutter. Again assuming that the required signal-to-clutter ratio

---

[1] This assumption does not account for the effect that the velocity response characteristics of the canceler have on the detection probabilities. For a treatment of this topic, see Wainstein and Zubakov [393, sec. 42].

[2] The additional loss due to the mean velocity component V_0 may be included in the loss factor L.

[3] Another example of the turbulence limited condition is for a chaff cloud that has a small vertical extent.

may be reduced by an amount equal to the clutter improvement ratio, the detection range expressions of Chap. 2 become

$$R^2 = \frac{\lambda^2 f_r^2 (\sin\psi)\,\sigma_T L}{2\pi^3 \theta_2 \varphi_2 (S/C)\,\sigma_0\,(\sigma_v^2 + V_0^2)} \qquad \tan\psi > \frac{\varphi_2 R}{c\tau/2} \tag{9-44}$$

$$R = \frac{\lambda^2 f_r^2 (\cos\psi)\,\sigma_T L}{8\pi^2 \theta_2 (S/C)(c\tau/2)\,\sigma_0\,(\sigma_v^2 + V_0^2)} \qquad \tan\psi < \frac{\varphi_2 R}{c\tau/2} \tag{9-45}$$

Equations (9-42) to (9-45) show that the power law of the single-canceler range equation may vary between one and four, depending on the condition of clutter. The maximum range in precipitation or in sea clutter may be calculated in the crosswind case by using a zero mean clutter velocity and in the up or downwind case by using V_0 from Chap. 6. As a result, the range coverage for MTI systems can be described by an eliptical pattern.

The previous range equations have been developed for the single-canceler system. The corresponding equations for the double-canceler can, using the clutter improvement factor I_2, be developed in a similar manner.

10 ENVIRONMENTAL LIMITATIONS OF CW RADARS

The continuous-wave (CW) radar is frequently used for detection and tracking of moving targets. In its simplest form a single sinusoid is transmitted, and the received signals are mixed with the transmitted carrier frequency. The existence of moving targets is determined from the *beat note* or doppler frequency shift f_d.

$$f_d = \frac{2v}{\lambda} \qquad\qquad (10\text{-}1)$$

where v is the radial velocity difference between the target and the radar (positive for closing geometries) and λ is the carrier wavelength. The advantage of this technique is its simplicity since simple headphones provide an efficient detector for the doppler beat note.

For general references see Skolnik [354, chap. 3], Povejsil et al. [293], and Vinitskiy [387].

As with virtually all radar waveforms, the angle of arrival of the target echoes can be determined with multiple receive apertures or with monopulse or conical scan receivers. On the other hand there is no target-range determination or resolution except with special geometries that occur with lunar or planetary observation radars. If range resolution is required, the transmit waveform must be modulated. Frequency modulation is discussed in Sec. 10.2, and binary-phase modulation is discussed in Secs. 12.2 and 12.3.

Since the CW receiver responds to echoes from all ranges, it is generally necessary to separate the *leakage* or *spillover* from the transmitter and signals from close-in clutter. This is usually accomplished by filtering out received signals at the carrier frequency and at the doppler frequency of the clutter. An alternate technique is to pulse the transmitter at a rate higher than twice the expected doppler frequency. This technique is called interrupted-CW or ICW to distinguish it from *pulse-doppler* or range-gated processors. ICW will be discussed in Secs. 10.2 and 10.5, and pulse-doppler processors will be discussed in Chap. 11.

Rather than delve into the wide variety of CW or FM-CW receivers that are described in the literature ([354, chap.3; 267; 388]), this chapter will emphasize the clutter limitations of CW radars. The equations and graphs of Secs. 10.3 through 10.6 combined with the clutter backscatter coefficients of Chaps. 6 and 7, should lead to quantitative computation of the detection performance of CW radars. It will be shown that even if the spillover of the transmit signal into the receiver can be reduced to a tolerable level, clutter signals will exist that are far in excess of the minimum detectable signal (MDS).

10.1 TRANSMITTER SPILLOVER AND NOISE LIMITATIONS

Perhaps the primary design problem in CW radar is transmitter-receiver isolation. Because they are both operating simultaneously, the receiver must reject the transmit signals and operate on the received ones. This direct coupling between the transmitter and receiver, variously called *leakage, spillover,* and *feedthrough,* can easily obscure target echoes. As an example, if the radar transmits 1 kW (60 dBm) of continuous power and the minimum detectable signal (MDS) is -130 dBm, there must be in excess of 190 dB of isolation at the target-doppler frequency. About

20-40 dB of this can be obtained with ferrite circulators between the transmitter and receiver of a single aperture system. If separate or polarized apertures can be utilized, 60-100 dB of isolation is obtainable.[1] When the geometry permits, such as in space tracking or planetary observation radars, the antennas can be widely separated or on opposite sides of a mountain to achieve even greater isolation. In addition to this isolation, additional rejection can be obtained with circuitry that samples the transmit signal and effectively *subtracts* a portion of it from the spillover signals. These specialized techniques can achieve additional isolation exceeding 60 dB [159, 270].

Even with these techniques the spillover signal power can exceed the MDS by 100 dB and cause severe dynamic range problems in the receiver. The problem often does not arise from the spillover alone, but from the AM and FM noise sidebands of the transmitter leakage that appear in doppler passband of the target. A high quality CW transmitter may have AM noise sidebands that are 90-120 dB below the carrier in a 1-kHz bandwidth and FM noise sidebands 70-95 dB below the carrier in the same bandwidth. These sidebands will either raise the MDS or appear as spurious false targets. Noise sideband limitations are discussed in [301; 139; 140; 159; 387, chap. 2] and will only be briefly summarized here. Transmitter noise will also appear on all clutter signals.

The permissible transmitter noise levels can in general be handled separately for AM and FM noise. However, there is no single parameter that can be specified that describes the transmitter stability requirements. The simplest requirement to consider is that of spurious signals separated from the carrier frequency (or the main beam clutter frequency for an airborne radar) by at least a doppler shift Δf_{min}. These signals could be interpreted as a target. As might be expected, the lower the value of Δf_{min} the more stringent are the transmitter stability requirements. For maximum sensitivity, the spurious signals in the doppler filter of interest must be below the noise level in that filter. For example, if the spillover (or clutter) is 80 dB above the rms noise, a sideband rejection ratio requirement R of 86 dB on the AM and FM noise independently will ensure that in the worst case of in-phase AM and FM components the resultant signal will no more than equal the noise level [140]. First consider the AM noise component. If M is the fractional amplitude modulation, there are two sidebands of amplitude

---

[1] Only 20-30 dB by polarization alone.

1/2M. The maximum allowable value of M for a given AM sideband rejection ratio R_{AM} is

$$M_{max} = 2R_{AM}^{\frac{1}{2}} \qquad (10\text{-}2)$$

For noise sidebands due to frequency modulation, the relationship is not as simple. In most cases the modulating frequency f_m is below Δf_{min} and the higher order sidebands must be calculated from the Bessel coefficients. Graphs are available in [140, fig. 3-2] for calculating the maximum deviation for a given modulating frequency f_m and sideband rejection ratio R. For single sinusoidal frequency modulation and a very small index of modulation

$$\frac{P_c}{P_{SB}} = \left(\frac{f_m}{\Delta f_{rms}}\right)^2 = 20 \log \left(\frac{f_m}{\Delta f_{rms}}\right) \text{ in dB} \qquad (10\text{-}3)$$

where P_c is the carrier and P_{SB} is the total power in both sidebands. Δf_{rms} is generally defined for a 1-kHz bandwidth as the rms deviation of a single sinusoidal frequency modulation at the center of a 1-kHz bandwidth that would equal the rms deviation of the total frequency modulation of the actual signal contained in the whole 1-kHz band [301]. At C or X band typical unstabilized triodes or solid-state osciilators have a Δf_{rms} of 2 to 5 Hz at $f_m = 10$ kHz. Nongridded two-cavity klystrons have a Δf_{rms} of 0.2 to 0.5 Hz. With passive stabilization Δf_{rms} can be kept below 0.1 Hz under the same conditions but at lower power output. At $f_m = 2$ kHz the values of Δf_{rms} are typically 4-10 times higher, resulting in considerably higher sidebands. For a given type of microwave source, the noise is almost independent of carrier frequency from 3-10 GHz.

The configuration of the receive signal processor also influences the sensitivity of AM and FM noise if limiting is used to achieve a constant false alarm rate (CFAR). Raduziner and Gillespie [301] have shown that limiting at IF, prior to doppler filtering, has a different effect on AM and FM noise on clutter echoes than does a synchronous detection or homodyne receiver followed by a high-pass filter, remodulation to IF, and then limiting before doppler filtering. When clutter limited, the homodyne system will tend to reduce the effect of FM components resulting from a single sinusoid or low-index complex frequency modulation entering the receiver from stationary clutter echoes. (There is some controversy on this subject.)

10.2   CW, FM-CW, AND ICW TRANSMISSIONS

It was shown in Sec. 8.3 on the environmental diagram that the doppler spectra of clutter echoes as seen by a stationary radar are often disjoint from the spectra of aircraft or missile echoes. As a result, there can be complete resolution of moving target echoes from clutter. Thus in theory the clutter rejection or improvement factor I is infinite. In practice, transmitter noise or receiver saturation is the limiting factor on target detectability. With an airborne CW radar the target spectrum may or may not overlap the clutter spectrum depending upon the geometry and closing velocities as discussed in Sec. 10.6.

The detection range of a CW radar can be written from Eq. (2-6) as

$$R^4 = \frac{\bar{P}_T G_T L_T A_e L_R L_p L_a L_s \sigma_t}{(4\pi)^2 K T_s b (S/N)} \tag{10-4}$$

These terms are the same as those defined in Sec. 2.1 with the exception that the average transmit power $\bar{P}_T$ is used rather than the peak power P_T and the doppler filter or *speedgate* bandwidth b is used for the noise bandwidth. This equation is obtained by making use of the general theorem given in Chaps. 1 and 8 which states that detectability depends on the transmitted energy and not on the details of the transmit waveform (see also [71, chap. 6]). The only modification to Eq. (10-4) that is necessary for CW radar computations concerns the requirement for a matched filter. For a sine-wave CW transmission and a stationary reflector, this would call for a zero bandwidth receiver. In practice, the filter bandwidth b must encompass at least the target spectral width due to fluctuation, target acceleration, antenna scanning modulation, time on target, and any frequency uncertainties in the receiver local oscillators, filters, etc. In current radars the filters generally have wider bandwidth than the absolute minimum, but there is a partial compensation for the loss in detectability with the use of postdetection filtering or integration. Typically, (S/N) for a CW radar must be at least 6 dB rather than the 13-dB value required for detecting steady targets with pulse radars.

If the CW transmission is truncated at time T_d as shown on Fig. 10.1A, a closer approximation can be made to a matched filter, but spillover and clutter problems are aggravated. The effect of truncation is to give all the spillover signals a $\sin^2(\pi f_d T_d)/(\pi f_d T_d)^2$ spectrum. Even if the product of the doppler shift and the dwell time $f_d T_d$ is large, clutter

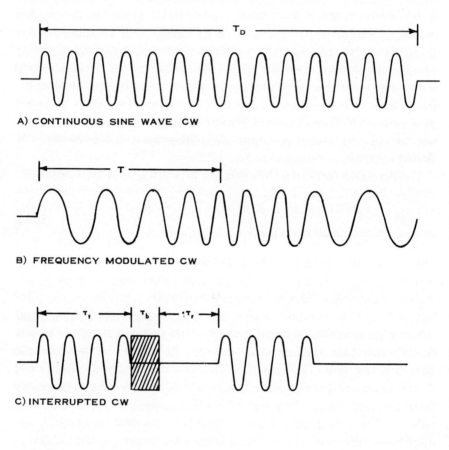

A) CONTINUOUS SINE WAVE CW

B) FREQUENCY MODULATED CW

C) INTERRUPTED CW

D) BINARY PHASE–CODED CW

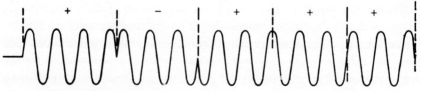

figure 10-1 *Waveforms for the general class of CW radar.*

signals that are often 50-100 dB above the target echo will have spectral components at f_d that are well above the noise in the filter bandwidth b. This effect will be expanded upon in Chap. 11 on pulse-doppler and burst waveforms.

FM-CW

While it is possible to truncate a CW transmission at a time T_d and estimate range to some fraction of $cT_d/2$, pulsed-CW transmission for greater than 1 msec has found little application except for planetary observation radars. By far the most common technique for determining target range is to frequency modulate a continuous transmission (FM-CW). The modulation waveform may be a linear sawtooth, a triangle or a simple sine wave. These modulations have the effect of broadening the spectrum of the transmit waveform and thus enhancing the ability to determine range.

If the modulation waveform is a linear sawtooth, the time-frequency relationships will be as in Fig. 10-2. A carrier frequency f_0 is modulated at a rate $\dot{f}_0$ and transmitted. The echo from a target at range R will occur after a time $T_t = 2R/c$ as shown on Fig. 10-2A. The received signal is mixed with the transmit signal and the difference or beat frequency f_b is extracted. If the target is stationary,

$$f_b = f_r = \dot{f}_0 T_t = \frac{2R}{c} \dot{f}_0 \qquad (10\text{-}5)$$

where f_r is the beat frequency due only to the target range ([354, sec. 3.3]). The beat note has a constant frequency except near the turn-around region of the sawtooth. If the carrier is modulated at a rate f_m with a frequency deviation Δf, the range of a target can be determined from

$$R_0 = \frac{cf_r}{4f_m \Delta f} \qquad (10\text{-}6)$$

If the target has a radial velocity v, the echoes will have a doppler shift $f_d = 2v/\lambda$. Then the beat frequency waveform appears as in Fig. 10-2B. The range frequency f_r can be determined from the average beat frequency [354]

$$f_r = \tfrac{1}{2} \left[f_b(\text{up-slope}) - f_b(\text{down-slope}) \right] \qquad (10\text{-}7)$$

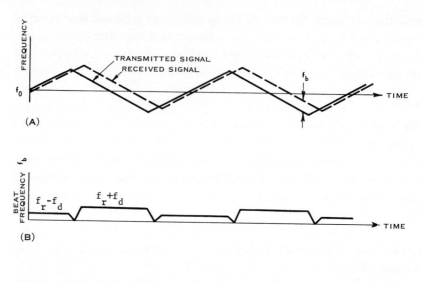

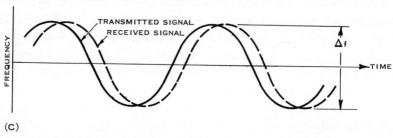

figure 10-2 *Frequency time relationships in FM-CW radar when the received signal is shifted in frequency by the doppler effect. (A) Transmitted (solid curve) and echo (dashed curve) for triangular modulation; (B) beat frequency for triangular modulation; (C) transmitted and received signals for sine-wave modulation. (After Skolnik [354].)*

The inverse of the frequency deviation can be considered analogous to an *effective* pulselength in pulse radar while the modulating frequency is analogous to the PRF [293].

If the carrier is sinusoidally modulated with a voltage waveform

$$e_t = E_T \sin\left(2\pi f_0 t + \frac{\Delta f}{2f_m} \sin 2\pi f_m t\right) \tag{10-8}$$

the received signal from a target at range R when mixed with the carrier frequency will yield a difference frequency voltage waveform [354, p. 89; 293]

$$e_b = kE_T E_r[\sin 2\pi f_0 T_t + \pi \Delta f T_t \cos(2\pi f_m t - \pi f_m T_t)] \qquad (10\text{-}9)$$

where k is a constant of proportionality, E_r is the received voltage from the target and $T_t f_m \ll 1$. The transmit and received waveforms are shown as Fig. 10-2C.

If the argument of Eq. (10-9) is differentiated with respect to time, the average beat frequency $\bar{f}_b$ over one-half a modulating cycle can be approximated by letting $\cos \pi f_m T_t \approx 1$. Then [354, p. 90]

$$\bar{f}_b = \frac{4Rf_m \Delta f}{c} = f_r \qquad (10\text{-}10)$$

The total spectrum of a carrier that is frequency modulated by a sine wave consists of a series of spectral lines that can be described by Bessel functions of the first kind and order 0, 1, 2, 3, etc. A reduction in spillover signals and close-in reflections can be obtained by the use of filters that only pass certain harmonics of the modulating frequency. For example, extraction of only those received frequencies near $3f_m$ will result in considerable spillover rejection at a penalty of 4-10 dB in signal-to-receiver noise power. The detection of targets up to a given range can be optimized by proper choice of the modulating frequency f_m and extraction of the receive signal components corresponding to the desired order of the Bessel function $J_n(f)$. It should be emphasized that frequency modulation will also smear the clutter spectrum over a wide range of frequencies. This may or may not be desirable depending on the total clutter power-to-signal ratio and the clutter location.

Target range can also be determined by phase modulation of the carrier. The most common waveform consists of simple binary phase modulation (0-180° phase shift) as illustrated on Fig. 10-1D. This technique will be described in Secs. 12.2 and 12.3.

Interrupted CW (ICW)

When a single antenna must be used for both transmitting and receiving, it is often impractical to achieve the desired isolation between the final transmitter stage and the receiver. To alleviate this problem and to eliminate the close-in clutter backscatter, the transmitter is often turned on or *pulsed* at a high rate for 10 to 60 percent of the total time. During this time τ_t and for a short *blanking period* τ_b the receiver is switched off to eliminate transients, spillover, and close-in clutter. The remainder τ_r of the interpulse period is used for reception of the target echoes. The timing is shown diagrammatically on Fig. 10-1C. Then the interpulse period T can be written as

$$T = \tau_t + \tau_b + \tau_r \tag{10-11}$$

then

$$d_t = \frac{\tau_t}{T} < 1$$

$$d_r = \frac{\tau_r}{T} < 1$$

$$d_b = \frac{\tau_b}{T} < 1$$

where d_t and d_r are the transmit and receive duty factors and d_b is the blanking duty factor.

It is assumed that the receiver has a single range gate and that the PRF = $1/T$ is greater than twice the highest target-doppler frequency. This type of transmission is often called *pulse-doppler* in airborne radar systems but is referred to in this text as Interrupted-CW or ICW to distinguish it from the multiple range-gate systems discussed in Chap. 11 on pulse-doppler and burst waveforms.

Since the echo from a distant target with unknown range can enter the receiver no more than 100 d_r percent of the time, there is an average loss in received signal power. This is generally referred to as *eclipsing* of the target echoes. One of the primary problems in the choice of an ICW waveform is to optimize the proportion of the interpulse period T devoted to τ_t, τ_b, and τ_r. Three types of optimization have been studied where it is assumed that the target range is much greater then $cT/2$:

1. The transmitter is peak power limited and it is desired to maximize the target signal-to-noise ratio S/N averaged over all possible target ranges for a given blanking time τ_b.

2. The transmitter is average power limited such that a large value of d_t will necessitate a reduction in peak transmit power. The optimization is again for maximum average S/N for a given blanking time τ_b.

3. The system sensitivity is limited by close-in clutter, and it is desirable to increase τ_b to blank out the close-in clutter. This optimization will maximize the signal-to-clutter ratio for a given amount of degradation in S/N.

The optimization of the average signal-to-noise ratio is not the same as the optimization of detection probability [139]. If the target is eclipsed during any given observation time, the loss in detectability is infinite. To alleviate this situation the PRF is often switched between two or three values during the time the antenna moves one beamwidth.

Peak Power-limited Case

For the peak power-limited radar Walcoff [393] has shown that the average receive power from the target can be written

$$\bar{P}_r \approx \frac{2}{3}\tau_r^3 + (1 - 2\tau_r - \tau_b)\tau_r^2 \quad \text{for} \quad \tau_t > \tau_r \tag{10-12}$$

$$\approx \frac{2}{3}\tau_t^3 + (1 - 2\tau_t - \tau_b)\tau_t^2 \quad \text{for} \quad \tau_t < \tau_r \tag{10-13}$$

Assuming that the receiver noise during the period τ_r is the limiting factor on detectability, the receiver noise power N can be written

$$N = KT_s b d_r \tag{10-14}$$

where K = Boltzmann's constant

T_s = system noise temperature as defined in Eq. (2-3)

b = doppler filter noise bandwidth

d_r = receiver duty ratio

By calculating $\bar{P}$ and N for various values of τ_b, it was found that the optimum ratio of τ_t/τ_r was 1.67 for all values of d_b. The normalized signal-to-noise ratio is shown on Fig. 10-3 as a function of d_t for several values of d_b. Since the minimum value of d_b is a function of the antenna, receiver layout, and spillover problems, it can generally be estimated prior to the choice of τ_t and τ_r. Once τ_b can be approximated, the other terms can be determined by letting $\tau_t/\tau_r = 1.67$.

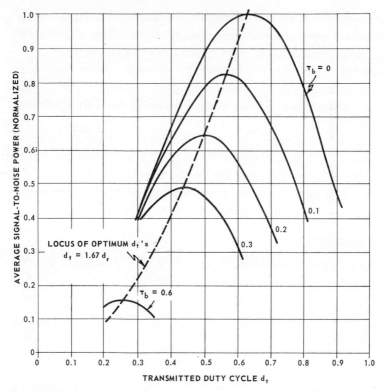

figure 10-3 *Signal-to-noise ratio in the receiver gate as a function of duty cycle and receiver blanking time. Peak power-limited case. (After Walcoff [393].)*

Average Power-limited Case

A second case of importance is when the transmitter average power is limited. This case also applies to preliminary design studies since transmitter cost for high-duty ratio systems is more dependent on average than peak power. Walcoff [393] has also calculated the average return power for this case.

$$\bar{P}_r \approx \frac{2\tau_r^3}{3\tau_t} + (1 - 2\tau_r - \tau_b)\frac{\tau_r^2}{\tau_t} \qquad \tau_t > \tau_r \qquad (10\text{-}15)$$

$$\approx \frac{2}{3}\tau_t^2 + (1 - 2\tau_t - \tau_b)\tau_t \qquad \tau_t < \tau_r \qquad (10\text{-}16)$$

The receiver noise power is the same as expressed in Eq. (10-14). The optimization of S/N leads to $\tau_t = \tau_r$ for all values of τ_b. In this case there is less loss in S/N than for the peak power-limited cases when τ_b is large, since the peak power can be increased indefinitely as τ_b is increased. One method of selecting τ_b is to draw graphs of clutter-to-signal power ratio P_c/P_s versus range to be shown in Figs. 10-6 through 10-8. From these graphs the location of the maximum clutter regions can be determined. Often τ_b can be extended in order to gate out much of this clutter at the price of a moderate reduction in receiver duty factor.

Clutter-limited Case

The clutter-limited case is difficult to evaluate in closed form. When the clutter is primarily the reflections from uniform rain close to the radar, the blanking period (in distance units) must be at least equal to the extent of the antenna near field to significantly reduce the clutter power. Thus,

$$\frac{c\tau_b}{2} > \frac{D^2}{\lambda} \qquad (10\text{-}17)$$

where D is the antenna diameter and λ is the carrier wavelength. This is necessary since, as shown in Sec. 10.3, only one half of the clutter power is returned from the near field of a single-aperture CW system.

If the clutter is primarily due to sea or land clutter, the choice of τ_b which maximizes S/C is strongly dependent on antenna height, main beam elevation angle, antenna sidelobes, and the potential use of a *clutter fence.*

10.3 TOTAL RAIN CLUTTER POWER FOR SINGLE ANTENNA CW RADAR[1]

In a CW radar the return power at the antenna from distributed reflectors (such as rain) is the sum of the powers from the individual reflectors at all ranges. In cases where the spatial distribution of the reflectors is relatively uniform and the cross sections are not grossly different, the sum may conveniently be replaced by an integral. If it is assumed that the rain extends into the radar and the usual radar equation

[1]The analysis in this section was carried out by P. J. Luke.

is used for the return from an element of volume, the integral diverges. In order to properly account for the return from rain very close to the antenna, it is necessary to replace the usual far field expression for the echo by the near or Fresnel field expression. This procedure leads to considerable mathematical difficulty. It is possible, however, to make some approximations and obtain a reasonably good result.

The power scattered back toward the radar by rain in an element of volume dv is (see Sec. 2.5 and [296]).

$$\frac{PG\,|f(\theta,\varphi)|^2}{4\pi\,r^2}\,\Sigma\sigma\,dv$$

where $\Sigma\sigma$ is the rain backscattering cross section per unit volume. Of this backscattered power the fraction received by the radar is $A_e\,|f(\theta,\varphi)|^2/(4\pi r^2)$, where A_e is the effective area of the antenna. The echo power received from volume element dv is thus

$$dP_r = \frac{PG\,|f(\theta,\varphi)|^2}{4\pi r^2}\,\Sigma\sigma\,dv\,\frac{A_e\,|f(\theta,\varphi)|^2}{4\pi r^2}$$

$$= \frac{PG\,A_e\,\Sigma\sigma}{(4\pi)^2}\,|f(\theta,\varphi)|^4\,\frac{dv}{r^4}$$

(10-18)

Noting that $G = 4\pi A_e/\lambda^2$ and $dv = r^2\,dr\,d\Omega$, where $d\Omega$ is an element of solid angle, the total rain return, assuming the rain is uniform, is given by

$$P_r = \int dP_r = \frac{P\Sigma\sigma}{4\pi}\left(\frac{A_e}{\lambda}\right)^2 \int_{r_1}^{r_2}\frac{dr}{r^2}\int |f(\theta,\varphi)|^4\,d\Omega$$

where r_1 and r_2 are the limits of the rain cloud in range, and the integration in angle is over the solid angle subtended by the rain cloud. The integration over r yields a factor

$$\frac{1}{r_1} - \frac{1}{r_2} = \frac{r_2 - r_1}{r_1 r_2}$$

For the integral over angles, let Ω_1 and Ω_2 be the equivalent solid angles for one-way and two-way transmission, that is,

$$\Omega_1 = \int_{4\pi} |f(\theta, \varphi)|^2 \, d\Omega \tag{10-19}$$

and

$$\Omega_2 = \int_{4\pi} |f(\theta, \varphi)|^4 \, d\Omega \tag{10-20}$$

since $|f(\theta, \varphi)| \leq 1$, $\Omega_2 < \Omega_1$. Therefore, if one lets $\Omega_2 = \eta^2 \Omega_1$, then η^2 will be a number less than one. From the definition of gain,

$$G = \frac{4\pi}{\displaystyle\int_{4\pi} |f(\theta, \varphi)|^2 \, d\Omega} = \frac{4\pi}{\Omega_1} \tag{10-21}$$

one finds $\Omega_1 = \lambda^2/A_e$; and the power returned from the rain is (assuming the lateral extent to be great compared to the beamwidth)

$$P_r = \frac{P\Sigma\sigma}{4\pi} \left(\frac{A_e}{\lambda}\right)^2 \left(\frac{r_2 - r_1}{r_1 r_2}\right) \eta^2 \left(\frac{\lambda^2}{A_e}\right) \tag{10-22}$$

$$= \frac{P\Sigma\sigma\lambda}{4\pi} \eta^2 r_0 \frac{(r_2 - r_1)}{r_1 r_2}$$

where $r_0 \equiv A_e/\lambda$.

If $r_2 - r_1$ is small compared to r_1, the received power is approximately (putting $r_2 - r_1 = h$)

$$P_r = \frac{P\Sigma\sigma\lambda}{4\pi} \eta^2 \frac{r_0 h}{r_1^2} \tag{10-23}$$

This equation is also applicable to a pulse radar with pulselength equal to $2h/c$. If r_2 is large compared to r_1, then Eq. (10-22) becomes

$$P_r \simeq \frac{P\Sigma\sigma\lambda}{4\pi} \eta^2 \frac{r_0}{r_1} \tag{10-24}$$

The factor η^2 may be computed if the pattern function $f(\theta, \varphi)$ is known. Probert-Jones [296] makes a calculation assuming a gaussian antenna pattern and obtains a value of

$$\eta^2 = \frac{\pi^2}{32 \ln 2} = 0.445$$

For rain extending into or nearer than the Fresnel-Fraunhofer transition point, we set r_1 equal to the range of the transition point (κr_0, where κ is a number of the order of one, the precise value depending on the shape and illumination of the antenna); and Eq. (10-22) or Eq. (10-24) gives the far field contribution to the total rain echo

$$P_r = \frac{P\Sigma\sigma\lambda}{4\pi} \left(\frac{\eta^2}{\kappa} - \eta^2 \frac{r_0}{r_2} \right) \tag{10-25}$$

and Eq. (10-24) becomes

$$P_r = \frac{P\Sigma\sigma\lambda}{4\pi} \frac{\eta^2}{\kappa} \tag{10-26}$$

Hansen [157, p. 40] lists transition ranges for four antennas, as follows:

1. Uniformly illuminated square: $r_t = D^2/\lambda$
2. Square with cosine tapered illumination: $r_t = 4D^2/(\pi^2\lambda)$
3. Uniformly illuminated circular aperture: $r_t = \pi D^2/(4\lambda\sqrt{2})$
4. Circular aperture, $(1 - r^2)$ tapered illumination: $r_t = \pi D^2/(8\lambda)$

Silver [350, chap. 6] gives gain factors for various antennas from which the effective area is obtained, respectively, namely

1. $A_e = D^2$
2. $A_e = 8D^2/\pi^2$
3. $A_e = \pi D^2/4$
4. $A_e = 3\pi D^2/16$

From these relations the value of κ for each of these antennas is found to be

1. Uniform, square: $\kappa = 1$
2. Tapered, square: $\kappa = 1/2$
3. Uniform, circular: $\kappa = 1/\sqrt{2}$
4. Tapered, circular: $\kappa = 2/3$

Near Field (Fresnel Region)

The electromagnetic field in the Fresnel region is quite different from that in the Fraunhofer region although there is a smooth transition from one to the other. This is shown in Hansen [157, p. 46] for a uniformly illuminated square aperture. The most significant feature for the present purpose is that the radiation is largely concentrated within a cylinder whose cross section is the aperture area until the distance from the aperture is approximately that designated as the transition range. This behavior is also indicated by the on-axis power density (shown in [157, fig. 19, p. 36 and fig. 22, p. 38]) which at large distance varies inversely as the square of the distance but in the Fresnel region oscillates about a constant value. It may also be noted that at the transition range κr_0 a cone with solid angle Ω_1 (i.e., the equivalent solid angle for one-way transmission) would intercept an area of $\kappa^2 A_e$. This is shown as follows: The area within the cone and on the surface of a sphere of radius κr_0 is $(\kappa r_0)^2 \Omega_1$. The definitions of Ω_1 and of the gain G yield

$$\Omega_1 = \frac{4\pi}{G} = \frac{\lambda^2}{A_e}$$

The definition of $r_0 = A_e/\lambda$ and the above relation between Ω_1 and A_e then give the area as

$$(\kappa r_0)^2 \Omega_1 = \kappa^2 \left(\frac{A_e}{\lambda}\right)^2 \left(\frac{\lambda^2}{A_e}\right) = \kappa^2 A_e$$

All of the above arguments are presented in support of the plausibility of the method used below for computing the echo power from rain in the Fresnel region.

It will be assumed that in the Fresnel region the radiation is confined to a cylinder of cross section equal to the effective area of the antenna and of length κr_0. Inside this cylinder the power density is uniform. The rain is assumed to be uniform and to fill the cylinder from range r_a to range r_b.

The power density (P/A_e inside the cylinder) may be compared to that in the far field to obtain the correspondence

$$\frac{PG \, |f(\theta, \varphi)|^2}{4\pi r^2} = \frac{P A_e \, |f(\theta, \varphi)|^2}{\lambda^2 r^2} \longrightarrow \frac{P}{A_e}$$

hence

$$\frac{A_e |f(\theta, \varphi)|^2}{4\pi r^2} \longrightarrow \frac{\lambda^2}{4\pi A_e}$$

The echo power received from a volume element dv from Eq. (10-18) is then

$$dP_r = \frac{P}{A_e} \Sigma \sigma \, dv \, \frac{\lambda^2}{4\pi A_e} = \frac{P \Sigma \sigma \lambda}{4\pi} \frac{\lambda}{A_e^2} \, dv \tag{10-27}$$

Integrating this gives the total rain return, namely,

$$P_r = \frac{P \Sigma \sigma \lambda}{4\pi} \frac{\lambda}{A_e^2} A_e (r_b - r_a)$$

$$= \frac{P \Sigma \sigma \lambda}{4\pi} \frac{r_b - r_a}{r_0} \tag{10-28}$$

If the rain extends to or beyond the transition point κr_0 the contribution to the echo power from the Fresnel region is

$$P_r = \frac{P \Sigma \sigma \lambda}{4\pi} \left(\kappa - \frac{r_a}{r_0} \right) \tag{10-29}$$

and if r_a is small compared to r_0

$$P_r = \frac{P \Sigma \sigma \lambda}{4\pi} \kappa \tag{10-30}$$

Total Rain-clutter Echo Power

Adding Eqs. (10-26) and (10-30), one has the power returned from rain extending from the radar to essentially infinity

$$P_r = \frac{P\Sigma\sigma\lambda}{4\pi}\left(\frac{\eta^2}{\kappa} + \kappa\right) \qquad (10\text{-}31)$$

The ratio of near field contribution to far field contribution is κ^2/η^2, the value of which is about one for the circular aperture with uniform illumination and about 2.2 for the square antenna with uniform illumination. Note that the total clutter power is independent of aperture size; and since $\Sigma\sigma$ for rain echoes is proportional to λ^{-4}, the overall frequency dependence varies as λ^{-3}.

For an interrupted CW radar the rain-clutter echo power can be determined by substituting for $r_1, r_2, r_a,$ and r_b in the preceding equations.

If extended rain clutter as seen by a stationary radar has a spectrum that is narrow compared to the doppler shift of the target and the modulating frequencies and indices ($m < 0.2$) of the transmitter noise are small, the amount of noise on the clutter signal can be approximated by

$$P_{rb} \doteq \frac{\overline{P}_T\Sigma\sigma\lambda}{4\pi}\left[\frac{1}{R_{AM}}\left(\frac{\eta^2}{\kappa} + \kappa\right) + \frac{1}{R_{FM}}\left(\frac{\pi^2 f_m \eta^2 r_0}{c}\right)\right] \qquad (10\text{-}32)$$

where P_{rb} is the transmitter noise power in the doppler filter of interest, R_{AM} is the AM noise sideband ratio of the transmitter at the doppler frequency of interest, and R_{FM} is the noise sideband ratio at that doppler frequency.

The theory derived in this section is compared with a series of experiments by Kiely [213] and illustrated on the upper portion of Fig. 10-4 for a K_a band (λ = 8-mm) CW radar. The total receive power relative to the transmit signal is plotted for various rainfall rates. The values of $\Sigma\sigma$ to obtain the theoretical curve were taken from Chap. 6. The antenna taper and efficiency were estimated. The disparity between the theory derived in this section and the experiments will require further study. The relative rain backscatter power using separate antennas will be summarized in the next section.

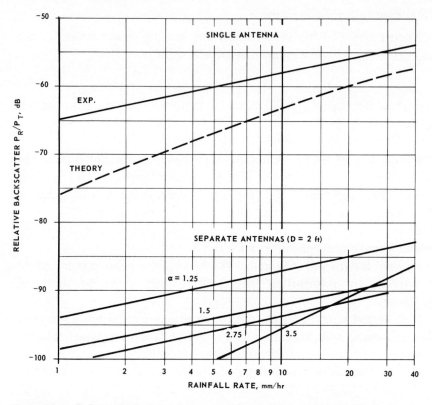

figure 10-4 *Relative backscatter power as a function of rainfall rate for K_a-band CW radar in uniform rain; single and separate antennas. (Experimental data from Kiely [213] with permission of the Controller of Her Majesty's Stationery Office.)*

10.4 RAIN CLUTTER POWER FOR SEPARATE TRANSMIT AND RECEIVE ANTENNAS

This section considers the backscatter power from a uniform rain surrounding a bistatic CW radar. The radar has separate circular transmit and receive antennas that are collimated (parallel beams) on some distant target. It is assumed that the near fields (Fresnel regions) of the two antennas do not overlap and that the total backscatter from the rain is primarily due to the overlap of the far field patterns of the antennas as shown on Fig. 10-5. Further assumptions for this section are

1. The antennas have the same sense linear polarization and uniform phase excitation.

2. Both apertures are large compared to a wavelength and small angle approximations can be made.

3. The rain is uniform everywhere.

4. Direct spillover and multiple scattering are negligible.

In a manner similar to the derivation in the previous section, Wild [405] considered the total clutter power in the receiver for separate equal diameter coplanar antennas. For uniformly illuminated apertures and neglecting attenuation, the relative backscatter power P_R/P_T can be approximated by

$$\frac{P_R}{P_T} = 0.022 \frac{\lambda \Sigma \sigma}{\alpha} \quad \text{uniform illumination} \tag{10-33}$$

where λ = carrier wavelength, m

$\Sigma \sigma$ = volume reflectivity, m^2/m^3

α = center-to-center spacing in antenna diameters as illustrated on Fig. 10-5, $\alpha \geq 1$

For uniform illumination (first sidelobe 17.5 dB down from the main lobe) about 93 percent of the total power is from the overlapping regions of the main beams. The region where the mainlobe of one antenna intercepts the first sidelobe of the other contributes 4 percent. The remaining sidelobe contributions are about 3 percent.

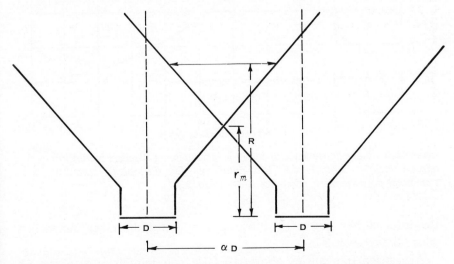

figure 10-5 *Antenna beam geometry for separate apertures of equal diameter D.*

The range dependence of the clutter is shown on Fig. 10-6. The ordinate is the incremental relative backscatter and the abscissa is an arbitrary normalized range in units of the minimum range where the beams overlap r_m.

$$r_m \approx \frac{\alpha D/2}{3.8\lambda/\pi D} = \frac{\alpha D^2}{2.44\lambda} \tag{10-34}$$

where D is the diameter of the uniformly illuminated apertures and r_m is slightly less than the transition range nr_0 between the near and far fields.

Three curves are shown on the graph. Curve A is the contribution of the mainlobes, B includes the addition of the main lobe to first sidelobe power, and C is the total power (in increments of R/r_m). From these curves the spectrum of the transmit FM noise which is received with the rain clutter can be derived. In addition curve C can be used to estimate

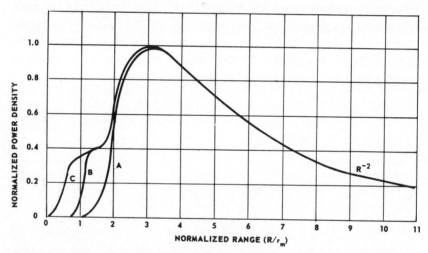

figure 10-6 *Range dependence of echo for two antenna CW radars in uniform rain. (A) Mainlobe contribution; (B) mainlobe plus first sidelobe; (C) total CW echo range power density. (Courtesy of Wild [405].)*

the total clutter power by integrating only for the range increments where the receiver is on.

At carrier frequencies above 10 GHz and rainfall rates above 16 mm/hr, the attenuation of the distant rain clutter signals may become significant. By expressing the two-way attenuation coefficient (from Chap. 6) in units of dB/r_m, Eq. (10-33) can be multiplied by an empirical attenuation factor to yield the relative backscatter for heavy rainfall

$$\frac{P_R}{P_T} = 0.022 \, (0.27)^{a^{0.64}} \frac{\lambda \Sigma \sigma}{\alpha} \tag{10-35}$$

where a is two-way rainfall attenuation in dB/r_m and is valid for $0 \le a \le 3$. Attenuation can be neglected for $a \le 0.2$ with less than 1-dB error unless the FM noise on distant clutter is important.

Wild [405] also made hand calculations for amplitude-tapered apertures with a $[b + (1 - K^2)^n]$ distribution for $b = \frac{1}{2}$, $n = 2$ where K is the radial distance from the center of the aperture [157, p. 66]. This results in first sidelobes 26.5 dB below the peak gain. In this case 99 percent of the reflected power comes from the main beam intersection. Neglecting attenuation, the relative rain backscatter power can be approximated for equal size apertures.

$$\frac{P_R}{P_T} = \frac{0.023 \, \lambda \Sigma \sigma}{\alpha} \quad \text{tapered illumination} \tag{10-36}$$

A limited number of calculations were also made for unequal circular apertures with uniform illumination. With a few tenths of a decibel error, Eq. (10-33) can be modified as

$$\frac{P_R}{P_T} = \frac{0.056 \, \mu \lambda \Sigma \sigma}{\alpha \, (2.7)^\mu} \quad \text{unequal apertures} \tag{10-37}$$

where D is the larger aperture diameter and μD is the smaller aperture diameter. Then $0 \le \mu \le 1$. Sidelobe contributions should yield only a slight increase in the relative backscatter.

The only widely reported experiments on the backscatter from bistatic CW radars is the work of Kiely [213] which is shown on the lower portion of Fig. 10-4. As with the single-aperture experiments, there is considerable disparity with the theory even when the rainfall attenuation is estimated. The theory predicts higher relative backscatter than the experiments with the separated antennas.

10.5 SEA- AND LAND-CLUTTER POWER FOR SURFACE ANTENNAS

While the computation of the magnitude and spectrum of sea- and land-clutter echoes from an airborne antenna can be approximated by the relationships of Sec. 10.6, it is difficult to present simple relationships for the clutter echoes from a surface radar. This is a result of the clutter echoes being a highly sensitive function of the sidelobe structure of the antenna. One successful method of computation inserts the best estimate of the antenna pattern into a digital computer along with an analytic function for the clutter backscatter as a function of depression angle $\sigma_0(\Psi)$. The clutter response from a large number of small areas on the surface is then integrated. The numerical results given in this section are the result of such a procedure. The emphasis is on the effect of beamwidth, sidelobe structure, and elevation of the centerline of the beam above the horizontal. The results are applicable to a CW, Interrupted CW (ICW), or pulsed radar.

In order to give physical significance to the results, a number of assumptions have been made.

1. The target is in all cases a 5-m² aircraft or missile at 50 nmi from the radar.

2. The radar is vertically polarized at 60 ft above a flat earth and has an average CW power of 1 kW or an ICW peak power of 5 kW.

3. The frequencies of interest are C band (λ = 6 cm) or X band (λ = 3 cm).

4. The mean sea- or land-clutter reflectivity can be represented by $\sigma_0(\Psi) = \gamma \sin \Psi$ for $0.001 < \Psi < 0.1$ radians where γ is a constant, independent of Ψ. This gives a better approximation to the reflectivity than would a constant value for σ_0 although the values in Chap. 7 are even better. For the computations $\gamma = -28$ dB, which roughly corresponds to the reflectivity from a sea state three at C band, a sea state two at X band or very flat terrain at either band.

5. Forward-scatter effects have been neglected. This should not be a major factor at C or X band.

6. Antenna spillover has been neglected.

7. The computation is performed in range increments of 1.2 μsec (0.1 nmi) and in azimuth increments of 0.004 radians. Clutter at azimuth angles beyond 20° did not influence the results.

8. Two basic antenna types are considered. The first is a small symmetrical antenna with a 3° beamwidth with -20-dB first sidelobes

and -30 dB for the next few sidelobes. The second is a measured pattern of a higher quality antenna with a 1.8° beamwidth at C band, -30-dB first sidelobes and the next sidelobes falling off rapidly. This latter antenna pattern is scaled for X-band transmission and also for a 3.6° beamwidth.

Two systems are considered; the first is a pure CW system, and the second is an ICW system with 1.2-μsec transmit time, 2.0-μsec clutter-decay time, and 1.8-μsec receive time. The pulse repetition frequency is thus 200 kHz.

It may be well to collect the approximate equations for the return signals to get a feeling for effects such as transmit frequency. The clutter return can be written for main beam illumination (neglecting loss terms).

$$P_c = \frac{P_T G_T G_R \lambda^2 (\sigma_0 A)}{(4\pi)^3 R_c^4} \tag{10-38}$$

where A = illuminated area

σ_0 = reflection coefficient = $\gamma \sin \Psi \approx \gamma h / R_c$ (Secs. 7.2, 7.11)

h = antenna height above the surface

R_c = range to center of clutter element

but $A = R_c \theta (c\tau/2)$ for small θ and low grazing angles

θ = two-way azimuth beamwidth, radians

τ = pulse duration or range-gate duration

Reducing the equation, one obtains

$$P_c = \frac{P_T G_T G_R \lambda^2 \gamma h \theta (c\tau/2)}{(4\pi)^3 R_c^4} \quad \text{main beam} \tag{10-39}$$

The return from a target of cross section σ_t and at range R_T is

$$P_s = \frac{P_T G_T G_R \lambda^2 \sigma_t}{(4\pi)^3 R_T^4}$$

and the received clutter-to-signal ratio is

$$\frac{P_c}{P_s} = \frac{\gamma h c\tau}{2\sigma_t} \left[\frac{R_T}{R_c} \right]^4 \tag{10-40}$$

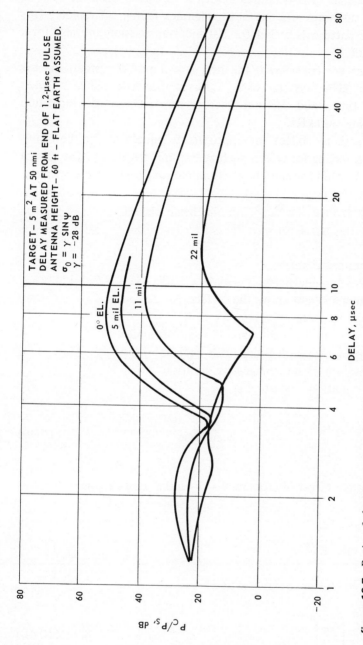

figure 10-7 *Ratio of clutter power to target echo power per 1.2 μsec of range delay, 1.8° beamwidth. The total clutter power for a CW system is −52 dBm for 0° beam elevation, −58 dBm for 5-mil elevation, −62 dBm for 11-mil elevation, −79 dBm for 22-mil elevation.*

By using γ rather than σ_0 the range term drops out of the equation for targets in the clutter. A small antenna height would seem to be advantageous, but it reduces the radar horizon and delays initial *acquisition* of targets.

The results to follow were computed using typical antenna patterns and calculated antenna gains at the target and clutter locations. Some typical results are presented in Fig. 10-7 and Table 10-1. Figure 10-7 illustrates the clutter-to-signal power ratio as a function of clutter range R_c for the 1.8° beamwidth high-quality circular antenna with elevation of the antenna centerline in milliradians as a parameter. The point at which the main beam hits the clutter is the *hump* at the center of the graph. The smaller hump at the left is due to the first sidelobe clutter power. The total clutter power for a CW transmission (the area under the curves) is also indicated along with the beam elevation. It can be seen that elevating the beam reduces the clutter signal much faster than the target signal although the signal-to-receiver noise power ratio is reduced. The total clutter signal for the ICW transmission is obtained by integrating at only those delay times when the receiver is on (1.8 μsec

TABLE 10-1   Clutter-to-signal Ratios for Various Antennas for h=20m, 5 m^2 Target at 50 nmi, γ=−28

Antenna size, frequency, and beamwidth	Elevation angle, mils	CW transmission			ICW transmission	
		P_C dBm	P_C/P_S, dB	Sensitivity loss, dB	P_C/P_S, dB	Sensitivity loss, dB
8 ft X band High quality $\theta = 0.9°$	0	−58	47	0	47	5.5
	5.5	−70	37	—	37	—
	11	−85	28	7.5	22	13.0
8 ft C band High quality 1.8°	0	−52	60	0		
	5	−58	54	—		
	11	−62	51	—		
	22	−79	41	7.5		
2 ft X band High quality 3.6°	0	−58	73	0		
	22	−68	64	—		
	44	−85	53	7.5		
2.5 ft X band Low quality 3.0°	0	——	82	—		

every 4.0 μsec). In this case there is a 5.5-dB loss *on the average* in target signal from the CW case for 0° elevation due to the receive duty factor of 36 percent. Results with several other antenna configurations are shown on Table 10-1. The first column gives the antenna type and frequency. The second column is the beam elevation in milliradians (17 mrad = 1°). The third column is the received clutter power P_c for $\gamma = -28$ dB. This factor is important to determine if there is receiver saturation. For rough seas add 10 dB to P_c and P_c/P_s and for typical wooded or hilly terrain add 20 dB. The clutter-to-signal power is given for both the CW and ICW cases. The sensitivity loss column accounts for the reduced power density on the horizon as a result of elevating the beam and the 36 percent receiver duty factor of the ICW system.

It can be seen that a reduction of the beamwidth by a factor of four reduces the clutter-to-signal ratio by about 20 dB, and elevating the beam by about two-thirds of a beamwidth achieves a similar reduction. The ICW mode is advantageous only with the elevated beam with the nine-tenths beamwidth.

It can be seen from the values of total clutter power that considerable doppler filtering is required even for the moderate environment that was assumed. The clutter problem does not vanish even if the centerline of the antenna beam is two to three beamwidths above the horizon. This is illustrated on Fig. 10-8 for the two antennas previously discussed. The total clutter power exceeds the target echo, but the problem is far less severe for high-gain antennas with low sidelobes.

10.6 CLUTTER SPECTRUM FOR AIRBORNE RADARS

Since the primary purpose of a CW radar is the detection and tracking of targets on the basis of their velocity separation from clutter, it is useful to calculate the doppler spectrum of these clutter echoes. If the radar is stationary, the relationships of Chaps. 6 and 7 should suffice although it may be necessary to include a computation of the spectra of antenna sidelobe echoes. The doppler spectrum at an airborne receiver is dominated by the platform motion.

This section will give the equations to compute this spectrum by assuming that the antenna is circularly symmetrical with a beamwidth of less than 0.25 rad. In addition, the clutter is assumed to be homogeneous with a negligible mean velocity compared to the platform motion. The

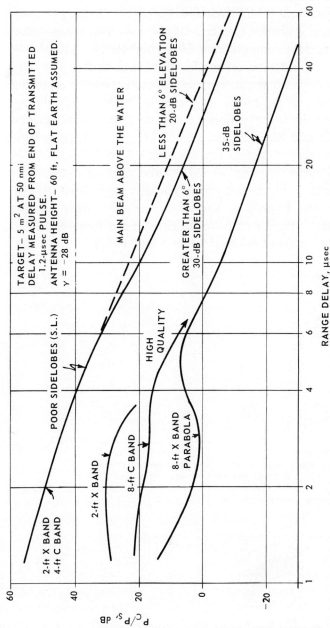

figure 10-8 *Ratio of sidelobe clutter power to target power for antenna beam elevation angles of 2-3 beamwidths and targets in the main beam.*

analysis applies to a CW or ICW radar where the target and clutter are many times the unambiguous range.

Following the procedures of Farrell and Taylor [112], the computation of the spectrum is divided into an antenna main lobe computation and one or more sidelobe computations. They computed that, outside of the first sidelobe, the sidelobe clutter spectrum is essentially flat. This occurs because there is no special correspondence between a particular sidelobe and a region on the surface of the earth. While this simplifying assumption is used in the equations of this section, Helgöstam and Ronnerstam [166] and Biernson and Jacobs [41, fig. B-1] give more detailed spectral shapes for the first few sidelobes.

The geometry is shown on Fig. 10-9. The airborne radar at height h is moving at a velocity V in the Y-Z plane. The dive angle ζ will arise in later discussions. The hyperbolas shown in the X-Y plane are the contours of constant doppler or *isodops*. The doppler frequency f_d of the echoes from the center of the main beam is then [112]

$$f_d = \frac{61\,V\,\cos\Lambda}{\lambda} \tag{10-41}$$

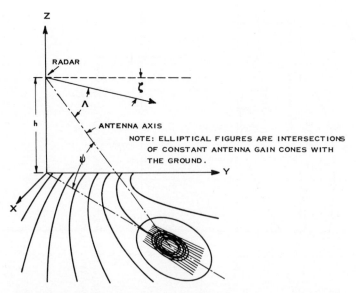

figure 10-9 *Antenna geometry for airborne radar observing surface clutter. (After Farrell and Taylor, IEEE [112].)*

where V is the platform velocity in feet/second, λ is the RF wavelength in centimeters, and Λ is the angle (radians) between the antenna axis and the radar platform velocity vector if $0 < \Lambda < \pi$. If both V and λ are in meters, the constant is equal to two. The maximum spectral density in watts/hertz at this center frequency neglecting radar loss terms L is

$$W_m(f_d) = \frac{10^{-8}\,\hat{P}_T\,d_t\,G_0^{\,2}\,\lambda^3\,\theta_1\,\sin\Psi\,\sigma_0(\Psi)}{h^2\,V\,\sin\Lambda} \qquad (10\text{-}42)$$

where d_t = transmit duty factor ($\hat{P}_T d = \bar{P}$ for CW radar)
$\quad G_0$ = peak antenna gain ($G_T = G_R$)
$\quad \theta_1$ = one-way half-power beamwidth, radians
$\quad \Psi$ = depression angle from the horizontal
$\quad \sigma_0(\Psi)$ = normalized backscatter coefficient at angle Ψ
$\quad h$ = antenna height above surface, ft
The shape of the main beam clutter spectrum is

$$W_m(f_d \pm \Delta f) = W_m(f_d)\ \mathrm{erfc}\left[\frac{0.0387\,\lambda\Delta f}{V\theta_1\,\sin\Lambda}\right] \qquad \text{w/Hz} \qquad (10\text{-}43)$$

where erfc is the complement of the error function.

These expressions were obtained [112] assuming that

1. The earth is flat. Determination of $\sigma_0(\Psi)$ from Chap. 7 at the true incidence angle rather than the depression angle will make this a better approximation.

2. The antenna main beam has a gaussian shape.

3. The reflection coefficient σ_0 is a constant over the main beam illuminated area.

4. The range R to various points in the illuminated area is essentially constant.

5. The angle from the velocity vector to the line of sight Λ is greater than the angle from the center of the beam to the first null. If this is not the case the spectrum of the clutter will most likely be disjoint from an inbound target.

The main beam clutter-to-signal power ratio can be calculated by dividing the product of Eq. (10-43) and the doppler filter bandwidth b by the target echo power.

The sidelobe clutter-to-noise ratio was also derived by Farrell and Taylor [112]. Assuming a flat doppler spectrum in the sidelobe regions,

$$\frac{P_{cn}}{P_s} = \frac{4.3 \times 10^{12} \hat{P}_T d_t^2 G_0^2 g^2 \lambda^3 \sigma_0 \cos\zeta}{\bar{F} h^2 V} \qquad \text{for} \quad \zeta \le 45° \qquad (10\text{-}44)$$

where P_{cn} = the clutter power in the nth *isodop* or nth doppler channel

ζ = the dive angle from the horizontal of the aircraft or missile, positive downwards, radians

$\bar{F}$ = receiver noise figure, power ratio

g = the normalized sidelobe power gain (if the sidelobes (one-way) are 30 dB below the main beam $g^2 = 10^{-6}$; also $G_0 g$ is the absolute sidelobe gain).

Clutter spectral densities for ICW transmission are given in [166, figs. 2 and 3].

Even if the target echo is well above the clutter, or if the clutter and target echo spectra are disjoint, there is always a possibility that a doppler filter will *lock-up* on the main beam clutter. This problem is often alleviated by looking at the highest doppler frequencies first if only inbound targets are of interest. An alternate procedure is to determine the spectral width of the echo since in general the clutter has a broader spectrum then the target. This latter method is sometimes called a *coherency check.*

11 PULSE-DOPPLER AND BURST WAVEFORMS

11.1 TERMINOLOGY AND GENERAL ASSUMPTIONS

A pulse-doppler radar combines the range discrimination capability of pulse radar with the frequency discrimination capability of CW radar by using a coherent pulse train, i.e., a train of pulses which are samples of a single unmodulated sine wave. For a fixed repetition rate, the spectrum consists of a set of lines with spacing equal to the repetition frequency. When a coherent pulse train is reflected by a moving object, the lines of the spectrum are doppler frequency shifted an amount proportional to the object's radial velocity. When a number of objects with different

For general references see Resnick [309], Rihaczek [319, 316, 317], Ares [7, 6], Barton [20, chap. 12], Galejs [130].

velocities are present, the resultant echo is a superposition of a corresponding number of pulse trains, each with its own doppler shift. A range gate is used to select only those pulse trains coincident in time (within a pulse width) with the pulse-train echoes from the target. A narrow-band filter following the range gate often selects only a single spectral line corresponding to a particular doppler shift, thus attenuating all those trains which pass the range gate but do not have the proper doppler shift.

Burst waveforms may be considered a special case of pulse doppler, with an implied limit of 50 to 100 pulses per antenna beam position. This distinction is somewhat related to radar function, with the truly continuous transmission of pulses (pulse-doppler) being more appropriate to tracking systems with long *dwell* times on the target. The burst waveforms are better suited to surveillance or acquisition systems, as the parameters are chosen for compatibility with time restrictions inherent in three-dimensional surveillance radars.

Therefore, the term *pulse doppler* will be used to describe systems with essentially continuous transmission of pulses, and the terms *burst waveform* or *pulse train* to describe transmissions of finite extent. A *pulse-doppler receiver* will be considered one that extracts the energy in a single PRF line of the spectrum, while a *comb filter receiver* extracts the energy in all the spectral lines. With proper design, the performance of both types of receivers should be equivalent.

The use of the term *burst waveform* will also imply removal of three restrictions generally attributed to pulse-doppler processes.

1. A burst waveform need not have all of the pulses radiated at the same power level, and the processor may have amplitude or phase weighting on a pulse-by-pulse basis.

2. A processor for burst waveforms need not have a physical range gate but may have its gating defined by the time of arrival of the processor output at a threshold.

3. A burst waveform need not have all of the pulses on the same carrier frequency.[1]

In this section several assumptions will be made about the waveform and the target being observed.

---

[1] The burst waveform with each pulse at a different carrier frequency or with only a few pulses per frequency will be discussed in Chap. 13 on linear FM and frequency coding.

1. The target velocity is not high enough so that its echo passes out of the range gate during the coherent integration time.[1]

2. The target is not accelerating or decelerating by an amount sufficient to spread the doppler echoes beyond the response of a single doppler filter. (See Chap. 5.)

3. The carrier frequency is much greater than the spectral width of an individual pulse.

4. The change in phase versus time due to target radial velocity is negligible during a single pulse.

5. The target can be considered as a point reflector or a few reflectors having an echo spectrum which is narrower than the doppler filter width.

It is to be noted that *pulse doppler* in this book will also be differentiated from multiple-pulse MTI or other multiple-pulse coherent systems which do not require that a range gate precede the doppler filtering. It will also be assumed that an attempt is made to make the receiver a matched filter to the pulse train. Deviations from these assumptions will be handled as efficiency factors L_s rather than by defining separate techniques.

The pulse-doppler waveform will consist of a train of pulses which are constant amplitude samples of a coherent carrier frequency. For convenience in definitions, the minimum number of pulses N will be $\geq$ 10 and the duty cycle low enough to separate this technique from *interrupted* CW, which has no range gating other than to prevent the receiver from saturating on the transmitted signal leakage or from close-in clutter returns (see Sec. 10.2).

The potential advantages of the pulse-doppler technique, which include many of the virtues of CW, pulse, and MTI systems, also suffer some of the limitations of each. Some of the advantages include:

1. The ability to measure range and velocity unambiguously over a predetermined region of the ambiguity plane in the presence of multiple targets.

2. The ability to reject unwanted echoes in either doppler or range domains or in both.

3. Coherent rather than incoherent integration of the returns from many pulses with the attendant reduction in the required (S/N) or (S/C) per pulse.

---

[1] If this is not the case, see Rihaczek [320] who discusses the errors involved and methods of compensation.

4. Less sensitivity than MTI systems to the mean velocity and spectral width of the unwanted clutter if the target velocity is separated from the center of the clutter spectrum by $> 1/T_d$, where T_d is the coherent integration time.

5. Much greater *spillover*[1] rejection and often greater *close-in* clutter rejection than with CW systems.

The limitations cannot as easily be categorized but are intimately tied to the choice of such parameters as carrier frequency, PRF, pulse width, and filter bandwidth. The only truly general limitation is that pulse-doppler receiver circuitry is usually more complex than that found in pulse, CW, or simple MTI receiver systems. Pulse doppler also requires a transmitter with a pulsed-power amplifier chain or at least a pulsed transmitter where the phase of each transmission is stored in some manner. This can be construed as a limitation of most coherent systems.

11.2 RANGE-DOPPLER LIMITATIONS

In the discussion on MTI systems, it was generally assumed that the interpulse period was greater than the required unambiguous range. If the resulting *blind speeds* presented a serious limitation, a staggered or dual PRF system was suggested. Ambiguous ranges were only considered if second-time-around clutter echoes *folded* into the range interval of interest. In most current search radars the time per beam position has been sufficient to allow long interpulse periods and still retain three or more pulses per beam position. With the advent of large phased arrays with long detection ranges and narrow antenna beams, the time per beam position is considerably reduced. Since the pulse-doppler technique implies many pulses per beamwidth, the range ambiguity problem becomes difficult to avoid.

The ambiguity problem is more acute in search systems for air defense where the PRF and pulselength must be chosen to accommodate a broad span of target ranges and radial velocities. The carrier frequency and PRF determine the doppler ambiguities (blind speeds) from the usual doppler equation

$$f_d = \frac{2v}{\lambda} = \frac{2vf_0}{c}$$

---

[1] Leakage of signals directly from transmitter to receiver.

where f_d = the doppler frequency shift
  λ = the transmit wavelength = c/f_0
  v = the echo radial velocity

For an unambiguous velocity response it is required that

$$f_d < \frac{\text{PRF}}{2} = \frac{1}{2T}$$

where T is the interpulse period. For an unambiguous range response,

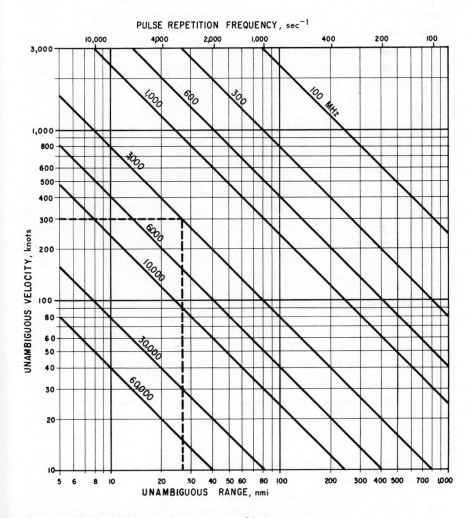

figure 11-1 *Unambiguous velocity vs. unambiguous range.*

$R_{max} \leq (c/2)(T - \tau')$ where τ' is the pulse duration.[1] The solution of both of these inequalities is shown in Fig. 11-1. The graph can be entered at the desired maximum unambiguous velocity with a horizontal line to the carrier frequency line. The projection of this intersection to the abscissa gives the maximum unambiguous range and the related pulse repetition frequency. Alternately the desired maximum range or PRF can be entered on the abscissa to determine the unambiguous velocity. The example shows that for ± 300-knot velocity coverage at S band (3,000 MHz), the unambiguous range is less than 27 nmi. For a finite number of pulses, the width of the spectral lines will slightly reduce these values.

It becomes obvious that for many short wavelength applications, unambiguous operation is impractical and parameters should be chosen by careful study of the ambiguity diagram.

11.3 AMBIGUITY DIAGRAMS FOR SINGLE CARRIER PULSE TRAINS

The ambiguity function described in Chap. 8 has appeared in many forms. It has proven useful in interpreting the performance resulting from the transmission of finite pulse trains in either multiple-target or clutter environments. In studies of the ambiguity function or diagram, it must be emphasized that the peak of the function is centered at the range and doppler of the target. Clutter and interference are then assumed to have the differential range and doppler shifts.

Constant Interpulse Period

The uniform pulse train consisting of N rectangular pulses of duration τ' and interpulse period T is the simplest and most widely used pulse-doppler waveform. Its great utility is based on the complete absence of time ambiguities in the region between the transmit pulses.[2] The ambiguity diagram of this waveform for $N = 9$ is shown in Fig. 11-2 (see Resnick [309]). The central response is labeled $\psi(0, 0)$ and the

---

[1] In this chapter τ' is used for pulselength where it might be confused with the time displacement variable τ.

[2] If all the targets in the environment are located within the unambiguous range $c(T - 2\tau')/2$ and are separated by at least $2\tau'$, there can be infinite resolution of their echoes. See Chap. 8 for terminology.

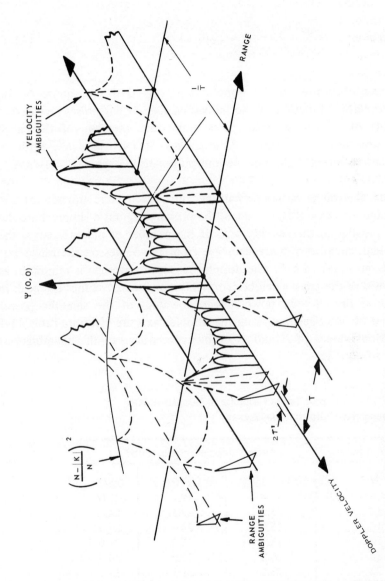

figure 11-2 *Ambiguity diagram for uniform pulse train.*

397

velocity response near the range axis where $\tau = 0$ is of the form for $\omega = 2\pi\nu \ll 1/T$

$$\psi(0,\omega) \approx \frac{\sin^2(N\omega T/2)}{N^2 \sin^2(\omega T/2)} \approx |\chi(0,\omega)|^2 \tag{11-1}$$

The range response on the doppler axis $\omega = 0$ is the square of the autocorrelation function of the individual rectangular pulses repeated at intervals of T. For a finite pulse train, the *envelope* of the range ambiguity peaks of period T decreases as $(N - |K|)^2/N^2$, with K representing integer values of the interpulse period over $-(N - 1) \le K \le (N - 1)$.

Since the range response can be zero for a large portion of the unambiguous range if τ'/T is small, the primary regions of interest are the ridges parallel to the velocity axis. If the number of pulses is large, the sinusoidal sidelobes rapidly decrease from the values on the range axis and at multiples of $1/T$. The doppler response envelope is minimum at $1/2T$, where the *peak* sidelobe is roughly equal to $1/N$ times the peak in amplitude or $1/N^2$ in power. *Average* values of the sidelobe power response at other regions along the doppler axis are given in Table 11-1 for various values of N and at various percentages of the unambiguous doppler frequency.

TABLE 11-1   Ratio of Target Response to Doppler-shifted Interference for Uniform Pulse Spacing

Number of pulses	Average doppler response, dB*		
N	$0.1/T, 0.9/T$	$0.3/T, 0.7/T$	$0.5/T$
8	-------	19.1	21.0
12	15.4	22.5	24.6
16	16.5	24.7	27.0
20	18.1	27.3	29.0
24	20.2	29.0	30.6
32	23.0	31.5	33.1

*The values in this table can be interpreted as the average improvement in clutter-to-signal ratio for *point* clutter at the target range but separated in doppler frequency. This illustrates the interference rejection of the matched-filter receiver for a uniform pulse train.

Staggered Pulse Trains

Since in the time frequency plane of the ambiguity diagram the area between range ambiguities has identically zero sidelobes, the uniformly spaced pulse train provides maximum freedom from extended clutter echoes. When the target is relatively small and the clutter is widely distributed or the interfering targets are large, this would be the best burst waveform to use if N can be made sufficiently large. Amplitude, phase, and pulse width tapering can be used to reduce the doppler sidelobes, and each individual pulse can be modulated to increase range resolution. These variations will be discussed in later sections.

The main difficulty with the uniform pulse burst is that large range ambiguities will be *close-in* unless the PRF is low, and doppler ambiguities will be *close-in* unless the PRF is high. The nonuniform burst provides maximum freedom from *significant* ambiguities and is most useful where the interfering targets are all of approximately the same cross section as the desired target. A small main lobe volume of the ambiguity function is desirable, as in other pulse trains.

To construct a suitable waveform, the interpulse period should be varied so that a maximum number of time sidelobe locations are generated. This requires an *almost uniform* stagger in multiples of the pulse width. Time axis sidelobes in the autocorrelation function are not allowed to overlap, and hence no sidelobe has an amplitude greater than $1/N$. The term *magic stagger* has been applied to this type of pulse train. In terms of the ambiguity surface $\psi(\tau, \omega)$ for this type of waveform, Resnick [309, p. 29] has shown that

$$\frac{\psi(0,0)}{N^2} \geq \psi(\tau, 0) \quad \text{for} \quad |\tau| \geq \tau'. \tag{11-2}$$

In other words, the central spike of the ambiguity function is at least N^2 greater than any zero doppler time sidelobe in the region outside twice the pulse width. He also showed that this inequality holds for all doppler. For the above type of burst, when the number of pulses N exceeds a *few dozen,* it is desirable that the average PRF be high enough so that the first doppler ambiguity at $f_d = 1/T$ is at a higher frequency than the highest doppler frequency of interest to minimize doppler ambiguities at the target range. The ambiguity function for the Resnick-type waveform is shown in Fig. 11-3. The apparent periodicity for the doppler response at zero range offset is somewhat misleading as will be

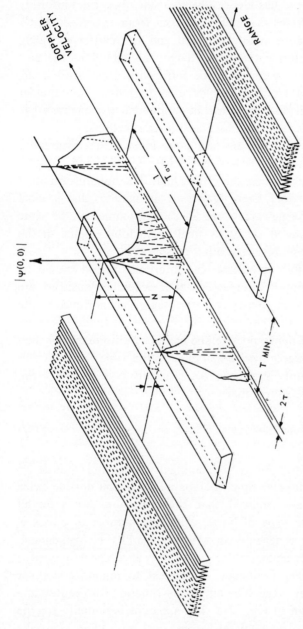

figure 11-3 *Envelope of the correlation function for a nonuniformly spaced pulse train of N pulses. (After Resnick [309].)*

explained subsequently. Resnick also showed that

$$\int_{-\infty}^{\infty}\int_{K\Delta-\tau'}^{K\Delta+\tau'} |\theta(\tau,\omega)|^2 \, d\tau \, \frac{d\omega}{2\pi} = \frac{1}{N^2} \tag{11-3}$$

where $\theta(\tau,\omega)$ = the signal correlation function [see Eq. (8-10)]

Δ = the minimum interpulse period

K = integer used to define discrete periods of the ambiguity surface

Since a sidelobe of the autocorrelation function is produced by each of the N pulses with the $N - 1$ other pulses, the total volume contained in these sidelobes is

$$\int_{-\infty}^{\infty}\int_{-\infty}^{\infty} |\theta(\tau,\omega)|^2 \, d\tau \, \frac{d\omega}{2\pi} = 1 - \frac{1}{N}, \quad \text{for} \quad |\tau| > \tau' \tag{11-4}$$

and the remainder of the volume is along the doppler axis within the strip $(\tau' \geq \tau \geq -\tau')$.

Since the pulse train consists of small deviations from a uniform pulse train, the close-in spectrum (near the range axis) resembles the $\sin NX/\sin X$ shape of the uniform pulse train. In that region, the clutter or interference rejection is comparable to the uniform pulse train. The clutter rejection for higher doppler frequencies is described in the section on clutter computations. Thus, this waveform merits use where major range ambiguities cannot be tolerated, and the maximum expected target doppler velocities are within the first few ambiguous regions. This is illustrated in Fig. 11-4 for a 32 pulse train where $\Delta = 55 \, \tau'$ and total duration is 3,346 τ'. The abscissa is the normalized doppler frequency in units of T_{av}^{-1} where T_{av} is the average interpulse period.

Nonuniformly spaced pulse trains having a uniformly low sidelobe level will be generated when the sequence of interpulse spacings is

$$\Delta_i = \Delta + \tau'[X + iq] \text{ modulo } (N - 1)$$

where $i = 0, 1, 2, \ldots, N - 2$; N is the number of pulses in the train $(N - 1$ is a prime number); and X, q are positive integers such that $N - 1 > q > 0$ (choose $q \approx (N - 1)/2$, but not $N/2$ or $(N - 2)/2$) and

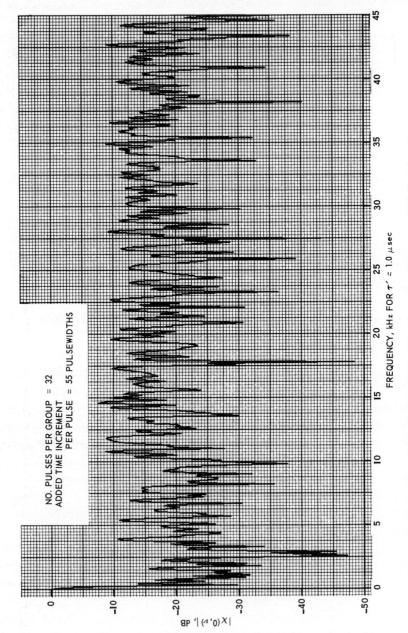

NO. PULSES PER GROUP = 32
ADDED TIME INCREMENT
PER PULSE = 55 PULSEWIDTHS

figure 11-4 Ambiguity function, $20 \log |\chi(0, \nu)|$ for 32 pulse train. (Courtesy of M. Davidson.)

$N - 1 > X \geq 0$. The duration of the train will be

$$T_d = (N - 1)\left[\Delta + \tau'\left(\frac{N - 2}{2}\right)\right]$$

The minimum interpulse spacing Δ, which is also the distance to the first time sidelobe, can be as low as

$$\Delta \approx \frac{N^{3/2}\tau'}{4}$$

and the duration-bandwidth product of the pulse burst can then be expected to be

$$T_d B \approx \frac{T_d}{\tau'} \approx \frac{N^{5/2}}{4}$$

For a pulse train with 32 pulses the maximum sidelobe level will be 30 dB down and the minimum $T_d B \approx 1,450$. It can be seen that as the number of pulses becomes large ($N > 100$) the length of the pulse train may become excessive. The duty factor may be lower than desired for target detectability or full transmitter utilization. Several of the Resnick trains are illustrated in Table 11-2.

TABLE 11-2 Tabulation of $T_d B$ and Sequence of Spacings for Small Values of N

N	$T_d B$ required	Sequence
2^3	1^1	1
3^3	3^1	1,2
4^3	6^1	1,3,2
5^4	11^1	3,1,5,2
6^4	17^1	1,3,6,2,5
7^3	33^2	4,7,5,8,6,3
8^4	35^2	7,3,6,2,12,1,4
9^4	45^2	2,8,14,1,4,7,6,3
10^4	55^2	1,5,4,13,3,8,7,12,2
12^4	85^2	2,4,18,5,11,3,12,13,7,1,9
14^4	127^2	5,23,10,3,8,1,18,7,17,15,14,2,4

[1] Optimum sequence ($T_d B$ minimum possible).
[2] Extensive but not exhaustive search.
[3] Resnick [309] and E. Paaske and V. G. Hansen, Note on Incoherent Binary Sequences, *IEEE*, vol. AES-4, pp. 128-130, January, 1968.

The second-order statistical properties of the target ensemble will undoubtedly limit the maximum duration of the waveform. Further, for the narrow-band approximation to be valid

$$T_d B \leq \left[\frac{2v}{c} \right]^{-1}$$

or for Mach-5 targets $T_d B \leq 90,000$ (see also Remley [308]).

11.4 AMPLITUDE, PHASE, AND PULSE WIDTH TAPERING OF FINITE PULSE TRAINS

The preceding sections on constant PRF and staggered pulse trains illustrated that the timing of the pulses in the waveform can be chosen to minimize the effects of localized clutter or interference. On the other hand, if there is extended clutter, the average improvement in the signal-to-clutter ratio (S/C) is limited to about the number of pulses in the waveform N. In both cases it was assumed that the transmitted pulses were constant in amplitude and width and the receiver consisted of a matched filter or set of filters.

It was shown (Chap. 8) that there are also *optimum* clutter filters which maximize the output S/C at a small penalty in the output S/N. An analogy to this optimization is the tapering of the transmit and/or receive apertures of a phased-array antenna wherein the angle sidelobes are reduced at the price of a small reduction in antenna gain and resolution.

One of the limiting factors in achieving clutter reduction by pulse-doppler techniques in narrow-beam rotating or phased-array radar systems is that the limited time allotted per beam position generally restricts the duration of the pulse train and hence the number of pulses. Since it is usually difficult to *transmit* a pulse train with pulses of varying amplitude, it seems best to first consider the optimization of the receiver for the constant amplitude uniformly spaced pulse train of finite extent.[1]

---

[1] As in all of this section, the effects of target acceleration will be neglected, and the targets will be assumed to approximate point reflectors. The range law of the clutter echoes is not included until Sec. 11-6.

Several examples will be given to illustrate that *Taylor, cosine, Hamming,* etc. amplitude weighting of the received pulse train can improve the signal-to-clutter ratio in many situations. It will also be shown that there are *optimum* complex (both amplitude and phase) weighting techniques that can yield even more improvement if the clutter range and doppler are known approximately. With all of these techniques, there is a slight penalty in signal-to-noise ratio SNR. A figure of merit for these receivers would be the efficiency L_s for a given doppler sidelobe reduction, defined as

$$L_s = \frac{\text{output signal-to-noise of weighted processor}}{\text{output signal-to-noise of the matched filter}}$$

Both the efficiency and the sidelobe reduction can be made quite high if the unwanted reflectors or targets occupy a small region of the ambiguity plane. If they are less than a pulselength from the target range and are within the first doppler ambiguity, the reduction in the doppler sidelobes (compared to the uniform amplitude case of Table 11-1) at the target range can be substantial as shown in Fig. 11-5 for a train of only 21 pulses. The particular taper was based on the Hamming function [373] and the loss in output SNR is only slightly over 1 dB while all doppler sidelobes are down by 41 dB or more.

The Hamming function is an efficient taper and can be represented in discrete form by its weighting function amplitude [92]

$$W(n) = K - (1 - K) \cos\left[\frac{\pi(2n - N - 1)}{(N - 1)}\right] \tag{11-5}$$

where n = the number of the pulse in the train
  K = a constant (equal to 0.54)
There is no phase weighting in this type of taper.

For $K = 0.54$, the theoretical peak sidelobe level is 42.8 dB below the central peak with a 1.3-dB loss in peak signal-to-noise ratio (see Sec. 13.9). An approximation for the efficiency with weighting is

$$L_s = \frac{\left[\sum_1^N A_n\right]^2}{N \sum_1^N A_n^2} \tag{11-6}$$

where A_n is the weighted amplitude of the Nth pulse.

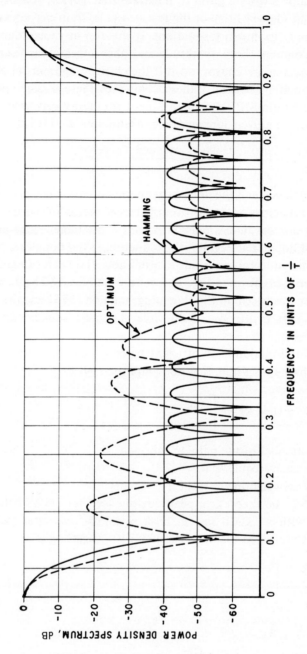

figure 11-5 *Power density spectra for Hamming weighting and for "optimum weighting" in the region of 0.5/T to 0.83/T for a 21 pulse train. (Derived by Rummler [330].)*

If resolution of targets with nearly the same velocity is desired, there is the additional penalty of the widening of the central doppler lobe by 40 to 50 percent. Taylor or Dolph-Chebyshev weighting functions [402, 373] can then be used to minimize the degradation in doppler resolution for a given sidelobe level. This will be discussed in another context in Chap. 13. One limitation of the *antenna-type* tapers is that they optimize for clutter rejection only at the target range, whereas the clutter often extends into the ambiguous range intervals. These *antenna* tapers are widely tabulated and usually do not involve phase weighting.

Significantly better efficiency and doppler sidelobe rejection can be achieved for specific doppler separations if the interference has a limited range extent. A specific example would be the detection and tracking of a reentry body where the echoes from the wake and tank fragments have comparable ranges and velocities but are not of interest. Unfortunately the determination of the optimum receiver weighting functions usually involves variational calculus techniques which are quite complex and require a digital computer for solutions for the longer pulse trains. Descriptions of optimization procedures can be found in Stutt [364, 365], Spafford [359], and Rummler [330, 331].

The value of these optimization techniques can be summarized as follows:

1. If the extent of the *clutter* in range and doppler is small and its location in the ambiguity plane is approximately known, substantial S/C improvements can be achieved over the use of a matched filter. This is achieved with generally less than 1-1.5 dB loss in efficiency.

2. As the range extent of the clutter increases to one-half the length of the total waveform and the doppler extent exceeds one-half the unambiguous velocity, the *optimum* receiver becomes less advantageous. The improvement I in S/C is limited to 5-16 dB over that of matched filters for pulse trains of 20-40 pulses.

3. If the range extent of the clutter is comparable to the length of the waveform, the optimization yields only 2.6 to 3.0 dB improvement above the matched filter [330]. An example of this will be shown in the section on truncated pulse trains.

4. The penalties of using optimization rather than matched filtering are small if the estimate of the clutter location is poor.

5. Optimization degrades the doppler resolution. This may or may not be a problem. The efficiency factors which are quoted here are dependent on the width and number of the doppler filters that are actually implemented.

6. The advantages of optimization are only obtained with tight phase and amplitude tolerances. Rummler [330] states that the results of the optimization (shown in Figs. 11-5 and 11-6) can only be obtained for an rms phase and amplitude uncertainty on each pulse of 1° and 0.2 dB respectively.

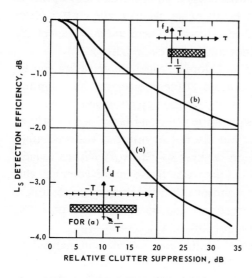

figure 11-6 *Detection efficiency vs. relative suppression with a 21 pulse train, clutter cloud distributed in doppler from $f_d = -5/6T$ to $f_d = -1/2T$, in range (a) from $n = -5$ to $n = 5$; (b) from $n = 0$ to $n = 5$. (After Rummler [330].)*

7. The optimization is dependent upon the clutter-to-noise ratio *and* the signal-to-noise ratio.

An example that illustrates the loss in detection efficiency for a given clutter reduction as a result of the use of optimum waveforms in moderate clutter extents is contained in Fig. 11-6. Two clutter extents are shown for a clutter velocity spread of from $1/2T$ to $5/6T$ Hz. The ordinate is the efficiency L_s as defined earlier, and the abscissa is the clutter suppression relative to that of a matched filter. For curve *a*, the range extent is 10 interpulse periods (five preceding and five following the target). For curve *b* the extent is five interpulse periods beyond the target ($n = 0$ to 5). It can be seen that even for moderately extensive clutter, significant relative improvement can be obtained with only 1-2 dB loss in efficiency. The zero range *cut* of the doppler response for about a 1.7-dB loss in efficiency (and the clutter extent given for case *a*) is sketched in Fig. 11-6.

A similar optimization is sketched in Fig. 11-7 from Spafford [357]. The clutter is assumed to extend throughout *all* the range ambiguities but was limited to between $0.275/T$ and $0.375/T$ in doppler. The zero range cut of the ambiguity diagram $\psi(0,\omega)$ is shown for the unweighted and *optimum* pulse trains. The loss in SNR with the optimum weighting is only about 0.1 dB. The zero range error cut is deceptive in that high

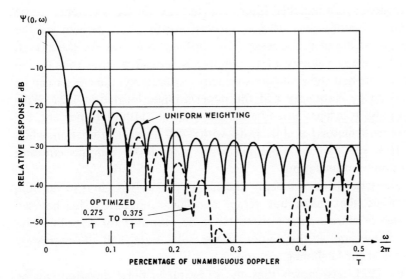

figure 11-7 *Power spectra at target range for 32-pulse transmission. (After Spafford [357].)*

rejection is shown in the desired doppler region. However, the doppler cuts at the distal range ambiguities (where the clutter is also assumed to exist) degrade rapidly, and the overall improvement is only 2.85 dB better than the matched-filter receiver for range extended clutter.

If the clutter extends throughout the $2N-1$ range ambiguities, Ares [7] has shown that the use of a constant envelope burst and the optimum mismatched receiver yields a S/C improvement that is within 3 dB of the constant envelope burst and its matched filter.

In some situations, where it is necessary to use a burst of only a few pulses, the signal-to-clutter improvement for constant envelope burst is inadequate for the environment. If the transmitter chain is sufficiently linear, by weighting the amplitude of the pulses in the transmission as a function of time, an additional improvement may be obtained. This is especially useful in range-extended clutter since the clutter echoes from the distal range ambiguities then have relatively lower power at the input to the receiver.

With the aid of a digital computer and nonlinear optimization routines, Ares [7] has computed the improvement in signal-to-clutter ratio over that of a single pulse for bursts of from four to eight pulses. Some typical results are given in Fig. 11-8 for a weighted burst transmission and a matched-filter receiver. The clutter is assumed to be extended in range and have a spectral width of 2 percent of the unambiguous doppler frequency. The optimization was for the case in which the mean velocity of the clutter is separated from that of the target by one-half the ambiguous velocity. Alternately, the clutter can be assumed to be stationary and the target doppler located at $1/2T$. With the six-pulse transmission and a clutter spectral width of 10 percent, the improvement factor I is still 26 dB with $1/2T$ doppler separation, while only 10.8 dB can be obtained with a uniform transmission and receiver-only optimization.

An alternate form of tapering the energy of a pulse train is to vary the duration of the pulses but retain constant amplitude. This has the advantage of allowing the transmitter to be saturated for all pulses. The widest pulses are at the center of the pulse train, and the narrowest pulses are at the beginning and end.

Deley [92] has shown that the variation in pulse duration can be chosen to optimize the clutter rejection at a given doppler or yield a flat region with low doppler sidelobes. He calculated a number of examples for 16- and 30-pulse bursts. An example for $N = 30$ is shown as Fig. 11-9. The ordinate is the relative Q function which is defined as

$$\text{Relative } Q(\omega) \;=\; \frac{\displaystyle\int_{-\infty}^{\infty} |\chi_N(\tau,\omega)|^2\,d\tau}{\displaystyle\int |\chi_1(\tau,\omega)|^2\,d\tau} \tag{11-7}$$

This is a measure of the doppler response of a waveform in clutter. The numerator is the integral of the magnitude squared of the ambiguity function of the N-pulse burst. The denominator is the integral of the magnitude squared of a single pulse. In deriving the figure, the single pulse considered was the *shortest* pulse in the train. The abscissa of Fig. 11-9 is the normalized doppler frequency in units of $1/T$.

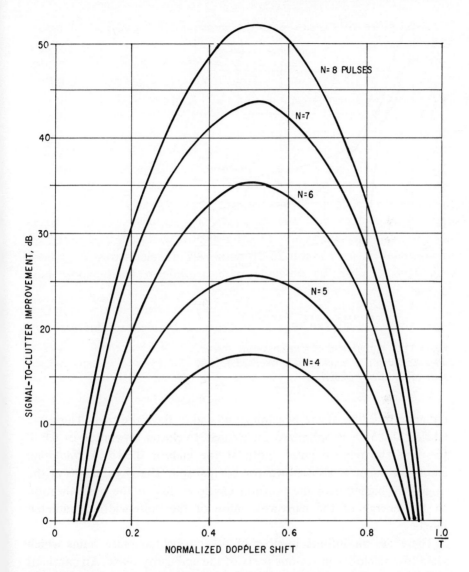

figure 11-8 *Improvement for N-pulse burst with weighted transmission and matched filter. Clutter spectral width = 0.02/T. (After Ares [7].)*

Four examples are shown including the *uniform* envelope 30-pulse burst. The two bursts with pulse-width variation are labeled flat Q and $Q(0.5/T)$. To obtain the flat Q doppler response from $0.2/T$ to $0.8/T$ the normalized pulse width is unity for the number 15 and 16 pulses

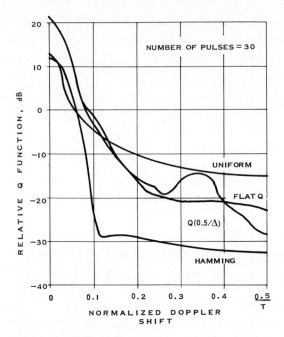

figure 11-9 *Q-functions of four uniformly spaced 30-pulse bursts as a function of normalized doppler shift. (From Deley [92].)*

reducing to 0.086 at the beginning and end of the pulse train. The curve labeled $Q(0.5/T)$ is optimized for a target-to-clutter separation of $0.5/T$ Hz, and the relative pulse width at the ends is 0.092. A *Hamming* amplitude taper is also shown for comparison. The Q functions at the origin are higher than the uniform envelope due to the normalization to the energy of the narrowest pulse of the pulse-width modulated bursts.

There are an infinite number of other weighted pulse trains which yield low sidelobes in various parts of the ambiguity plane. An excellent compilation of phase- and amplitude-weighted uniformly spaced pulse trains has been prepared by Spafford [358].

11.5 BLOCK DIAGRAMS FOR PULSE-DOPPLER RECEIVERS

There are many pulse-doppler receiver configurations for both search and tracking systems. The choice of configuration is based on the function of the radar. Search radars usually require more sophisticated circuitry than tracking radars since a greater number of range gates and doppler filters are needed to *find* the target. Discussions of the choice of carrier frequency are contained in the section on the ambiguity diagram and the section on clutter computations.

If the purpose in using pulse doppler in a tracking radar is primarily to determine target velocity and if the clutter environment is not severe, the range and angle circuits may involve only incoherent (envelope) processing. Similarly, there may be coherent range and doppler processing and incoherent angle processing to reduce the total amount of circuitry. A discussion of these *hybrid* coherent pulse trackers can be found in Barton [20, pp. 385-389].

The discussion that follows will assume that both coherent range and doppler processing are needed to reduce clutter as well as integrate the energy from many pulses.

Range Gates and Doppler Filters at IF

A block diagram of a pulse-doppler receiver whose doppler processing is accomplished with narrow-band IF filters is given in Fig. 11-10. Waveform parameters are shown at the top of the figure with the square pulses representing coherent samples of a constant frequency carrier. The received signals are mixed to a convenient first IF, filtered to approximately the inverse of the pulse width, and then mixed again to a second or third IF which is convenient for the final doppler band-pass filtering which occurs after range gating. The matched filtering for an individual pulse determines the design of the first band-pass filter and the range gate. In this section a *matched filter* will imply optimization with respect to receiver noise unless called a *clutter matched filter*. This will be discussed in the section on detection range computations and in Sec. 14.8. The use of double conversion in the diagram is meant to illustrate two points. While it will be shown that a pulse-doppler receiver can closely approximate a matched filter for a coherent pulse train, the receiver in Fig. 11-10 is rather inefficient in the utilization of the received signal energy. This energy is contained in an entire series of spectral lines separated by the pulse repetition frequency. In this figure, as in most receiver schemes, each narrow band-pass filter after the range gates responds to only a single spectral line.[1] The high total electronic

[1] The alternate use of comb filters will be discussed later in this section. Also see Galejs [130], George and Zamanakos [134], and Flesher [121].

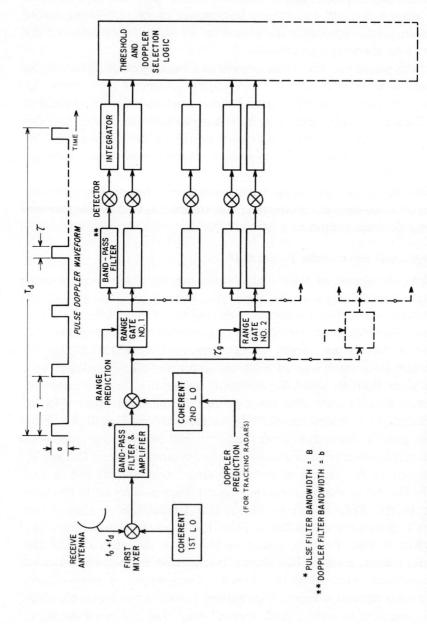

figure 11-10 *Pulse doppler waveform and basic receiver block diagram.*

414

gain required to make up for this signal energy loss usually implies the use of more than one intermediate frequency. This gain may be distributed in ways other than in the figure. The second reason for using a multiple conversion receiver is to simplify the filtering processes and remove the doppler images. The spectrum of a continuous pulse train with interpulse period T and N approaching infinity is shown as the series of impulses centered about the carrier frequency f_0 on Fig. 11-11. The decrease in the height of the spectral lines away from f_0 results from multiplication by the spectrum of a single pulse in the train. If the pulse length is 1 μsec and the interpulse period is 200 μsec, there are about 400 significant lines in the spectrum.

The dotted lines in the figure represent the echoes from a point-source target approaching the radar at a doppler frequency of $f_d = 2v/\lambda$. The gaussian-shaped spectra near the transmit spectral lines represent clutter echoes with zero mean velocity but with finite width due to turbulent effects.

If the receiver is to coherently integrate many of the target-echo pulses, the doppler band-pass filter must have a width $b \ll 1/T$. In the example of a 5-kHz PRF, the bandwidth b would be about 100 Hz for 50-pulse coherent integration. With other than high-quality crystal filters, the carrier frequency after the final intermediate frequency would have to be under 1 megacycle to keep the Q of the filters within practical bounds. If a single conversion receiver were used at an IF of 500 kHz, the pulse-filter bandwidth ($\approx$ 1 MHz) would become meaningless and the spectral lines and interference near the carrier frequency plus 1 MHz, 1.5 MHz, etc. would *fold* into the desired spectral region. With the double conversion technique illustrated, the first IF could be about 10 MHz with a pulse-filter bandwidth of 1 MHz, and the second IF could be of the order of 0.5 MHz.

In the example, as in several of the implementations known to the author, the pulse out of the range gate contains less than a full cycle of the second intermediate frequency. This need not involve a loss in performance since the phase of each pulse return is such that the ringing in the *correct* narrow-band filter will increase with each return pulse. A stringent requirement placed on the range gate is that the switching transients will not integrate in the filter. It should be noted that there is no restriction that the PRF be constant. Partially because of the switching problem, the *homodyne* mixer configuration, which is discussed in the next section, is becoming more popular.

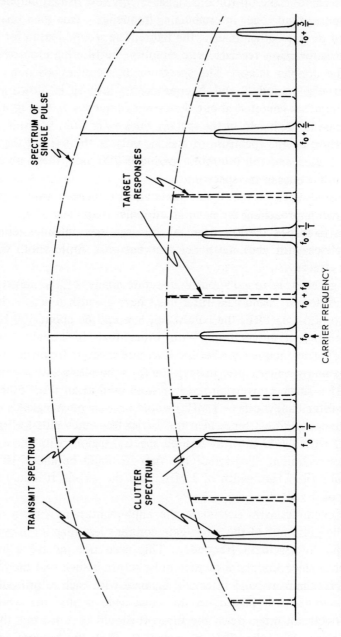

figure 11-11 *Spectrum of returns from a coherent pulse train.*

Labels within figure:
SPECTRUM OF SINGLE PULSE
TRANSMIT SPECTRUM
TARGET RESPONSES
CLUTTER SPECTRUM
CARRIER FREQUENCY

$f_0 - \frac{1}{T}$ f_0 $f_0 + f_d$ $f_0 + \frac{1}{T}$ $f_0 + \frac{2}{T}$ $f_0 + \frac{3}{T}$

The number of range gates and doppler filters is dependent on the prior knowledge of target location and radial velocity. In a tracking radar, only two range gates may be needed: an *early* and a *late* gate straddling the predicted range. Each of these gates could be followed by two or three doppler filters *straddling* the predicted radial velocity. In an acquisition mode or for a search radar the number of range gates and velocity filters may fill the entire unambiguous range including all possible doppler frequencies up to $1/T$. Since the doppler response is symmetrical around the PRF lines, the bank of filters may cover either the region between one pair of lines or may straddle a single line.

The minimum number of filters for *all doppler* coverage is approximately equal to the number of pulses desired for coherent integration N_c with the width of each filter b equal to $1/(N_c T)$ or $1/T_d$. Additional filters tend to reduce the loss when a target echo straddles two doppler filters but does not give additional unambiguous information. If the filters are not perfectly rectangular, there is a loss of response in the crossover region between the filters. On the other hand, a rectangular band-pass filter is not a matched filter for a uniform pulse train.

After the doppler filtering, the sinusoidal output of the filters is envelope or square-law detected and is often stored or integrated with a capacitor. The integration at this point is incoherent for all of the pulses after the effective time constant of the doppler filter. The reduction in per pulse (S/N) below that required for coherent integration of N_c pulses can be approximated by determining N/N_c and entering the appropriate curves of Marcum and Swerling [244] (see Chap. 3). If the signal-to-noise ratio prior to the detector resulting from the coherent integration of N_c pulses is much greater than unity, the loss from incoherent integration is small.

Homodyne or Zero-IF Processor

Several limitations of the narrow-band IF filter configuration for pulse-doppler receivers have led to the zero-frequency IF (homodyne) configuration. These limitations include:

1. The coherent integration time is fixed at the value of the original design and cannot be adapted to the spectrum of different types of targets or better knowledge of the target velocity.

2. The range-gate switch is not easily designed when there is only about one cycle of the second IF during the gate time.

3. With phased arrays or other rapidly scanning antennas, the signals in crystal filters cannot be *dumped* instantaneously to permit coherent integration to start at a new location in space.

A basic block diagram of the homodyne pulse-doppler receiver is shown in Fig. 11-12. The one-step conversion from microwave RF to zero frequency is not necessary but can be accomplished with sufficient preamplification prior to the single sideband mixer. The high gain prior to conversion compensates for the poor noise figure of diode mixers near zero output frequency and is the only point in the system where the signal amplification is common to all range gates and doppler filters. The mixers are of the single sideband type so as to attenuate the doppler images and to provide the inphase I and quadrature Q components of the RF pulse signal. Using a signal representation basically that of Rice,

$$f(t) = I \cos(\omega t) + Q \sin(\omega t)$$

where ω is the carrier frequency in radians. The mixer outputs are in quadrature (90° phase separation) and the phase and amplitude of each RF pulse are contained in the amplitude of the instantaneous I and Q components. These components are bipolar pulses where $(I^2 + Q^2)^{1/2}$ is the magnitude of the RF pulse. $\mathrm{Tan}^{-1} - (Q/I)$ is the phase angle of the RF pulse. These bipolar pulses then allow reconstruction of the doppler signal. If the PRF is constant and is greater than the highest target doppler, the doppler frequency determination is unique. Otherwise the ambiguities in doppler must be tolerated or resolved by other measurements such as range rate.

The transmit pulses are usually much shorter than a cycle of the doppler, therefore the range gating follows a filter which is matched to the transmit pulse width. The determination of which range gate and doppler filter contain the target signal (or if a signal exists) can be performed by several methods. The dc integration shown in Fig. 11-12 is one of the simplest to consider. If the doppler prediction is exactly correct in one of the range-gate-doppler filter channels, all of the outputs from the train of echo pulses in its I channel will have the same polarity, and those in the Q channel will also be unipolar.[1] The I and Q dc integrators will simply add the returns from each pulse and after squaring

---

[1] Both channels are necessary since it is possible for the signal to exist in only one channel (Secs. 12.6 and 14.8).

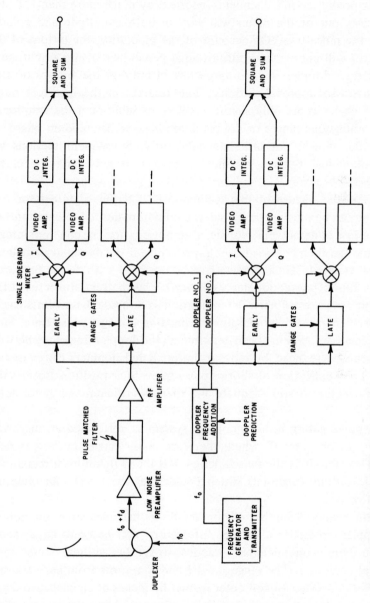

figure 11-12 *Zero IF (homodyne) pulse-doppler receiver.*

419

and summing will yield the coherent summation of N_c pulses. If the doppler prediction in a channel is incorrect by much more than $1/T_d$ Hz, the pulses out of the mixer will vary in both amplitude and polarity during the pulse train. On account of the bipolarity, the output of this integrator will approach an infinitesimal percentage of the amplitude in the *correct* doppler as N approaches infinity if the integrator time constants also approach infinity. The residues in the incorrect range-doppler channels are only significant if N_c is small or if the samples are at the ambiguous ranges or dopplers. Since noise has random phase and amplitude, it will not integrate coherently. Subsequent sections will place numerical bounds on the integration of noise and the clutter residues.

One of the virtues of this technique is that the dc integrator time constant can be much greater than the minimum desired integration time without loss in detectability or signal-to-noise ratio. The coherent integration time can then be varied over a wide range if the dynamic range limits are not violated. Thus, targets can be *acquired* with fewer but wider doppler filters (poorer doppler resolution). When better doppler resolution or more accurate velocity determination is desired, these filters can be effectively narrowed by simply transmitting a longer pulse train. Since approximate velocity has been determined, the doppler frequency between filters can be reduced by moving the predicted dopplers closer to each other. The application of shorter pulse trains to acquisition assumes that clutter rejection is still adequate and there is sufficient energy for detection.

The range gating in a homodyne system is performed at microwave frequencies or at an IF where there are many carrier cycles per range gate. This simplifies the switch design, but the switches must remain free of a dc output (pedestal) which could integrate to give an undesired output.

As the required number of doppler filters becomes large, the number of doppler predictions, mixers, integrators, etc. becomes large; and it may be more economical and practical to synthesize these operations in a digital computer. The range-gated I and Q returns from each transmit pulse are converted to computer format by means of an analog-to-digital converter A/D and stored in a digital memory. After the storage of N_c returns, each stored return is sequentially rotated in phase in a manner analogous to the *predicted doppler and mixer* combination. After vector summation of *each* predicted doppler for each range gate, the maximum output is compared to other range-gated dopplers or to a preset threshold. This type of processing will be discussed further in Sec. 14.8.

The bipolar video signals after the mixers can also be made to undergo progressive rotations in a video doppler *resistor matrix* where the various range-gate positions form the column inputs, and the doppler rotations are the row outputs.

Tapped Delay-line and Comb-filter Processors

If the number of pulses to be coherently integrated is relatively small, the *tapped delay-line implementation* can be used. The portion of this configuration after the mixing processes is shown in Fig. 11-13. The delay lines are matched to the intervals between pulses with T_i corresponding to the time between the transmission of the first and second pulses, etc. The time between the last two transmit pulses to be coherently integrated is then given as $T_{(N-1)}$. This manner of doppler decoding can be used for a fixed nonuniform pulse train, with the special case of constant PRF resulting in equal lengths for all the delay lines.

The summing bus directly below the delay lines provides the zero doppler output if the delay lines are all an integral number of wavelengths at the intermediate frequency. The time delays or phase shifters correspond to the radial distance the target moves between pulses. The arbitrary constant K is shown to indicate that the phase shifts must be adjusted to the individual spacings between pulses if the interpulse period is not constant. The conversion from time delay to phase shift must be made on the basis of target motion in carrier wavelengths. The value of ϕ is often chosen such that $N\phi = 2\pi$ radians, which yields contiguous doppler coverage with a 4-dB *notch* between doppler filters for a uniform transmission. Smaller values of ϕ will reduce the depth of the notch but will increase the response of a target in adjacent doppler filters. The outputs of this processor have the bandwidth of the individual pulses. Coherent integration results from the vector voltage addition of the target echoes in the appropriate channel while noise adds powerwise.

What has been described is often called a *doppler matrix* or phasing matrix. Its utility is limited by delay-line limitations and the requirement for a large number of components for large numbers of pulses. One of its advantages is that there is no physical range gating, target range being determined by the time of appearance of echo pulses at doppler outputs.

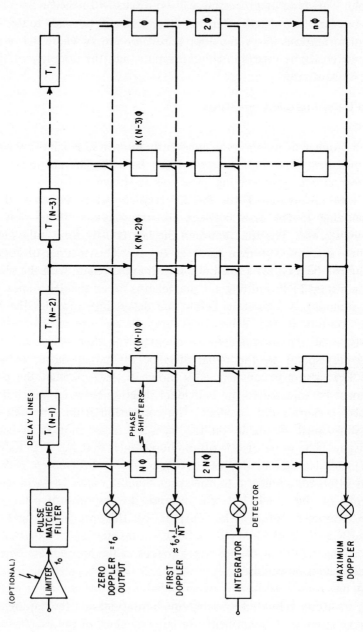

figure 11-13 *Tapped delay-line receiver for coherent pulse train.*

A *zero range* reference occurs when the first transmit pulse appears at the output of delay line T_1. Both quartz and magnetostrictive [226] delay lines have been used for this technique. A parallel configuration is common for quartz lines rather than the series of lines shown in Fig. 11-3. Delay-line limitations are discussed in Resnick [309].

If the number of pulses is large enough to achieve CFAR action, a limiter may be placed ahead of the delay lines. Then the normalized maximum output power is obtained from the coherent or voltage sum of N_c pulses (N_c^2 in power). The average noise power level is less than this value by a factor of N_c, the number of pulses coherently integrated.

Since this configuration responds to any set of narrow-band signals associated with all the PRF lines (see Fig. 11-11), it is one of a class of *comb filters*. While it is functionally the same as a range-gated parallel filter bank or a homodyne device it utilizes all of the signal's spectral components. This type of processor should require less amplification than those that use a single spectral line. However, the losses of most types of delay lines exceed the signal increase obtained from utilizing all of the spectral lines.

Other versions of the comb filter are based on the use of delay lines and feedback techniques. Examples are shown in Fig. 11-14 with the delay-line length equal to the interpulse period and the gain A equal to the delay-line loss. The similarity of Fig. 11-14A to an MTI system becomes obvious if the amplifier has a gain of minus one. A null in the response occurs at zero doppler and multiples of the PRF. Coherent integration occurs only at odd multiples of one-half of the PRF. This variation is discussed fully in the MTI section.

Figure 11-14B is more appropriate to a pulse-doppler system if the loop gain is near unity. In this case the response is maximized at the predicted doppler and at the predicted doppler plus multiples of the PRF. The response characteristics of feedback filters including higher order feedback networks are given in [38, chap. 3; 121, 130]. In general, the feedback filter may be either at IF or video.

The advantages of this feedback technique include the need for only one delay line, the ability to observe the signal in many range gates without additional circuitry, and, alternately, the ability to perform postdetection (incoherent) integration of a pulse train by integration of the envelope-detected returns. Disadvantages include the necessity for stringent requirements on loop gain, bandwidth, and spurious responses. To coherently integrate many pulses, the loop gain must be very close to

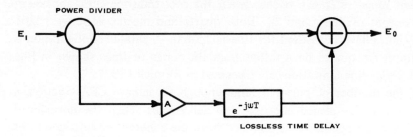

(A) FEED–FORWARD COMB FILTER

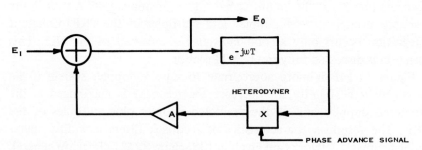

(B) FEEDBACK COMB FILTER

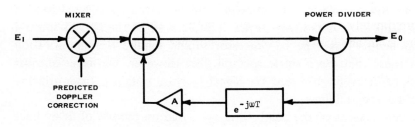

(C) COHERENT MEMORY FILTER

figure 11-14 *Comb-filter processors.*

unity without incurring oscillation at any frequency in the passband. As shown in Fig. 11-14*B*, coherent integration can occur only at a single *predicted* doppler frequency that is corrected to within $1/N_c T$ before entering the loop.

The block diagram shown in Fig. 11-14C has been used to extract doppler information, and is commonly known as the Coherent Memory Filter.[1] It has been shown [60] that if the phase advance signal is

$$e_d = \cos\left(\frac{2\pi}{T}\right) t$$

the output of the device will be an approximation to the Fourier transform of the input signal.[2] The output envelope of the device will have a response which is narrower than the input pulse by roughly $1/N_c$, and its location in time with respect to the input envelope is a function of the doppler shift of the input pulse train. The doppler resolution is approximately $1/(N_c T)$ Hz. If the range of a pulsed return is unknown, there is an ambiguity of range and doppler similar to that for an FM signal. For this reason, this type of dynamic filter is more commonly used with continuous CW input signals.

11.6   RANGE COMPUTATIONS FOR PULSE-DOPPLER RADAR

The computation of detection range for a pulse-doppler search radar does not differ substantially from that of the simple pulse radar if the dependence on energy is kept in mind. The pulse radar range equation from Chap. 2 was given as

$$R^4 = \frac{\hat{P}_T G_T L_T A_e L_R L_P L_a L_c L_s \sigma_t}{(4\pi)^2 K T_s B_N (S/N)} \tag{11-8}$$

For a single transmit pulse and approximately a matched filter, the noise bandwidth $B_N \approx 1/\tau'$ where τ' is the pulse width. Since a pulse-doppler system coherently integrates many (N_c) pulses, it is almost sufficient to multiply the numerator of the range equation by N_c if the transfer characteristic of the narrow-band filters following the range gate

---

[1] Registered trademark of Federal Scientific Corporation, New York [60].
[2] This is a close approximation to single sideband modulation of $1/T$ cycles per recirculation.

is such that they optimally integrate N_c pulses, i.e., the narrow-band filter must also be matched to the total transmission time T_d.

The simplest processor to consider is the tapped delay line (Fig. 11-13). In this configuration N_c pulses are added voltagewise ($N_c{}^2$ in power), and N_c noise samples are added with random phase (N_c in power). The (S/N) required for a given probability of detection and false alarm probability is thus reduced by N_c after allowance is made for the appropriate number of false alarm opportunities.

In any pulse-doppler system, the false alarm probability must be based on the product of the number of range gates n_g and the number of doppler filters K per range gate. The number of opportunities for false alarms is not necessarily greater than for a simple pulse radar since the number of independent samples at the output of the filter is proportional to the filter bandwidth b. For a given velocity coverage, the filter bandwidth decreases as the number of doppler filters increases. Thus, in most cases the number of potential false alarms is independent of the number of doppler filters.

Another version of the pulse-doppler range equation can be derived from the following relations:

$$d_t = \frac{\overline{P}}{\hat{P}} \tag{11-9}$$

where d_t = transmit duty factor
$\overline{P}$ = average transmit power
$\hat{P}$ = pulse transmit power (average power during pulse)
Also

$$d_t = \frac{\tau' N_c}{T(N_c - 1)} = \frac{N_c}{B_N T(N_c - 1)} \tag{11-10}$$

where T = average interpulse period
N_c = number of coherently integrated pulses
Eliminating the duty factor from Eqs. (11-9) and (11-10),

$$\frac{\hat{P}}{B_N} = \frac{\overline{P} T(N_c - 1)}{N_c} \tag{11-11}$$

A good approximation for the noise bandwidth of the *matched* doppler filter is

$$b \approx \frac{1}{T_d} = \frac{1}{(N_c - 1)T}$$

(11-12)

where T_d is the coherent transmit time of pulse train. Then substituting for $(N_c - 1)T$ in Eq. (11-11) yields

$$\frac{\hat{P}}{B_N} \approx \frac{\bar{P}}{bN_c}$$

(11-13)

An alternate development by Bussgang et al. [57] yielded $d_g B_N$ instead of bN_c where d_g is the receiver duty factor. These results are almost equivalent if the receive range gate is comparable to the pulselength.

If Eq. (11-13) is then substituted in the range equation after the numerator has been multiplied by N_c to account for the coherent integration, the alternate pulse-doppler range equation becomes

$$R^4 = \frac{\bar{P}_t G_T L_T A_e L_R L_P L_a L_c L_s \sigma_t}{(4\pi)^2 K T_s \, b(S/N)}$$

(11-14)

It must be emphasized that (S/N) is the per pulse signal-to-noise ratio in the doppler filter for the desired probability of detection and false alarm probability.

Combined Coherent and Incoherent Integration

In many practical cases, the doppler-filter bandwidth is deliberately made greater than the matched bandwidth for the number of available pulses in the transmission. This occurs where the target spectrum is broader than $1/NT$, where the target is accelerating or decelerating in the time NT sufficiently to broaden the spectrum beyond $1/NT$, or where the construction of the optimum number of doppler filters is too costly. In these cases, coherent integration exists for N_c pulses; and if the detected output of the doppler filters is linearly integrated, there is incoherent integration of N/N_c *bursts.* This latter value is entered into the incoherent integration curves of Chap. 3. The signal-to-noise requirements of various proportions of coherent and incoherent integration versus the number of doppler filters K are also plotted by Steinberg [38, chap. 4] where it is assumed that the doppler filters are contiguous and cover the entire spectrum between the PRF lines. The number of doppler filters K is then

$$K = \frac{\text{PRF}}{b} = \frac{1}{bT} = \frac{N_c}{bT_d} \quad \text{for large } N_c$$

but $bT_d \approx 1$ for contiguous matched filters so that $K \approx N_c$. Similar graphs of probability of detection versus normalized range for various amounts of postdetection integration are given in Bussgang [57]. Sample computations show that the use of Marcum's or Bussgang's curves gives results within a few tenths of a decibel.

Since the doppler filters may cover only a portion of the spectrum between the PRF lines and little improvement in sensitivity is obtained for $K > N_c$, the number of coherently integrated pulses N_c is felt to be more representative than K and will be used in subsequent discussions.

Deviations from Ideal Integration

The calculation of the efficiency factor L_s for pulse doppler must account for some potential added losses:

1. The deviation of the pulse matched filter from the optimum value.

2. The range gate being too narrow, too wide, or not centered on the target return. This loss should not always be directly added to number one above since pulse *stretching* by the pulse matched filter partially compensates for minor errors in the range gating.

3. The doppler matched filter shape (or bandwidth) may deviate from the optimum.

4. The target response may straddle more than one doppler filter or fall in the *notches* between them [226, p. 379].

5. There may be a significant doppler dispersion loss [308] if $2T_d v/r'c \geq 1$.

6. Doppler *image noise* may be *folded* onto the desired doppler spectrum.

7. Phase- and time-delay errors reduce the sensitivity.

8. FM and AM transmitter noise imposed on nearby clutter may exceed receiver noise (see Sec. 10.1).

The characteristics of the individual pulse and the doppler filters deserve special attention. Analysis of the improvement obtained from coherent integration for various filter shapes has been reported by North [269], George and Zamanakos [134], Galejs [130], and others. For the case of a uniform pulse train of rectangular pulses and a uniform video comb filter with an idealized square passband, it has been shown [134] that

$$\text{Improvement} \;=\; (N_c L_s) \;=\; 10 \log(0.45\,N_c) \text{ in dB, if } r' < \frac{T}{4} \qquad (11\text{-}15)$$

The assumption here is that the input noise is white, and the rectangular doppler filters cover the main lobe of the $(\sin x/x)^2$ energy density spectrum (~ 90 percent of the energy falls within the filter). The assumed filter bandwidth $b \approx 2/N_c T$ Hz is somewhat wider than the optimum.

By altering the gains of the *teeth* of the comb filter to *match* the amplitudes of the lines of the uniform pulse-train spectrum, an additional improvement of 2.1 dB is obtained. If, in addition, each *tooth* of the filter is given a $(\sin x/x)^2$ response, the North filter is obtained whose improvement is 4.3 dB greater than Eq. (11-15). The improvement in this case ≈ 0.8 dB greater than N_c, presumably due to the assumptions of an overly wide band pass for the single pulse and an idealized integrator.

Galejs [130] has extended this analysis to physically realizable comb filters and to the case where the pulse-train envelope is in a $|\sin x/x|$ form which would result from an antenna beam scanning past a target. Using the same general assumptions as George, he obtained the (S/N) power improvements, above $10 \log 0.45\, N_c$, for practical filters and rectangular pulse trains. For rectangular pulses the improvement was 3.8, 1.7, and 1.8 dB for the optimum, cascaded delay line, and feedback filters, respectively. The exact values quoted here are for special cases. They should not be used without study of the detailed assumptions in the references, including the discussions on output sampling time.

It is fairly common to overlap both the range gates and doppler filters since as much as 5-6 dB loss in detectability *can* occur in each case if the target signal falls in the *notches*. In practice, without a deliberate overlap, the average loss in each is only 1 to 2 dB due to the time smearing of the individual pulse filter, the inherent overlap of physical doppler filters, and the smearing of the doppler response of the target.

11.7 CLUTTER COMPUTATIONS[1]

Perhaps the most difficult part of predicting the performance of a pulse-doppler radar is the calculation of the rejection of extended clutter echoes. The difficulty arises because the dominant clutter may not be at the target location and may extend over a considerable portion of the ambiguity plane. The differing range dependencies of targets and clutter preclude the use of a single chart of equations. The general pulse-doppler range equation for uniform clutter will be derived in this section with some indication of how simplifications can be made. It will be shown that in many cases the right members of the range equation can be factored into the *range in clutter* for a pulse radar and a clutter attenuation CA factor for the pulse doppler.

The periodic character of the pulse-train spectrum causes doppler ambiguities at multiples of the repetition frequency. Thus, undesired targets (such as rain, chaff, sea or land clutter) may produce a large response at or near the doppler frequency of the desired echo thereby obscuring it. The use of a high pulse repetition frequency to reduce the doppler ambiguities is usual in the design of pulse-doppler radars but aggravates the problem of range ambiguities. With a high PRF, echoes from many different ranges may be received simultaneously so that even though the target might be in a clear region, clutter elsewhere along the antenna beam can cause interference. The peak echo power from clutter at the output of the receiver range gate, neglecting antenna sidelobes, will be

$$\hat{P}_c = \frac{\hat{P} G A_e L}{(4\pi)^2} \Sigma\sigma \int \frac{|f(\theta,\phi)|^4 L_a}{R^4} \, dV \tag{11-16}$$

where G = antenna gain at beam center

A_e = effective area of the receiving aperture

L = transmitter and receiver loss factors combined

$\Sigma\sigma$ = clutter scattering cross-section density (usually expressed in square meters per cubic meter for volume clutter)

$f(\theta,\phi)$ = antenna amplitude pattern which is a function of the angular coordinates θ, ϕ measured from the axis of the beam

---

[1]The majority of this section was prepared by Dr. P. J. Luke of the Applied Physics Laboratory.

L_a = attenuation factor due to the transmission medium (a function of range)

R = range to the volume element dV

$dV = R^2 \, d\Omega \, dR$

$d\Omega$ = element of solid angle

For volume-extended clutter (e.g., rain, chaff) the integral with respect to the solid angle may be evaluated immediately, giving [296]

$$\int |f(\theta, \phi)|^4 \, d\Omega \;=\; \frac{\pi \theta_1 \phi_1}{8 \ln 2} \;=\; 0.57 \, \theta_1 \phi_1$$

where θ_1 and ϕ_1 are the one-way half-power beamwidths. The echo power from volume extended clutter is then

$$\hat{P}_c \;=\; \frac{\hat{P} \, G \, A_e \, L}{(4\pi)^2} \, (\Sigma \sigma) \, (0.57 \, \theta_1 \phi_1) \int \frac{L_a \, dR}{R^2} \tag{11-17}$$

The integral with respect to R is taken over all ranges from which simultaneous or overlapping echoes are received.

Following the range gate, the signal is filtered by the doppler filter, envelope detected, and compared with a threshold. In order to determine the clutter energy at the detector input, we compute the spectral density of the clutter echo at the input to the filter and from that the clutter energy at the filter output. A similar computation yields the target-echo energy at the detector. The ratio of these two is the signal-to-clutter ratio for the particular range gate and doppler filter being considered, which are the ones having the most target-echo energy.

Since the clutter return consists of echoes from many small reflectors distributed at random in space and, hence, with random phases, the spectral density for the clutter will be the sum of the spectral densities for the individual returns. The expected value of the resultant spectral density (with respect to both the spatial distribution and the frequency or velocity distribution) integrated over the filter bandwidth yields the mean clutter energy at the detector input.

Constant Interpulse Period—Clutter Response

An echo consisting of a uniform train of N pulses with constant amplitude a_i and doppler shift f_i relative to the filter center frequency has an energy spectral density

$$
\begin{aligned}
E_i(f) = \tfrac{1}{2} a_i^2 \tau_i^2 \bigg\{ & |U_i(f - f_i)|^2 \frac{\sin^2[\pi(f - f_i)NT]}{\sin^2[\pi(f - f_i)T]} \\
& + |U_i(f + f_i)|^2 \frac{\sin^2[\pi(f + f_i)NT]}{\sin^2[\pi(f + f_i)T]} \bigg\}
\end{aligned}
\tag{11-18}
$$

where T is the interpulse spacing, τ_i is the pulse width of the echo after range gating (which may truncate the pulse), and $\tau U(f)$ is the single-pulse spectrum (defined in this manner so that $U(f)$ is dimensionless, e.g., for a rectangular pulse $U(f) = \sin \pi f \tau / \pi f \tau$). After truncation by the range gate the echo from the ith reflector has a single-pulse spectrum $\tau_i U_i(f)$.

The frequencies f_i are distributed over some interval which is usually small relative to the pulse-repetition frequency but may be greater than $1/NT$. Thus, the factor $|U_i(f \pm f_i)|^2/\sin^2 \pi(f \pm f_i)T$ does not vary appreciably over this interval and may be given the constant value $U_i(f \pm f_c)^2/\sin^2 \pi(f \pm f_c)T$ when averaging $E_i(f)$ with respect to the distribution of f_i. The frequency f_c is an average with respect to the distribution. Noting that the average of the factor $\sin^2 \pi(f \pm f_i)NT$ is $1/2$, the average spectral density is

$$
E_{ci}(f) = \tfrac{1}{2} a_i^2 \tau_i^2 \left\{ \frac{|U_i(f - f_c)|^2}{2 \sin^2 \pi(f - f_c)T} + \frac{|U_i(f + f_c)|^2}{2 \sin^2 \pi(f + f_c)T} \right\}
$$

This equation is valid only if $f - f_c > 1/NT$. Since f_i and f_j are independent, this is the spectral density for all the returns.

The total energy in the bandwidth of the filter is

$$
E_{ci} = \int_0^{b/2} E_{ci}(f)\, df = \tfrac{1}{4} a_i^2 \tau_i^2 \int_{f_c - b/2}^{f_c + b/2} \frac{|U_i(f)|^2}{\sin^2 \pi f T}\, df
\tag{11-19}
$$

where b is the filter bandwidth. Since b is small compared to $1/\tau_i$, $U_i(f)$ is essentially constant throughout the range of integration. Taking

$|U_i(f)|^2$ evaluated at f_c outside the integral and evaluating the remaining integral, noting that $a_i^2/2$ is the peak pulse power, and summing over all reflectors, one obtains for the total clutter energy out of the filter

$$E_c = \left[\frac{\hat{P} G A_e L}{(4\pi)^2} \right] [(\Sigma\sigma)(0.57\,\theta_1\phi_1)] \left[\int \frac{L_a \tau_i^2\, dR}{R^2} \right] \left[\frac{|U(f_c)|^2 K}{T} \right] \quad (11\text{-}20)$$

where

$$K = \frac{1}{\pi} \left(\frac{\sin \pi b T}{\cos \pi b T - \cos 2\pi f_c T} \right)$$

The integral with respect to range is approximately

$$\int \frac{L_a \tau_i^2\, dR}{R^2} \approx \tau^2 \frac{c}{2} \left(\tau_g - \frac{\tau}{3} \right) \sum_k \frac{L_a}{R_k^2} = \tau^2 \frac{c \tau_e}{2} \sum_k \frac{L_a}{R_k^2}$$

where τ = pulselength, $\tau_e = \tau_g - \tau/3$ if $\tau_g \geq \tau$

τ_g = range-gate length

R_k = range to the kth ambiguous range

If $\tau_g < \tau$, interchange τ_g and τ. Thus, finally one has for the total clutter energy

$$E_c = \frac{\hat{P} \tau G A_e L}{(4\pi)^2} \frac{\tau}{T} (\Sigma\sigma) \left(\frac{c \tau_e}{2} \right) (0.57\,\theta_1\phi_1) \left(\sum_k \frac{L_a}{R_k^2} \right) |U(f_c)|^2 K \quad (11\text{-}21)$$

Target-echo Response

The energy spectral density for a target echo consisting of a train of N pulses with amplitude a and doppler shift f_s relative to the filter center frequency is

$$E_s(f) = \frac{1}{2} a^2 \tau^2 \left\{ |U(f - f_s)|^2 \frac{\sin^2 \pi (f - f_s) N T}{\sin^2 \pi (f - f_s) T} \right.$$

$$\left. + |U(f + f_s)|^2 \frac{\sin^2 \pi (f + f_s) N T}{\sin^2 \pi (f + f_s) T} \right\}$$

If the range gate does not match the pulse (so that part of the pulse is eclipsed), τ should be replaced by $\epsilon\tau$ where $0 < \epsilon < 1$. The signal energy out of the doppler filter is then

$$E_s = \int_0^{b/2} E_s(f)\,df = \frac{1}{2}a^2\tau^2|U(f_s)|^2 \int_{f_s-b/2}^{f_s+b/2} \frac{\sin^2\pi f NT}{\sin^2\pi fT}\,df \quad (11\text{-}22)$$

The filter bandwidth b is generally much less than the PRF $1/T$, and f_s is less than $b/2$. Otherwise the signal is in another filter, and one should examine that filter. Therefore, $\sin(\pi fT)$ may be approximated by πfT.

The integral in Eq. (11-22) is then

$$\int_{f_s-b/2}^{f_s+b/2} \frac{\sin^2\pi f NT}{\sin^2\pi fT}\,df = \int_{f_s-b/2}^{f_s+b/2} \frac{\sin^2\pi f NT}{(\pi fT)^2}\,df$$

$$\qquad\qquad (11\text{-}23)$$

$$= \frac{N}{\pi T}\int_{\pi NT(f_s-b/2)}^{\pi NT(f_s+b/2)} \frac{\sin^2 x}{x^2}\,dx = \frac{NY}{T}$$

where the function Y represents the fraction of the energy of a single line contained in the filter bandwidth. The signal energy is then

$$E_s = \frac{\hat{P}_r G A_e L}{(4\pi)^2}\left(\frac{\tau}{T}\right)\sigma_t L_p \frac{L_a}{R_t^4}|U(f_s)|^2 NY$$

where R_t = target range

$\quad\;\; L_p$ = pattern loss in case the target is not centered in the beam

With N pulses integrated, the coherent integration on the dwell time is $T_d = NT$.

The ratio of signal energy to clutter energy at the output of the doppler filter containing the target signal is thus

$$\frac{E_s}{E_c} = \left[\frac{\sigma_t}{(\Sigma\sigma)(c\tau_e/2)}\right]\left[\frac{L_p}{0.57\,\theta_1\phi_1}\right]\left[\frac{(L_a/R_t^4)}{\sum_k(L_a'/R_k^2)}\right]\left[\frac{|U(f_s)|^2}{|U(f_c)|^2}\frac{NY}{K}\right] \quad (11\text{-}24)$$

or, naming the factors,

$$\frac{E_s}{E_c} = [\text{echo power}][\text{pattern factor}][\text{attenuation}][\text{spectral factor}]$$

For a duty ratio of $\tau/T \le 0.01$, the single-pulse envelope functions $U(f_s)$ and $U(f_c)$ differ by a small fraction of one percent (i.e., both signal and clutter are near the center of the single-pulse spectrum). Hence the ratio $U(f_s)^2$ to $U(f_c)^2$ is essentially unity. The function Y is always < 1 but generally > 0.5.

A number of approximations can be made to make Eq. (11-24) more tractable. A reasonable average value for Y is about two-thirds, the gate length $\tau_g \approx \tau$, the atmospheric attenuation term L_a can be neglected, and $L_p \approx 0.57$. Then Eq. (11-24) simplifies to

$$\frac{E_s}{E_c} = \left[\frac{\sigma_t}{(\Sigma\sigma)(c\tau/2)\,\theta_1\phi_1}\right]\left[\frac{1}{R_t^2 \sum\limits_k (R_t/R_k)^2}\right]\left[\frac{N}{K}\right] \qquad (11\text{-}25)$$

Clutter Echoes Only at the Target

If the clutter is small in extent compared to the interpulse period (less than $cT/2$) and located only at the target, then the energy ratio is

$$\frac{E_s}{E_c} = \left[\frac{\sigma_t}{(\Sigma\sigma)(c\tau/2)\,\theta_1\phi_1}\right]\left[\frac{1}{R_t^2}\right]\left[\frac{N}{\dfrac{1}{\pi}\left(\dfrac{\sin\pi bT}{\cos\pi bT - \cos2\pi f_c T}\right)}\right] \qquad (11\text{-}26)$$

It should be noted that the first two brackets of Eq. (11-26) are the same as the equation for the simple pulse radar in volume clutter. Thus, the terms in the final bracket constitute the clutter attenuation or improvement factor for a train of N pulses using a close approximation to a matched-filter receiver. The detection range equation for clutter at the target range is then obtained by rewriting Eq. (11-26)

$$R_t^2 = \left[\frac{\sigma_t}{(\Sigma\sigma)\,(c\tau/2)\,\theta_1\phi_1(S/C)} \right] \left[\frac{N}{\dfrac{1}{\pi}\left(\dfrac{\sin \pi bT}{\cos \pi bT \,-\, \cos 2\pi f_c\, T} \right)} \right] \qquad (11\text{-}27)$$

Numerical values for the second bracket were given in Table 11-1.

Extended-clutter Echoes

For clutter which extends quite close to the radar (uniform rain, etc.) the sum in the denominator of the third bracket of Eq. (11-25) is dominated by the term for the nearest ambiguous range R_1 and the signal-to-clutter ratio is then

$$\frac{E_s}{E_c} = \left[\frac{\sigma_t}{\Sigma\sigma(c\tau/2)\,\theta_1\phi_1} \right]\left[\frac{R_1^2}{R_t^4} \right][L_a(R_t - R_1)]\left[\frac{N}{K} \right] \qquad (11\text{-}28)$$

where $L_a(R_t - R_1)$ is the attenuation between the target and the first clutter region. The significance of this equation is that for volume clutter which is close to the radar at range R_1, the signal-to-clutter ratio varies inversely as R_t^4 rather than as R_t^2 as in unambiguous pulse radars. The first ambiguous clutter range can be approximated by using the largest integer N_1 which satisfies

$$R_1 \approx \left[R_t - N_1\!\left(\frac{cT}{2} \right) \right] > 0 \quad \text{if} \quad \frac{c\tau}{2} \ll \frac{cT}{2} \qquad (11\text{-}29)$$

The approximation breaks down if R_1 is in the near zone of the antenna or if an STC circuit is used to reduce the near-in clutter. The maximum clutter echo power was derived in Chap. 10 on CW radars.

For the intermediate case where the clutter extends at least several interpulse periods but not to the radar (the target may or may not be within the clutter), the sum in Eq. (11-25) may be approximated by

$$\sum_k R_k^{-2} \approx \frac{(N_2 - N_1 + 1)}{[R_t + (2N_1 - 1)\,cT/4][R_t + (2N_2 - 1)\,cT/4]} \qquad (11\text{-}30)$$

where $R_t + N_1\ cT/2$ = smallest ambiguous clutter range

$\qquad R_t + N_2\ cT/2$ = largest ambiguous clutter range

$\qquad N_2 - N_1 + 1$ = number of ambiguous ranges from which clutter returns are simultaneously received

Both N_1 and N_2 may be negative as long as $R_t + (2N_1 - 1)\,cT/4$ is positive. If this term is near zero range, the approximation to obtain Eq. (11-30) is poor; and the system is most likely dominated by the first clutter echoes. In this case it will usually suffice to use Eq. (11-29).

Optimum Value of Filter Bandwidth

In order to determine the optimum bandwidth for maximizing S/C, let $b = X/NT = X/T_d$ for large N, where X is a constant factor. The last factor in Eq. (11-24) becomes

$$\frac{NY}{\dfrac{1}{\pi}\left[\dfrac{\sin(\pi X/N)}{\cos(\pi X/N) - \cos 2\pi f_c T}\right]} \qquad (11\text{-}31)$$

Referring to Eq. (11-23), it is easily shown that the function Y is a monotonically increasing function of X which approaches the value one for $X = N$. For X decreasing, Y becomes asymptotic to X. Examination of the denominator of the above expression shows it to be also a monotonic increasing function of X which, however, does not approach a constant value for large X but in fact increases indefinitely as X/N approaches $2f_c T$. At this point the approximations used to obtain Eq. (11-20) are no longer valid. For small X the denominator is proportional to X/N.

The value of the fraction is thus proportional to N^2 for small X and decreases as X increases. This suggests that the best bandwidth to use is the narrowest obtainable. This is true when clutter is the only interference to consider. In the presence of broad-band noise, however, reduction of the bandwidth much below the reciprocal of the signal duration T_d seriously degrades the signal-to-noise ratio. Also, if the target fluctuates, little improvement in signal-to-noise ratio will be obtained by reducing the bandwidth below the spectral width of the fluctuations.

The *best* bandwidth is then obtained by taking X to be of the order of one, making the signal-to-clutter ratio proportional to N^2. The constant of proportionality depends on f_c. If b is replaced by $1/NT$ in the last bracket of Eqs. (11-27) through (11-31), this term can be written

$$\left[\frac{N}{\frac{1}{\pi}\left(\frac{\sin \pi bT}{\cos \pi bT \, - \, \cos 2\pi f_c T}\right)} \right] = \left[\frac{N}{\frac{1}{\pi}\left(\frac{\sin (\pi/N)}{\cos (\pi/N) \, - \, \cos 2\pi f_c T}\right)} \right] \qquad (11\text{-}32)$$

If, in addition, N is at least 30, $\sin (\pi/N) \approx \pi/N$, $\cos \pi/N \approx 1$, and the bracketed term becomes

$$N^2 [1 \, - \, \cos 2\pi f_c T] \approx \text{clutter attenuation} \qquad (11\text{-}33)$$

As expected, the clutter attenuation term is maximized at $f_c T = \frac{1}{2}$. Then for $bT = 1/N$ and N large, the maximum clutter attenuation for a doppler separation of $f_c T \approx \frac{1}{2}$ (averaged over dopplers in that region) can be approximated by

$$\text{CA}_{\max} \approx 2N^2 \quad \text{near the optimum doppler separation} \qquad (11\text{-}34)$$

Alternately for an average over all doppler velocities (except for the ambiguous regions) the improvement factor I is N^2.

Random Interpulse Period

In extended clutter, staggering of the pulses alters the location of ambiguous clutter responses but does not alter the number of ambiguous ranges from which clutter returns are received. Therefore, a good approximation is that the integral with respect to R in Eq. (11-17) is unchanged by the staggering. The spectrum of a train of staggered pulses differs from that of a regularly spaced train except in the immediate vicinity of the central line, resulting in an increase in the clutter energy in the bandwidth of the nonambiguous doppler filters. The function K that appears in Eqs. (11-20) through (11-28) must be replaced by a function K' for the staggered pulse train.

The exact form of the function K' is dependent on the nature of the staggering. In the case where the jth pulse of a train is shifted by a random amount ϵ_j from the position it would have in a regularly spaced pulse train, the function is

$$K' = [1 \, - \, |\varphi(f)|^2] Nbt + |\varphi(f)|^2 K \qquad (11\text{-}35)$$

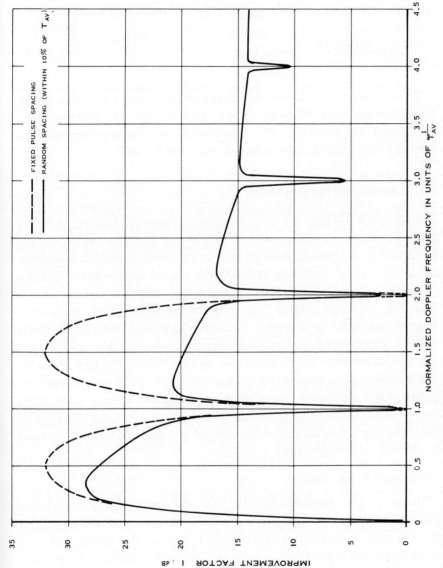

figure 11-15 *Pulse-doppler point clutter improvement factor.*

where $\varphi(f)$ is the characteristic function of the probability distribution of ϵ_j. If ϵ_j may have any value from $-\frac{1}{2}\alpha T$ to $+\frac{1}{2}\alpha T [\alpha < 1 - (\tau/T)]$ with uniform probability, the characteristic function for large N is

$$\alpha(f) = \frac{\sin \pi \alpha f T}{\pi \alpha f T}$$

The relative improvement in signal-to-clutter ratio I is shown in Fig. 11-15 for the case $NbT = 1$, $\alpha = 0.2$, and $N = 32$. The abscissa is the normalized doppler shift $1/T_{av}$. For comparison, the improvement for a constant PRF pulse doppler is also shown (dashed curve).

11.8 TRUNCATED PULSE TRAINS

The previous section on clutter computations emphasized that the *close-in* clutter at range R_0 is often the limiting factor in the performance of pulse-doppler systems in range extended-clutter environments. For most detection criteria, the overall improvement factor I contains a term of the form R_0^2/R_t^4 for uniform volume clutter extended in range with a high PRF transmission. Thus, the overall improvement may be negligible for targets at a range of many interpulse periods. The type of clutter that is of most concern in this description is assumed to be extended in range and separated from the target velocity but confined in doppler extent. Rihaczek [316, 317] and others have shown that a uniformly spaced pulse train is desirable since most of the energy under the ambiguity surface is concentrated in finite locations on the axes. The following discussion is based on uniform interpulse spacing of uniform amplitude pulses.

Single-burst Pulse Train

It can be seen from Eq. (11-30) that the summation of the clutter echoes from many ranges provides a limit to the signal-to-clutter ratio improvement in partially extended clutter. Ares [5] has shown that for a given target-to-clutter velocity ratio and interpulse period T there exists for each target range R_t a value of N that maximizes the signal-to-clutter ratio. This *optimization* exists only for clutter which extends for all the $2N - 1$ ridges of the ambiguity diagram. The explanation for this optimization is that for small interpulse periods, the close-in clutter at range R_0 is competing with target signals from distant ranges; and it is the echoes from the later transmitted pulses which constitute the greatest portion of the clutter entering the receiver. The optimum number of pulses found by Ares is

$$N \approx \frac{1}{\sqrt{3}} \left(\frac{R_t}{cT/2} \right) \tag{11-36}$$

and the optimum clutter attenuation is $CA_{opt} \approx 2/3N$, the improvement factor then being

$$I_{opt} \approx \frac{2}{3} N = \frac{2R_t}{3\sqrt{3}\,(cT/2)} \tag{11-37}$$

It can be seen that the improvement factor increases with target range. The optimum doppler velocity occurs at approximately $1/2T$ as expected. If the target range and doppler frequency are approximately known, as in an acquisition or tracking radar, the *optimum* number of pulses and their spacing can be chosen. The target detection range for uniform range-extended clutter can then be written from Eq. (2-30)

$$R_{opt} = \frac{2L_c L_s' \sigma_t}{3\sqrt{3}\,(\pi/4)\,\theta_1\phi_1(c\tau'/2)\,(cT/2)\,(S/C)\Sigma\sigma} \tag{11-38}$$

It can be noted that the required (S/C) is only first-order dependent on target range and that both short pulselengths and small interpulse periods are desired. Since a short interpulse period can only result from optimization with respect to a high doppler frequency, the interpulse period term $cT/2$ in the denominator only shows the desirability of having a large velocity separation between the targets and the clutter.

If the interpulse period is not fixed by other considerations, a double optimization involving the period and the number of pulses can be performed. An example is given in Table 11-3. For a given range R_t (in microseconds) and target-doppler frequency f_d, the optimum number of pulses is shown. Also shown on the table is the length of the burst $T_d \approx 2R_t/3c$ and the doppler resolution b in kHz.

The improvement factor for these same cases is shown in Table 11-4. The improvement factor is in decibels above the signal-to-

TABLE 11-3 Optimum Number of Uniformly Spaced Pulses for Extended Clutter

Target range R_t, μsec	Dwell time T_d, μsec	Doppler resolution b, kHz	N_{opt} for doppler frequency offset, kHz										
			4	5	8	10	12	15	20	25	30	35	40
400	231	4.33	2	2	4	5	6	7	9	12	14	16	18
450	260	3.85	2	3	4	5	6	8	10	13	16	18	21
500	289	3.46	2	3	5	6	7	9	12	14	17	20	23
550	318	3.15	3	3	5	6	8	10	13	16	19	22	25
600	346	2.89	3	4	6	7	8	10	14	17	21	24	28
650	375	2.67	3	4	6	8	9	11	15	19	23	26	30
700	404	2.48	3	4	7	8	10	12	16	20	24	28	32
750	433	2.31	3	4	7	9	10	13	17	22	26	30	35
800	462	2.17	4	5	7	9	11	14	18	23	28	32	37
900	517	1.94	4	5	8	10	12	16	21	26	31	36	42
1,000	577	1.73	5	6	9	12	14	17	23	29	35	40	46
1,100	635	1.58	5	6	10	13	15	19	25	32	38	44	51
1,200	693	1.44	6	7	11	14	17	21	28	35	42	49	55
1,300	751	1.33	6	8	12	15	18	23	30	38	45	53	60
1,400	808	1.24	6	8	13	16	19	24	32	40	49	57	65
1,500	866	1.16	7	9	14	17	21	26	35	43	52	61	69
1,600	923	1.08	7	9	15	18	22	28	37	46	55	65	74

clutter ratio that would result from a single pulse of the train, and it only applies to an extended-clutter situation. The values shown are slightly better than would be indicated by Eq. (11-37) for certain numbers of pulses (see also [211]).

Ares also considered receiver amplitude weighting of the truncated pulse train and showed that there was little additional improvement. The explanation for this is that the ambiguous clutter returns at other than the target range do not receive the symmetrical weighting which is necessary to reduce the doppler sidelobes. Figure 11-16 gives examples [5] of the improvement obtained with a constant amplitude and a *Taylor*-tapered train of 16 pulses for various target ranges. It is drawn for the case of uniform rain clutter which has a narrow spectrum but is extended in range until the radar antenna beam reaches a 35,000-ft altitude. The improvement increases slowly from 40 to 100 nmi, and then as the beam extends above the rain, the improvement increases more rapidly. The additional improvement resulting from tapering the received waveform is less than 3 dB at the optimum velocity and almost trivial at other velocities, even with a 40-dB taper. Thus, significant improvement over the uniform case can only occur with complex weighting of the transmitted waveform as well as of the receiver waveform as discussed in Sec. 11.4.

Geometric or Burst Programming[1]

A pulse-doppler radar operating at high PRF will have blind ranges and close-in clutter problems if the transmit pulse train is not turned off immediately prior to and during the target-echo receive period. By use of such a turn-off procedure, reduction of close-in clutter echoes can be achieved but at the expense of reducing the total number of pulses on the target during the coherent integration time. A *burst repetition frequency* (BRF) is thus created that is a function of target range and the time allowed for nearby clutter to diminish to a tolerable level. This

TABLE 11-4   Optimum Improvement Factor, I, dB, for Uniformly Spaced Pulses and Extended Clutter

Target range R_t, μsec	Doppler frequency offset, kHz										
	4	5	8	10	12	15	20	25	30	35	40
400	1.3	1.3	4.3	5.2	6.0	6.7	7.8	9.0	9.7	10.3	10.8
450	1.3	3.0	4.3	5.2	6.0	7.3	8.2	9.4	10.3	10.8	11.5
500	1.3	3.0	5.2	6.0	6.7	7.8	9.0	9.7	10.5	11.3	11.9
550	3.0	3.0	5.2	6.0	7.3	8.2	9.4	10.3	11.0	11.7	12.2
600	3.0	3.0	6.0	6.7	7.3	8.2	9.7	10.5	11.5	12.0	12.7
650	3.0	4.3	6.0	7.3	7.8	8.7	10.0	11.0	11.9	12.4	13.0
700	3.0	4.3	6.0	7.3	8.2	9.0	10.3	11.3	12.0	12.7	13.3
750	3.0	4.3	6.7	7.8	8.2	9.4	10.5	11.7	12.4	13.0	13.7
800	4.3	5.2	6.7	7.8	8.7	9.7	10.8	11.9	12.7	13.3	13.9
900	4.3	5.2	7.3	8.2	9.0	10.0	11.5	12.4	13.2	13.8	14.4
1,000	5.2	6.0	7.8	9.0	9.7	10.5	11.9	12.9	13.7	14.3	14.9
1,100	5.2	6.0	8.2	9.4	10.0	11.0	12.2	13.3	14.0	14.7	15.3
1,200	6.0	6.7	8.7	9.7	10.5	11.5	12.7	13.7	14.5	15.1	15.6
1,300	6.0	7.3	9.0	10.0	10.8	11.9	13.0	14.0	14.8	15.5	16.0
1,400	6.0	7.3	9.4	10.3	11.0	12.0	13.3	14.2	15.1	15.8	16.4
1,500	6.7	7.8	9.7	10.5	11.5	12.4	13.7	14.6	15.4	16.1	16.6
1,600	6.7	7.8	10.0	10.8	11.7	12.7	13.9	14.9	15.6	16.4	16.9

[1] Material for this section courtesy of A. Chwastyk.

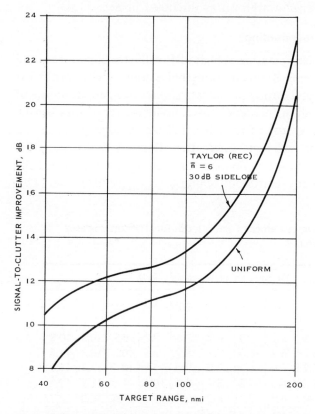

N=6,   T= 23.5 μsec

RAIN UNIFORM FROM 0 TO 35,000 ft
TARGET AT OPTIMUM VELOCITY
3° ELEVATION BEAMWIDTH − CENTER AT 1.5°

figure 11-16   *Signal-to-clutter improvement vs. target range*
for 16 pulse train and uniform rain. (From Ares [5].)

frequency is nominally equal to $1/(2T_t - T_{cd})$ where T_t is the target echo time and T_{cd} is the *clutter-decay time.*

In the case of surface radars, it is often necessary to impose the requirement for a clutter-decay time to prevent large clutter returns from regions much closer than the target range from entering the range gate. These returns may be from either the main lobe or sidelobes of the antenna. For uniform clutter, increasing clutter-decay time must be compromised with respect to the number of pulses transmitted. The functional relationship between target range, clutter decay, and the BRF is shown in Fig. 11-17. Here, the maximum percentage of pulses is taken to occur when transmit on and off times are equal; that is, the clutter-decay time is zero.

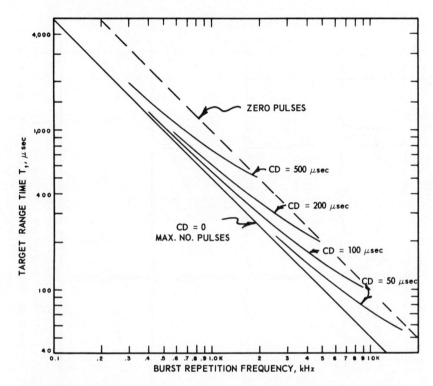

figure 11-17 *Target range delay vs. burst repetition frequency for various clutter-decay times.*

The adaptive control of the BRF in response to changes in target range and the range extent of the nearby clutter is termed *geometric programming*. Some experimental effects of using geometric programming are shown on the range profiles of Fig. 11-18. The equipment used was a high PRF (150-kHz) digital implementation of a pulse-doppler radar. In all plots, the zero doppler channel is displayed (normalized to the largest signal) to illustrate the land-clutter response. In Fig. 11-18A no geometric programming was used. This mode maintains maximum energy on the target (a water tower in this case) but suffers from ambiguous clutter folding onto the target range. Figure 11-18B shows

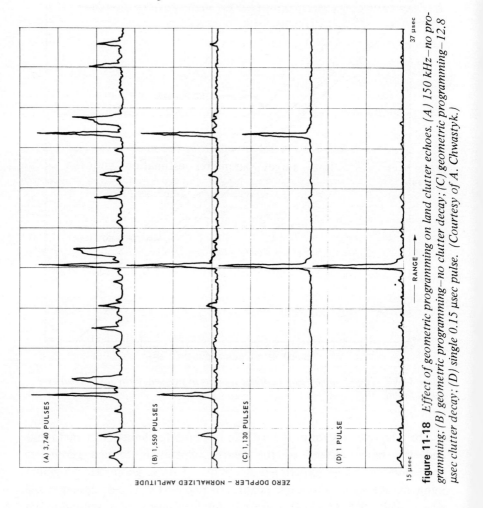

figure 11-18 *Effect of geometric programming on land clutter echoes. (A) 150 kHz—no programming; (B) geometric programming—no clutter decay; (C) geometric programming—12.8 μsec clutter decay; (D) single 0.15 μsec pulse. (Courtesy of A. Chwastyk.)*

the effect of geometric programming but without any clutter-decay time again allowing nearby clutter to fold onto the target region. In Fig. 11-18C local clutter has been removed by keeping the receiver off an additional 12.8-μsec interval after each burst. Since the system has a constant PRF, range ambiguities still occur beyond the target. In Fig. 11-18D a single-pulse transmission is shown for comparison. With this nonambiguous waveform, the return energy from the target is relatively low (no integration) and cannot yield any doppler resolution.

The modulation of the transmit waveform with burst timing creates spectral components at the burst repetition frequency and its harmonics. The main doppler lobe width (the velocity resolution) remains essentially unchanged, but a pair of additional spectral lines due to the burst modulation are less than 4 dB down from the main lobe. The spectral lines can affect both target parameter estimation and the clutter performance of the system. The worst case for clutter rejection occurs when the clutter-doppler frequency is separated from the target-doppler frequency by the burst repetition frequency. In this case, the clutter residue will often integrate from burst to burst, leaving a substantial residue at the end of the dwell time. Increased coherent integration time will not improve this condition. A sketch illustrating this condition is shown in Fig. 11-19 where the burst spectral line of the clutter appears at the target-doppler frequency.

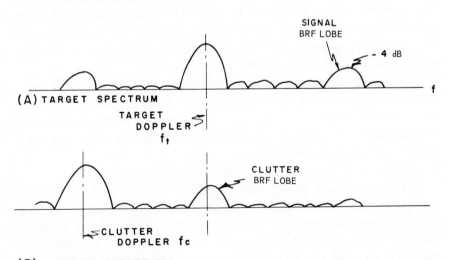

figure 11-19 *Spectra of target and clutter illustrating effect of burst spectral lines.*

11.9 FREQUENCY AGILE PULSE TRAINS

The Resnick method of staggering of pulse trains (Sec. 11.3) was shown to have a correlation function which maintains a maximum time-residue level of $1/N$ except on the doppler axis. This requires that the $\frac{1}{2}N(N-1)$ spacings between *any* pairs of pulses all be different multiples of the pulselength. However, the time correlation function residues in the neighborhood of the doppler axis (for range offsets $< \tau$) increase at values of doppler frequency which are higher than the first main doppler ambiguity which is due to the pulse repetition frequency. Resnick [309] showed that these residues reach an average level of $1/\sqrt{N}$, normalized to the value of the correlation function at the origin. This was illustrated in Fig. 11-3. Thus, the Resnick waveform yields good time resolution at all dopplers of echoes from objects separated in range but mediocre doppler resolution of echoes originating at the same range, but which are considerably separated in doppler frequency.

When a large amount of multiple-target resolution or interference rejection is desired and the location of the clutter or interference is not known, the number of pulses in the waveform must be increased. Since the total length of the pulse train will vary about as $\frac{1}{2}\tau N^2$, there may be insufficient time per beam position to use large values of N, especially in a narrow beamwidth surveillance radar. This problem will be aggravated if the desired detection range is large and the minimum interpulse period Δ is substantial. In a surface radar the *close-in* clutter may completely preclude target detection in the ambiguous range intervals. In any case, finding suitable receiver storage mechanisms (delay lines, etc.) becomes much more difficult for processing long pulse trains.

Several techniques are available to extend the unambiguous range when the longer pulse trains are impractical. These techniques are based on using *frequency agile* trains which are composed of pulses on several different carrier frequencies, separated such that spectra of the signals on the various carriers do not overlap.

There are at least three classes of this type of waveform:

1. The incoherent pulse train where all pulses which could give ambiguous responses are on different frequencies. No attempt is made to maintain coherency between the various pulses in the train or between the frequencies transmitted. Since the relative phases of the pulses are then arbitrary, only incoherent processing can occur (see Sec. 14.1).

2. The frequency-coherent pulse train where each pulse is on a different carrier frequency, but coherency *is* maintained between all the transmitted pulses. The properties of this waveform and the matched-filter implementation will be discussed in Chap. 13.

3. The pulse train has several phase-coherent pulses per frequency interleaved with the pulses from the other carrier frequencies. The pulse spacing may be uniform or nonuniform.

The third example is of interest in this section. The signal processor for reception of the return signals can take at least two forms. The coherent processor in Fig. 11-20*A* can be used where the pulses on each frequency are coherent with those on all the other frequencies as well as those on the same frequency. The discussion of this technique is also included in the second on frequency coding.

The incoherent frequency processor shown in Fig. 11-20*B* logically follows the discussion of the Resnick-type pulse train since Kaiteris and Rubin [199] have shown that most of the desirable properties of the former can be maintained with frequency agile *magic N* trains when incoherent processing of the outputs on each frequency is performed. In addition, the length of the staggered pulse train often can be reduced to achieve a given doppler residue level or, at least, ambiguous clutter responses can be reduced.

Following the development of [199], consider the transmission of a pulse train of N pulses with a total length of $(N - 1)\overline{T}$ where the spacings of *any* pair of pulses are different and $\overline{T}$ is the average interpulse spacing. This train is then divided into m subtrains each having length of approximately $(N - 1)\overline{T}/m = T_d/m$. If the processor outputs on each of the m frequencies are added (postdetection integration), the resulting ambiguity function of total waveform after normalization (the *video ambiguity function*) has about the same relative residues off the doppler axis as the Resnick pulse train; and the residues on the time axis itself are never greater than $1/N$. The ambiguity function of the incoherent summation is simply the addition of the magnitudes of the ambiguity functions of the waveform on each frequency. The sidelobes of each component ambiguity function occur at different locations on the ambiguity plane; therefore, with respect to these sidelobes the ambiguity functions are disjoint and the following expression applies:

$$|\psi_j(\tau, \omega)| \cdot |\psi_k(\tau, \omega)| = 0 \quad \text{for} \quad j \neq k \quad \text{and} \quad |\tau| > \tau' \qquad (11\text{-}39)$$

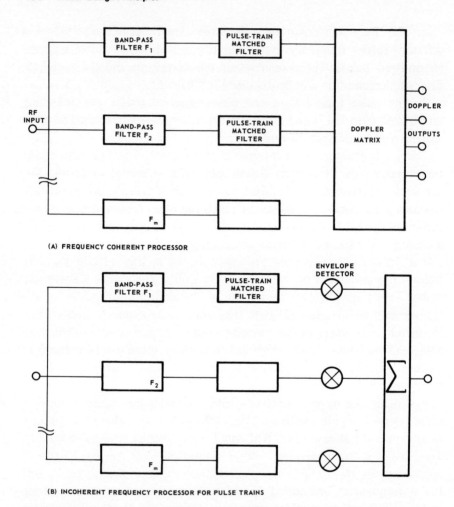

(A) FREQUENCY COHERENT PROCESSOR

(B) INCOHERENT FREQUENCY PROCESSOR FOR PULSE TRAINS

figure 11-20 *Processors for frequency agile pulse trains. (A) Frequency coherent processor; (B) incoherent frequency processor for pulse trains.*

where $\psi_j(\tau, \omega)$ is the ambiguity function of the waveform on the jth carrier frequency. If the m frequency bands are disjoint the video ambiguity function is

$$\psi_v(\tau, \omega) = \frac{1}{m} \sum_{k=1}^{m} |\psi_k(\tau, \omega)| \tag{11-40}$$

Since there has been no general restriction placed on the choice of which pulses will occur on which frequency band, the doppler response for $|\tau| < \tau'$ cannot be described in detail. If maximum doppler resolution is desired, the pulses on a given frequency should be distributed throughout the pulse train. The penalty for this higher resolution is that many *matched* doppler filters must be instrumented for each carrier frequency. If poorer doppler resolution is acceptable, all the pulses on a given carrier can be transmitted in succession and then the group of pulses on another frequency, etc. For a given Resnick pulse train, this will introduce more closely spaced range ambiguities but still only of amplitude $1/N$.

12 PHASE CODING
TECHNIQUES

The next two chapters will elaborate on the two primary methods of obtaining range resolution by means of *pulse compression*. This chapter will expand on phase coding of a single-frequency carrier while Chap. 13 will emphasize pulse coding by shifting the carrier frequency during the waveform. Since the *compressed pulse* is generally the output of a matched filter, the properties of the output waveforms (in the absence of a doppler shift) can be discussed in terms of the autocorrelation function of the transmit signal. When there is a doppler shift the complete Ambiguity Function must be discussed. The various forms of phase coding (or phase modulation) provide an excellent basis for general studies of signal processing and provide an intuitive grasp of the

For general references see Elspas [105], Golomb [143, 142], Ristenblatt [20, chap. IV], Chandler [64], Sakamoto et al. [332].

significance of the *Ambiguity Diagram*. The effects of the time-bandwidth product of the transmission are emphasized since phase coding can be utilized for products as low as three and up to the hundreds of thousands, and there are several practical implementations of the matched filter for stationary or moderate-speed targets as well as for high-speed targets.

While many of the concepts in this chapter are also applicable to amplitude coding, the emphasis will be on the transmission of a constant amplitude sinusoidal carrier which is divided in time into N equal segments of duration τ'. The majority of the material will be on *binary* phase coding where the sine wave in each segment can be either 0 or 180° from an arbitrary reference. This will be expanded in Sec. 12.10 to include *polyphase* waveforms where each segment can have any one of M possible phases.

Phase coding can be further divided into two general classes depending on the length and periodicity of the code. The *continuous code* has wide application in communication systems to convey messages and in continuous wave (CW) radars to provide time or range resolution. A *single pulse* whose carrier is phase coded is often used in pulse-radar systems to increase range resolution and accuracy when energy requirements dictate a pulselength substantially in excess of the desired resolution. Coded pulses (or coded words) are also used for clutter or interference rejection. The classes of phase coding have also been described as *periodic* and *aperiodic* where periodic refers to the repetition of a code. The aperiodic code would then include the special case of a continuous transmission with *random binary coding* of the phases of individual segments as well as a short-coded word. Obviously there are hybrids between the classes.

12.1 PROPERTIES OF WAVEFORMS WITH RANDOM BINARY CODING

Consider a carrier divided in time into N segments each of which is coded either + or − as shown in Fig. 12-1. This will be called a binary-coded waveform. A *plus* corresponds to the nominal carrier phase and a *minus* corresponds to a 180° phase shift. The amplitude of the kth segment of the waveform will be denoted by a_k. Each segment will be assumed to have unit amplitude and one of two phases.[1] The effect of the carrier

---

[1] For example, by flipping a coin and letting heads equal a plus and tails a minus.

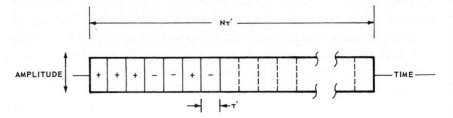

figure 12-1   *Representation of binary phase coding.*

frequency will be neglected until the complete ambiguity function is discussed.

In accordance with this notation, the temporal autocorrelation function of this waveform can be conveniently written in discrete form

$$\phi(m) = \sum_{k=1}^{N} a_k a_{k+m}$$

where the integer index m steps over the domain $-(N-1) \le m \le (N-1)$. Each step corresponds to a shift in a_{k+m} of one segment duration τ'. By describing the autocorrelation function only for discrete steps of τ', the determination of its value for various codes will result in integer values approximating the triangular waveforms that result from an exact autocorrelation. The discrete procedure simplifies the calculation and drawing of the autocorrelation function on the time axis. The loss in accuracy is slight since the detailed shape of the ambiguity function depends on the response of the matched filter.

The autocorrelation function is always a maximum at $\tau = 0$ and is equal to N.

If the polarities of the values of a_k are chosen randomly and N is large, the autocorrelation function near the origin is triangular as shown in Fig. 12-2. The power spectrum of a random sequence, according to the Wiener-Khintchine theorem, is the Fourier transform of its autocorrelation function. The shape of the power spectrum is then approximately that for a single rectangular segment of length τ' or

$$G(f) = \frac{\sin^2(\pi f \tau')}{(\pi f \tau')^2}$$

This type of power spectrum is shown in Fig. 12-3*C* for a 255-segment pseudo-random binary code whose starting point was varied from code to code. While not truly a random code, the first nulls appear at a frequency of $1/\tau'$ from the carrier, and the first *sidelobes* are about 13 dB below the peak.[1] The code sequence shown in Fig. 12-3*D* was used to reverse the phase of a microwave carrier. These photos of the display of a microwave spectrum analyzer have a logarithmic amplitude scale (10 dB/division).

The values of the autocorrelation function away from the origin (range sidelobes) on the time axis can be computed by numerous techniques but can easily be visualized by setting up the following tables where illustratively the code consists of the first seven segments of Fig. 12-1. For a one-segment offset

$$
\begin{array}{cccccccc}
 & + & + & + & - & - & + & - \\
\hline
 & & + & + & + & - & - & + & - \\
\hline
\phi(m = 1) = & & 1 & 1 & -1 & 1 & -1 & -1 & \quad = 0, \ \text{first side peak}
\end{array}
$$

The logic used above assumes that the occurrence of two +'s or −'s (a *match*) is unity and a + and a − equal − 1. Thus, in this case, the value of the autocorrelation function at a one-segment offset is zero. Similarly, a two-segment offset would yield

$$
\begin{array}{cccccccc}
 & + & + & + & - & - & + & - \\
\hline
 & & & + & + & + & - & - & + & - \\
\hline
\phi(m = 2) = & & 1 & -1 & -1 & -1 & 1 & \quad = -1, \ \text{second side peak}
\end{array}
$$

It has been shown [64] that the amplitude of the autocorrelation function for offsets exceeding τ' (range sidelobes) will fall between ±

[1]The dc component for this sequence is lower than indicated on the figure (see Golomb et al. [143, p. 77]).

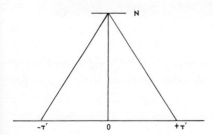

figure 12-2 *Autocorrelation function of a random sequence near the origin.*

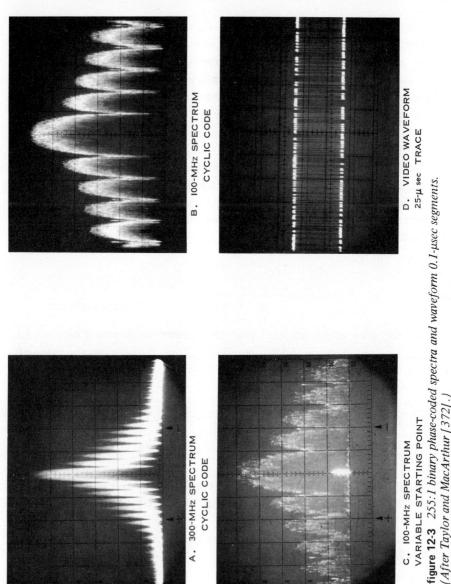

B. 100-MHz SPECTRUM
CYCLIC CODE

D. VIDEO WAVEFORM
25-μ sec TRACE

A. 300-MHz SPECTRUM
CYCLIC CODE

C. 100-MHz SPECTRUM
VARIABLE STARTING POINT

figure 12-3   *255:1 binary phase-coded spectra and waveform 0.1-μsec segments. (After Taylor and MacArthur [372].)*

$0.667\sqrt{N}$ with a 0.5 probability as $N \approx \infty$. The rms sidelobe level is about $0.7\sqrt{N-1}$. It will be shown in later sections that by proper choice of code both the average and peak output amplitudes for the *mismatched* case can be held to a lower level and hence yield better clutter rejection.

Before describing specific codes and their applications, it is worthwhile to state the properties of a long code where the choice of the polarity of each segment is made randomly [142].

1. The number of pluses (or ones) is approximately equal to the number of minuses (or minus ones).

2. Consecutive segments of the same polarity (or runs) occur frequently. About half of these have a length of one, one-fourth have a length of two, one-eighth have a length of three, etc.

3. The value of the ambiguity function drops rapidly from the origin on both the time and doppler axes and constitutes a relatively simple implementation of a *thumbtack* ambiguity function.

The sidelobes on the time axis are more easily controlled and the bulk of the following discussion will emphasize the properties for zero doppler shift.

12.2 BINARY PERIODIC SEQUENCES

While the binary periodic codes or binary sequences have greater application in communications than in radar, they have several properties that make them useful in CW radar (especially when the transmit and receive antennas are separated to alleviate the spillover problem). The first property of importance results from the fact that the ratio of the peak of the matched-filter output (the autocorrelation function) to the value at other regions on the time axis of the ambiguity function increases faster than the square root of the number of segments in the transmission. An important extension of this phenomenon is that there are specific periodic codes where the amplitude of the autocorrelation function has a constant value of minus unity on the time axis for the length of the period except at the origin. The peak amplitude of the autocorrelation function has a height equal to the number of segments in the period. This important property of *linear maximum-length sequences* means that the rejection of undesired *interference* is proportional (in amplitude) to the length of one period of the sequence.

The class of radar waveforms which have this desired property of high rejection of undesired echoes on the time axis was first reported about 1955 by Elspas [105], Golomb [142], Zierler [411], and several other investigators at M.I.T. There are many names for the maximal-length sequences depending on their applications in communication systems. While there are other sequences of this nature, the terms *maximal-length binary shift-register sequences, m-sequences* or *pseudo-random sequences* will be considered synonymous. The spectrum of one of these sequences of length 255 was shown in Fig. 12-3*A* and *B*. A shift-register encoder for these sequences is shown as Fig. 12-4.

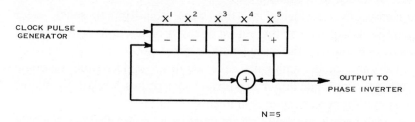

figure 12-4 *Shift-register encoder with modulo 2 adder for period of 31 segments.*

As distinguished from random binary sequences, the following properties are attributed to pseudo-random sequences [142, 143, 64].

1. The number of segments of plus ones (+) in each period of the sequence is within one of the number of segments of minus ones (−). This is sometimes called the balance property.

2. In every period, half of the runs have length one, one-fourth have length two, one-eighth have length three, etc. (the run property).

3. The initial condition in the shift register determines the starting point of the code. The condition of all *pluses* is forbidden.

4. The last stage in the shift register must be connected to the feedback circuit and there must be an even number of feedback taps.

5. Each period has length $2^n - 1$. This is the longest possible period for an *n*-stage register.

6. In the absence of doppler shift, the autocorrelation function has two levels. Its amplitude at the origin ($r = 0$) is equal to the length of a period N. For all offsets other than multiples of one period the magnitude of the function is unity. This can be written

$$\phi(m) = \sum_{k=1}^{N} a_k a_{k+m} = \begin{cases} N & \text{if } m = 0 \\ -1 & \text{if } 0 < m < N \end{cases} \qquad (12\text{-}1)$$

The notation of $- - - - +$ in the shift register of Fig. 12-4 corresponds to the state of the individual binary elements (flip-flops, etc.) and the particular sequence of pluses and minuses is denoted the *initial condition.* The time duration of the segments is determined by the clock generator frequency. The proper choice of the connection of the Modulo two adder determines whether the sequence will be maximal length. It has been shown [143] that the maximum length of a period will be $2^n - 1$ since if the initial condition of all *plus* were to occur in the shift register, the subsequent stages will remain all plus. Adders with multiple feedback paths are also possible and the various allowable connections[1] are shown in the literature [105, 143, 38]. Other sequences with similar autocorrelation functions can also be generated but without the convenient shift and add property of the binary shift-register sequences [64, 144].

A block diagram of a CW radar system with pseudo-random phase coding could look like Fig. 12-5 if the target range is known within a fraction of $\pm r'$. A CW signal from the RF generator is reversed in phase according to the polarity of the shift-register encoder. The 0-180° coded CW signal is then amplified and transmitted. Upon reception the target echoes are mixed to a convenient IF and demodulated. If the time delay of the echo is exactly known, a CW signal appears at the narrow-band filters. Range and doppler tracking circuits will be shown later in the chapter.

12.3 AMBIGUITY FUNCTION FOR PSEUDO-RANDOM SEQUENCES

It would be desirable if the low value of the autocorrelation function for pseudo-random sequences which was obtained on the time axis (except at the origin) was also obtainable throughout the ambiguity plane. This would create a *thumbtack* ambiguity surface with a height (before normalization) of N^2 at the origin and a plateau of unity height extending in range for the length of the period and throughout the doppler region. Unfortunately, this is not the case, and it has been shown that the squared magnitude of the ambiguity function for these sequences can be written as [105, 64, 125][2]

---

[1] As an example, if the Modulo two adder were connected to the fourth and fifth stage of the shift register the output would not be maximum length.

[2] The segments are all assumed to be contiguous phase modulations of a high carrier frequency.

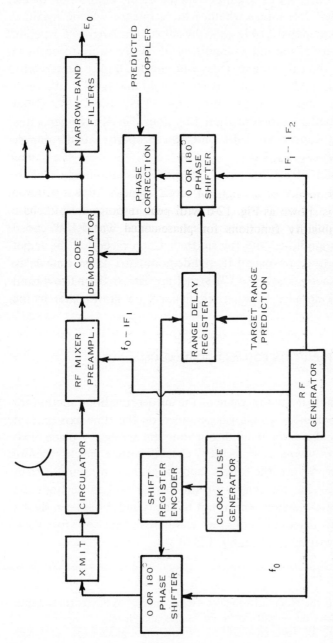

figure 12-5 *Block diagram of phase-coded CW radar when approximate target range and doppler are known.*

$$|\chi(r, s)|^2 = \left| \sum_k a_k \, a_{k+m} \left(\exp \frac{2\pi j}{N} \right)^{ks} \right|^2 \tag{12-2}$$

$$= N^2 \text{ for } m = 0 \; (\text{Mod } N), s = 0$$

$$= 0 \quad \text{for } m = 0 \; (\text{Mod } N), s \neq 0$$

$$= 1 \quad \text{for } m \neq 0 \; (\text{Mod } N), s = 0 \; \text{Mod} \frac{1}{N\tau'}$$

$$= N + 1 \text{ elsewhere}$$

Mod N corresponds to the periodicity in range and s is the doppler shift. The region where the value is $N + 1$ is often called the *plateau region* and Persons [280] has calculated the ambiguity function for a pseudo-random sequence of a length of seven segments. The square root of this function is shown as Fig. 12-6 with peak response normalized to unity. Other ambiguity functions for phase-coded waveforms can be found in Sakamoto et al. [332].

The central peak of height N^2 corresponds to the target response, and all range and doppler shifts are measured from its location. The doppler axis is completely free of ambiguities as long as the code contains many periods. The ambiguities along the range axis are unity height except at the code period. The so-called plateau region can be seen to consist of a series of ridges parallel to the range axis and at multiples of the rate of the periodic repetition.

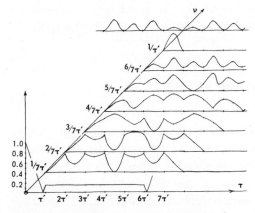

figure 12-6 *Square root of the ambiguity function for pseudo-random sequence (N = 7). (From Persons [280].)*

It can be seen from the figure that if the length of the period is sufficient to place the major range ambiguities beyond a range where clutter or undesired targets can occur,[1] excellent clutter rejection can be obtained for small doppler shifts. If in addition the period $N\tau'$ is short such that the doppler difference between desired and undesired targets is less than $1/N\tau'$, there will be virtually complete resolution of the desired target. If the radar parameters are such that both of these constraints cannot be met, there is a choice between competing against high-volume ambiguous range spikes or the ridge parallel to the range axis at a doppler shift of $1/N\tau'$. In many ways this waveform resembles a pulse-doppler waveform with period $N\tau'$.

If the integration time of the narrow-band doppler filters in Fig. 12-5 is short, the values given for the regions other than the correlation period will fluctuate since the desirable properties of these codes only apply for long sequences. Chandler [64] has placed some limits on this fluctuation.

Periodic Barker sequences [98] for $N = 11$ and 13, which are derived from codes described in Sec. 12.5, also have excellent reduction of sidelobes in the absence of doppler shift. However, they degrade rapidly for moderate velocity targets. Periodic codes with more than two possibilities of phase per segment have also been reported [165, 127]. A detailed study of *ternary* codes was made by Tompkins [371] where the coding includes +, −, and 0. The zero corresponds to the absence of a segment. Properties of ternary codes were tabulated up to $N = 18$ for those cases in which the periodic correlation for $m \neq 0$ or N is zero in the absence of a doppler shift. In the regions off the range axis, the codes do not seem attractive.

12.4   MAXIMUM-LENGTH BINARY SHIFT-REGISTER CODED WORDS

Maximum-length binary shift-register *sequences* have been shown to have the longest period for a given number of stages and to have desirable autocorrelation functions in the absence of doppler shift. For maximum-length binary phase-coded *words,* the side peaks on the time axis are not as low as for the words with polyphase coding (Sec. 12.10) or for the relatively short Barker codes (Sec. 12.5). On the other hand, they are

---

[1]Or at any range if the clutter doppler is sufficiently separated.

preferable to randomly coded words; and due to this property (and the ease of generation and decoding), they have achieved great popularity for word lengths of $N = 31$, 63, 127, etc. In this section it will be seen that there are several maximum-length binary shift-register codes for each length $2^n - 1$, and there are $2^n - 1$ starting positions for each of these. Within this multiplicity of codes, there are those which have minimum values of either the maximum peak value or rms value of the side peaks. There are also codes with especially low values of the side peaks near the origin (see Sec. 12.5). Use of these *best* codes can yield desirable performance in both multiple-target and clutter environments.

Maximum-length binary codes can be generated by sensing and logically operating on the state of predetermined stages of a n-bit shift register and entering the resultant $+$ or $-$ into the first-bit stage of the register. If the register is *clocked* every τ' units of time, a nonrepeating sequence of $2^n - 1$ units of length τ' will result. The bits used in the feedback path are the *one* coefficients of irreducible polynominals *modulo two*. An example for degree five, length 31 can be written in binary notation as 100101.[1] The polynominal which this denotes is (see [372, 64])

$$(1) x^5 + (0) x^4 + (0) x^3 + (1) x^2 + (0) x^1 + 1$$

The constant *one* in every polynominal refers to the closing of the loop to the first bit in the register. This was shown diagrammatically in Fig. 12-4.

In addition to the first five properties of maximum-length sequences in Sec. 12.2, the following properties can be attributed to maximum-length binary codes:

1. There are $2N - 1$ side peaks in the autocorrelation function and this function is symmetrical about the origin.

2. The number of maximal-length codes (including mirror images) for an n-stage shift register is given by $\phi(2^n - 1)$ [20, p. 29; 143, p. 24] where $\phi(m)$ is Euler's phi function. If $2^n - 1$ is a prime number, the number of codes is $(2^n - 2)/n$. If $2^n - 1$ is factorable into prime numbers denoted P_i, then the number of codes is

$$\frac{\phi(2^n - 1)}{n} = \frac{[2^n - 1](P_1 - 1)(P_2 - 1)(P_3 - 1)\dots}{n(P_1)(P_2)(P_3)\dots} \tag{12-3}$$

[1] It is often written in octal notation as 45_8. The mirror image is also irreducible and generates the reverse code.

Each prime P_i is used only once even if it appears more often.

3. The algebraic sum of all the autocorrelation functions for all the starting points of a given code is $(2^n - 1)^2$ at the origin and $[(N - k)/N](2^n - 1)$ for each segment k away from the origin in either direction (see Sec. 14.7).

4. The ambiguity function of this waveform is symmetrical about both the time delay and doppler shift axes.

A summary of the desirable codes and their starting positions is given in Table 12-1 after Taylor [372]. The first column is the degree of the polynomial (number of register stages) and the code length. The second column is the polynomial in octal notation. The feedback conditions can be determined by drawing the shift-register diagram and superimposing the polynomial in binary form on the stages. The polynomial always has an extra *one* (one bit more than the number of stages) which refers to closing the loop to the first bit of the register. All of the *one* bits in the register are fed to logic circuitry in synchronism with the clock, and the resulting output is entered into the first bit of the shift register. The output for the particular case of Fig. 12-4 is taken from the bit as shown. The third column lists the lowest maximum sidelobe level that can be obtained with its code when a suitable starting position is used. These starting positions are tabulated in the fourth column in the decimal equivalent form of the required binary state of the register. Column five shows the lowest rms sidelobe[1] amplitude of the codes, and column six shows the decimal equivalent of the initial conditions in the register to achieve this level. In some cases there were too many pertinent starting conditions to list, and only the total number of these is shown.

After degree eight, only a portion of the code properties are available, and only the total number of maximal-length codes and the code with the lowest peak value are shown. None of the mirror images are shown; therefore, the total number of desirable codes is actually twice that contained in the table. Further information on longer codes can be obtained from Roth [327] and Braasch and Erteza [50] or from tables of irreducible polynomials.

Roth's paper also contains an extensive table through degree nine for determining the minimum number of Mod two adders and the

---

[1]The mean value of the residues of these codes is 0.5. The values in the table are the rms deviations from this mean except for lengths three and seven.

TABLE 12-1   Maximum-length Binary-code Properties

Degree (number of stages) and length	Polynomial, octal	Lowest peak sidelobe amplitude	Initial** conditions, decimal	Lowest rms sidelobe amplitude	Initial conditions, decimal
1 (1)	003*	0	1	0	1
2 (3)	007*	−1	1,2	0.707	1,2
3 (7)	013*	−1	6	0.707	6
4 (15)	023*	−3	1,2,6,8 10,11,12	1.39	2,8
5 (31)	045*	−4	5,6,26,29 (9 conditions) 2,16,20,26	1.89 1.74 1.96	6,25 31 6
6 (63)	103*	−6	1,3,7,10 26,32,45,54 (9 conditions) (9 conditions)	2.62 2.81 2.38	32 35 7
7 (127)	203*	− 9	1, 54	4.03	109
	211*	− 9	9	3.90	38
	235	− 9	49	4.09	12
	247	− 9	104	4.23	24,104
	253	−10	54	4.17	36
	277	−10	14,20,73	4.15	50
	313	− 9	99	4.04	113
	357	− 9	15,50,78,90	4.18	122
8 (255)	435	−13	67	5.97	135
	453	−14	(20 conditions)	5.98	254
	455	−14	124,190,236	6.10	246
	515	−14	54	6.08	218
	537	−13	90	5.91	90
	543	−14	(10 conditions)	6.02	197
	607	−14	(6 conditions)	6.02	15
	717	−14	124,249	5.92	156
9 (511)	(24 codes)	−19 (1743 polynom.)		∼8.0	
10 (1,023)	(30 codes)	−29 (3023 polynom.)			
11 (2,047)	(88 codes)				
12 (4,095)	(72 codes)				
13 (8,191)	(315 codes)				

After Taylor and MacArthur [372].
*Only single Mod-two adder required.
**Mirror images not shown.

corresponding feedback connections for a given polynomial. The number of adders is minimized by using intermediate feedback connections.

For some cases in which it is not convenient to use a code of length $2^n - 1$, Delong [93] has found a number of desirable binary codes which cannot be simply generated with a shift register. By means of an extensive computer search, he has found binary codes of length 31, 45, 85, and 99 with peak sidelobe amplitudes of three, four, seven, and seven, respectively. Above a code length of 100, the best maximum-length binary codes appear to have lower peak sidelobes than nonmaximum-length codes.

12.5 PERFECT WORDS AND CODES WITH CLEAR REGIONS

In multiple-target environments, it may be significant that the distribution of the time sidelobes of the binary phase-coded words is different from that of linear FM pulse compression. The time sidelobes of linear FM are maximum immediately adjacent to the main lobe and decrease with distance from the main peak unless some unusual form of tapering is used. In contrast, the time sidelobes of the phase-coded words tend to be more random and are often quite low adjacent to the main lobe. The average sidelobe level does not decrease until about $N/2$ side peaks from the main lobe are reached. The sidelobe magnitude of course finally decreases to unity at Nth side peak.

This section will show that there are also optimum codes for radar pulses whose autocorrelation functions have "valleys" similar to binary sequences along the range axis. The general assumption to be made in this section is that the pulselengths are short enough such that the radial velocity of the target does not cause a significant phase change during the length of the pulse. This limit can be expressed by

$$Tf_d = \frac{2vT}{\lambda} \ll \frac{1}{2} \tag{12-4}$$

or

$$vT \ll \frac{\lambda}{4}$$

where λ = the transmit wavelength

v = the target radial velocity

$T = N\tau'$ = the pulselength

Violation of this inequality will either necessitate a bank of doppler matched filters or cause a loss in (S/N) or (S/C).

There is a class of binary phase-coded pulses whose autocorrelation functions over one period (finite time autocorrelation) have only two nonzero levels. These codes are called *perfect words* and include the Barker codes [17]. Using the same notation as in the previous section, the known codes are shown in Table 12-2.

TABLE 12-2 Perfect Binary Words of Finite Length

Code length, N	Perfect words
2	+−
3	++−
4	+++−, ++−+
5	+++−+
7	+++−−+−
11	+++−−−+−−+−
13	+++++−−++−+−+

The autocorrelation function of these words with zero doppler shift can be written

$$\phi(m) = \begin{cases} N & \text{if } m = 0 \\ 0 \quad \text{or} \quad \pm 1 & \text{if } m \neq 0 \end{cases}$$

It has been shown [379] that there are no other *binary* codes with this property from $N = 13$ to $N = 6,084$ and that it is unlikely any exist for N greater than 6,084.

These codes are only perfect in the time domain (unknown range, zero doppler shift) as it has been shown by Key et al. (see [125, p. 108]) that the output degrades rapidly in the presence of a doppler shift. The ambiguity diagram shown in that reference indicates that these codes are not desirable even if a bank of matched filters is used for all possible dopplers. Interference and clutter return from ranges and velocities other than the target *fold* into the target cell and obscure the target return signal.

The limitation of a maximum length of 13 segments is a serious one if automatic detection is desired. A limiter (at IF) can be inserted ahead of the matched filter to normalize the output response to a point target to 13 units (in amplitude). However, there is a significant probability that the sum of the 13 samples of the side peaks plus noise will exceed threshold levels of typical search systems. The noise output can be reduced by widening the prelimiter bandwidth, but this will be ineffective in the presence of clutter which has essentially the same spectrum as the target return. The value of the rms sidelobe levels resulting from coded words (which are needed for distributed clutter computations) was given in Sec. 12.4.

One way of achieving longer codes from the Barker codes is to phase code within each segment of one Barker code with another Barker code. This has been called *Barker squared* or *combined Barker* coding. The properties of such codes were calculated by Hollis [177]. He combined a Barker code of length four with the code of length 13 in two ways. When each bit of the 13-bit word was coded into four bits, the zero doppler autocorrelation function of the waveform yielded four side peaks of amplitude 13 located at range offsets of ± 1, ± 3 segments and 12 peaks of amplitude four. When each bit of the four-bit word was coded into 13 bits, the same number of side peaks of amplitude greater than unity appeared, but the location of the side peaks of amplitude 13 occurred at offsets of ± 13 and ± 39 segments. The main peak of the autocorrelation function in both cases was 52. The first combination may be useful if expected inference is considerably separated in range from the target.

When an unwanted target or small regions of clutter are separated by an appreciable portion of the pulse envelope, the linear FM[1] or chirp waveform is less susceptible to side-peak interference than phase-coded words. On the other hand, there are certain binary phase-coded words which are desirable for resolving closely spaced targets or observing missiles in the presence of tank fragments or decoys. There is no apparent method of finding these codes other than observation of a number of autocorrelation functions. For example, the code 453_8 (octal) of length 255 with initial condition 21 has the following time sidelobes adjacent to (and symmetrical about) the main peak:

$$0, -1, 0, -1, 0, 1, 0, 1, 2, 1, 2, -3, -2, -1, 2, 1, 2$$

[1] With a *tapered* receiver (see Chap. 13).

Thus, within eight segment lengths of the main peak, the ratio of the main lobe to side peak amplitude is 255 to 1 or 48 dB. A number of other codes have similar characteristics for six or seven segment offsets. To obtain this degree of discrimination the decoder must be highly linear and have a large dynamic range.

12.6 DECODING TECHNIQUES—ALL-RANGE DECODERS

One of the advantages of the phase-coding technique is that a number of types of simple, efficient, and flexible decoders can be built. This flexibility is an important factor since the particular code can easily be changed from radar to radar or even from pulse to pulse in a single radar, as contrasted to most pulse-compression techniques. In general all of the decoders of this section will approximate matched-filter or correlation detectors.

This discussion will be divided into parts with the functional breakdown indicated below:

Type	General radar function	Typical doppler coverage
All-range decoders		
a. Analog	Surveillance—short-coded words	None
b. Digital	Surveillance—long-coded words	None
Cross correlators	Tracking—words or sequences	Limited doppler

Decoders have also been built in matrix form [226] and many more variations are used in communication receivers.

Encoding techniques will not be discussed except to state that it is apparent that encoding directly at the microwave carrier frequency is by far the most satisfactory method for binary or four-phase encoding. Multiple, low-power diode crystal switches with good isolation are now available with switching speeds of about 5 nsec. Microwave "magic tees" or "rat races" provide antiphase outputs, and "hybrids" provide 0 and 90° phase shifts over bandwidths in excess of 20 percent of the carrier frequency. This type of encoder was used to generate the 0.1 μsec per segment code of Fig. 12-3D.

If little prior information is available about the loop propagation time from the radar to the target, the receiver must be implemented to provide a matched filter for a large set of time delays. This is the typical surveillance or *search* configuration. The class of matched filters where the filter's output will reach its maximum value for *any* target range will be called the *all-range decoders.* In the following discussion it will be assumed that a phase-coded word is transmitted, and the doppler shift is negligible or at least known.

Analog All-range Decoders

A basic block diagram for the analog decoder is shown in Fig. 12-7 for $N = 7$. It is assumed that the input is at the RF carrier and that each delay element is an integral number of wavelengths at that frequency. This is often accomplished by inserting phase trimmers at each tap of the delay line. Obviously, the matched filter for the pulse segment must also be centered at that frequency. The easiest way to visualize that this oversimplified diagram is a matched filter is to assume that all the elements are bilateral and then consider what appears at the input when an impulse is applied to the output. The summing amplifier then acts as a power divider, and a pulse of length $\approx N\tau'$ will appear at the input with the code reversed in time. The necessary conditions for this filter to be truly *matched* are described in Sec. 12.8. Reference [332] describes the loss incurred when using single-tuned or rectangular filters.

The advantage of this matched filter over the correlation decoder shown in Fig. 12-5 is that knowledge of target range is not needed. The

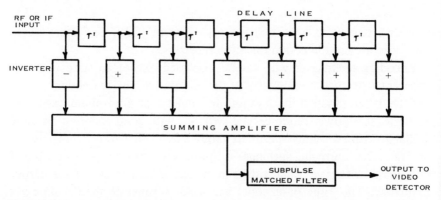

figure 12-7 *Matched filter for binary-coded words. (Polarities shown are for 7-bit Barker code.)*

implementation in Fig. 12-7 or its digital equivalent is thus useful for search radars in which the number of segments in the code is sufficiently small so that the bandwidth of the N-cascaded delay lines will pass a rectangular pulse of length τ' with little distortion. Attenuation in the delay lines can be compensated for by varying the gain of the inverters prior to the summing amplifier.

The bandwidth of the delay line presents a practical problem in that its N-cascaded stages[1] must have an overall bandwidth $\geq 1/\tau'$ (the inverse of the segment duration), thereby requiring each stage to have a bandwidth $\gg 1/\tau'$. Thus, the total time-bandwidth product of the tapped delay must be at least $(N\tau')(1/\tau') \approx N$. This is not a practical problem with quartz or glass ultrasonic delay lines. With lumped-parameter delay lines the requirement that $N > 31$ becomes a more difficult design problem, especially since they are generally low-pass networks. With low-pass delay lines, the bandwidth must be at least $f_2 + 1/2\tau'$ where f_2 is the intermediate frequency. The number of sections n for a lumped-parameter *m-derived* or *constant k* delay line would then vary as

$$n_s \doteq [K(\text{time-bandwidth})]^{1.5} \approx \left[K(N\tau)\left(f_2 + \frac{1}{2\tau'}\right) \right]^{1.5} \qquad (12\text{-}5)$$

Numerical values for the constant and more exact values for delay-line design can be found in Millman and Taub [250, pp. 291-298]. In addition, the impedance matching problem of *tapping* the lines at IF presents additional problems, and individual gain adjustments are usually necessary at each of the taps. It can be appreciated that only short codes yield practical numbers of delay-line sections.

While ultrasonic lines are usually preferable, the high insertion losses of the transducers and temperature stability problems for long total delays of the order of 0.1 msec need to be taken into account. The additional stable amplification and temperature control circuitry required can be somewhat costly. For short delays ($< 3\mu$sec) a helical center conductor or a wide-band semirigid coaxial cable has been found to be a simple delay mechanism. While magnetostrictive delay lines have

---

[1] Actually only $N - 1$ delays are required, but this will be neglected in the discussion. Also the time delay is only constant for about 60 percent of the bandwidth for lumped-parameter delay lines.

also been used, the distortion that they induce tends to degrade the waveform at the decoder output.

An alternate matched filter for decoding phase-coded words is shown in Fig. 12-8. This is the zero intermediate frequency (dc-IF) or homodyne configuration where bipolar video is fed into each delay line. The local oscillator for the pair of mixers is at the transmitted frequency f_0, and the relative phases of this oscillator at the mixers are in quadrature. The mixer outputs are thus the *in-phase I* and *quadrature Q* components of the echoes and contain no carrier frequency as such.[1] In practice both mixers are constructed as a single-sideband mixer whereby the appropriate pair of outputs is obtained with only a single local oscillator input. Minor doppler shifts (if known) can be added to the LO frequency to prevent degradation of the matched-filter output. Since the total signal may appear in either or both outputs and the resultant amplitude is $e_0(t) = (I^2 + Q^2)^{1/2}$, both channels are necessary to avoid an average 3-dB loss in signal-to-noise ratio. Another way of stating this is that since the phase of the target is unknown, the received signal may appear in the channel that was not constructed [392].

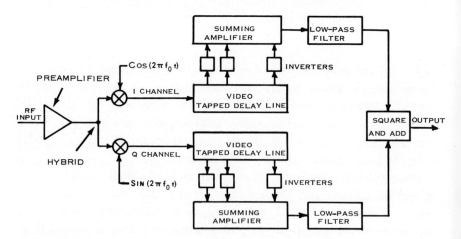

figure 12-8 *Bipolar video matched filter for phase-coded words.*

While this approach requires twice the number of delay-line elements, it often requires less hardware than the decoders at IF. This results from

---

[1]The general configuration is discussed in Chap. 14.

a combination of two effects which were shown in Eq. 12-5. First, the delay-line bandwidth for the video waveform need be only one-half that of the band-pass waveform; and second, no IF carrier need be passed through the lines. If lumped-parameter delays are used, the number of sections for both lines is then

$$n_s = 2[K(\text{½ time-bandwidth})]^{1.5} = 2\left[K\left(\frac{N}{2}\right)\right]^{1.5} \tag{12-6}$$

As an example consider a code of 32 0.1-μsec segments which can be decoded either with a tapped IF decoder at 30 MHz or with a tapped homodyne decoder. If m-derived delay sections are used, the IF delay in the systems must have an overall bandwidth in excess of 35 MHz. The ratio of the number of sections for the two techniques is then

$$\frac{n_{\text{IF}(s)}}{n_{\text{homodyne}(s)}} \doteq \frac{[(3.2)(35)]^{1.5}}{2[16]^{1.5}} \doteq 9 \tag{12-7}$$

With lower compression ratios requiring less distributed delay elements, the difference is less dramatic.

The delay tolerances are also less critical for the dc-IF but the physical realization of a wide-band squaring and summing circuit is not trivial. An excellent approximation to $(I^2 + Q^2)^{1/2}$ is to use the larger of $|I| + \text{½}|Q|$ or $|Q| + \text{½}|I|$ with a simple logic circuit to decide which is larger.[1] An even simpler approximation is to use $|I| + |Q|$ or the larger of the two. With this simplification the loss in detectability is less than 0.65 dB for $P_D = 0.5$ and $P_N = 10^{-6}$ for the steady- or fluctuating-target cases. For lower signal-to-noise ratios (< 10 dB) the loss can usually be neglected. Table 12-3 shows the losses compared to the sum of squares resulting from the various approximations. The losses are slightly different for steady and fluctuating target models [238, 392]. A final point in favor of the dc-IF is that it lends itself to digital decoding or digital processing of the decoder outputs.

Digital Decoders

When binary or quadrature (four-phase) codes are transmitted, a pair of shift registers can be used as a digital *tapped delay-line* pulse

[1] This will reduce the loss in detectability to about 0.2 dB.

TABLE 12-3   Signal-to-Noise Penalty for Using Single Homodyne Channel or Sum of Magnitudes instead of Square-law Detector, dB

A. Penalty, dB, for Rayleigh fluctuating targets* (case 1 or 2) as compared to $(I^2 + Q^2)^{1/2}$

False alarm probability	$P_D = 0.2$		$P_D = 0.5$		$P_D = 0.9$	
	$\|I\|$	$\|I\| + \|Q\|$	$\|I\|$	$\|I\| + \|Q\|$	$\|I\|$	$\|I\| + \|Q\|$
10^{-1}	1.6	0.2	3.4	0.2	9.1	0.2
10^{-2}	2.2	0.3	3.6	0.3	9.9	0.4
10^{-4}	2.4	0.4	4.0	0.4	10.5	0.5
10^{-6}	2.5	0.5	4.4	0.6	10.7	0.6
10^{-8}	2.5	0.6	4.6	0.7	10.8	0.7

B. Signal-to-Noise penalty for steady targets,* dB

False alarm probability, P_N	$P_D = 0.2$		$P_D = 0.5$		$P_D = 0.9$	
	$\|I\|$	$\|I\| + \|Q\|$	$\|I\|$	$\|I\| + \|Q\|$	$\|I\|$	$\|I\| + \|Q\|$
10^{-1}	1.6	0.15	2.4	0.15	10.4	0.25
10^{-2}	1.5	0.30	2.7	0.35	12.0	0.40
10^{-3}	1.5	0.35	2.8	0.40	12.8	0.55
10^{-4}	1.5	0.45	2.8	0.55	13.2	0.70
10^{-6}	1.4	0.50	2.8	0.65	13.7	0.75
10^{-8}	--	0.50	--	0.70	---	0.85

*Averaged over all possible values of target phase. Greatest of $\|I\|$ or $\|Q\|$ gives the same loss as $\|I\| + \|Q\|$.

compressor. An *all-range* implementation of this form of digital pulse compression has been built for maximum-length binary codes of up to 255 segments [372] and is illustrated in Fig. 12-9. The shift-register encoder is at the upper left, and the I-Q (homodyne) mixer is at the lower left of the diagram. The low-pass filters approximate matched filters for the segments of the code. The IF limiter is needed only to ease dynamic range problems and reduce the effect of impulse-type noise since the I and Q polarity *samples* yield the same effect. The samplers are normally *clocked* at the bit rate of the code, and only the polarity of the bits in each channel is shifted down the respective signal registers.

At the time of transmission, the encoded signal is stored in the *code register* with the polarity of each stage representing the polarity of the

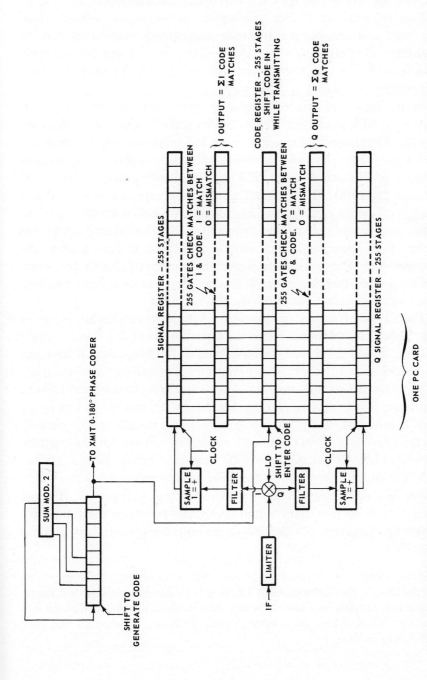

figure 12-9 *Digital pulse compressor using 0-180° binary phase-coded modulation. (After Taylor and MacArthur [372].)*

corresponding segment of the code. Several unique features of this decoder are worth noting. First, any binary code of length 255 can be transmitted and stored prior to reception of the received echoes. Thus, the code can be changed for each transmitted pulse if desired. In addition, any shorter maximum-length code can be stored if lower range resolution is desired. Also, if only one code is to be used, the code register can be replaced by fixed wiring.

As the quantized target echoes propagate down the signal registers, their polarity is continuously matched with the stored code in the code register. The number of *matches* in the 255 gates between the signal and *code register* is continuously summed. When there is perfect alignment of a target and the stored word, 255 matches should appear in the I or Q summer (or both). The summation of the agreements is combined in a *greatest of* circuit yielding the larger of $|I|$ or $|Q|$. This summation is fed to a threshold circuit where the threshold level is predetermined by the desired false alarm rate. Since there is broad-band limiting prior to the decoder, the false alarm rate should be constant (CFAR) for noise, jamming, and other uncorrelated echoes. Target range is determined by stopping a counter when the threshold is exceeded.

The CFAR effect and an example of clutter reduction are shown in Fig. 12-10.[1] At the upper left is the uncompressed (all +'s in the code) *A-scope* output from land-clutter echoes at the beginning of the trace. The backscatter from a localized rainstorm is at the center of the trace. In all of the figures, the transmitted pulse length is the same and is about 3 percent of the illustrated trace length. As the number of bits in the code is increased, the decoder output decreases rapidly for the *spatially* uncorrelated rain clutter and with moderate rapidity for the partially correlated land-clutter echoes. As the number of bits in the code reaches 255, the clutter is reduced to the noise level, demonstrating the CFAR action. If a point target echo of sufficient energy were present, it would always reach the maximum output (equivalent to voltage addition) while distributed targets would be held to the noise level (equivalent to power addition).

---

[1] See Chap. 4 for discussion of CFAR. In that chapter it was shown that the desired effect can only be achieved for large time-bandwidth products. With the configuration described here, the number of stages should be $\geq$ 63 when low false alarm probabilities are desired.

There is some signal processing loss in the digital processor illustrated in Fig. 12-9. The combination of limiting and quantization causes about a 2-dB loss in efficiency. About 1 dB of this is the loss that is inherent in a CFAR processor. There is an additional fractional decibel loss for using the *greatest of* circuit. This loss was tabularized in Table 12-3. In addition there is about a 2.40-dB average *range cusping* loss if the full 255 code is used since the target echo may straddle adjacent bits in the

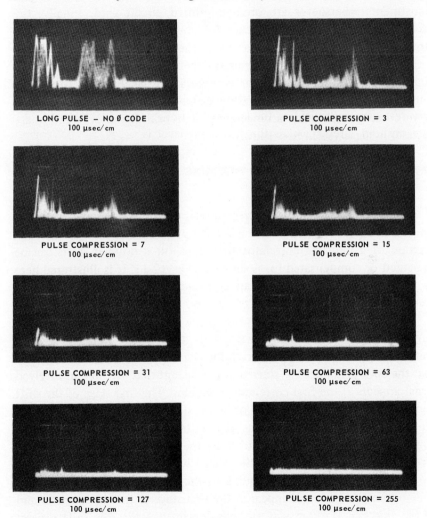

figure 12-10 *Detected video out of digital pulse compressor from rainstorm. (After Taylor and MacArthur [372].)*

registers. This loss is about 1.2 dB if the number of stages in the I and Q registers is twice the number of segments in the code [300]. The range-cusping loss will be discussed in Chap. 14.

Some advantages of this type of processor that were not mentioned previously include freedom from adjustment and the inherent reliability of microelectronic digital circuitry. The unit that was described uses microcircuits. Eight entire stages of the decoder not requiring any adjustments whatsoever are on one printed circuit card. Finally, the maximum-length code transmitted by one radar will not cause false targets to appear on another radar which uses a different code.

Since there are two signal registers there are effectively four phase quantization levels and a similar configuration can be used to decode four-phase codes. A two-bit code register is required with somewhat more complicated gating arrangement. A hybrid digital MTI and phase decoder is suggested in Sec. 14.6.

12.7 CROSS CORRELATORS AND TRACKING TECHNIQUES

If the target range is approximately known, one or more cross correlators can be used to acquire and track the target echoes. The basic block diagram of a multiple correlator that is useful if doppler shifts can be neglected or compensated for is shown in Fig. 12-11. It is illustrated here in the I-Q configuration, but an equivalent IF processor can also be constructed with band-pass filters. In the homodyne processor the I and Q components of the signal appear at a succession of *correlator-integrators,* each of which corresponds to a fixed range delay. To consider the action of single correlator channel, it is simplest to assume that the code generator[1] is started at the instant that the bipolar received signal appears at the first correlator integrator. The first stage of the shift register will be either in the plus or minus state corresponding to the first bit of the code, and will energize either the a or b line to the correlator-integrator c/i. As shown in the insert, this will either pass directly or invert the bipolar video signal while preserving its amplitude. At a subsequent time corresponding to the one-segment duration of the code, the second *bit* will appear at the first stage of the shift register and at the first c/i. If there was an inversion in the code, the shift register

---

[1] The range delay to the target will be neglected temporarily.

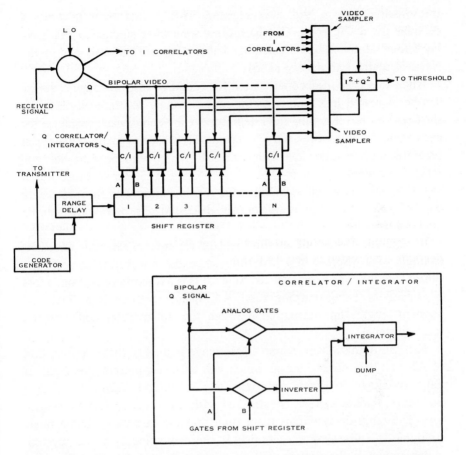

figure 12-11 *Functional block diagram of range-gated correlator. (Courtesy of M. Davidson.)*

would energize the opposite gate of the c/i, and the signal applied to the integrator would have the same polarity as the first bit. Thus, if the code is entirely aligned, the received voltage of all N bits will be integrated. A similar integration process will occur for signals in the I channel. At the other integrators, cross correlations of the codes will appear and the polarities of the segments will be almost random. At the end of the N segments the *sidelobes* will appear on the other integrators. At the termination of the N-segment correlation code the levels in the integrators will correspond to the side peaks of the autocorrelation waveforms as discussed in previous sections. Obviously the integration will continue if the code is repeated as a sequence both in the transmission and in the shift register. This circuit then becomes a decoder for a CW code. Configurations somewhat similar to this have been used at the Naval Electronics Laboratory in California and at the Naval Research Laboratory [156].

When this type of decoder is used for coded words, the range coverage can be increased indefinitely by running the code continuously into the shift register but dumping the integrators at the end of each sequence for each stage. In this case, each c/i periodically observes many ranges each separated by Nr'. The video sampler must then be sequenced to yield the range dependent signals (to the $I^2 + Q^2$ circuit) similar to those which would appear on a conventional range display. In any of these alternate configurations the various, previously discussed approximations can be made to $I^2 + Q^2$.

In sum, the basic configuration and its variants represent an excellent example of a discrete matched filter since the integrators perform the filtering on bipolar video signals. Since this is a discrete operation, a loss in sensitivity (range-cusping loss) will occur if there is a fractional segment length time mismatch between the shift-register code and the received signal.

The determination of coarse target range is merely the determination of which c/i receives a signal which will exceed a present threshold. If range is known to within approximately $\pm$ one code segment, a simple fine-range tracker can be constructed with only two correlator integrators. The range-delay circuit is set such that the target's expected range straddles the first two c/i circuits. If the prediction is good, approximately one-half the voltage will be integrated on each circuit. For any small errors in estimation the voltage will be split proportionally. Thus, a range-error signal can be derived from the ratio of the integrated voltages. This decoder is a version of an *early-late gate* or *split gate* range tracker. If short-coded pulses are used, the I and Q signals are gated into their respective integrators only for the predicted duration of the pulse envelope.

An alternate technique of deriving the range-error signal is to invert the polarity of the $I^2 + Q^2$ output from the first c/i and add it to the $I^2 + Q^2$ output of the second c/i. The difference of these signals versus error in estimating range ΔR has the appearance of a range discriminator characteristic. Such a characteristic is shown in Fig. 12-12, curve a (neglecting the *self noise* of the range sidelobes of the code). The difference signal can be filtered and used to vary the range-delay circuit with one polarity increasing the range-delay prediction and the other signal decreasing it for subsequent input signals. The amplitude of the difference signal when normalized to the total input signal power determines the amount of correction. The use of a limiting amplifier prior to the final mixer is one such form of normalization.

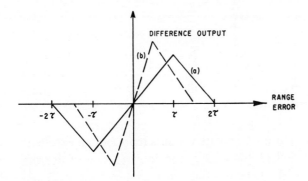

figure 12-12 *Early-late integrator difference output vs. range error (unfiltered).*

A somewhat simpler version of the range tracker for phase-coded waveforms is shown in Fig. 12-13. In this form the delayed code is used to switch the phase of the local oscillator LO signal to decode the received signals. The received signal is amplified (and sometimes limited), divided into the *early* and *late* channels, mixed to dc, and integrated. A time delay of the order of the segment length is inserted to form the early channel. In alternate configurations, the delay may be placed in the appropriate LO line or may be obtained from the outputs of successive states of the shift register which forms the delayed-code generator. The shift-register outputs are used to switch a pair of phase inverting switches in two separate LO lines similar to Fig. 12-8. This final variation is sometimes called the *delay-lock tracking discriminator* [360, 100].

In all of these decoders, one channel is typically delayed by one-half to two segment lengths from the other, with the variations between them dependent upon the nature and location of the delay element. The choice of the delay time[1] τ'/K is a compromise between the error sensitivity of the range circuit, the *pull-in* range and the efficiency or threshold level required for small signal-to-noise ratios [251]. If the

[1]K is a constant which is usually between one-half and two.

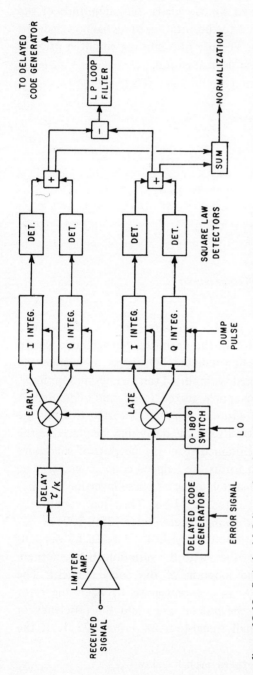

figure 12-13 Switched LO linear decoder and range tracker, range determined from contents of delayed-code generator.

delay time between early and late channels is less than τ', the discriminator characteristic will have a steeper slope as in curve b of Fig. 12-12, but a small *pull-in* range ($\sim \pm 1.5\tau'$).[1] As the delay is increased, the range-error sensitivity will decrease but a greater pull-in range is achieved. Discussions of the optimum choice of parameters for the generalized radar split-gate tracking process are available in texts such as Barton [20, pp. 357-365] and Mityashev [251, chap. 2], but descriptions for tracking of phase-coded signals are primarily limited to the communications field.

Spilker [360] has analyzed the range tracking of maximum-length *sequences* using the delay-lock discriminator technique. The block diagram would be somewhat similar to Fig. 12-13 with the limiter included.[2] The error signal output of the loop filter is used to vary the clock rate of the delayed-code generator. He showed that the mean-square delay error due to white gaussian receiver noise can be represented by

$$\sigma_n^2 = 2.12 \, (\tau')^2 \left(\frac{N}{N+1} \right)^2 \frac{p_0 N_0}{S} \quad \text{for} \quad |\epsilon| < \tau', \ K = \tfrac{1}{2} \qquad (12\text{-}8)$$

where σ_n^2 = variance of time-delay error

τ' = segment duration

N = number of segments in the period $2^n - 1$

S = average input signal power during sequence

N_0 = input noise power spectral density

p_0 = loop filter frequency constant [100]

ϵ = the error in the range prediction

This equation can be expressed in another form for large N and with both I and Q channels [136]

$$\frac{\sigma_n}{\tau'} = \left(\frac{N_0 B_n}{S} \right)^{\tfrac{1}{2}} \quad \text{for} \quad K = \tfrac{1}{2} \quad \text{or} \quad \frac{\sigma_n}{\tau'} = \left(\frac{N_0 B_n}{2S} \right)^{\tfrac{1}{2}} \quad \text{for} \quad K = 1 \ (12\text{-}9)$$

for $S/(N_0 B_n)$ above a threshold value and where B_n is the loop noise

---

[1]In practice, the matched filtering will smooth out these curves and extend them in the range-error direction.

[2]He used only a single channel (rather than an I-Q configuration) in his analysis.

bandwidth. The signal power for a given delay error and $K = 1$ is thus one-half that for $K = \frac{1}{2}$.

As expected, this equation shows that the rms range error is directly proportional to the segment length and inversely proportional to the square root of the signal-to-noise ratio. It was also demonstrated experimentally that for large values of N the threshold for the system to *lock on* occurs at $\sigma_n = 0.3\,\tau'$ and the threshold signal-to-noise ratio for stable tracking in the absence of transients is [360, 100]

$$\text{Filter output SNR}_{\text{threshold}} = 22.2 \left(\frac{N}{N + 1} \right)^2$$

$$= 13.5 \text{ dB for large } N,\ K = \frac{1}{2},\ \text{single channel} \quad (12\text{-}10)$$

For binary maximum-length sequences, the range error due to the range sidelobe contributions is generally negligible. In the case of coded words, the error may be significant for very fine tracking. An excellent comparison of different methods of range tracking of binary sequences is given by Gill [136].

12.8 PHASE-CODED WORDS—NOISE AND CLUTTER PERFORMANCE

The performance of radar systems using phase-coded waveforms in the presence of noise, clutter, or undesired targets is not difficult to calculate for binary-coded words. In computing detection range, either of two methods gives a close approximation to the correct answer for the matched-filter receiver competing with receiver or jamming noise.

1. In the denominator of the range equations, the noise bandwidth term B_N can be made equal to the equivalent matched-filter bandwidth of the *uncoded* pulse.

2. The noise bandwidth term B_N can be made equal to the matched-filter bandwidth of the *segment* length, and then the number of segments N must be placed in the numerator of the range equations to account for the coherent integration of the segments.

In both cases the required (S/N) for detection must be determined by using a desired false alarm number n' which is based on the number of *segments* in the observation range.

The losses due to deviations from a matched filter L_s are not easy to formulate concisely. Some simple examples include:

Limiting loss: If the signals are limited at RF or IF before decoding, the loss in detectability will be approximately 1 dB if N is greater than about 63, for $P_D \approx 50$ percent, $P_N \approx 10^{-6}$ (see Chap. 4).

Quantization loss: If the decoder is all digital (see Sec. 12.9) as in the polarity-coincidence detector, the quantization loss will be approximately 1 dB for large N in addition to the limiting loss (see I. Jacobs [187]). If there are two bits in *both* the I and Q channels, the total limiting and quantization loss will reduce to 0.5 dB [300].

Range cusping loss: If there is a time mismatch between the stored waveform which is to be correlated with the target echoes, there will be a sampling or *range-cusping* loss. If the time mismatch is one-half of a segment length, the target signal in both of the adjacent range outputs will be reduced by 6 dB. The average cusping loss is only about 2-3 dB and depends on the matched-filter bandwidth and the type of postdetection filtering or integration. This loss is about 1.2 dB for two samples per τ'.

Velocity compensation loss: If the target range changes an appreciable portion of a carrier wavelength during the time duration of the entire pulse of N segments, the summation of the N segments will not yield N^2 at the matched-filter output. If there are no *doppler* filters, the output will be proportional to the voltage summation of the N segment vectors each differing slightly in phase

$$e_0 \sim P_0^{1/2} \sim \sum_{n=1}^{N} \cos[2\pi f_0 (t - t_d) + \pi a_n + (N - n) 2\pi f_d \tau'] \quad (12\text{-}11)$$

where P_0 = power output
$\quad\quad t_d$ = time delay of transmitted pulse
$\quad\quad a_n$ = pulse code = 0 for a *plus* bit and 1 for a *minus* bit
$\quad\quad f_d$ = doppler frequency
$\quad\quad \tau'$ = segment duration

As an example, for the case of a target with a radial velocity of 1,000 fps (Mach 1) at a carrier of 3 GHz, the doppler frequency is about 6 kHz. If the segment length is 3.3 μsec, the phase shift per segment is about 0.13 radians. If there are only 10 segments in the code, the loss in signal power is only about 0.5 percent. If the code length is 50 segments, the loss in signal power will be substantial (~ 4 dB). A graphical technique to determine this loss is given by J. S. Jacob [186]. (See also Sakamoto [332].)

Phase-coded Words—Clutter Performance

The performance of the phase-coded words in a distributed clutter environment can be estimated from the simple pulse radar equations for clutter (which were given in Chap. 2) if the segment length is used to determine the resolution cell rather than the envelope of the transmitted pulse. The alternate forms and primary correction factors to these equations are now discussed.

In *discrete* clutter or in multiple target environments the radar cross section of the clutter can be used directly in computation of the signal to clutter ratio S/C. In simplified form

$$\frac{S}{C} = \frac{\sigma_t(CA)}{\sigma_c} \quad \text{predetection}$$

where σ_c is the radar cross section of the undesired scatterers. The square of the ratio of the number of segments to the time sidelobe peaks or the rms sidelobe level affects the clutter attenuation factor CA. For the Barker codes, this ratio is simply N^2 for the peaks and $2N^2$ for the rms sidelobe level. The choice of the peak sidelobe level or the rms sidelobe level must correspond to the choice of the signal-to-clutter ratio (S/C) to be used in the denominator of the clutter equations. Figure 12-14 shows that there is a 4- to 5-dB main peak-to-highest sidelobe advantage by using the codes with the lowest peaks. Polynomials and starting points corresponding to the better codes were given in Table 12-1.

In a distributed clutter environment where the range extent of the clutter is greater than the transmitted pulselength $N\tau'$, the segment length ($c\tau'/2$ in distance units) is used to determine the resolution cell. A loss term must be entered to account for the time sidelobe contributions. Since there are $2(N-1)$ time sidelobes, the *mean-square* sidelobe value must be multiplied by $2(N-1)$ and compared to N^2 at the peak.[1] A linear processor is assumed. Table 12-4 has been calculated (see [298])

---

[1] It is assumed that the transmit pulselength is small enough compared to the target range to neglect the range-dependent terms.

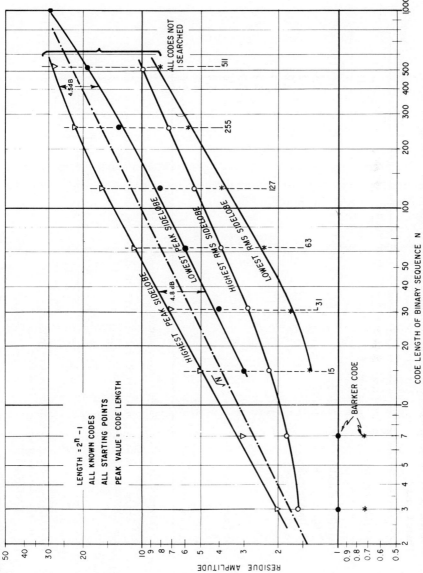

figure 12-14 *Maximum and minimum values of binary phase-coded residues.*

TABLE 12-4 Range Sidelobe Clutter Contributions for Barker and Maximum-length Binary Phase Codes

Code length, N	Least rms* sidelobes	$1 + \dfrac{2(N-1)}{N^2}$ (mean-square sidelobes)	Degradation L_S dB
3 (Barker code)	0.707	1.22	0.88
7 (Barker code)	0.707	1.12	0.51
13 (Barker code)	0.707	1.07	0.3
15	1.4	1.24	0.94
31	1.7	1.19	0.75
63	2.4	1.18	0.70
127	3.9	1.24	0.93
169 (Barker squared)†	– –	1.15	0.6
255	5.9	1.27	1.05
511**	8.0	1.25	1.0
511 (Noise codes)	~16.0	~2.0	~3.0

*Computed for discrete code shifts of one segment. The mean value is −0.5.

**Not all codes searched.

†It can be derived from results in [178] that for combinations of Barker codes the rms sidelobe contribution increases the total clutter power by the factor

$$1 + \frac{N-1}{N^2}\left[\frac{L_o^2(L_i-1)+(L_i-1)(L_o-1)+L_i^2(L_o-1)}{L_o\,L_i-1}\right]$$

where L_i = number of segments in the *inner* code
L_o = number of segments in the *outer* code
$N = L_i L_o$

from Table 12-3 to show this degradation for the maximum-length codes. This table was computed for discrete one-segment length offsets of the autocorrelation functions. A slightly different result would be obtained if correlation between adjacent sidelobes of the autocorrelation functions were included [300]. Several manual calculations have shown that the difference in computation technique yields a minor error, and the degradation column in the table is slightly pessimistic. It can be seen that for the better codes the degradation is only about 1 dB and is almost independent of code length. An entry is shown for 511-segment noise codes which illustrates that the sidelobe degradation in (S/C) with distributed clutter is about 3 dB for these codes [300]. From Fig. 12-14 it can be seen that the poorest codes give a few decibels of additional degradation for small values of N.

In summary it can be seen that the better binary codes give an improvement in S/C which is only about 1 dB less than the pulse-compression ratio. The peak range sidelobes are somewhat larger than those of the tapered linear FM waveforms in Chap. 13, but the performance in distributed clutter is comparable.

12.9 SIDELOBE SUPPRESSION OF PHASE-CODED WORDS

In multiple-target environments, or where there are large undesired reflectors (point clutter), it is often desirable that the time sidelobes of the autocorrelation function of the binary phase-coded pulses be reduced to as low a level as possible. Otherwise the time sidelobes of one large target may appear as a smaller target at another range. Key, Fowle, and Haggarty [212, 125] have demonstrated that weighting networks can be designed to reduce these sidelobes to an arbitrarily low level.

As an example they use the 13-bit perfect word (Barker code for $N = 13$) whose autocorrelation function has a spectrum

$$\Phi(\omega) = (\tau')^2 \left[\frac{\sin(\omega\tau'/2)}{\omega\tau'/2} \right]^2 \left[12 + \frac{\sin(13\,\omega\tau')}{\sin\omega\tau'} \right] \qquad (12\text{-}12)$$

It was shown that the desired part of the spectrum can be extracted from the signal received from a point target with negligible doppler shift by a filter whose frequency characteristic is

$$H(\omega) = \frac{13}{12 + [\sin(13\omega\tau')/\sin\omega\tau']}$$

For the Barker code they derived two weighting functions to approximate Eq. (12-12), the first *of 13th order* and the second of *25th order.*

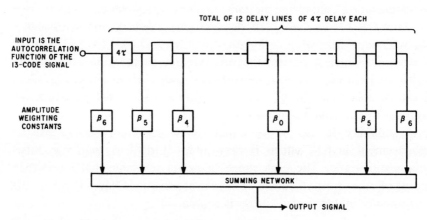

figure 12-15 *Weighting network which controls amplitudes of the close-in sidelobes of the Barker 13 code. (After Fowle [125].)*

The 13th-order weighting network, which eliminates the first six sidelobes, is shown in Fig. 12-15; and the peak-to-residual sidelobe ratios are given in Table 12-5.

The weighting network is placed after the matched filter and causes only about a 0.25-dB loss in signal detectability for the higher order network. The reduction in sidelobes much below the 32.4-dB level does not seem practical due to probable phase errors in the circuitry, transient effects, and the degradation due to slight target-doppler shifts.

TABLE 12-5   Sidelobe Reduction of a 13-segment Barker Code

Type of weighting function	$\dfrac{\text{Peak signal}}{\text{Residual peak sidelobes}}$	$\dfrac{\text{Peak signal}}{\text{rms sidelobes}}$
None	13 (22.3 dB)	25.3 dB
13th order	41.9 (32.4 dB)	
25th order	189 (45.6 dB)	

In distributed clutter, the above technique has minimal value for Barker codes since their rms sidelobes are 25 dB down without weighting. The loss in detectability due to the unweighted sidelobe contributions of the clutter is only 0.3 dB.

Another technique to reduce the time sidelobes of a finite length binary sequence has been reported by Erickson [107]. He has found codes of length 2^n which have a complementary code whose autocorrelation function will cancel the autocorrelation function of the original code except at the origin. Thus, if the two codes can be transmitted simultaneously and their matched-filter outputs added vectorially,[1] there will be no residues. A practical problem exists with this approach in a clutter environment. If the two transmitted codes are separated in frequency, there will be decorrelation of the clutter echoes and little cancellation. If the codes are separated in time and the clutter is extensive, the residues will be temporally or spatially decorrelated. The primary use of this technique would be to prevent the time sidelobes of a large target (or large point clutter) from obscuring a smaller target.

12.10 POLYPHASE-CODED WORDS

The *perfect words* yield high autocorrelation peak-to-sidelobe ratios of $N:1$ in amplitude but only up to $N = 13$. The maximum-length codes can be used for many discrete lengths up to $N = \infty$. While they are easy to generate and decode, the ratio of amplitude of the central peak to largest side peak remains close to $\sqrt{N}$. In cases where the word length must be greater than $N \approx 30$ from accuracy, resolution, or automatic detection considerations, there are some additional *good* codes if the restriction of 0-180° phase shifting is removed [93, 127, 128].

Higher order or *polyphase*-coded words can be generated by coding each segment into one of M phases.[2] Then in the case of Frank codes [128] the length of the word becomes M^2 or N. These codes for $M = 3$ through 8 are shown in Table 12-2 where the numbers correspond to phase shifts of $2\pi/M$ radians from an arbitrary phase reference. The codes for $M = 3$ found by Delong [93] are labeled b and c. The code length can be the square of any integer, and the properties to be described have

---

[1] In the absence of doppler shift.

[2] M is used here for the number of possible phases rather than the N in Frank's paper to be consistent with the notation of the previous sections.

been verified up to $M = 13$. The generation of these codes is described in the reference.

The maximum sidelobe level is also shown in the table, and it can be seen that the amplitude ratio of the central peak N to the peak sidelobe is considerably greater than $\sqrt{N}$ and approaches $3\sqrt{N}$ for $M = 3$ to 12. This may be important for surface radars that must cope with land clutter or in multiple-target environments and may justify the additional expense of encoding and decoding into many phases. The important properties of polyphase codes can be summarized as follows:

1. The autocorrelation function is symmetrical about $\tau = 0$. Each set of autocorrelation side peaks for $\tau > T/N$ or $\tau < T/N$ (negative offsets) is also symmetrical (for zero doppler shift).

2. The autocorrelation function equals zero at multiples of M segments.

3. In all cases studied, the amplitude of the largest side peaks is the vector sum of $M/2$ unit vectors (if M is even) or $(M + 1)/2$ unit vectors (if M is odd) where the vectors are separated by angles of $2\pi/M$ radians.

4. Other starting points than the illustrated sequence of all zeros (except for mirror images) seem to result in higher side-peak amplitudes.

5. In the presence of doppler shift, the autocorrelation function *degrades* at a slower rate than for binary codes, but the peak shifts its position rapidly. For $M = 4$, a doppler shift of M cycles ($4 \times 360°$) per sequence length will shift the main peak by four segment lengths (see Sec. 8).

6. The codes have zero periodic correlation in the absence of doppler shift [165]

$$\phi(m) = 0 \qquad m \neq 0, N, 2N, \text{etc.}$$

Property number (5) implies that the ambiguity function is somewhat similar to a chirp or linear FM waveform. This can be seen by observing the code for $M = 8$ in Table 12-6. After the initial sequence of zeros, the phase versus time is monotonically increasing (Mod 2π) for the first eight segments of the code. After that the phase increases by $4\pi/M$, $6\pi/M$, $8\pi/M$, etc., and the general phase variation approximates

$$d\theta \approx \alpha t^2$$

where t is time in units of the segment length and α is a constant. Since

TABLE 12-6 Polyphase Codes for N = 9 to N = 256

M	N	Peak sidelobe amplitude	Code sequence in phase steps of $\frac{2\pi}{M}$
3	9	1.0	(a) 0,0,0;0,1,2;0,2,1 (b) 0,1,1;2,1,2;1,1,0 (c) 0,0,2;2,0, 0;1,0,1
4	16	1.4	(a) 0,0,0,0;0,1,2,3;0,2,0,2;0,3,2,1
5	25	1.6	(a) 0,0,0,0,0;0,1,2,3,4;0,2,4;1,3;0,3,1,4,2;0,4,3,2,1
6	36	2.0	(a) 0,0,0,0,0,0;0,1,2,3,4,5;0,2,4,0,2,4;0,3,0,3,0,3;0,4,2,0, 4,2;0,5,4,3,2,1
7	49	2.25	(a) 0,0,0,0,0,0,0;0,1,2,3,4,5,6;0,2,4,6,1,3,5;0,3,6,2,5,1,4; 0,4,1,5,2,6,3;0,5,3,1,6,4,2;0,6,5,3,2,1
8	64	2.6	(a) 0,0,0,0,0,0,0,0;0,1,2,3,4,5,6,7;0,2,4,6,2,0,4,6;0,3,6,1,4,7, 2,5;0,4,0,4,0,4,0,4;0,5,2,7,4,1,6,3;0,6,4,2,0,6,4,2;0,7, 6,5,4,3,2,1
9	81	~2.8	
10	100	~3.1	
12	144	~3.9	
16	256	~4.7	
16*	256	28	
16**	256	13	
16†	256	26	

After Frank [128], Delong [93], and Queen [299].
*Four-phase approximation to Frank code (rms sidelobes = 5.51).
**Four-phase approximation to good binary codes (rms = 5.99).
†Four-phase approximation to linear FM (rms = 6.81).

a second-order change of phase versus time is a linear frequency shift, the similarities to linear FM become obvious. Polyphase coding can be considered as a discrete phase approximation to discrete frequency coding, and this analogy will be used to introduce Chap. 13. This property can be advantageous for single-pulse radars in the presence of a *small* doppler shift since there is little loss in sensitivity or clutter rejection as compared to pseudo-random codes. On the other hand, a doppler shift will be interpreted as a range error and the ambiguity diagram will have high diagonal ridges as in linear FM.

Also included in the table are some four-phase approximations to several pulse-compression waveforms that have desirable properties.

Four-phase waveforms are receiving considerable interest since digital decoders are relatively easy to implement. While the four-phase approximations to the Frank $M = 16$ waveforms do not have outstanding range sidelobe properties, Queen [299] has shown that the peak range sidelobe is about 19 dB down from the peak and the rms sidelobes are about 33 dB down. The four-phase approximation to linear FM that was studied yields a peak range sidelobe which is 21 dB down from the value at $\tau = 0$.

13 LINEAR FREQUENCY MODULATION AND FREQUENCY CODING

This chapter will include a discussion of the most widely used forms of pulse compression, those which fall into the general class of frequency coding. Since both frequency and phase modulation are forms of angle modulation, the discussion will begin with an example relating frequency and phase coding. The field of linear FM pulse compression will be briefly summarized, but most of the material will concentrate on the broad class of waveforms which can be generated with discrete frequency techniques and the corresponding matched filters. The discussion will conclude with descriptions of several *hybrid* frequency-coded systems, including multi-frequency pulse trains and stepped plus linear FM waveforms.

13.1 MULTIPLICITY OF FREQUENCY-CODING TECHNIQUES

It was shown in the chapter on signal processing concepts and waveform design that the basis of improved range resolution is some form of modulation or coding within the envelope of the transmit pulse which widens the signal spectrum. While binary or 0-180° phase coding accomplished this same end, it should not be categorized as frequency modulation.

It is interesting to note that the Frank polyphase codes (Sec. 12.10) are actually a phase approximation to a frequency code. If the Frank code for 64 segments is divided into segments 1-8, 9-16, 17-24, etc. and the spectrum is calculated for each set of eight, a series of spectra is obtained as shown in Fig. 13-1. The total spectrum of the 64-segment pulse could then be considered as that resulting from a set of frequency-coded pulses with the dashed line approximating the spectrum envelope sin x/x shape.[1]

Two conclusions can be drawn from this figure. First, the more or less linear frequency variation versus time makes this waveform somewhat ambiguous in the range-doppler plane of the ambiguity diagram. A single doppler channel receiver will be tolerant to greater doppler shifts with this waveform than with a binary phase-coded waveform though not as tolerant as Linear FM. Second, the *tapering* of the spectrum envelope leads to low pulse compression time sidelobes. Typical values for those sidelobes were shown in Chap. 12; the subject of range sidelobe reduction will be explored further in Sec. 13.10. Thus, the distinction in terminology between frequency and phase coding is primarily dependent upon the number of phase segments in the transmit waveform and their ordering.

While the generation of linear FM or *chirp*[2] waveforms has in the past been based on the use of passive dispersive delay lines, most of the description in this section will be based on active generation of the desired waveform by discrete frequency techniques. In this variation, the transmit spectrum is formed by the summation of a coherent series of short duration RF pulse segments, selected from a *comb* of coherent frequencies. This departure from the conventional description of chirp systems [214] can be justified in a number of ways:

1. The description of the properties of chirp techniques is by far the best documented of the radar signal processing techniques; and rather than repeat earlier analyses, the reader is referred to the excellent articles by the personnel of Bell Laboratories (Klauder and Price [214]) and those of the Sperry Rand Corporation [39, 71, 274]. The original FM

---

[1] Details of the overall spectrum can be found in Cook and Bernfeld [71].

[2] A common term for linear FM radar signals coined by B. M. Oliver of the Bell Laboratories in 1951.

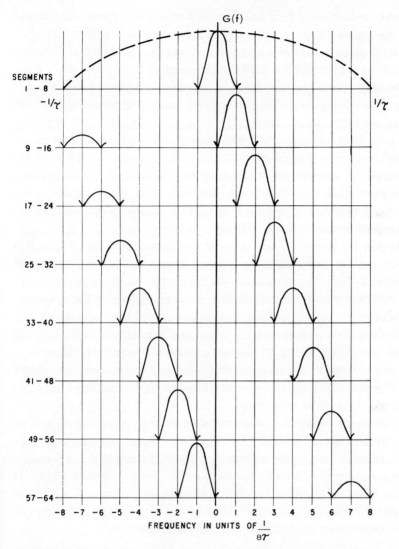

figure 13-1 *Spectra of component parts of 64-segment polyphase code (only main lobe and one sidelobe of each component shown).*

concepts are contained in patents which date back to 1940 and are described in these references.[1]

2. It can be shown that virtually all of the desired properties of linear FM are achievable by transmission of discrete frequency segments. These

---

[1] Darlington's patent (No. 2,678,997) described *chirp* by means of linearly stepped frequencies as well as dispersive techniques.

include the reduction of time sidelobes, the ability to improve target resolution and signal-to-clutter ratio, the ability to achieve greater detection
range for a given peak transmitter power, and the capability of achieving
automatic detection with a hard-limited receiver [351, 271, 402, 260].

3. With discrete transmission of frequencies, it is possible to utilize an
approximation to linear FM and accept the range-doppler ambiguity with
perhaps slightly more hardware than is needed for dispersive
techniques. However, accepting this penalty yields a more flexible means
of waveform generation. For example, it is possible to *scramble* the
ordering of the frequencies to eliminate the FM ambiguity and replace it
with a thumbtack-type ambiguity function. This would eliminate the
major range-doppler uncertainty but yield a general increase in the
sidelobe level throughout the ambiguity plane.

4. It will be shown in Sec. 13.8 that higher overall time-bandwidth
products can be achieved with discrete frequency systems than with
conventional dispersive systems, utilizing narrow-band receiver components. An example of a matched-filter processor in which the frequency
segments are not linearly ordered or even contiguous is given in Sec. 13.13.

5. With radar signal processing becoming increasingly digital, it will
become increasingly easier to convert a number of narrow-band received
waveforms, rather than a single analog broad-band waveform, into digital
form. The preceding discussion is not intended to demean the utility and
simplicity of linear FM waveforms.

The summary of general characteristics of chirp-type signals and receivers given in Secs. 13.3 and 13.4 extends the ambiguity diagram discussion in the signal processing section. It was noted there that a rectangular
spectrum is generally desirable for maximizing clutter rejection [243]. In
addition, a rectangular time function is the simplest to transmit with most
types of transmitter power amplifiers. Section 13.3 contains a description
of how both of these goals can be obtained simultaneously.

13.2   LINEAR FM PULSES (CHIRP)

By far the most widely implemented technique for pulse compression is
the use of linear frequency modulation during the pulse.[1] Many of the

---

[1]Nonlinear frequency modulation will not be further discussed except as a form of
discrete frequency coding. The interested reader can pursue the subject in [71, 305],
along with the discussion that was included in Sec. 8.7.

goals of waveform design can be attained with relatively simple dispersive devices. The desired transmission energy can often be obtained by the choice of the requisite pulselength, while the desired resolution can be independently specified by the frequency deviation. Improved performance over uncoded pulses is obtained in distributed clutter environments, and the receiver can be a *constant false alarm-rate* device to cope with broad-band noise interference.

Unfortunately, chirp is not a universally desirable waveform, hence the emphasis in previous sections on the multitude of other waveform modulations. The lack of unambiguous range and velocity resolution, discussed in the chapter on signal processing concepts and waveform design, sometimes results in inadequate performance in *sorting out* targets, land- and sea-clutter echoes, decoys, and dense rain and chaff echoes. It would be a sharp contrast with the related field of communications if a single waveform with a *single* relatively simple matched filter were adequate for all environments.

In the next paragraphs, chirp will be discussed in its simplest forms. More extensive analysis and block diagrams will be limited to multi-frequency processors to avoid duplication in a well-documented subject.

13.3 GENERATION AND DECODING WITH DISPERSIVE TECHNIQUES

In the design of a simple chirp system the first parameter to be chosen is the length of the pulse envelope, given the peak transmitter power. This determines the pulse energy and allows the prediction of the detectability of a given target model at a given range. Second, the desired resolution determines the bandwidth of the transmission and therefore the performance in a distributed clutter environment.

One class of techniques for generating a long-duration FM signal is the so-called passive one illustrated in Fig. 13-2. An IF pulse of the approximate length of the desired compressed pulse is generated and inserted into a dispersive delay device. Since the compressed pulse for a linear FM waveform has a sin x/x time envelope, it can be shown that the input to a linear dispersive delay should have a time waveform [71]

$$ f_1(t) = \frac{\sin(\mu T t/2)}{(\mu T t/2)} \cos\left(\omega_{IF}\, t\right) $$

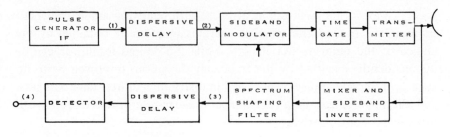

(A) GENERATOR AND DECODER

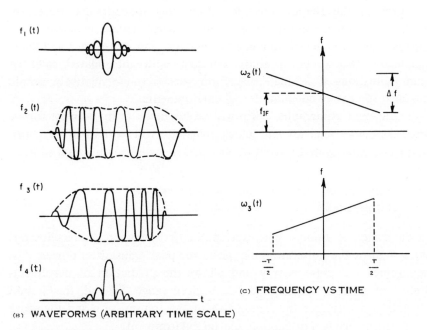

(B) WAVEFORMS (ARBITRARY TIME SCALE)

(C) FREQUENCY VS TIME

figure 13-2 *Passive system for linear FM pulse compression.*

where $\mu = \Delta\omega/T$, the desired frequency slope at the dispersive delay output (the rate of frequency sweep)

T = the duration of the transmit pulse envelope

$\omega_{IF} = 2\pi f_{IF}$ = the radian center frequency of the dispersive delay line

After traversing the dispersive line the waveform is "up-mixed" (converted) to the desired carrier. Filtering out unwanted sidebands, the transmitted waveform can then be written as [39]

$$f_2(t) \approx \cos\left[\omega_0 t + \frac{\mu t^2}{2}\right], \quad \text{for} \quad -\frac{T}{2} \leq t \leq \frac{T}{2}$$

where ω_0 is the transmit carrier frequency. The generating waveform $f_1(t)$ can be obtained by passing a narrow pulse (radian bandwidth $> \Delta\omega$) through a rectangular band-pass filter of width $\Delta\omega/2\pi = \Delta f$. The filter must have a linear phase characteristic.

The dispersive delay device can be composed of cascaded all-pass networks [71, 195], dispersive ultrasonic delay lines [111, 71], waveguide operated near cut-off [183], or optical devices. For purposes of this discussion we assume whatever device is used has negligible attenuation over the bandwidth Δf and a linear time delay versus frequency. The waveform at its output is shown diagrammatically as $f_2(t)$, its frequency-versus-time characteristic as $\omega_2(t)$. It should be noted that there is actually a certain amount of amplitude ripple versus time at the transmitter input. With the usual saturation in the transmitter, there will consequently be waveform distortion at the receiver output. However, this distortion has a small effect with large compression ratios.

The sideband modulator is used to convert the FM signal to the desired carrier frequency, and a time gate is usually employed to form the rectangular transmit waveform. Upon reception, the echo is mixed with a local oscillator which yields the sideband inverse to that utilized on transmission [302]. By this technique the frequency-versus-time characteristic is inverse with respect to the transmitted signal as illustrated by $f_2(t)$ and $f_3(t)$. This technique permits the use of identical dispersive delays in the transmit and receive lines. In practice, the received signal is often inserted in the same dispersive delay as was used for transmission, thus reducing hardware and tolerance requirements.

The spectral shaping filter is used to reduce the -13-dB time sidelobes to a tolerable level at the cost of a slight reduction in signal-to-noise and signal-to-distributed-clutter ratios and at the further cost of a small increase in compressed pulse width. This subject will be covered in a slightly different context in Sec. 13.9. The detected waveform shown as $f_4(t)$ has a $|\sin x/x|$ form near its central lobe in the absence of spectral shaping. For a large compression ratio the waveform will be similar to that to be shown in Fig. 13-7. The peak signal amplitude is increased by $T\Delta f$, and the pulse width is reduced by approximately the same factor.

An alternate chirp generation technique that utilizes *active* generation of the FM waveform is to apply a sawtooth or ramp modulation waveform to a voltage-controlled oscillator or a square-law modulation to a phase-controlled oscillator [71, 39]. Small variations in the modulation function can be injected to compensate for distortions elsewhere in the system. If target range and range rate are approximately known as in a tracking system, a similar ramp may be used for *decoding* in the receiver. If target range is not known, a dispersive delay matched filter is usually used in the receiver. Other techniques for approximating a linear FM waveform are discussed in Secs. 13.10 and 13.12, and an excellent summary of dispersive devices is found in Cook and Bernfeld [71, chaps. 12, 13, and 14]. Table 13-1 summarizes the compression ratios and range sidelobe levels that have been reported in the open literature for various techniques. The first column in the table lists the technique with some rough bounds on the loss in the dispersive line L, the maximum delay T_{max}, and the typical frequency range F in MHz. The succeeding columns give the compression ratio, the center frequency of the dispersive device, the experimental compressed pulse width, the peak sidelobe level, and the reference from which the data was extracted or estimated. In most of these examples slight variations in pulse width can be traded for different peak sidelobe levels, but the sidelobe increase due to transmitter distortion (not included in the table) will generally preclude achieving peak sidelobe levels of better than -35 dB. An example of the waveforms of a high compression-ratio passive linear FM system is shown in Fig. 13-3. The peak sidelobes are -40 dB for the unit.

13.4 DISTORTION EFFECTS ON LINEAR FM SIGNALS

The effects of amplitude and phase distortion on pulse-compression systems are best described by paired-echo analysis techniques [39, 214, 71] as follows in brief. In the frequency domain the distortion has been related to the transfer admittance of the system [214]

$$Y(\omega) = A(\omega) \exp(jB\omega)$$

The terms can be expanded into Fourier series of n terms

$$A(\omega) = a_0 + \sum_n a_n \cos(nc\omega)$$

TABLE 13-1 Characteristics of Typical Linear FM Pulse Generation and Compression Circuits

Technique	Compression ratio, $T\Delta F$	Center frequency, MHz	Compressed pulse width, μsec	Peak sidelobe, dB below peak	Reference
(1) Lumped constant dispersive lines	50	13.5	0.10	35	[183]
	~130	13.5	0.077	31	[183]
(2) Dispersive strip delay lines	~110 (80)**	60	~0.18 (.25)	30	Andersen*
	118 (72)	30	~0.36 (.56)	40	Andersen*
F(steel) 5–45 MHz	(64)	15	0.8	28	[111]
F(aluminum) 1–5 MHz	35	5	0.8		Brew*, [71]
L~10–40 dB	~500	2.0	~1.3		Brew*
(3) Diffraction dispersive delay lines—perpendicular	64	30	0.18–(.25)	36	[111]
	400 (250)	45	0.05–(.08)	35	[111]
$T_{max} \approx 225\mu sec$	400	60	0.07+		Andersen*
$F = 20$ to 60 MHz	1,000	100	0.025		Andersen*
$L = 15$ to 60 dB					
(4) Diffraction dispersive delay lines—wedges	64	30	~0.02	28	[111]
	250	500	~0.004		Andersen*
$T_{max} \approx 65\mu sec$		(Sapphire)			
(5) Discrete frequency	1,000	15	1.0	~23	[188]
	1,024	30	0.012	~30	
(6) Waveguide near cut off	~120	2,775	~0.009	12–18	[183]
(7) YIG	125	2,880	0.014	>13	[71]

*Reported by Andersen Laboratories, Bloomfield, Connecticut and by R. D. Brew and Co., Concord, New Hampshire.
**Parentheses indicate values with weighting functions.

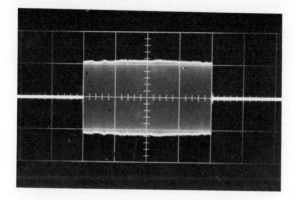

A. EXPANDED PULSE, 40 μsec

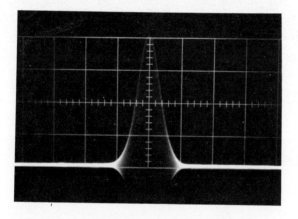

B. COMPRESSED PULSE, 0.58 μsec

HORIZONTAL SCALE: , 1 μsec/cm

figure 13-3 *Transmit and receive waveforms of dispersive linear FM pulse-compression system. (A) Transmit waveform, 30 MHz IF; (B) compressed pulse, – 40-dB sidelobes. (Permission of Andersen Laboratories.)*

$$B(\omega) \;=\; b_0\omega \;+\; \sum_n b_n \sin(nc\omega)$$

where for linear FM signals c = complex constant

$a_n \cos(nc\omega)$ = amplitude distortion terms

$b_n \sin(nc\omega)$ = phase distortion terms

The system is considered ideal when $a_n = b_n = 0$, or the distortion terms are zero.

For small phase deviations ($< 25°$), it is sufficient to consider only the $n = 1$ term. A graph of the highest paired range sidelobes for various peak phase and amplitude errors is shown in Fig. 13-4 [214, 39, 273, 71]. It can be seen that for -40-dB range sidelobes, the amplitude must be held to ± 2 percent and the phase to within $\pm 1.2°$. These results assume a sinusoidal type of error versus time or frequency.

The results for random errors are more difficult to compute. Some limited experimental results involving only phase distortion for $T\Delta f = 50$, uniform amplitude (no weighting); and 10, 20, and 30° of rms phase error[1] are shown in Fig. 13-5. The far-out sidelobes for 20 and 30° error increase to -13.5 and -12.2 dB below the main peak [283].[2] The corresponding decreases in the main lobe amplitude were 1 and 2 dB, and the increases in the main lobe width are 15 and 19 percent. The ordinate is linear in amplitude.

The effect of transmitter tube phase ripple b_1 on paired-echo sidelobes has been analyzed by Liebman (see Cook [71, p. 395]) and is approximately

$$b_1 = K\left(\frac{\Delta E}{E}\right)\phi \quad \text{or} \quad L\left(\frac{\Delta I}{I}\right), \quad \text{degrees}$$

$\quad\quad = K$ (voltage ripple ratio) (electrical length of device in
$\quad\quad\quad$ degrees) or
$\quad\quad = L$ (current ripple ratio)

For klystrons $K = 1/2$, for TWTs $K = 1/3$ and for crossed-field amplifiers $K = 1/25$. For triode amplifiers $L = 50$ and for amplitrons $L = 40$. It should be kept in mind that three or four crossed-field devices must usually be cascaded to equal the gain of a single klystron or TWT.

13.5 SPECTRUM OF A COMB OF FREQUENCIES

Consider the pulse waveform containing N equal-amplitude rectangular time segments of length τ each on a different frequency f_n. The

---

[1] Derived with a Monte Carlo procedure using tables of random normal deviates.

[2] The far-out sidelobes are somewhat exaggerated since the triangular weighting of a matched-filter output (autocorrelation of the transmit pulse envelope) was not included in the experiment.

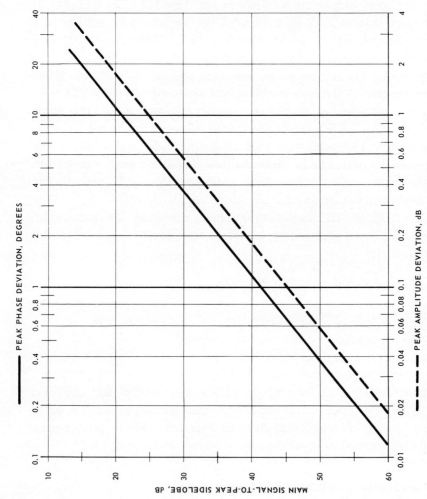

figure 13-4 *Peak sidelobe of linear FM signals with phase and amplitude distortion (paired-echo theory). (After Klauder et al. [214], copyright 1960, American Telephone and Telegraph Co. by permission.)*

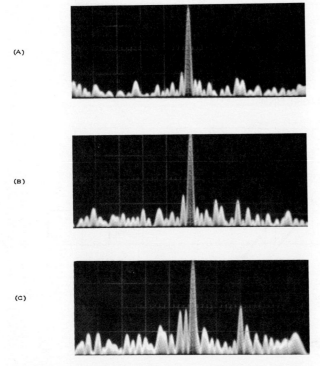

(A)

(B)

(C)

figure 13-5 *Typical compressed-pulse waveform (amplitude) for linear FM signal and random phase errors, $T\Delta F = 50$. (A) $20°$ phase error (std. dev.); (B) $30°$ phase error; (C) $40°$ phase error.*

envelope of this waveform is illustrated in Fig. 13-6.4. The time origin will be taken as the midpoint of the first segment but otherwise conforms to an analysis by O'Neill [273]. The waveform of the nth segment or subpulse is given by

$$v_n(t) = A \exp j2\pi (f_n t + \phi_n) \qquad \left(n - \frac{3}{2}\right)\tau < t < \left(n - \frac{1}{2}\right)\tau \qquad (13\text{-}1)$$

where A = segment amplitude

$\quad\quad\quad f_n$ = frequency of the nth segment

$\quad\quad\quad \phi_n$ = phase of the nth segment

$\quad f_n - f_{n-1} = 1/\tau$ (see Sec. 13.6)

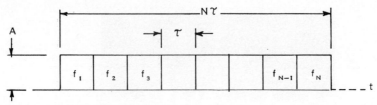

A) LINEAR STEPPED-FREQUENCY PULSE

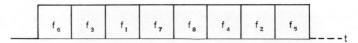

B) SCRAMBLED FREQUENCY-CODED PULSE

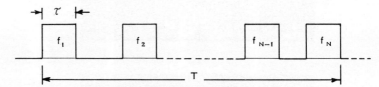

C) MULTIFREQUENCY PULSE TRAIN (SHOWN FOR LINEAR FREQUENCY)

figure 13-6 *Discrete frequency waveforms.*

The Fourier transform of the segment can be written

$$V_n(f) = A \int_{(n-3/2)\tau}^{(n-1/2)\tau} \exp j2\pi (f_n t + \phi_n) \exp(-j2\pi ft)\, dt \qquad (13\text{-}2)$$

This can be shown to be equal to [273]

$$V_n(f) = A\tau\, \frac{\sin \pi [(f_0 - f)\tau + n]}{\pi [(f_0 - f)\tau + n]} \exp j\left[2\pi\left(f_0 - f + \frac{n}{\tau}\right)(n - 1)\tau\right] \qquad (13\text{-}3)$$

where f_n is replaced by $f_0 + n\Delta f$ for a frequency spacing of Δf. Now let $\phi_n = n\phi_0 = 0$ (see Sec. 13.6). Then for N subpulses and after removing the carrier term, the spectrum of the modulation of the entire pulse becomes

$$V(f) = A\tau \sum_{n=1}^{N} \frac{\sin \pi (f\tau - n)}{\pi (f\tau - n)} \exp j[2\pi (f\tau - n)(1 - n)] \qquad (13\text{-}4)$$

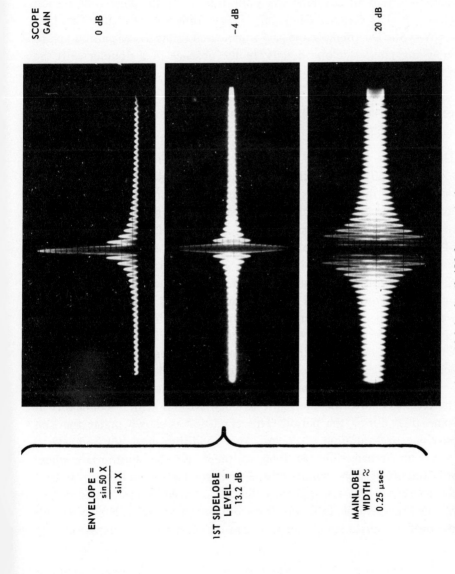

SCOPE GAIN

0 dB

−4 dB

20 dB

ENVELOPE = $\dfrac{\sin 50\, X}{\sin X}$

1ST SIDELOBE LEVEL = 13.2 dB

MAINLOBE WIDTH $\approx$ 0.25 μsec

figure 13-7 *Uniform weighting sidelobe details (50 frequencies).*

It will be shown in later sections that the envelope of the compressed pulse out of the matched filter for this waveform has a sin x/x shape in the neighborhood of its main lobe. The width of the compressed pulse at the 4-dB points is equal to τ/N for uniform weighting of the frequencies. Since the transmit envelope has a duration of $N\tau$, the pulse-compression ratio $\approx N^2$. The matched-filter output waveform is illustrated in Fig. 13-7 for $N = 50$. The oscilloscope gain is increased in the lower photograph to show the range sidelobe details. In the absence of a doppler shift the scrambled-frequency waveform of Fig. 13-6B will yield the same output.

13.6 WAVEFORM ANALYSIS FOR DISCRETE FREQUENCIES

The rigorous analysis of discrete-frequency pulse compression loses touch with physical significance unless some restrictive assumptions are made about the relations among the various parameters, such as the spacing between frequencies Δf, the subpulse or segment length τ, and the phase of each of the subpulses ϕ_n. Figure 13-8 is the time waveform of the envelope of the *coherent* summation of eight continuous sinusoidal carriers spaced by exactly Δf in frequency. The bandwidth of the RF waveform is $N\Delta f$ and is assumed to be much less than the mean or carrier frequency f_c. The spectrum is periodic in frequency, and as would be expected, the time waveform is also periodic with a repetition period of $1/\Delta f$. A typical transmit time waveform is shown in Fig. 13-9. With an even number of frequencies, a center *carrier frequency* does not actually exist as any one of the individual frequencies.[1]

The necessary condition for compression of the energy is that all of the phasors (shown on the figure as small arrows) achieve colinearity at some time during the period. This condition is shown under the first envelope peak at time $t = 2n/\Delta f$. This condition need not be made to occur on transmission as long as there are the appropriate phase adjustments in each channel of a receiver such as shown in Fig. 13-10. If the phases are not adjusted for colinear vector addition at some instant of time, the periodic time waveform will appear noiselike. However, with the proper adjustment the phasors can be made to peak up periodically

---

[1] From an *analytic signal representation* point of view involving Hilbert transforms there would be no need to limit the discussion to a narrow-band comb spectrum or to designate a specific carrier frequency.

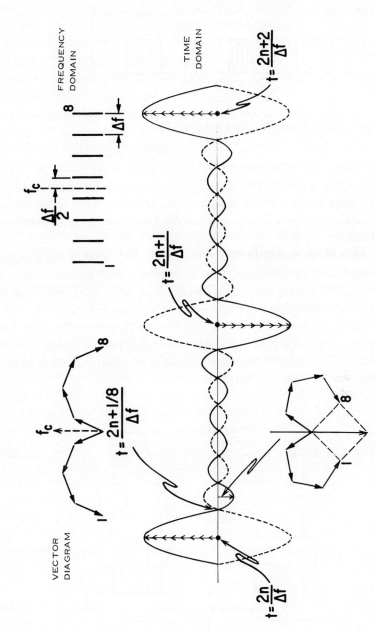

figure 13-8 *Eight-frequency compressed-pulse details.*

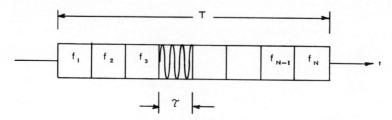

figure 13-9 *Transmit waveform for stepped chirp.*

for all time, providing the generated comb frequencies are truly coherent. This is essentially the definition of coherence for multifrequency systems. This condition can be made to occur if all of the frequencies are derived from a single stable oscillator either by mixing or multiplication.

The alternating polarities of the phasors in Fig. 13-8 at the envelope peaks occur with an even number of frequencies. With an odd number of frequencies, the center frequency can be considered the *carrier* and the phasors at the envelope peak always have the same polarity. In either case, if the mean frequency of the comb is much greater than the total spectral width, there will be many RF cycles under the envelope, and the exact time of occurrence or polarity of the *mathematical* peak is of little importance in the detected output.

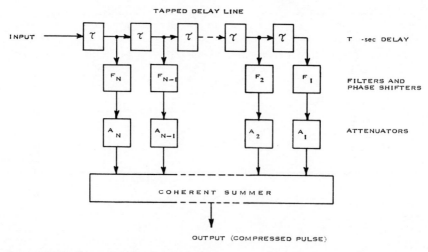

figure 13-10 *Stepped chirp matched filter.*

A result of these assumptions is that the phase ϕ_n of each of the N-frequency segments can be assumed to be zero. While this is a mathematical convenience, the results are valid as long as the phase responses in the individual channels of Fig. 13-10 are equal and opposite in sign to the phases of the corresponding received frequencies.

The ensuing discussion will be based upon the pulse sequential transmission of the frequencies, and the CW-derived periodic time waveform envelope discussed above will not be quite relevant. The envelope of the transmission on each frequency, which will be referred to as the *subpulse* and which has a duration of τ seconds, now determines what portion of that frequency is time selected for coherent summation with the other frequencies of the comb.

Before proceeding further it is desirable to make the assumption that the frequency spacing is equal to the inverse of the pulselength $\Delta f = 1/\tau$ and the subpulse envelopes are rectangular. This has the desirable effect of keeping the mathematics reasonable, but more importantly it yields the best waveform properties for most practical situations. The validity of the preceding statement can be illustrated by considering the matched-filter outputs when $\tau \Delta f \neq 1$.

If $\tau \Delta f > 1$, there will be significant range ambiguities at time $t = |1/\Delta f|$ as seen from the autocorrelation function in Fig. 13-11. It can also be seen that when $\tau \Delta f = 1$, the nulls of the autocorrelation function of the subpulse envelope tend to suppress these ambiguous peaks. When $\tau \Delta f < 1$, the subpulse filters F_n shown in Fig. 13-10 will overlap; and the contribution of each channel to the coherent summation will contain

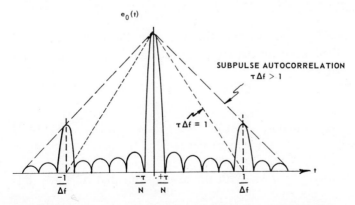

figure 13-11 *Autocorrelation function of frequency-coded pulse.*

spurious signals. The details of the output waveform for $\tau \Delta f < 1$ will depend on the ordering of the frequencies in the pulse envelope.

While there may be special cases where $\tau \Delta f$ is made slightly greater than unity to attain somewhat better range resolution and accuracy, the remainder of this section will be limited to discussions of systems with $\tau \Delta f = 1$.

The transmitted waveform for a linearly stepped frequency waveform can be written

$$v(t) = \sum_{n=0}^{N-1} [u(t - n\tau) - u(t - [n + 1]\tau)] \cos(\omega_0 + n\Delta\omega) t \qquad (13\text{-}5)$$

where $\omega_0 = 2\pi f_0$, lowest frequency in the transmission comb

$\Delta\omega$ = frequency spacing = $2\pi \Delta f$

$u(t)$ = unit step function = $\begin{cases} 1, & \tau \geq 0 \\ 0, & t < 0 \end{cases}$

N = number of frequencies in transmission

The phase and time origin is taken to be the leading edge of the first subpulse. It can be shown that the matched filter has an impulse response which can be written [402]

$$h(t) = v(-t)$$

$$= \sum_{n=0}^{N-1} [u(-t - n\tau) - u(-t - [n + 1]\tau)] \cos(\omega_0 + n\Delta\omega) t \qquad (13\text{-}6)$$

If the matched-filter delays and filters are in the inverse order of transmission and the delay to the target is neglected,[1] the matched-filter output can be written

$$e_0(t) = \sum_{n=0}^{N-1} A_n{}^2 \exp j(\omega_0 + n\Delta\omega)t \qquad (13\text{-}7)$$

in the region $(N - 1)\tau < t < N\tau$. With the assumption that $\Delta f = 1/\tau$, this

[1] That is, only differential delays with respect to target location are formulated.

becomes the only region of interest. Equation (13-7) is the finite sum Fourier series approximation to the matched-filter output, neglecting the detailed shape of the filters in each channel.[1] A_n is the weighting function of the nth frequency segment, and it will be assumed equal to unity until Sec. 13.9. Equation (13-7) can then be factored into

$$e_0(t) = \exp(j\omega_0 t) \sum_{n=0}^{N-1} \exp(jn\Delta\omega t) \tag{13-8}$$

By means of the following identity

$$\sum_{n=0}^{N-1} X^n = \frac{1 - X^N}{1 - X} \tag{13-9}$$

and letting $X = \exp(j\Delta\omega t)$, the summation in Eq. (13-9) becomes

$$\sum_{n=0}^{N-1} \exp(jn\Delta\omega t) = \frac{1 - \exp(jn\Delta\omega t)}{1 - \exp(j\Delta\omega t)} \tag{13-10}$$

Equations (13-8) and (13-10) can then be combined and manipulated so that finally

$$e_0(t) = \exp j \left[\omega_0 + \frac{(N-1)\Delta\omega}{2} \right] t \left(\frac{\sin N(\Delta\omega/2)t}{\sin(\Delta\omega/2)t} \right) \tag{13-11}$$

To account for the autocorrelation of the rectangular segment envelope as was illustrated in Fig. 13-11, Eq. (13-11) should be multiplied by the triangular function $\tau[1 - (|t|/\tau)]$ for $|t| \leq \tau$, and zero elsewhere.

The term $\omega_0 + [(N-1)\Delta\omega/2]$ is the mean frequency of the comb or *carrier* frequency and can be factored out. Thus, the second term determines the compressed-pulse envelope. The nulls in the envelope occur when $\sin[N(\Delta\omega/2)t] = 0$ or when

$$N \frac{\Delta\omega}{2} t = \pm m\pi, \text{ where } m \text{ is any integer}$$

---

[1] An analysis which includes the effects of practical filters is included as Appendix C of [273].

The first sidelobe peak is reduced from the main lobe by 13.46 dB when $N \geq 50$ and is 13.06 dB down for $N = 8$. This envelope has the $\sin NX / \sin X$ shape and is similar to that illustrated in Fig. 13-11. At one-half the distance between the nulls, the linear envelope response is down by 4 dB and

$$\text{Compressed pulse width}_{(4 \text{ dB})} = \frac{1}{N\Delta f} = \frac{\text{transmit envelope width}}{N^2} \quad (13\text{-}12)$$

The pulse-compression ratio is then slightly greater than N^2 with the customary definition of 3-dB pulse widths.

Thus, the compressed-pulse matched-filter output for a stepped linear FM waveform has almost the same shape as that for a continuous linear FM waveform for large compression ratios.

13.7   CAPABILITIES FOR EXTREME BANDWIDTHS WITH DISCRETE FREQUENCY SYSTEMS

In Secs. 13.1, 13.2, and 13.3 it was indicated that among the requirements imposed on very wide-band processors, many imply large time-bandwidth products. The time-bandwidth product is typically proportional to the square of the number of segments in a frequency-coded pulse, and the signal processing bandwidth is proportional to the number of frequencies. As a general rule, it is preferable to use discrete frequency coding instead of linear FM when the time-bandwidth product is in excess of 250 and the signal bandwidth is over 100 MHz. The previously mentioned problems with dispersive lines also tend to favor the discrete techniques for wide bandwidths and $T\Delta f$'s of 150 or more.

There are three other major factors that favor the use of the discrete frequency techniques. The first is the ability to eliminate the range-doppler ambiguity of linear FM. It will be seen that the matched-filter response, tapering effects, receiver sensitivities, etc., *do not* depend on the order in which the frequencies are transmitted. In simple terms, if the price is paid to implement matched filters for *all* expected target dopplers, a *thumbtack* ambiguity function can be obtained. While there will be an increase in the number of low-level residues, there will be no range-doppler confusion between targets of comparable size. A related benefit to *scrambling* the frequencies is that the system will not respond to interfering FM-like signals generated by other radars.

The second and perhaps the most significant advantage of multiple-frequency channels is that the components in each channel need only have a bandwidth of $1/N$ times the total processing bandwidth. This is illustrated in the waveform pictorial and receiver block diagram for the *stepped-chirp* frequency processor shown in Fig. 13-10. It is assumed that all subpulses or segments have the same amplitude upon transmission. The particular waveform shown in Fig. 13-9 is stepped linearly in frequency. After mixing to the desired RF, the signal is transmitted, received, and mixed down to a convenient intermediate frequency to become the input of the matched filter of Fig. 13-10. The processor delays τ are equal to the segment length of the transmission.

As shown, the filters F_n are placed inversely in time to the transmission frequencies f_n. If the filter bandwidths are approximately $1/\tau$, a nearly matched receiver is obtained. A description of the waveforms and spectra will be given in subsequent sections, but ordinarily the bandwidth of each channel is only $1/N$ times the total signal bandwidth. The phase and amplitude response to the received frequency segments is controlled in the individual channels. This eliminates the requirement for components having phase and amplitude linearity across the total transmission bandwidth. In this simplified diagram, the coherent summer is simply a voltage adder, which is then followed by an envelope detector. Channelization, as will be discussed later, also facilitates the use of digital processing.

The third advantage of discrete systems is that channelization allows the capability for selective limiting in automatic detection or CFAR receivers. As was mentioned in Chap. 4, strong CW interference anywhere in the total signal bandwidth can *capture* a limiting receiver for FM signals. This interference will cause suppression of the target echo signals and prevent the matched-filter output from crossing a preset threshold. In a channelized system, CW interference will only suppress the target signal in one of N channels and the target signal power output will be $(N - 1)^2/N^2$ times the output without the interference. If N is greater than about 10, target detection will not be prevented with a fixed threshold system for CW interference which is 20-30 dB above the per channel echo signal power. This will be illustrated in Sec. 13.11.

In summary, there is virtually no limit to step frequency processor time-bandwidth product capability. Systems of this nature have been built with bandwidths of several hundred megahertz and with time-bandwidth products of 100,000.

13.8 RESOLUTION PROPERTIES OF FREQUENCY CODED PULSES

The virtues of wide-band waveforms have been well explored analytically, but it is always useful to show available experimental results. Figures 13-12, 13-13, and 13-14 are typical outputs from two similar discrete-frequency processors each with a coherent bandwidth of $\approx$ 80 MHz. These signal processors will be described in Secs. 13.11 and 13.12. In one of the systems, the transmission consisted of eight frequency segments of τ = 0.13-μsec duration with 10-MHz frequency separation. In the second system, the subpulse length was 0.4 μsec, and the frequency separation Δf was 2.5 MHz. In each system there was also an incoherent output which was simply the summation of the detected envelopes on each frequency. The resolution of the incoherent channel was 0.13 μsec in the first processor and 0.4 μsec in the second. The compressed-pulse output of each system was $\approx$ $1/N\Delta f$ $\approx$ 12.5 nsec in duration at the $-$ 4-dB points.

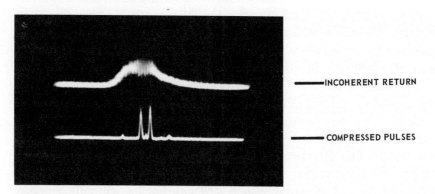

INCOHERENT RETURN

COMPRESSED PULSES

figure 13-12 *Resolution of 6-in. spheres (25-ft separation). Trace length = 500 ft, no. of frequencies = 8, no. of pulses = 32, cross section = 0.2 ft^2, range = 2,000 yd, power = 2 kW, pulse width = 0.13 μsec, compressed = 0.012 μsec.*

The resolution of the echoes from two 6-in. metallic spheres is illustrated in Fig. 13-12. They were suspended from a weather balloon 2,000 yards from the radar. The A-scope photo was taken when they were 25 ft apart, but they were also resolved with less than 8-ft radial separation. The upper trace showing the incoherent output illustrates that resolution was not possible with 0.13-μsec (65 ft) resolution. The two small range sidelobes on either side of the echoes from the spheres resulted from having $\tau\Delta f$ slightly greater than unity. (See Sec. 13.6.)

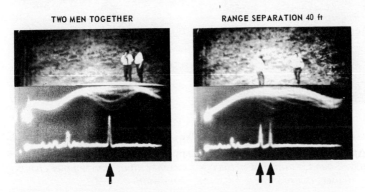

figure 13-13 *Resolution of human targets in ground clutter. (Center of each photo shows incoherent echoes; bottom traces show compressed pulses. Trace length = 500 ft, range = 300 yd, bandwidth = 80 MHz, bandwidth = 2.0°, antenna elevation = -3°.*

Figure 13-13 similarly shows the resolution of two men standing in a grassy field. The upper half of each photo is from a TV system boresighted with the antenna. In the left-hand photo, when the men were standing side by side, they could not be resolved; but their reflected echoes did stand out from the local ground clutter. The fluctuation of the incoherent output (center of the photos) was primarily due to the ground-clutter contributions at the left of the trace. With the two men separated in range by 40 ft, the resolved echoes at the bottom of the right-hand photo are evident. They not only could not be resolved in the incoherent output, but could not be detected in the ground clutter. While the improvement in target-to-clutter ratio was evident in the grassy areas, the echoes from the men did not stand out very well in a nearby wooded area unless they walked continuously in one direction.

The third set of photos in Fig. 13-14, illustrates that rainstorm echoes do not in general yield compressed pulses. In other terms, the echoes from distributed clutter are uncorrelated (not coherent) between frequencies which are separated by at least $1/\tau$ (see Secs. 6.5 and 7.6). The photo on the left is of the PPI of a conventional incoherent S-band

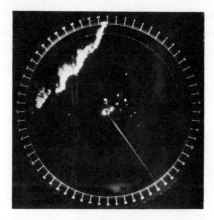 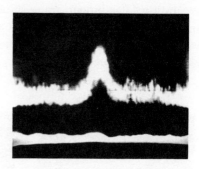

figure 13-14 *Radar returns from heavy rainstorm. (Left) S-band PPI of rainstorm; range scale = 120,000 yd; April 30, 1963, beamwidth = 2°. (Right) Upper trace = single-frequency C-band return; lower trace = coherent output zero doppler channel; trace length = 60 μsec; transmit pulse = 0.4 μsec; bandwidth = 80 MHz.*

search radar when a narrow, but intense rainstorm was 40-50 miles from the radars. The photos on the right are again A-scope photos of the detected output of the multifrequency radar which was pointing at about 290° azimuth. The upper trace is again the incoherent output showing intense returns extending for about one mile. The lower trace is the coherent output, and other than noise, nothing is visible. The absence of an increased noise level at the rain location results from the CFAR action of hard limiting on each frequency, which will be discussed later in this chapter. If there had been a single point reflector, the compressed pulse from it would have extended up to the baseline of the incoherent trace on the scope screen.

The three photos illustrate that high resolution, clutter reduction, and a constant false alarm-rate receiver can be achieved in practice as well as in theory. While these results were obtained with discrete frequency coding, they could also have been obtained with phase coding or linear FM.

13.9 SIDELOBE REDUCTION BY AMPLITUDE WEIGHTING[1]

Amplitude weighting of the frequency comb is frequently used to reduce the range sidelobes present in the output of a pulse-compression system.

[1]Most of the analysis and experimental work for this section was carried out by D. M. White [402]. While the analysis is based on discrete frequency systems, the
(Footnote continued on next page.)

The first sidelobe resulting from a uniformly weighted comb was shown to be ≈ 13.5 dB below the main lobe peak. The remaining sidelobes decay monotonically and symmetrically about the main lobe. By amplitude weighting the frequencies the sidelobes, in theory, may be reduced any desired amount.

In addition to the desirable sidelobe reduction, amplitude weighting of the frequencies has undesirable effects on the output waveform, including main-lobe widening, and output signal-to-noise ratio degradation. These effects are measured as deviations from the $\sin Nx / \sin x$ envelope waveform, obtained from a uniformly weighted frequency comb. The reason for the main-lobe broadening is deduced intuitively from the Fourier transform reciprocity relationship. When the spectral shape is made narrower, the time domain waveform becomes broader and vice versa. A bell-shaped or tapered spectrum would then be expected to correspond to a wider time domain waveform than would a flat or unweighted spectrum of the same total frequency extent. Therefore, the resultant waveform in the time domain will be broader for an amplitude-weighted comb than for the uniformly weighted comb. Amplitude weighting may be introduced into a pulse-compression radar either entirely at the transmitter, entirely at the receiver, or at both simultaneously. When the amplitude weighting is imposed equally on both the transmitted waveform and on the corresponding gains of the receiver channels, the system is still considered *matched.* For a matched system, there will be no decrease in the signal-to-noise ratio due to weighting. However, when the amplitude weighting is introduced either at the transmitter or the receiver alone, the system is *mismatched,* with a resulting degradation in the signal-to-noise ratio.

The most feasible way to amplitude weight is at the receiver alone. This is mainly due to the necessity of operating the transmitter at its efficient peak-power limit point. Amplitude weighting solely at the receiver is accomplished by proper adjustment of the filter gains or by the insertion of the proper value attenuators in each of the parallel channels. It can be conveniently maintained, due to the accessibility of the components and the low power levels involved. For these reasons it is henceforth assumed that weighting is performed solely at the receiver at the expense of a lower output signal-to-noise ratio.

results are in general applicable to dispersive pulse-compression techniques. With proper interpretation, the analysis also applies to tapering of the pulse-train waveforms for reduction of doppler sidelobes.

The frequency response of a matched filter is adjusted to maximize the output signal-to-noise ratio and requires that the amplitude response of the matched filter be equal to the amplitude spectrum of the input signal. The amount of mismatch is measured as a degradation from the optimum signal-to-noise ratio normally produced with a matched filter.

The loss in signal-to-noise ratio resulting from the use of weighting at the receiver alone is characterized by the loss factor L_s given by [373]

$$L_s = \frac{(S/N) \text{ weighted}}{(S/N) \text{ matched}} = \frac{[\int_T w(t)dt]^2}{T \int_T w^2(t)dt} \tag{13-13}$$

where $w(t)$ = the weighting function

T = the processing time interval or the length of the transmitted waveform

Equation (13-13) states that (S/N) is degraded by a factor L_s times that obtained in the absence of *matched* weighting. For discrete frequency amplitude weighting of an N-frequency radar, L_s becomes

$$L_s = \frac{\left[\sum_1^N A_n\right]^2}{N\left[\sum_1^N A_n^2\right]} \tag{13-14}$$

where A_n are the weighted amplitudes, for the N frequencies. Equation (13-14) is derived from the following relations:

$$S/N_{\text{matched}} \approx N \tag{13-15}$$

and

$$S/N_{\text{weighted}} \approx \frac{\left[\sum_1^N A_n\right]^2}{\sum_1^N A_n^2} \tag{13-16}$$

The basis of Eqs. (13-15) and (13-16) is that for coherent summation, signal components add as voltage levels while noise components add as power levels.

The problem in selecting an appropriate amplitude weighting function for a pulse-compression system is to find what finite spectrum shape can produce a desired time waveform under some condition of optimization. A common criterion for optimization is to find a weighting function which will provide a minimum main-lobe widening for a specified sidelobe level.

The amplitude weighting functions that will lower the sidelobes in the output time waveform all have a characteristic bell or tapered shape. When this function is applied to the comb, the frequency in the center will have the maximum amplitude; and all other frequencies will have decreasing amplitudes symmetrically about the center. The output waveform for the more familiar functions, such as cosine and gaussian weighting, can be obtained from the Fourier transform relationship between the input and output of pulse-compression systems. To evaluate the output $e_0(t)$, it is convenient to allow Eq. (13-7) to approach the integral form of the Fourier transform. This is performed by rewriting it as follows [391]:

$$e_0(t) = \frac{1}{2\pi} \sum_{n\omega=-W}^{W} a_n(n\omega) \exp(jn\omega t) \Delta(n\omega) \qquad (13\text{-}17)$$

where $2W$ is the total radian bandwidth. This is derived by (1) removing the carrier $[\omega_0 + (N-1)(\Delta\omega/2)]$; (2) changing the variable $n\Delta\omega$ to $n\omega$; and (3) changing the amplitudes A_n into the distribution function $a_n(n\omega)$,

$$a_n(n\omega) = \frac{A_n(2\pi)}{\Delta(n\omega)}$$

Then as $\Delta(n\omega)$ approaches zero, Eq. (13-17) becomes

$$e_0(t) = \frac{1}{2\pi} \int_{-W}^{+W} a(\omega) \exp(j\omega t) \, d\omega \qquad (13\text{-}18)$$

This Fourier transform relationship describes the output $e_0(t)$ of a pulse-compression receiver with an amplitude response $a(\omega)$. As an example, it will be used to find the output waveform for cosine weighting.

The weighting function for cosine weighting with the carrier removed is

$$a(\omega) = \cos \frac{\pi\omega}{2W}$$

where $a(\omega)$ equals unity at the center of the spectrum. Using Eq. (13-18), the output waveform can be expressed by

$$e_0(t) = \frac{1}{2\pi} \int_{-W}^{W} \cos \frac{\pi\omega}{2W} \exp(j\omega t)\,d\omega \tag{13-19}$$

Performing the integration, the envelope of the output waveform is

$$e_0(t) = \frac{2W}{\pi^2 - 4t^2W^2} \cos Wt \tag{13-20}$$

The maxima of Eq. (13-20) occur approximately midway between the nulls. The nulls occur when

$$t = \frac{k\pi}{2W}, \quad k = 3, 5, 7, \dots$$

The peak of the first sidelobe occurs when $k = 4$ and $t = 2\pi/W$. The ratio of mainlobe to first sidelobe level from Eq. (13-20) is then

$$\frac{e_0(t = 0)}{e_0(t = 2\pi/W)} = 15 = 23.5\,\text{dB} \tag{13-21}$$

The sidelobe structure has been verified experimentally with a multi-frequency synthesizer called a DELRA (Delay Line Range Analyzer) [260, 402], which is in turn a variation of a Coherent Memory Filter

[40] built for this author by Federal Scientific Corporation. As used in the following figures, it simulates a 50-channel stepped-frequency generator and receiver. Figure 13-15*A* shows amplitude versus frequency in 50 discrete steps for cosine weighting. In Fig. 13-15*B*, the

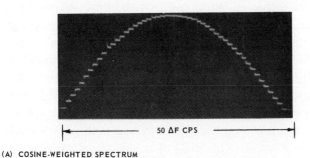

50 ΔF CPS

(A) COSINE-WEIGHTED SPECTRUM

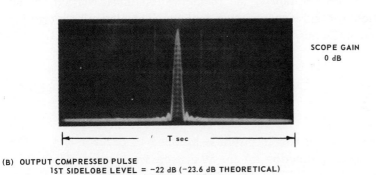

SCOPE GAIN
0 dB

T sec

(B) OUTPUT COMPRESSED PULSE
    1ST SIDELOBE LEVEL = −22 dB (−23.6 dB THEORETICAL)
    MAIN LOBE WIDENING = 56%

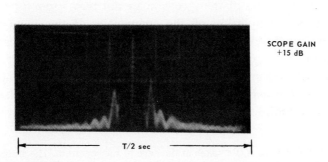

SCOPE GAIN
+15 dB

T/2 sec

(C) EXPANDED OUTPUT

figure 13-15 *Sidelobe reduction with cosine amplitude weighting.*

compressed-pulse amplitude is shown with about $-$ 22-dB first sidelobes as compared to the theoretical value of $-$ 23.6. Figure 13-15C shows the main-lobe region expanded in both dimensions to show the rapid decay of the sidelobes.

Several other amplitude tapers have been used to obtain lower range sidelobes than can be achieved with cosine weighting. As another example, the sidelobe structure for Dolph-Chebyshev [92] tapering for 30-dB sidelobes was obtained on the synthesizer and is shown as Fig. 13-16b and c. DELRA is an analog device with a certain limit of accuracy. Thus, ideally, the sidelobes should be equal in amplitude. As can be seen, this expected result is only approximated. The sidelobe structure for uniform weighting is illustrated as a reference at the top of the figure. While this taper is not practical for antennas or dispersive lines due to the large values required at the ends of the *spectrum*, there is no problem in approaching the theoretical limits for discrete frequency systems.

One other class of tapers, the Hamming functions, warrant special consideration for pulse-compression systems. The Hamming weighting function in continuous form is described in general terms by [373]

$$G(\omega) = a + (1 - a) \cos\left(\frac{\pi\omega}{W}\right) \quad 0 < a < 1 \qquad (13\text{-}22)$$

where the center[1] of the spectrum is assumed to occur at $\omega = 0$. The resultant weighting function for the 50 discrete frequencies and $a = 0.54$ is shown in Fig. 13-17A. The distinguishing feature between this weighting function and the cosine function is that the amplitudes of the frequencies at the edges of the spectrum are not zero. From Eq. (13-22) for $\omega = \pm W$, the amplitude of the frequencies at the edges of the spectrum are equal to 0.08 (compared to a value of one at the center of the spectrum).

The output waveform shown in Fig. 13-17 may be described approximately by the Fourier integral used to find the envelope for Hamming weighting. The result for any value of a is

---

[1] In some references $\omega = 0$ is taken at the edge of the spectrum. The amplitude at the edge of the spectrum is $a - (1 - a) = 2a - 1$. This is often referred to as the *pedestal*.

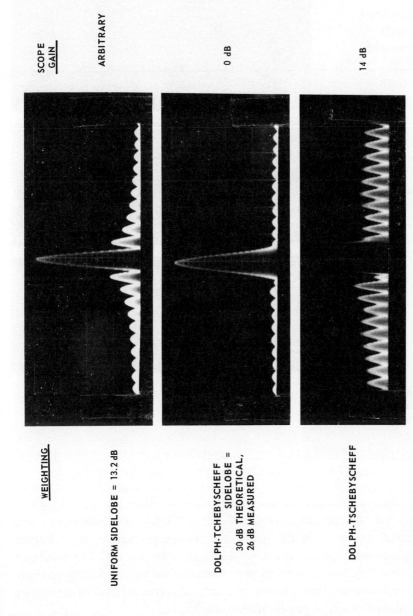

ARBITRARY

0 dB

14 dB

WEIGHTING

UNIFORM SIDELOBE = 13.2 dB

DOLPH-TCHEBYSCHEFF
SIDELOBE =
30 dB THEORETICAL,
26 dB MEASURED

DOLPH-TSCHEBYSCHEFF

figure 13-16 *Dolph-Chebyshev weighting sidelobe suppression (40 frequencies).*

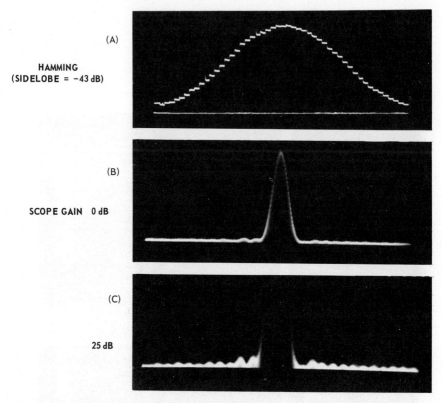

(A)

HAMMING
(SIDELOBE = −43 dB)

(B)

SCOPE GAIN 0 dB

(C)

25 dB

figure 13-17 *Hamming function weighting sidelobe suppression (50 frequencies).
(After [403].)*

$$e_0(t) = \frac{a}{2} \frac{\sin Wt}{Wt} + \frac{1-a}{2W(1-t^2)} \cos Wt \qquad (13\text{-}23)$$

The theoretical maximum sidelobe level for $a = 0.54$ is -42.8 dB which is beyond the dynamic range of the DELRA. The sidelobes were measured to be ~ 40 dB below the main-lobe peak. The main-lobe widening over that for uniform weighting is 46 percent. The Hamming weighting function, due to its very low sidelobe level and comparatively low main-lobe widening, serves as a very effective method of amplitude weighting in practical pulse-compression receivers.

Some of the properties of various weighting functions are summarized in Table 13-2 [373, 402, 71]. The Hamming class of weighting functions

TABLE 13-2 Sidelobe Suppression of Frequency-coded Waveforms

Weighting function	a	Time mismatch, %	Main-lobe broadening Theor.	Main-lobe broadening Exp.**	Processing loss, dB	Highest sidelobe, –dB	Sidelobe decay function*
Unweighted		0	1.00	1.00	0	13.2	$1/t$
Cosine		0	1.56		1.0	23.6	$1/t^2$
Hamming	0.54	⎰ 0	1.47	1.46	1.34	42.8	$1/t$
		⎱ 5	1.52		1.42	36.6	$1/t$
		10	1.55		1.47	32.2	$1/t^3$
(Cosine-squared)	0.50	0	1.59	1.70	1.76	31.4	$1/t$
Hamming	0.52	0	1.55		1.54	36.1	$1/t$
Hamming	0.56	0	1.45		1.16	37.7	$1/t$
Hamming	0.58	0	1.41		1.01	34.2	$1/t$
Dolph-Chebyshev		0	1.20		0.80	30.0	1
Dolph-Chebyshev		0	1.35	1.40	1.20	40.0	1
Cosine-cubed		0	1.88		2.38	39.1	$1/t^4$
Taylor $\bar{n} = 5$		0	1.34	1.44	1.1	34.0	$1/t$
$\bar{n} = 6$		0	1.41	1.7	1.23	40.0	$1/t$
$\bar{n} = 6$		10	—		—	28.0	$1/t$
$\bar{n} = 8$		0	~1.50		1.46	47.5	$1/t$
$\bar{n} = 8$		10	—		1.66	34.5	$1/t$

*Decay function far from compressed-pulse location.

**Compression ratio = 50:1 (experimental).

is characterized by the parameter a. The main-lobe broadening and processing loss are referenced to uniform weighting. The lobe decay function describes the rate of decrease of the sidelobes beyond the peak sidelobe. The theoretical and experimental values. apply for $N \geq 40$ frequencies, and some additional degradation is to be expected for smaller values of N [305].

The *time mismatch* column is not directly applicable to short-duration stepped-frequency waveforms, but is pertinent to linear FM when there is a large doppler shift. It is also significant in clutter computations for pulse-doppler waveforms. The value in the rows for 5 and 10 percent mismatch show the degradation when the weighting function is displaced from the received signal by that amount. The most significant effect is the increase in the range sidelobes. The column on processing loss can also be used for the loss in signal-to-clutter ratio for the case where clutter has an extent greater than twice the transmit pulselength ([71, pp. 341-343]). The losses for tapered FM are about equal to the losses for the better binary phase codes shown in Chap. 12 on phase coding.

In order to maintain low-range sidelobes, careful consideration must be given to the detailed band-pass characteristics of each channel [273]. In practice it has been possible to obtain -40-dB sidelobes by very careful adjustment of receivers employing 35-50 frequencies. In a practical operating system, -35 dB is a limit with most analog configurations.

13.10 COHERENT MULTIPLE-FREQUENCY GENERATION

This section contains a brief description of techniques for generating a coherent set or *comb* of frequencies. In addition to the requirements for stability and spectral purity at each frequency, there is imposed a requirement for coherence between the frequencies. In simple terms, coherence will occur if at a given time the phase of any of the frequencies can be determined relative to the phase of one of the other frequencies, plus a knowledge of the frequency separation.

The form of multifrequency generator that is most easily related to pulse compression is obtained when the comb of frequencies is generated from a train of *impulse-type* functions. As an example, assume that it is desired to generate a comb of N frequencies around a microwave carrier

f_0 with each frequency separated by Δf. As was stated earlier, these frequencies can be used to achieve a pulse-compressed waveform of width $1/N\Delta f$. For convenience, let the center of the microwave spectrum f_0 be a harmonic of the frequency separation Δf. Referring to the block diagram of Fig. 13-18, the stable oscillator output is at frequency Δf and is divided into two paths. At the left, it is multiplied to f_0 by means of varactor or diode frequency multipliers. If at least 30 dB reduction of the temporal sidelobes is part of the system requirement, the signal should be well filtered at this point. The stable oscillator output is also used to trigger an *impulse* generator which may be a blocking oscillator, snap diode, or other monostable pulse circuit. The requirement at this point is that the impulse generator output be a pulse which is shorter in duration than $1/(N\Delta f)$ and is repeated at a rate equal to Δf.

The microwave CW signal from the multiplier f_0 is fed to a high-speed microwave switch where it is switched *on* for the duration of the impulse. The output of the switch is then a train of short RF pulses whose spectrum can easily be shown to be a comb of frequencies centered around f_0 and spaced by Δf. The envelope of the set of spectral lines is the transform of one of the impulses.

Temporarily neglecting the dispersive line, the comb of frequencies is amplified and then inserted into a filter bank with N discrete narrow-band filters. The bandwidth of these filters must be $\ll \Delta f$ and have good rejection of the adjacent frequencies.[1] The attenuators can be used to compensate for the unequal amplitudes of the spectral lines at the output of the microwave switch or to taper the spectral lines for transmission. The multipole microwave switch is energized by the transmit frequency selection logic and may be programmed linearly with frequency or with any other pattern. Additional switching circuitry may be used to derive a convenient set of local oscillators for the receiver.

One of the limitations of this technique is that the output of the frequency multiplier is not used very efficiently since the switch is *on* only a small fraction $1/N$ of the time. In addition, the output of a typical linear power amplifier would be used only a small portion of the time. Since the maximum output of the power amplifier must be divided into at least N outputs, the power out of any filter is less than $1/N^2$ times the saturation output of the amplifier. One technique to obtain more efficient use of the available CW energy at frequency f_0 is to insert a

---

[1] The Q of the filters must be much greater than $f_0/\Delta f$.

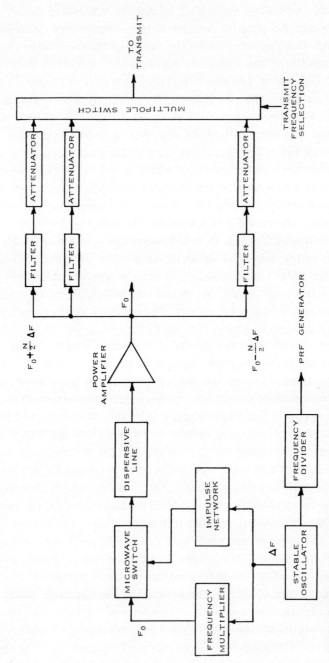

figure 13-18 *Impulse-derived multifrequency generator.*

dispersive device ahead of the power amplifier. This device will distort the impulse shape, reducing its peak amplitude, and thereby spreading its energy in time. The maximum delay in the dispersive line must be less than $1/\Delta f$. One convenient method of obtaining the dispersion is to use a length of waveguide which has a cut-off frequency just below $f_0 - (N/2)\Delta f$. In one experimental model with 11 frequencies, this technique of increasing the waveform duty cycle increased the power out of each of the filters by 8 dB since the power amplifier was used almost continuously.

Line spectra with a spread of over 600 MHz have been obtained with the above impulse generator techniques. A variation results from replacing the frequency multiplier and switch with a stable power oscillator which can be frequency modulated, such as a backward-wave oscillator. Alternately, f_0 can be inserted into an amplifier with an FM capability such as a helix modulated traveling-wave tube. A sine wave at frequency Δf is used to frequency modulate the oscillator or amplifier. The resulting amplitude distribution of the spectral lines is determined by the Bessel function expansion for the chosen frequency deviations. The difficulty with this technique is that the relative amplitudes of the spectral lines are very nonuniform, and for large N, some of the filters will have little power output.

As was implied in the preceding description, the power output of each filter may not be adequate to supply both transmitter and local oscillator signals. In addition, it may be difficult to filter the adjacent spectral lines if $\Delta f/f$ is less than 0.002. A solution to both of these problems is to phase-lock a high-level oscillator to each of the spectral lines. Since it is relatively easy to phase-lock with small signals and oscillator circuits having loop bandwidths that are much less than typical values of Δf, the filter bank and attenuators may be eliminated. The oscillator may be composed of transistors, varactor-tuned ceramic triode oscillators, klystrons or some of the newer forms of diode oscillators. The oscillators act both as narrow-band filters and power amplifiers.

The comb of frequencies can also be obtained with so-called frequency synthesis methods which use extensive combinations of mixers and multipliers with or without phase-locked oscillators. It is also possible to construct coherent frequency generators with minor modifications to commercial instrumentation equipment such as the Hewlett-Packard Company's family of frequency synthesizers. A unit of this type was built by Stanford University [114] to obtain a 100-kHz frequency ramp with a 1-sec duration. The Hewlett-Packard Model 5100A-5110A was switched in increments of 1 Hz, every 10 μsec for a 1-sec period. The time-bandwidth product is then 100,000. Since all of the frequencies in the synthesizer are derived from the same stable oscillator, the only additional requirement is to switch between frequencies when these frequencies have identical phase. With suitable modifications of the switching circuitry in the frequency unit so as to obtain switching within 0.1 cycle, the spurious signal level was held to -55 dB.

13.11 PULSE-COMPRESSION DECODERS AND LIMITER EFFECTS

In this section a matched-filter processor for a short random stepped-frequency transmission will be described. If the doppler shift of the target echo is much less than $1/N\tau$, a random ordering of frequencies can be used and only a single doppler channel need be implemented in the receiver. With linear stepping of the frequencies on the other hand, the tolerance to doppler shift is much higher (see Fig. 8-8). A matched-filter receiver for large doppler shifts will be described in the next section.

The tapped delay-line pulse-compression system of Fig. 13-19 is an expansion of Fig. 13-10. The transmit waveform is a contiguous sequence of eight constant amplitude, 0.1-μsec pulses with a frequency spacing Δf of 10 MHz. In a particular system of interest, the comb of frequencies was generated by modulating a carrier at several gigahertz with an impulse-type function. This is represented by the signal generator block. These frequencies are sampled with an eight-pole microwave diode switch, amplified, and then transmitted. The received target echoes are amplified in a low-noise traveling-wave tube and then mixed to a convenient intermediate frequency. Since the spectral width is 80 MHz, an intermediate frequency IF centered at 567 MHz was chosen to allow about 14 carrier cycles to occur between the nulls of the compressed-pulse envelope. With a lower IF, detection of the 12.5-nsec pulses would become more difficult. With a higher IF, the matched-filter delay-line losses would become excessive.

The signals are then injected into a tapped delay line consisting of eight sections of 0.140 in. OD semirigid coaxial cable. Each section is 67 ft long (1.5 nsec/ft delay), is quite stable, has only a 5-dB insertion loss, and coils onto a spool occupying the volume of about a 3-in. cube. The delay line *taps* are *Stripline*-type directional couplers with the coupling

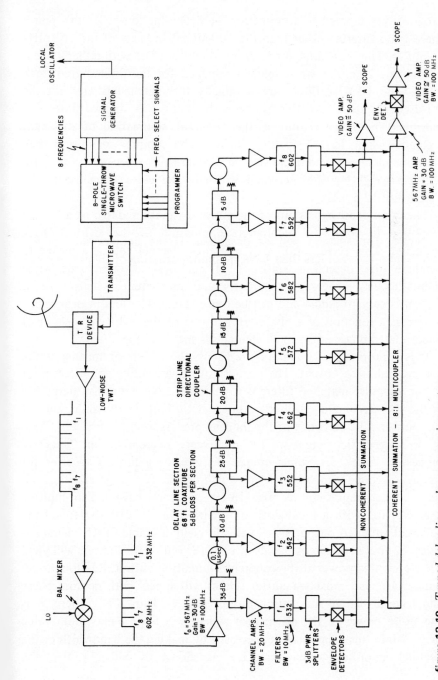

figure 13-19 *Tapped delay-line pulse-compression system.*

535

progressively tapered to account for the delay-line losses. At this point, there is no filtering other than the 100-MHz band-pass of the 567-MHz amplifier. The individual channels are then amplified with either a linear or limiting amplifier whose gain is adjusted for the desired frequency taper. The 10-MHz filters are nominally the matched filters[1] for the 0.1-μsec segments. At this point each channel is divided into two outputs. The main output is the coherent summation of the eight segments which have been time aligned. The compressed pulse is then envelope detected and amplified in a wide-band video detector. With uniform weighting the measured width of the pulse was close to the theoretical 12.5 nsec.

The other eight signals out of the power dividers are detected and then amplified to form the incoherent summation of the eight 0.1-μsec segments. This serves both as a reference channel and as a means to compare the detection statistics of complex target returns when simultaneously processed coherently and incoherently. This type of system is easy to adjust and can be quite stable. The loss in peak signal power as a function of a phase error in one of the eight channels or of uncorrelated phase errors in all channels is shown in Fig. 13-20.

It is appropriate at this point to discuss the effects of limiting on the compressed-pulse output. The purpose of using limiting amplifiers prior to the subpulse matched filters is to maintain a constant false alarm rate at the detected matched-filter output and to limit the dynamic range of the output. The statistics of the limiting process were discussed in Chap. 4 on automatic detection by nonlinear, sequential, and adaptive processes. There are some unique features of a limited receiver for frequency-coded signals which are worth noting. If N^2 is greater than 64, the channel amplifiers of Fig. 13-19 can each have a hard-limiting characteristic with a bandwidth of slightly over $1/\tau$ (10 MHz); and the coherent summed output signal will still yield desirable CFAR characteristics. Saturated coherent signals of power K out of the limiters will yield KN^2 times this power at the coherent summation output. Random noise, rain, chaff, and most sea-clutter echoes will appear with random phase and have a relative summation output of KN. Thus S/N', where N' is the noise power in the absence of signal, will have a maximum value of N. While this value may be increased by making the bandwidth of the channel amplifiers wider than $1/\tau$, as shown by Silber [349], the

---

[1] In this system the conjugate phase shift adjustment is in the signal generator.

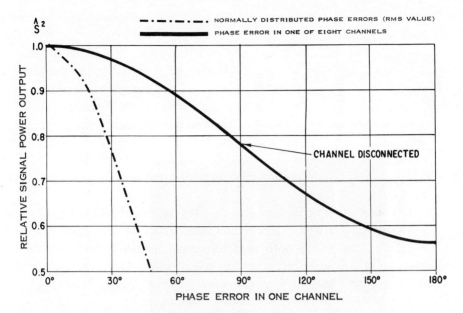

figure 13-20 *Relative signal power loss vs. phase errors.*

immunity of the system to strong CW interference will then decrease. Strong CW signals will *capture* several of the overlapping channel limiters and suppress significant portions of the target echo.

When the factor of strong CW interference can be ignored, there are advantages in using a prelimiter bandwidth greater than $1/\tau$, as illustrated in Fig. 13-21. A 30-MHz IF signal out of a channel limiter is shown along with the eight-frequency pulse-compressed output. The prelimiter bandwidth is 10 MHz for the 0.1-μsec subpulse. The upper set of pictures was taken with the signals in each channel just at the limit level. The compressed-pulse output was square-law detected and typical of the system at that time. In the center photos, the signal level was increased by 13 dB. The output of the limiter has a greater duration $(\tau > 1/\Delta f)$ due to the rise and fall portions of the subpulses receiving additional amplification while the center portion of each subphase is held constant by the limiter. The detected compressed pulse shows that ambiguous outputs at ± 0.1 μsec from the peak are visible and could be confused with other targets.[1] With 26-dB limiting, the limiter output is doubled in

[1]It should be recalled that the coherent summation of the ungated comb frequencies has periodic peaks every 0.1 μsec.

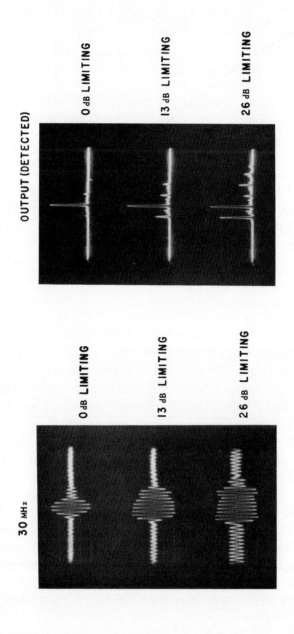

30 MHz

0 dB LIMITING

13 dB LIMITING

26 dB LIMITING

OUTPUT (DETECTED)

0 dB LIMITING

13 dB LIMITING

26 dB LIMITING

TRACE LENGTH = 1 μ sec

figure 13-21 *Effect of narrow-band limiting on coherent output. Transmit pulse = 0.1 μsec; limiter bandwidth = filter bandwidth.*

extent and several peaks are apparent at the detector. Tapering to reduce these range sidelobes would be completely ineffective. Some of the extraneous signal is also due to leakage through the transmit switches.

A similar system with $\tau = 0.4$ μsec and a limiter bandwidth of 10 MHz does not exhibit this problem. As can be seen in Fig. 13-22, the compressed pulse is not altered and there is negligible increase in the range sidelobes for 55 dB of limiting. Since the prelimiter bandwidth is $4/\tau$,[1] there was a negligible lengthening of the pulses. The dynamic range of this particular system was in excess of 65 dB.

A somewhat surprising result was also obtained in the incoherent summation. With the prelimiter bandwidth of only four times the channel filter bandwidth, the dynamic range of the incoherent summation was marginally acceptable for visual target detection with the eight-channel system, and was more than adequate for automatic detection with a 32-channel system.

13.12 HYBRID MULTIFREQUENCY SYSTEMS—PULSE TRAINS

Most of the processors described heretofore have been more or less in their basic form. To meet specific operational requirements it is often necessary to combine several techniques involving MTI, the use of pulse trains, and complex modulations for coding purposes into what will be called hybrid techniques. Combinations of discrete and linear FM come under this heading. It will also be convenient to discuss the digital processing of FM signals within this context. The first system to be described is a signal processor for the multifrequency pulse-train waveform which was shown in Fig. 13-6C.

In many radar defense systems the requirement for a short system reaction time is extremely critical. One of the major problem areas is the transfer of the range and radial velocities of targets detected by the *search* system to the tracking radar or, alternately, to the track portion of a multifunction radar. If the tracking system is of the pulse-doppler type and a wide range of target velocities is anticipated, it is desirable that the search system determine at least the approximate radial velocity of the target in a single look. The preceding discussions of frequency-coded waveforms and processors did not discuss problems attendant to

---

[1] The limiter amplifier itself has a bandwidth much in excess of this.

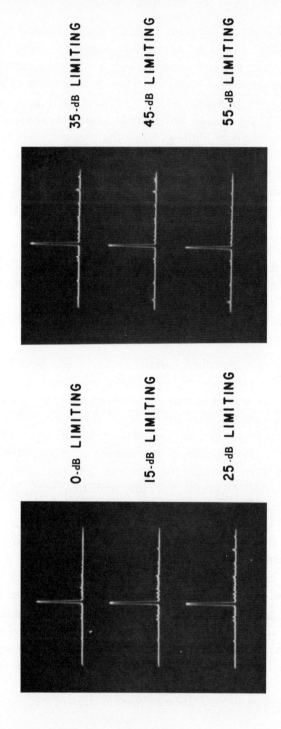

35-dB LIMITING

45-dB LIMITING

55-dB LIMITING

0-dB LIMITING

15-dB LIMITING

25-dB LIMITING

TRACE LENGTH = 1.0 μ sec

figure 13-22 *Effect of hard limiting on detected coherent output; transmit pulse = 0.4 μsec; limiter bandwidth = 4 × postlimiter bandwidth.*

resolving target range and velocity simultaneously. The range-doppler ambiguity for linear continuous or stepped FM was accepted in order to maintain receiver simplicity, or the waveforms were short enough so that uncertainty in target velocity caused only a small error in range determination. (See Sec. 8.6.)

In the following description, the frequency segments of a coded pulse are separated into a pulse train with a total duration of T such that the velocity resolution $\approx 1/T$ is adequate to meet the velocity designation specification. In addition, the transmission bandwidth $N\Delta f$ is made large enough for the desired range resolution $\approx 1/N\Delta f$ as well as for enhanced detection of targets in the presence of distributed clutter (rain, chaff, etc.). The following discussion of the pulse-train waveform and processor is based upon the original ideas of Dr. John B. Garrison, as described in his patent for a Radar Search System.[1]

The theory of the operation can best be explained in connection with a hypothetical version of the processor shown in Fig. 13-23. The figure shows a wide-band radar receiver connected to a length of nondispersive waveguide with a multitude of radiating elements whose spacings are properly chosen. Consider a radar transmitting a coherent train of short pulses whose times of transmission are matched to the spacing of the radiating elements. If this waveform is reflected from a stationary point target and the echo received and fed through an amplifier into the delay line, there will be an instant of time when all of the pulses are aligned at the radiating elements. If the pulses are samples of a coherent oscillator, a *beam* is formed perpendicular to the delay-line antenna and exists for the duration of the pulse. With a stationary target the phase front across this antenna will be parallel to the antenna elements regardless of the transmit frequency if the distance to each feed is equal to an integral number of wavelengths at the transmit frequency. A microwave pickup horn placed at the location shown for *zero doppler* will receive the coherent summation of the individual pulses. In the case when multiple carrier frequencies are to be used and these frequencies are integral multiples of a basic frequency f_1, the spacing of the feeds must be an integral multiple of the wavelength corresponding to f_1. Thus, a single set of feeds can be used for a number of frequencies.

If the target is moving towards the radar at a constant velocity, the reflected signals will have undergone a linear change in phase versus

[1] Patent no. 3,309,700, March 14, 1967.

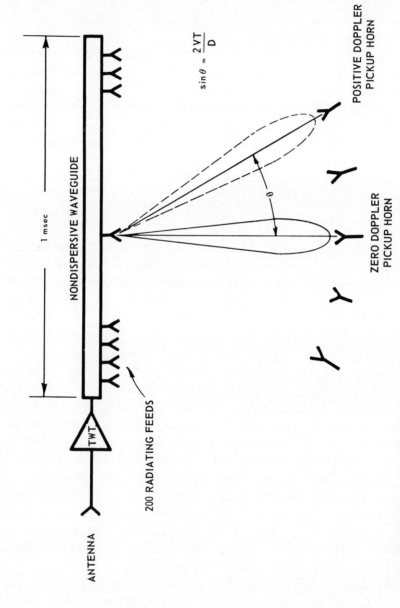

ANTENNA

1 msec

NONDISPERSIVE WAVEGUIDE

200 RADIATING FEEDS

$\sin\theta = \frac{2VT}{D}$

POSITIVE DOPPLER PICKUP HORN

ZERO DOPPLER PICKUP HORN

figure 13-23 *Hypothetical pulse-train processor.*

time. Upon reception, the phase front across the delay-line antenna will follow the dashed line of the figure, and an *antenna beam* will be formed at the *positive doppler* pickup horn. The sine of the angle θ with respect to the perpendicular to the *antenna* is proportional to the radial velocity of the target. If a large number of pickup horns are placed as shown in the figure, they will each respond to a given target velocity. Negative dopplers will appear on the left in the figure. The relationship for the beam steering angle is simply

$$\sin\theta = \frac{2vT}{D} \tag{13-24}$$

where v = target radial velocity

D = length of the delay line

The doppler resolution is roughly equal to $1/T$ in hertz.

What has been described up to this point is a multiple-frequency variation of a single-frequency pulse-doppler receiver. With a single frequency and a constant PRF, there would be range ambiguities every T/N time units as expected with single-frequency pulse trains. Obviously, suppression of the major ambiguities can be obtained by staggering the pulses for both single- and multiple-frequency cases and the corresponding taps on the delay line.

Transmitting multiple frequencies gives an interesting result in that pulse compression is obtained with reflections from point targets. As has been explained in conjunction with Fig. 13-23 and Eq. (13-24), the beam-pointing angle can be made independent of the carrier frequency. If the proper phase relationships are maintained, a compressed pulse of duration τ/N, as described in previous sections, will appear. The compressed pulse will occur at a time corresponding to the target range (plus a constant) and appear at the doppler pickup horn corresponding to the target radial velocity. If the order of the frequencies is *scrambled* and the PRF is made much greater than the expected target doppler, there will be no major ambiguities in the range-doppler plane. However, it can be shown that if the frequencies are linearly increasing or decreasing, the beam will *sweep* across the pickup horns during the pulse.

The single waveguide delay line at microwave frequencies is not practical for at least two reasons. First of all, for useful doppler resolution, the waveguide would be many miles long. Secondly, the placement of the pickup horns would result in an extremely cumbersome system. The block diagram in Fig. 13-24 was implemented to scale the system to a reasonable size. The time delay was obtained by a parallel system of ultrasonic delay lines at IF. The doppler and range determination remains at the original carrier frequency. To accomplish this, the received signal is amplified and divided into N separate frequency channels by a bank of band-pass filters. Each frequency segment is down mixed to a convenient IF frequency and then delayed by $T - t_n$ in the ultrasonic delay lines, where t_n is the time of transmission of the nth frequency. After the delays the signal in each channel is up mixed with the local oscillator back to the original carrier frequency. If the modulator to convert back to the carrier is not of the single-sideband type, an image signal at $f_n + 2f_{IF}$ will also appear and must be removed by a narrow-band filter.

The delay lines need be accurate in length only to about one-tenth of a pulse width.[1] The phase shifter in each channel is used as a vernier to adjust the delay to an integral number of carrier wavelengths for that channel. The pulses at the output of each phase shifter are then a delayed replica of each target echo.

One technique of resolving target velocity is to connect the radiating horns on the right of Fig. 13-24 to a constrained microwave *velocity lens* which performs the function of the waveguide delay-line antenna in Fig. 13-23. Since the interpulse delays have been removed and all echo envelopes appear simultaneously at the radiating horns, these horns need only be spaced $\lambda/2$ apart. Thus, one dimension of the constrained lens is only $N\lambda/2$, where λ is the carrier wavelength. The second dimension of the lens is dependent on the distance that the target moves during the dwell time (a Mach-4 target moves 2 ft in 0.5 msec). A sketch of a typical constrained *geodesic* lens is shown in Fig. 13-25 for 32 frequencies. Aluminum cover plates constrain the energy as in a waveguide structure, and the unequal curvature of the location of the elements compensates for spherical aberrations. The input horns are at the top, and the doppler outputs are at the bottom of the photo. Similar devices have been constructed with printed circuit microwave matrices or with a *Butler* matrix of *hybrids*.

Since individual pulse filtering is performed in each channel and all point-target echoes add voltagewise at the corresponding velocity

---

[1]The delay stability of the delay lines must be held to a few degrees of phase at the delay-line frequency, not at the original carrier frequency.

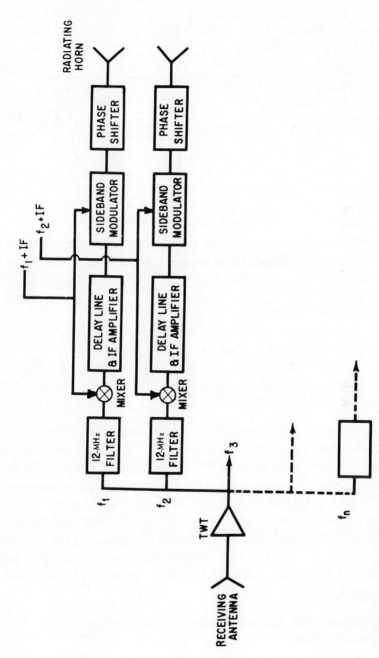

figure 13-24 *Processor for multifrequency pulse train.*

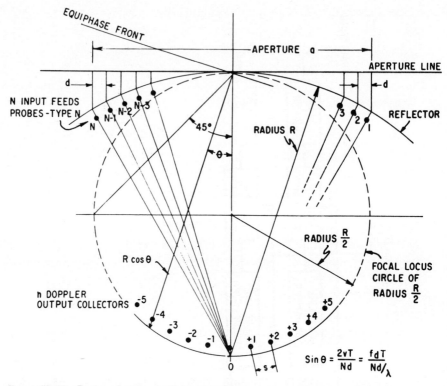

figure 13-25 *Design sketch, air dielectric matrix.*

(doppler) output, while noise or distributed clutter in each channel adds powerwise, the processor can be shown to be a matched filter. Hard limiters can be placed in each IF channel to obtain the CFAR action discussed earlier. The term *doppler outputs* was used rather loosely as the velocity lens is actually a true time delay device and is not a doppler approximation. In this sense, it does not have the bandwidth limitations of a *doppler-corrected* receiver as discussed by Remley [308] and Cook and Bernfeld [77, pp. 287-293]. Alternately, an *acceleration lens* can be constructed where the output ports only respond to constant acceleration targets.

13.13 AMBIGUITY DIAGRAMS FOR STEPPED-FREQUENCY WAVEFORMS

Ambiguity diagrams of frequency-stepped waveforms can take many forms depending upon the ordering of the frequencies and the time spacing between the segments. The major contours of the ambiguity diagram for the waveforms of Fig. 13-6 are shown in Fig. 13-26. The frequency steps in all three cases are separated by $1/\tau$. The diagram at the top is for contiguous stepped-frequency approximation to a linear FM pulse. The primary difference between this contour diagram and the chirp diagram of Fig. 8-3 is the result of the poor response of stepped FM to signals with a doppler shift of $1/2\tau$. At this point the stepped compressed pulse is considerably reduced in amplitude from the value at the origin. This effect is not important for short pulses where $1/N\tau \gg f_d$.

The effect of scrambling the order of the frequencies within the pulse envelope is shown in the center diagram. The energy that appeared along the diagonal ridge at $1/\tau$, $2/\tau$, etc. is broken up and forms a *pedestal* of dimensions $1/\tau$ by τ. There is also a lower *spillover pedestal* in the range dimension extending from $|\tau|$ to $|N\tau|$. This results from the far-out spectral components of each segment appearing in the *wrong* frequency channels. The lower level pedestal will be called *secondary sidelobes* if it occurs mostly along the range axis or *secondary ambiguities* if it extends throughout the ambiguity plane [180].

The contour diagram of a constant PRF linear stepped FM pulse train appears in Fig. 13-26C. Both the properties of linear FM and those of a pulse train are combined. As with all constant interpulse period pulse trains, there are clear regions parallel to the doppler axis. The major ambiguities appear at multiples of the pulse period T along a diagonal ridge. This is characteristic of linear FM signals as opposed to the nonskewed matrixlike appearance of single-frequency pulse trains.

The variations possible with frequency-coded pulse trains are almost infinite in that the phase, frequency, and time of transmission of each pulse can be varied. The waveform can be tapered in time and/or frequency to reduce the sidelobes. Howard [180] has studied the resultant variations in the ambiguity diagram of numerous variations of interpulse timing. Several other specific examples are available in the literature [316, 71, 211].

The central region of the range-doppler response has been determined experimentally for the waveform and processor described in Sec. 13.12. The ordering of the 32 frequencies and the deviation in pulse spacing from a fixed PRF were obtained by means of pseudo-random codes. The deviation from the nominal interpulse period was such that there were never more than four of the 32 pulse envelopes overlapping at the *distal* ambiguities. The pulselength was one-fiftieth the normal interpulse

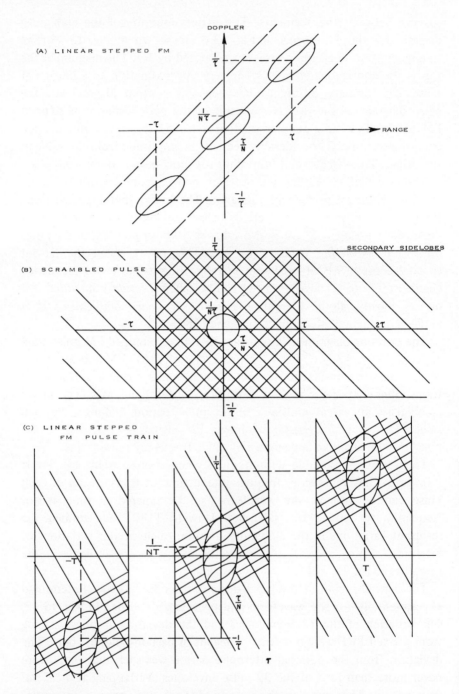

figure 13-26 *Ambiguity contours for discrete frequency waveforms.*

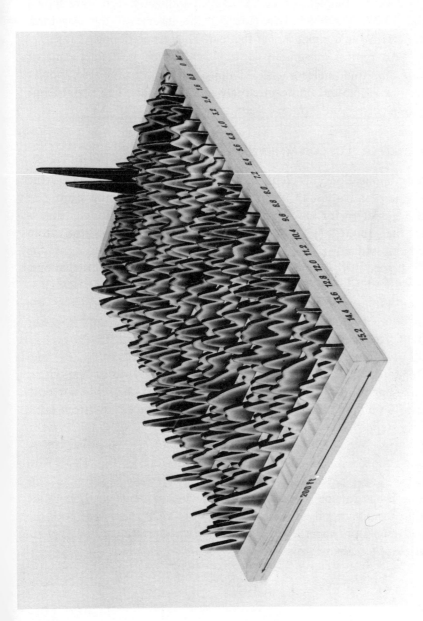

figure 13-27 *Amplitude of ambiguity region ($\pm \tau/2$) for 32-frequency stepped pulses. Twenty doppler cuts spaced by $1/2T$ are shown.*

549

period. Twenty doppler *cuts* of the central ambiguity region are shown in Fig. 13-27. There are two cuts per the inverse of the waveform duration with the doppler axis at the rear. The second doppler cut shows the -4-dB central response peak at $1/2T$, which is characteristic of constant amplitude pulse trains. The range sidelobes for the zero doppler cut are not visible but conform closely to $\sin Nx/\sin x$. The 200-ft range dimension corresponds to the 0.4-μsec pulse.

13.14 STEPPED PLUS LINEAR FM WAVEFORMS

In most applications of contiguous stepped FM to short pulses, the discrete approximation to linear FM does not result in any significant signal degradation for low doppler shifts. However, if the doppler shift is greater than about $1/(4N\tau)$, there will be a range sidelobe at a time τ from the compressed pulse. This can be inferred from the periodicity of the diagonal ridge in the ambiguity contour Fig. 13-26A. This problem may be alleviated by frequency modulating the individual segments as well as stepping the carrier frequency. The transmit waveform becomes the summation of a frequency-stepped signal and a sawtooth FM, the resultant signal being an excellent approximation to linear FM.

This technique has been studied both analytically [278] and experimentally [188], and results in a pulse-compression system with a high tolerance to large doppler shifts. In other terms, the *fill-in* sawtooth reduces the amplitude of the paired-echo effect that results from using only the stepped waveform.

Very large time-bandwidth products ($>$ 1,000) can be processed by replacing the band-pass filters in the block diagram of Fig. 13-19 with dispersive lines matched to the FM signal in each segment. Range sidelobe levels can be reduced by the previously mentioned amplitude tapering techniques. If -40-dB sidelobes are to be achieved, both the stepped oscillators and the *fill-in* sawtooth waveform must have a phase deviation of less than $1.15°$.

14 HYBRID PROCESSORS, CORRELATORS, AND INCOHERENT TECHNIQUES

The six preceding chapters discussed signal processing techniques in their basic forms. In many instances the performance that can be obtained does not yield adequate range or velocity resolution, clutter rejection, etc. In addition, the analog configurations that have been used in the past may not have sufficient stability or reliability to meet the system availability requirements, and it may be desirable to take advantage of the recent advances in digital micrologic circuitry. As a result of both factors there is a trend toward *hybrid* processors which are defined as combinations of two or more techniques. These may be composed of both analog and digital processor configurations. This chapter will discuss a few of the numerous possibilities of hybrid processors and also introduce some additional digital signal processor implementations. An alternate title for this chapter could be Trends in Signal Processing.

14.1 POSTDETECTION INTEGRATION

Postdetection integration of radar echoes is a form of signal processing that is common to many types of radars. The simplest configurations would include the phosphor integration of detected signals on a PPI display and the *smoothing* circuits in tracking radars. It was shown in Chap. 3 that with postdetection integration, the detectability of repetitive target echoes in a noiselike background can be substantially improved, especially for fluctuating targets. The Marcum and Swerling analyses [244] quantified the reduction in the required signal-to-noise ratio[1] for a square-law detector followed by a linear integrator for noise that is spatially and temporally uncorrelated. An estimate of the (S/N) reduction for *colored* or partially correlated noise was also given in Chap. 3 where it was shown that significant improvement was obtainable even in colored noise or clutter *if* the degree of correlation was known.

The descriptions of target and clutter echoes in Chaps. 5, 6, and 7 included the conditions under which these echoes could be considered correlated or uncorrelated. In most radar systems there is insufficient observation time in any single beam position to achieve temporal independence. Thus the emphasis placed on frequency decorrelation of clutter echoes is justified in that frequency agility is a means of ensuring that the echoes from distributed clutter are uncorrelated.

Decorrelation of the echoes from a complex target is desirable in a clutter environment as well as in a noise-only environment. An average of multifrequency target echoes will approximate the mean RCS (see Sec. 5.9). This averaging of independent samples essentially reduces the probability that there is a null in the RCS at the aspect of interest with a single carrier frequency transmission. While postdetection integration of a multifrequency transmission does not reduce the total clutter-echo power, the variance of the integrated sum is considerably reduced. Since detectability is proportional to the ratio of the target echo to the variance of the total noise power rather than the mean noise power, the improvements in detectability can approximate those illustrated in Chap. 3. The decorrelation of the target echo suggests the use of the *rapid fluctuating* Swerling case two or four target models rather than the more pessimistic cases one or three.

---

[1] The notation (S/N) is used for the required predetection power ratio for a given P_D and P_N.

Incoherent Pulse Compression

One type of signal processor based on postdetection processing is sometimes called an *incoherent pulse compressor.* A processor of this type was included in Fig. 13-19 where both the coherent and incoherent forms of integration were shown. The incoherent detected pulse width was τ with a pulse-compression ratio equal to the number of frequencies N. The coherent pulse width was τ/N, and the compression ratio $\approx N^2$. When the finer resolution of the coherent system is not required, comparable detection performance is often obtained in both receiver noise and distributed clutter environments with the incoherent system for $N \leq 8$.

MTI Followed by Incoherent Integration

The MTI processors of Chap. 9 may be further refined by squaring their residues and averaging them (postdetection integration). This combination has the property that when the clutter is highly turbulent and the MTI improvement I is poor, the postdetection integration gives additional improvement. When the clutter is highly correlated and the MTI improvement is good, the incoherent processor has little effect. This may be understood by examining the correlation coefficient of the MTI residue $\rho_r(t)$. This is formulated by

$$R_r(kT) = \overline{[f(t) - f(t + T)][f(t + kT) - f(t + T + kT)]} \qquad (14\text{-}1)$$

where R_r is the correlation function (not normalized) of the MTI residue, $f(t)$ is the input clutter signal, and T is the interpulse period. By carrying out this operation and normalizing, the residue correlation coefficient ρ_r may be expressed in terms of the input clutter correlation ρ by[1]

$$\rho_r(nT) = \frac{2\rho(nT) - \rho[(n + 1)T] - \rho[(n - 1)T]}{2[1 - \rho(T)]} \qquad (14\text{-}2)$$

When the residue signal is squared, the result is a signal having a dc component plus a fluctuating component. The correlation coefficient of the fluctuating component $\rho_s(kT)$ is given by [87, chap. 12]

---

[1] This is essentially a special case of an analysis by Urkowitz [382].

$$\rho_s(kT) = \rho_r^{\,2}(kT) \tag{14-3}$$

By using the values of $\rho_s(kT)$ in conjunction with Eq. (3-10), one may determine the reduction in noise variance as a result of the postdetection integration.[1] The results of this procedure are shown in Fig. 14-1. This figure illustrates the reduction in clutter fluctuation power for the single canceler (the improvement factor) and for the integrator (the incoherent processing gain) as a function of the standard deviation of the clutter spectrum. The gain for the hybrid processor is approximately their sum.

[1] This procedure was also described recently by Hall and Ward [155] giving the equivalent number of independent samples for the squared MTI residue:

$$N_I = \frac{N^2}{N + 2 \sum_{k=1}^{n-1} (N - k)\,\rho_r^{\,2}(kT)}$$

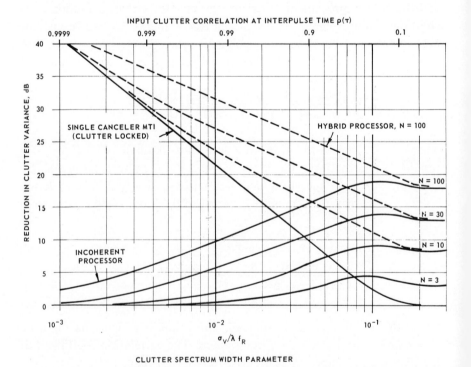

figure 14-1 *Clutter performance improvement for MTI followed by postdetection integration of N residues.*

The ordinate is not exactly the improvement factor I for the hybrid processor since the distribution function of the integrated residue of targets plus noise has not been determined. In addition the postdetection integration *loss* as discussed in Chap. 3 must be accounted for. The incoherent processing gain for higher order cancelers has also been evaluated in a manner similar to the preceding. The results are almost identical to those for the single canceler except for the asymptotic behavior as $\sigma_v/\lambda f_r \to 1$.

Postdetection Integration in Coherent Systems

Most coherent processors including CW and pulse-doppler systems have some degree of postdetection integration. In many cases it is only the *smoothing* in range or angle tracking circuits. However, there are specific instances where integrators are used prior to the threshold circuits. An example of this type was described in Sec. 11.9 on frequency agile pulse trains. There are numerous other techniques for improving performance following coherent processors, but a general analysis is not obvious. The graphs in Chap. 3 should give a first-order approximation to detectability improvement in noise or clutter.

14.2   DIGITAL INTEGRATORS AND QUANTIZATION LOSSES

The discussions in Chap. 3 and in the previous section assumed that the postdetection integrator was linear. This section will include some results of digital postdetection integration showing that the losses in detectability due to *quantization* or encoding into digital form can be quite small. The simplest case to consider is a coincidence detection procedure studied by Schwartz [337] and Harrington [160]. With this technique the detected video signals from each range gate are applied to a threshold circuit. If signal plus noise (or noise alone) crosses the threshold a *one* is entered into a shift register or counter for that range gate. There is then a counter for each range gate in the instrumented range. The echoes from successive transmitted pulses are subjected to the same procedure, and threshold crossings from these pulses are entered into the corresponding counters. If a decision must be made after K pulses (using Schwartz' notation), the contents of each register (each range gate) are examined to determine if there have been n out of K threshold crossings. It was shown that there is an optimum value of n and an optimum threshold setting for a given K and false alarm probability P_N. For $10^{-10} \leq P_N \leq 10^{-5}$ and $0.5 \leq P_D \leq 0.9$ the optimum value of n for a steady target is [337]

$$n_{\text{opt}} \approx 1.5\sqrt{K} \qquad (14\text{-}4)$$

Optimum values for fluctuating targets are lower [409]. While choosing the exact optimum value is not very critical, this value is several decibels better than using $n = 1$ or $n = K$. The efficiency of this process is illustrated for $P_N = 10^{-10}$ on Fig. 14-2. The ordinate is the number of pulses considered and the abscissa is the per pulse (S/N) for the assumed P_D and P_N. The quantization loss or required increase in (S/N) is about

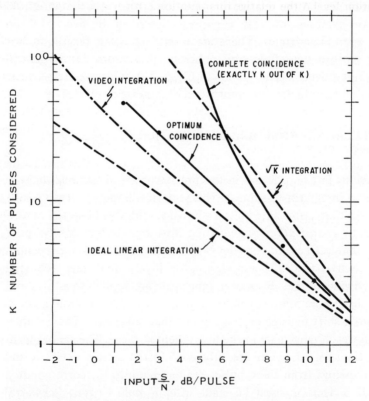

figure 14-2 *Comparison of coincidence detection with integration methods (nonfluctuating target). $P_N = 10^{-10}$, $P_D = 50$ percent. (After Schwartz [337].)*

1.4 dB above the linear video (postdetection) integrator for $K > 10$. For $P_N = 10^{-5}$ the required increase is slightly less. For optimum detection of small steady-target signals ($S/N \leq 1$ per pulse) noise alone would cross the threshold about 20 percent of the time [160, 96].

One problem with the coincidence technique is that when the *noise* level is increased due to instabilities, ECM, or clutter, the number of *detections* will increase. Thus, the threshold level should *adapt* to the mean-square noise in each region as was described in Chap. 4.

If the threshold is replaced by a multilevel analog-to-digital converter A/D, the quantization loss should be reduced. While in fact this is the case, the loss in S/N is a function of the number of levels, the dynamic range of the A/D, and the ratio of the maximum or saturation value of the A/D, E_m to the rms noise σ_n. If E_m/σ_n is large, there will be only a small error due to saturation of the A/D. However, for a fixed quantization level Δ the relative quantization error may be large for small signals.

For a given number of quantization levels there is then a trade-off between the loss when large signals and noise saturate the A/D and the loss when the gain is reduced prior to the A/D so that noise plus small signals does not cross the first threshold of the A/D a significant percentage of the time. The first condition results in a degradation in signal fidelity for large signals and the second results in a loss in detectability for small signals.

There have been a number of studies [216, 404, 246] of the required increase in S/N as a function of the number of quantization levels and of the value of E_m/σ for large numbers of pulses integrated. The quantization level is defined here as $\Delta = 2E_m/L$ where L is the number of levels. Assuming that the rms input noise is greater than or equal to Δ, the quantization noise can be treated as an addition of independent random noise with zero mean and mean square equal to $\Delta^2/12$ [404, 39]. Under this assumption the degradation in input (S/N) (in decibels) is

$$\epsilon = 10 \log \left[1 + \frac{E_m^{\,2}}{3\Delta^2 \sigma_n^{\,2}} \right] \tag{14-5}$$

Thus, the mean-square noise at the input, and not S/N, determines the loss. The degradation at the integrator output is believed to be similar to Eq. (14-5) which is then an upper bound on the degradation in detectability for small noise levels. Many calculations were performed on a digital computer with $E_m/\sigma_{(max)} = 3$ where $\sigma_{(max)}$ is the maximum value of the noise. With small signals and 250 pulses integrated, Eq. (14-5) was verified for E_m not more than 40 dB above the rms noise σ_n and 64, 128, or 256 quantization levels. When σ_n^2 is 30 dB below E_m^2, the degradation ϵ for small signals is 0.3, 0.08, and 0.02 dB for 64, 128, and 256 levels, respectively [216].

The less obvious advantage of multiple quantization levels is the ability to adapt to changing noise levels. It is assumed that the digital word for each pulse at each range gate is added to the contents of the shift register and the register contains several more bits than the A/D. At the end of the integration period the mean value of a number of range-gate registers (spatial average) is determined (see Sec. 4.10). A *target* decision is made only for a particular shift register whose content is well in excess of the spatial mean. This is a form of the *adaptive threshold*.

While multiple quantization levels are not extremely important for detectability in noise, this is not the case for coherent clutter rejection. Section 14.4 shows that quantization errors on large clutter signals (clutter quantization noise) can severely degrade MTI systems which require large improvement factors I.

Range-cusping and Collapsing Losses

The previous discussion assumed that the detected target signal was sampled at its maximum value and the target was stationary.[1] In practice the *samples* may occur prior to or after the peak of the matched-filter output and result in a *range-cusping loss*. As an example, assume that there is a rectangular pulse of duration τ and that the sampling rate is once every τ seconds. If the peak of the matched-filter output falls midway between the samples, a 4.5-5.0 dB higher (S/N) will be required to detect a target with $P_D = 0.5$ to $P_D = 0.99$ and $P_N = 10^{-5}$ [291]. This was calculated for a 6-out-of-32-pulse coincidence detector. This reduces the required number of counters in the instrumented range at the price of a collapsing loss.

Portner [291] studied the efficiency of coincidence detection of n out of 32 samples. A few results are shown on Table 14-1 for $P_N = 0.4 \times 10^{-5}$. Portner found that with the target at the optimum position in the

---

[1]This is also an implicit assumption in most detection studies.

gate there was only about a 1-dB loss compared to an ideal linear postdetection integrator. Similar results for other examples are given by DiFranco and Rubin [96]. This was for a steady target P_D = 0.9 and n_{opt} was about 15. For a fluctuating target (case two) the loss was about 1.8

TABLE 14-1   Coincidence Detection of n out of 32 Pulses for Overall P_D = 0.9, P_N = 0.4 $\times$ 10^{-5}

Type of target		Gate size in pulselengths		
		1	3	10
Steady*	n_{opt}	15	14	12
	(S/N), dB	2.5	4.2	5.7
	P_1, %	14	4	0.8
Fluctuating, case two	n_{opt}	8	6	4
	(S/N), dB	3.5	4.8	5.9
	P_1, %	4	0.4	0.04

*For a steady target and a linear video integrator, (S/N) is 1.45 dB per pulse. For a case two fluctuating target $(S/N) \approx$ 1.8 dB. P_1 is the percentage threshold crossings per pulselength.

dB and n_{opt} was about eight. When the gate size was increased to 10 times the pulselength, the losses increased to 5.7 to 5.9 dB and n_{opt} was reduced. Note also that for fluctuating targets the desired percentage threshold crossings of noise alone P_1 is much less than previously indicated (see Worley [409]).

Another significant loss that must be considered in high range resolution systems results from the target moving through several range gates during the integration process. Consider an example using 32 pulses, one-bit quantization of the detected signal, and where the target has a velocity of four pulselengths during the integration period. If the target instrumented range is divided into zones that are four pulselengths in extent, the detectability of the moving target is 3-5 dB poorer than if the target were stationary. The higher loss occurs if the worst initial position for the target is assumed. If the gate or zone size is increased to eight pulselengths the velocity loss is 4 dB and is essentially independent of target initial position. The 4-dB loss includes the loss from an overly wide gate. Note that the losses from an inadequate number of samples (range-cusping loss), an overly wide range gate (collasping loss), and target velocity are not usually additive.

14.3 DIGITAL MTI PROCESSORS

In addition to the environmental limitations of MTI processors discussed in Chap. 9, there have been numerous problems in the construction and maintenance of the hardware. The relative complexity of the equipment has resulted in most systems having only a single delay line with a fixed interpulse period. Recently, the rapid advances in A/D conversion rates and the rapid decrease in the cost of digital storage have led to a large number of highly flexible digital MTI processors.

In a typical digital MTI the delay lines are replaced by digital memories (primarily core storage). Since it is most convenient to store amplitude, the homodyne or dc-IF receiver is the simplest to implement. In this configuration the RF (or IF) signals are mixed to zero carrier frequency with a *single-sideband* or *quadrature* mixer whose two-video output represents the in-phase I and quadrature Q components of the signal [110, chap. 4]. This type of receiver was discussed in Secs. 11.5 and 12.6. A typical block diagram for digital MTI is shown as Fig. 14-3. The local oscillator is divided into two paths such that there is a 90° differential phase shift at the two mixers whose video outputs form the I and Q channels. These signals are sampled and converted to digital form at high speed.[1] The transfer characteristic of the A/D converters is shown as Fig. 14-4. Notice that the quantized values in an I-Q system do not directly represent either amplitude or phase and the total number of bits

---

[1] At this time an entire eight- to ten-bit conversion is practical in about 0.1 μsec.

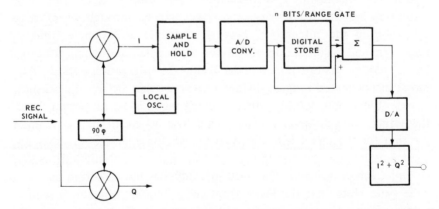

figure 14-3 *Block diagram for digital single-canceler MTI (I and Q configuration).*

in both channels is not numerically equal to quantization *levels* in either amplitude or phase. This will be discussed further in Sec. 14.5. In addition, saturation of the *A/D* converters is not the same as that in an RF or IF limiter prior to the quadrature mixers. While it is not necessary

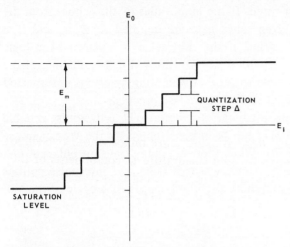

figure 14-4 *Transfer characteristic of A/D converter with two bits plus sign (redundant zero).*

to include both the *I* and *Q* channels to have an operable MTI, it is desirable to avoid the *average* detectability loss of about 3 dB in noise and clutter. See Table 12-3 and [238, 392]. If there is only one channel, the system will have *blind phases* (Chap. 9).

In the figure the bipolar video echo signal in each channel is sampled at least once per pulse width and converted to a digital word.[1] The digital word for each range gate is held in storage until the echo from the second pulse (from the same range gate) is converted to digital form. The difference between the two signals is then converted back to analog form and combined with the quadrature channel. As was shown in Table 12-3, summation of the magnitudes of the *I* and *Q* residues is a good approximation to a square-law detector. With sufficient steps (bits) in the quantization, the digital MTI is equivalent to the ideal vector canceler of Chap. 9.

---

[1] If the sampling rate is not high enough, there will be a *range-cusping* loss as described in the previous section.

The time when the stored echo from any range gate is withdrawn from memory is only a function of the timing of the logic circuits. Thus the choice of interpulse period is limited only by transmitter flexibility. In addition, the echoes from any number of transmit pulses can be stored before subtraction takes place. If the memory is large enough, multiple cancelers can be implemented. If the amplitudes of the echoes from the first four pulses for a given range gate are denoted A, B, C, and D, a four-pulse canceler will result if the subtractor is replaced by a logic network that computes $A - 3B + 3C - D$ [230]. Further shaping of the response function is possible with other weighting functions as suggested in Chap. 9. Digital *feedback cancelers* are also practical.

As a result of inherent flexibility of a digital MTI, a five-pulse variable interpulse period (VIP) MTI has been constructed by the Westinghouse Electric Corporation [230]. When there are many transmit pulses or *hits per beamwidth* in a slowly scanning search radar, the interpulse periods can be slowly varied to eliminate the blind speeds. If a cycle of the VIP

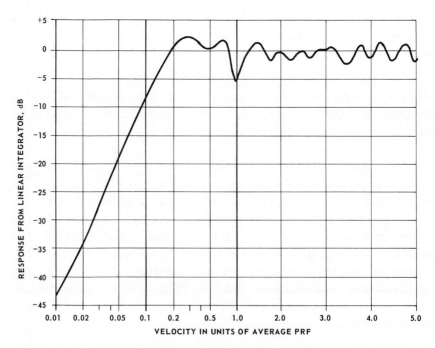

figure 14-5 *Computed response of five-pulse canceler with PRF varied by ± 15 percent (8 steps). (After Linder and Kutz [230].)*

pattern is completed (four period lengths of five pulses each) a typical composite response appears as in Fig. 14-5. The blind velocities are virtually eliminated with only a 1.6-dB average decrease in improvement factor I for echoes that have velocities of 0.02 times the blind velocity.

14.4 LIMITING AND QUANTIZATION ERRORS IN DIGITAL MTI[1]

When a signal is quantized, its exact value is uncertain; that is, it appears as though additional *quantization* noise was added to the system. Clutter cannot be canceled to a level below this basic *noise* limitation. The abrupt clipping imposed on a signal as it exceeds the saturation level causes a signal distortion. This results in additional frequency components which pass through the MTI system. These two effects are analyzed separately in the following analysis which ignores any losses associated with finite sampling in the range domain (see Sec. 14.2).

Quantization Errors

Signal and clutter may not be analyzed separately in a nonlinear process such as quantization. The *signal* must be defined as the incremental increase in output power when a signal is present at the input compared to that when it is not. Using this definition, the signal-to-clutter ratio at the output of the digital MTI processor (Fig. 14-3) is

$$\frac{P_{s0}}{P_{c0}} = \frac{\overline{[(E_1 + C_1 + \epsilon_1) - (E_2 + C_2 + \epsilon_2)]^2} - \overline{[(C_1 + \epsilon_3) - (C_2 + \epsilon_4)]^2}}{\overline{[(C_1 + \epsilon_3) - (C_2 + \epsilon_4)]^2}}$$

(14-6)

P_{s0}, P_{c0} = output signal and clutter power, respectively

E_1, E_2 = input signal voltage for the first and second pulse

C_1, C_2 = input clutter voltage for the first and second pulse

ϵ_1, ϵ_2 = quantization errors when signal and clutter are present

ϵ_3, ϵ_4 = quantization errors when only clutter is present

This equation may be considerably simplified by making some assumptions

---

[1] This section was prepared by J. P. Reilly.

concerning independence: $\overline{(E_1 - E_2)(C_1 - C_2)} = \overline{(E_1 - E_2)(\epsilon_1 - \epsilon_2)} = \overline{(C_1 - C_2)(\epsilon_1 - \epsilon_2)} = \overline{(C_1 - C_2)(\epsilon_3 - \epsilon_4)} = 0$. Defining σ_ϵ^2 as the variance of the quantization error results in

$$\frac{P_{s0}}{P_{c0}} = \frac{\overline{(E_1 - E_2)^2} - \overline{2(\epsilon_1\epsilon_2 - \epsilon_3\epsilon_4)}}{\overline{(C_1 - C_2)^2} + 2\sigma_\epsilon^2 - 2\epsilon_3\epsilon_4} \tag{14-7}$$

Defining ρ_ϵ as the correlation coefficient of the quantization error for clutter alone and $\rho_{\epsilon s}$ as that when both clutter and signal are present, Eq. (14-7) becomes

$$\frac{P_{s0}}{P_{c0}} = \frac{\overline{E_0^2} + 2\sigma_\epsilon^2[\rho_\epsilon(T) - \rho_{\epsilon s}(T)]}{2[P_{ic} - \overline{C_1 C_2} + \sigma_\epsilon^2(1 - \rho_\epsilon(T))]} \tag{14-8}$$

where $\overline{E_0^2} = \overline{(E_1 - E_2)^2}$, P_{ic} is the input clutter power, and T is the interpulse time. The improvement factor is obtained by dividing Eq. (14-8) by the input signal-to-clutter power ratio. Using the fact that the average signal gain for the single canceler is 2, the improvement factor is then

$$I_1 = \frac{1 + (\sigma_\epsilon^2/E^2)[\rho_\epsilon(T) - \rho_{\epsilon s}(T)]}{1 - \rho(T) + (\sigma_\epsilon^2/P_{ic})[1 - \rho_\epsilon(T)]} \tag{14-9}$$

where $\rho(T)$ is the correlation coefficient of the clutter signals. Equation (14-9) illustrates that the quantization error correlation $\rho_\epsilon(T)$ has a *quieting* effect on the noise power caused by σ_ϵ^2. However, except for clutter having an interpulse correlation coefficient unusually close to unity or for quantization levels approaching the rms value of the clutter signals, $\rho_\epsilon(T)$ is very small as illustrated by Fig. 14-6 [52]. Neglecting the correlation factor ρ_ϵ,

$$I_1 \simeq \frac{1}{1 - \rho(T) + \sigma_\epsilon^2/P_{ic}} \tag{14-10}$$

This analysis may be extended to include the double canceler. The details are quite similar to the steps used to analyze the single canceler

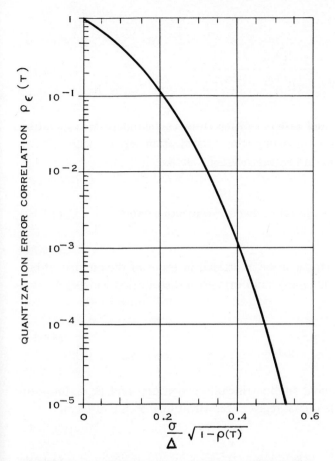

figure 14-6 *Quantization error correlation coefficient.* Δ = *quantization increment,* σ = *rms amplitude before quantization,* $\rho(T)$ = *correlation coefficient of input signal. (From Brennan and Reed,IEEE [52].)*

except that there are considerably more terms to contend with. The result is

$$
I_2 = \frac{1 + \dfrac{\sigma_\epsilon^2}{\overline{E^2}}\left\{\dfrac{4}{3}[\rho_\epsilon(T) - \rho_{\epsilon s}(T)] - \dfrac{1}{3}[\rho_\epsilon(2T) - \rho_{\epsilon s}(2T)]\right\}}{1 - \dfrac{4}{3}\rho(T) + \dfrac{1}{3}\rho(2T) - \dfrac{\sigma_\epsilon^2}{P_{ic}}\left[\dfrac{4}{3}\rho_\epsilon(T) - \dfrac{1}{3}\rho_\epsilon(2T)\right] + \dfrac{\sigma_\epsilon^2}{P_{ic}}}
$$

Again, neglecting the quieting effects of the quantization correlation results in

$$I_2 \simeq \frac{1}{1 - \frac{4}{3}\rho(T) + \frac{1}{3}\rho(2T) + (\sigma_\epsilon^2/P_{ic})} \qquad (14\text{-}11)$$

Equations (14-10) and (14-11) differ from the standard MTI equation only by the factor involving the quantization error, which is an additional limitation on the improvement factor

$$I_1 = I_2 = \frac{P_{ic}}{\sigma_\epsilon^2} \quad \text{limitation due to quantization errors} \qquad (14\text{-}12)$$

The quantization error may be expressed in terms of the number of bits of quantization by assuming a uniform error distribution which yields

$$\sigma_\epsilon^2 = \frac{E_m^2}{12(2^{n-1} - 1/2)^2} \qquad (14\text{-}13)$$

where n is the number of bits (including sign bit) and E_m is the saturation voltage of the digital register. Substituting into Eq. (14-12),

$$I_1 = I_2 = \frac{P_{ic}}{E_m^2} 12(2^{n-1} - 1/2)^2 \qquad (14\text{-}14)$$

This limitation on improvement factor, illustrated in Fig. 14-7, applies when the signal is less than E_m. For greater signals, the effects of clipping must be considered.

Losses due to Clipping

Clipping occurs when the quantized signal exceeds the saturation level of the register, which results in a broadening of its spectrum and a narrowing of its correlation function. The value of the correlation coefficient after clipping is desired since this quantity can be used to define MTI performance through Eqs. (9-14) and (9-15).

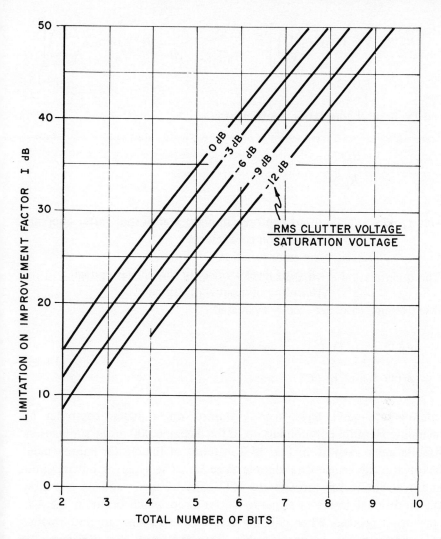

figure 14-7 *Improvement factor limitation imposed by quantization errors (non-redundant zero).*

The expression for the nonlinear transformation of an autocorrelation function has been derived in a number of places, but it is difficult to evaluate for the most general case when both signal and clutter are present. However, when the signal-to-clutter ratio approaches zero, the equations are greatly simplified. The expression for the correlation function of gaussian distributed noise alone is given by Van Vleck [386]:

$$R_0(T) = \rho_i(T)\left[\text{erf}\left(\frac{b}{\sqrt{2}}\right)\right]^2 + \frac{2}{\pi}\sum_{n=3,5,\cdots}\frac{\rho_i(T)}{n!}\left[H_{n-2}(b)\exp\left(\frac{-b^2}{2}\right)\right]^2$$

$$(14\text{-}15)$$

and the desired correlation coefficient is

$$\rho_0(T) = \frac{R_0(T)}{R_0(0)}$$

where $\rho_i(T)$, $\rho_0(T)$ = input (before clipping) and output (after clipping)
  correlation coefficients
  $\text{erf}(x)$ = the error function
  b = clipping level divided by rms clutter voltage
  $H_n(x)$ = Hermite polynomial of degree n

Two extreme cases are readily evaluated

$$\rho_0(T) = \rho_i(T) \qquad \text{for } b \to \infty$$

$$(14\text{-}16)$$

$$\rho_0(T) \frac{2}{\pi}\sin^{-1}[\rho_i(T)] \qquad \text{for } b \to 0$$

Intermediate cases have been evaluated on a digital computer by including Hermite polynomials up to the degree 6,000 in the summation. Results are presented in Fig. 14-8 in terms of the clutter improvement factor for the single and double canceler. These curves illustrate the improvement factor through the MTI canceler alone and do not show any additional losses in signal-to-clutter ratio which occur in the *A/D* conversion process. When clipping is severe, two results are noteworthy: MTI performance is considerably degraded, and the difference in performance between the single and the double canceler is much smaller than that for a linear system. One should interpret these results with caution. The improvement factor is defined in terms of the average residue power and does not indicate the distribution function which has a critical effect on the detection probabilities. The distribution function for small clipping levels will differ markedly from that for large clipping levels.

These clipping losses pertain when a video signal is limited and the clutter-to-signal ratio is large. Limiting on the IF waveform is somewhat

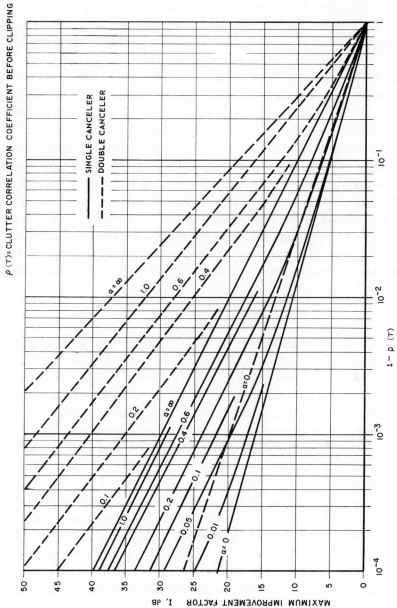

figure 14-8 *Maximum improvement factor for clipped waveforms.*

569

different. If two video channels in quadrature phase are individually limited and vectorially summed, the resultant vector is constrained to a square. In IF limiting, the vector is constrained to a circle inscribed in the square. When the limit level is exceeded, the MTI residue vector lies along the square in one case and along the circle in the other. It is difficult to predict exactly what the performance difference will be. Experiments have been reported by Shrader [344] that indicate performance degradations from limiting which are consistent with the video limiting losses of Fig. 14-8. For example, a dual canceler system using IF limiting had a theoretical scanning fluctuation improvement limitation of 40.5 dB from linear theory. In practice, this was found to be 23 dB. When the system was operated in a single canceler mode, the performance was found to degrade by an additional 7-10 dB.

Ward and Shrader [397] give results of a similar analysis and simulation showing that MTI improvement factor degradation as large as 20 dB is not uncommon in a three-pulse canceler. The four-pulse canceler is typically only 2 dB better in the presence of limiting clutter.

14.5 MATRIX MTI

In conventional range-gated digital MTI, the digital representations of two successive echoes from the same range bin are subtracted. This technique has achieved popularity partly because it is a direct analog to conventional MTI. This section describes an alternate method of designing a digital MTI system which could make it easier to alter the velocity response curve and to *clutter lock* to the mean velocity of the clutter. Also, digital MTI in this form could be combined with digital pulse compression with less total storage than if they were built separately.

Consider the left-hand portion of the digital MTI shown in Fig. 14-9. The echo from each range cell is converted into two quadrature vector components of the signal, of amplitudes I and Q, respectively. The total signal amplitude is then $(I^2 + Q^2)^{1/2}$ and the phase angle ϕ is $\tan^{-1} - (Q/I)$. To visualize the *matrix MTI* technique, consider for a moment that there is only one-bit quantization in each quadrature channel[1] denoted by I^+ or I^- and Q^+ or Q^-. Then $I_1^+ Q_1^+$ represents a signal echo vector from the first transmitted pulse that occurs in the upper

---

[1]Two-bit quantization total.

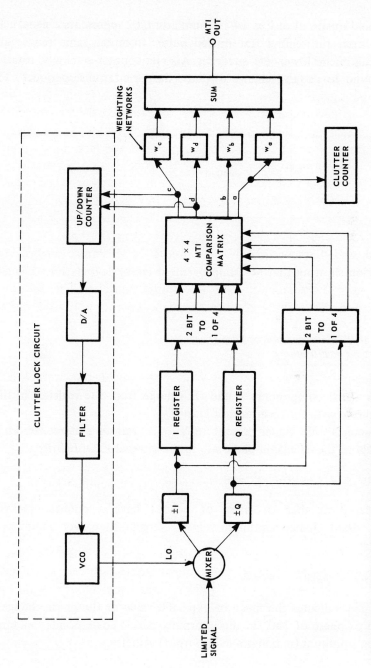

figure 14-9 *Matrix MTI with clutter lock and weighting.*

right-hand quadrant of Fig. 14-10. Similarly $I_2^- Q_2^-$ represents a signal echo vector from the second transmitted pulse (from the same range gate) occurring in the lower left quadrant. Also let counter-clockwise rotation correspond to positive change of phase during interpulse period T. The

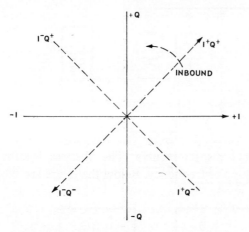

figure 14-10 *Representation of I and Q vectors (two-bit quantization).*

bipolar signal components I_1 and Q_1 emerge from the registers in time coincidence with the reception of I_2 and Q_2.

If there is only signal present $(S/N \gg 1)$, various statements can be made about the change in phase $d\phi_T$ in the interval T. Ordinarily,

$$90° \leq d\phi_T \leq 270° \qquad \mathrm{mod}\, 2\pi$$

If a large number of pairs of echoes have a counter-clockwise two-quadrant change (optimum velocity target), then over a number of periods[1]

$$\overline{d\phi_T} = +180° \qquad \mathrm{mod}\, 2\pi$$

where $\overline{d\phi_T}$ indicates the mean or expected value of the phase change. A change of phase of 180° in the interpulse period corresponds to a target signal at optimum (maximum MTI output) velocity.

[1] The initial phase of target or clutter echoes is uniformly distributed over 360°.

In a similar manner, when a large number of pairs of successive returns are quantized into $I_1^+Q_1^+$ and $I_2^+Q_2^+$, the expected value

$$\overline{d\phi}_T = 0 \qquad \text{mod } 2\pi$$

The best estimate in this case is that the echo is from a stationary object or a target moving at the blind speed.

Since each echo pulse can be located in one of four quadrants, there are 16 possible combinations of phase change for two-pulse MTI which can be represented by a 4 × 4 matrix. The *comparison* matrix block of Fig. 14-9 represents a digitally implemented matrix of decision elements capable of sorting the 16 responses into various categories.

Figure 14-11 illustrates the relationship between the change in vector direction and the classical MTI response curve. The abscissa is the change in phase in time T or doppler frequency. Below the figure are the 16 possible combinations of I and Q for two successive pulses. The four combinations in set a directly below 0° have no measurable phase change and are assumed to be clutter signals. Set b has an expected $\overline{d\phi}_T$ of 180°, and the returns are assumed to be targets. This set could also originate for $d\phi_T = -180°$. Thus of the 16 possible combinations, four can be attributed to targets and four to clutter.

The remaining eight combinations are indeterminate. Set c may result from a slowly receding target or clutter or, alternately, from an inbound target with $\overline{d\phi}_T = 270°$. In a similar manner a combination of vectors falling in set d may be a slowly approaching target or a higher speed receding target. There are four outputs of the matrix corresponding to regions a, b, c, and d.

In the presence of targets, clutter fluctuations, and noise the matrix outputs can only be handled statistically. The radar must then register many hits per beam or utilize a coded waveform to provide a large number of independent samples. The MTI response curve of such a statistical processor can be measured by first assigning weights to the matrix outputs. Let the occurrence of a pulse pair in set a be given a weight of zero, in set b a weight of unity,[1] and in sets c and d a weight of about 0.7. Then, if a large number of pairs of pulses with uniformly random initial phases are injected into the system and the weighted

---

[1] This is denoted by W_b and W_d on Fig. 14-11.

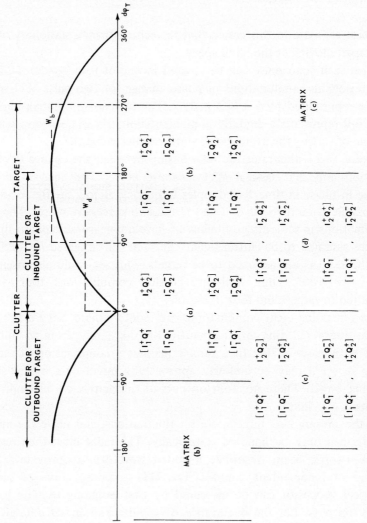

figure 14-11 *Equivalence of change of phase vector to MTI.*

outputs integrated (summed) at each value of $d\phi_T$, the overall response curve (integrated output versus $d\phi_T$) will appear like the conventional MTI response shown in Fig. 14-11.

It might at first be expected that with quantization the overall response curve would be a steplike approximation to the smooth curve. However, the smooth curve is obtained because of the *crossover* effect associated with the random starting phases. This effect is illustrated in Fig. 14-12 wherein the pair of successive signal vectors marked C straddles a phase quantization level. The vector pairs A and B, with other starting phases but the same differential phase $d\phi_T$, are within the quantization levels. Statistically, the ratio of crossing to noncrossing

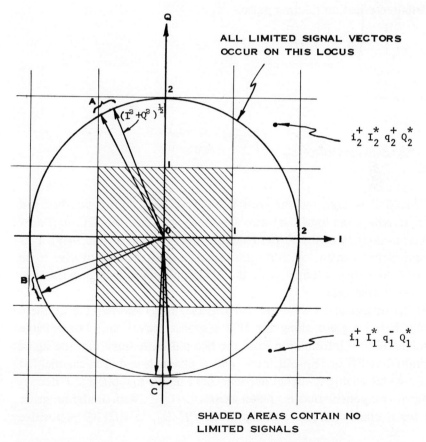

figure 14-12 *Vector representation of four-bit I-Q quantization (two bits I, two bits Q).*

vector pairs will vary smoothly as a function of $d\phi_T$ so as to yield the smooth overall MTI response curve.

With only one-bit quantization in each channel, it should be clear that clutter reduction can only be significant with considerable integration after the logic. The integration can be thought of as giving additional independent samples which effectively increase the number of bits.

If the echoes from N range gates of a rainstorm with a significant mean velocity are subjected to this logic, the mean radial velocity can be estimated by subtracting the number of outputs in region d from region c and dividing by a constant related to the number N of range gates observed. While the variance of a single estimate of $\overline{d\phi}_T$ is large, it reduces by the number of range gates. For two-bit quantization and uniformly distributed phase on the first pulse,

$$\overline{d\phi}_T = \frac{[(d) - (c)]\, 90°}{N}$$

and

$$\text{Standard deviation } \overline{d\phi}_T = \left(\frac{2}{3N}\right)^{\!\!\frac{1}{2}} (45°)$$

Thus if N is large and the clutter extensive, $\overline{d\phi}_T$ can be *smoothed* and used to adjust the local oscillator frequency to place the MTI *null* near the mean doppler of the clutter (see the upper portion of Fig. 14-9). This is equivalent to an analog *clutter-locked* MTI. Clutter does not have to be present to detect a target, i.e., the MTI null will still align with zero velocity in that case.

If the quantization were increased to two bits I and two bits Q, there would be 12 regions along the MTI response curve[1] and 144 possible combinations of information from the two pulses. Assume that the signal is limited at RF or IF, split into I and Q channels and each channel fed to a two-bit analog-to-digital decoder. Let i be the sign bit, and I^* denote a signal component that is greater than $(1/2)\,(I_{\max})$. With similar notation for the Q channel, a phase change from $i_1^+ I_1^* q_1^+ Q_1$ to $i_2^+ I_2^* q_2^+ Q_2^*$ is positive

---

[1] Quantization separately into I and Q is not directly equivalent to phase quantization. If quantization were in phase only, there would be 16 regions.

with any value between 0 and 60° and, therefore, has an expected change $\overline{d\phi}_T$ =30°. This will be denoted by the letter b. The notation for other phase changes is shown in Fig. 14-13A.

The matrix for deciding the phase change sector is shown as Fig. 14-13B. The phase locations for the first pulse are shown vertically and the intersections then show the phase change. The twelve possible phase locations for the second pulse are numbered and are drawn across the top of the matrix. For example, the occurrence of two threes corresponds to zero net phase change and is most likely clutter.

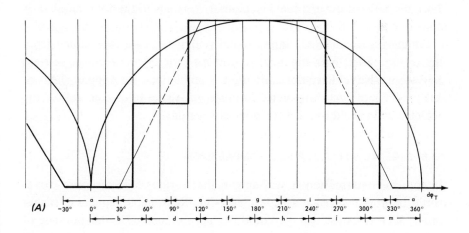

(A)

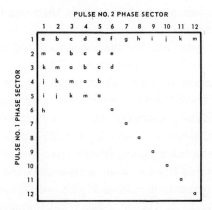

(B)

figure 14-13 *Matrix MTI with two-bit quantization of I and Q. (A) Response curve for 12 phase change regions; (B) MTI logic matrix.*

The response function can be weighted to give any desired shape within the limits of the quantization. A phase change in regions e, f, g, h, and i can be assigned a value of two. Regions, c, d, j, and k can be assigned unity and regions a, b, and m assigned zero. For large numbers of pulses the response curve will approximate a feedback MTI whose shape is dependent upon the weighting. This shape function may be desirable for weather or precipitation clutter. For clutter with narrow spectra such as land echoes, regions c, d, j, and k could be assigned a value of two. The mean velocity of the clutter can be determined by subtracting the occurrences of b from m. The clutter-to-noise ratio can be determined from the ratio of occurrence of regions a, b, and m to the total number of pulses.

At high S/N and S/C, target radial velocity can be determined from the region of $d\phi_T$ with the most counts. As with any coherent pulse-train waveform, the velocity determination is ambiguous for doppler shifts of $1/T$, $2/T$, etc. At low signal-to-noise or signal-to-clutter ratios the velocity will spread over several adjacent regions.

14.6 DIGITAL MTI PLUS PULSE COMPRESSION

In the previous section it was shown that various MTI response curves could be obtained using a matrix decoder. Reasonable replicas of higher order MTI response curves could be approximated if the number of quantization steps *or* the number of samples was sufficiently large. In this section it will be shown that the use of a coded waveform allows implementation of fewer quantization levels than would be needed if simple pulses were used.

One of the major problems in the design of a digital MTI is the requirement for wide dynamic range. If the MTI is to follow a linear FM or phase-coded pulse-compression system, the A/D converter must be able to handle targets and point clutter which are 20 dB above the *average* clutter level.[1] The average clutter may in turn be 40-50 dB above receiver noise. On the other hand, if the MTI precedes the pulse-compression subsystem, echo signals from targets and point clutter will be *stretched* in time (and reduced in amplitude); and it may not be necessary to handle signals in excess of the average clutter level for time-bandwidth products of 100 or more. Additional reduction of the

[1] See Sec. 14.4 on limiting and quantization errors.

dynamic range can be achieved with sensitivity time control (STC) circuits.

Another problem that may occur with a pulse-compression receiver preceding an MTI was pointed out by Shrader [344]. The pulse-compression sidelobes may not be stable in time and thus may not cancel in the MTI portion of the receiver.

One possible combination of MTI and pulse compression is shown in Fig. 14-14. The output of the matrix MTI described in the preceding section is in the form of weighting constants which are indications of the velocity of the radar echoes. In this figure the echoes from the first transmission are stored digitally, and no signals are sent to the PC decoder. The first and second transmissions are considered to be identical phase-coded waveforms of N segments with each corresponding segment denoted by I_{1n}, Q_{1n} or I_{2n}, Q_{2n}. The echoes from each segment of the second transmit pulse (I_{2n}, Q_{2n}) are compared with the stored echo from the first pulse (I_{1n}, Q_{1n}) in the block marked MTI matrix. The quantization shown in the figure is two bits each for I and Q. The MTI matrix outputs are then inserted into the weighting network where the weighting constants shown in the previous section are applied individually to I_{2n} and Q_{2n}. Thus, if $I_{2n} = I_{1n}$ and $Q_{2n} = Q_{1n}$, the best estimate for this segment is that the echo is from a stationary target and zero weighting (no signal) is sent to the PC decoder. If there is a $180°$ phase shift between I_{1n} and I_{2n}, a maximum signal is fed to the decoder. This should equal a multiplication by two in voltage. Intermediate values of phase change should give intermediate weighting. Notice that if amplitude weighting is used on both I_{2n} and Q_{2n}, there is no change in the phase of each segment and the only change in the decoded PC waveform is a change in amplitude and some degradation in the sidelobe regions.

The total clutter attenuation or improvement factor with only two-bit I and two-bit Q processing may not be outstanding unless the waveform contains many segments. However, there should be relatively few false alarms due to strong point-clutter echoes. Better performance should obviously be obtained with more quantization levels in the matrix MTI.

While the illustration was for phase-coded signals, there is no reason to preclude using the technique with linear FM signals as long as the sampling rate is higher than the signal bandwidth. Also a multilevel digital decoder can be used for the pulse-compression portion as well as the *digital* MTI portion of the receiver.

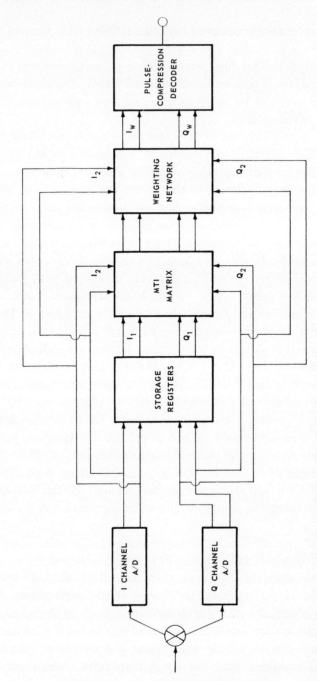

figure 14-14 *Matrix MTI plus digital pulse-compression decoder.*

14.7   A COMBINED PHASE-CODE AND PULSE-DOPPLER
SYSTEM

One of the limiting factors in unweighted phase-coded pulse-compression systems is the relatively high *time sidelobes* or *residues* that occur as compared to those generated when heavily tapered chirp or multifrequency pulse-compression techniques are used. Barker binary codes having residues in the time domain which are down from the main lobe by the number of segments N are available for several code lengths, but only up to $N = 13$. Fairly extensive computer studies of pseudo-random codes of length 511 have yielded lowest peak residues of amplitude 19 and rms values of 8-10 compared to a main lobe amplitude of 511 (see Chap. 12).

In some radar situations, especially those involving precision trackers, it is desirable to have a large time-bandwidth product and low residues near the main lobe; but moderate ambiguities in range and velocity can be tolerated if they are sufficiently distant from the expected target position and velocity. The multiple-target case is the best example of a need for low sidelobes since the residues of nearby targets may affect range and angle accuracy.[1]

One way to achieve low-range sidelobes is to transmit a pulse train with each pulse having a phase code of length 31. Figure 14-15 shows the decoder output for a single pseudo-random code of this length. This is the *finite time autocorrelation function* of the coded pulse. The residue levels are typical and not impressive for use in a multiple-target environment. This correlation function was plotted for a particular starting point. Similar autocorrelation functions can be obtained for the 30 other starting positions. The algebraic sum of the entire set of decoded pulses is shown in Fig. 14-16. Because of the simple binary phase structure of the waveform this figure also represents the coherent summation of this set of 31 compressed pulses [261].

The shape of the function shows that the residue peak amplitude is below the peak by 31 $\times$ 31/30. In other words, *rejection* is greater than 30 dB. The summation seems to approximate the properties of the periodic maximum-length sequences described in Chap. 12. The other

---

[1]Uniformly distributed clutter is not particularly bothersome in this situation. It has been shown that the time sidelobes of the better maximum-length binary phase codes cause only a 1-dB loss in signal-to-clutter ratio.

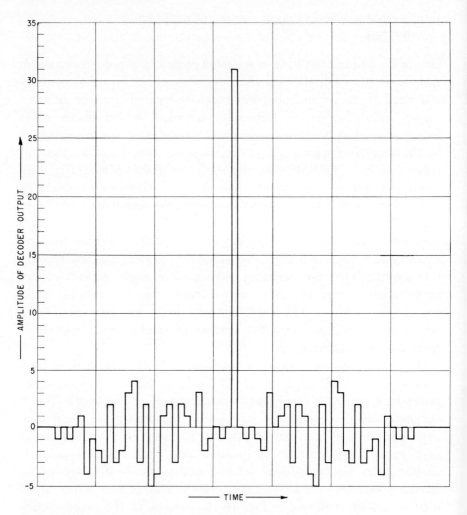

figure 14-15   *Video response of pulse-compression output.*

two maximum-length codes of 31 segments give the same result. On the other hand, other random codes did not yield this result.

Although not obvious, the residues all have the same 180° out-of-phase relationships with the main lobe. This, plus the smoothness of the rise and fall, suggests that even the residue could be eliminated. One method would be to transmit a 32d pulse of the same length without any internal phase code. The return from this pulse would be correlated with itself and form a triangular correlation function of the same extent as the residues in Fig. 14-16. The output of this correlator would be added to the summation of the original 31 pulses thereby subtracting from the residues but adding to the peak lobe. The energy content of the resultant pulse would be comparable to that which created the residues of the first 31. Thus, the ratio of the resultant peak to the residues could be as much as about 1,000:1 (60 dB) in amplitude. Instrumentation would undoubtedly be the limiting factor.

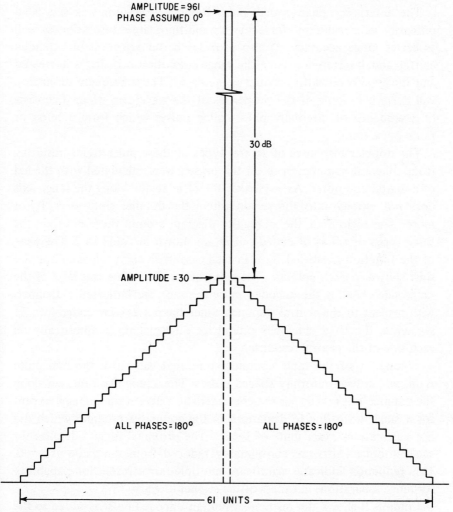

AMPLITUDE = 961
PHASE ASSUMED 0°

30 dB

AMPLITUDE = 30

ALL PHASES = 180°   ALL PHASES = 180°

|← —————————————— 61 UNITS —————————————— →|

figure 14-16 *Summation of matched-filter outputs of 31 starting positions of five-bit pseudo-random code.*

Since no restriction has been placed on the time separation of the coded pulses, the interpulse spacing can be held constant and multiple-doppler filters instrumented for the usual form of pulse-doppler processing. If the problem is merely one of multiple closely spaced targets, this may be adequate. If the interfering signal also includes weather or clutter, it may be desirable to use a nonuniform spacing to minimize the effect of the constant PRF such as suggested by J. Resnick [309] or W. Rubin [199] (see Chap. 11).

The internally phase-coded pulse-doppler waveform was suggested primarily as a technique for achieving multiple-target resolution as well as better range accuracy. It also provides better rejection of extended clutter and interference since the range resolution cell size is decreased and the time-bandwidth product is increased. The subsequent discussion will delve into some of the properties of the waveform when it consists of a sequence of internally phase-coded pulses which form a burst or finite pulse train.

The doppler responses of several types of these pulse trains, resulting from coherent summation of all the pulses, were calculated with the aid of a digital computer. As expected, the clear region along the range axis does not extend into the region where the doppler shift is $1/2T_d$ or more. The regions of the ambiguity diagram around the (τ, ν) origin for three types of trains of coded pulses are shown in Table 14-2. The peak of the function is labeled 0 dB in each case, with eight values of doppler shift shown in each column. The value at $\tau = 0$ and the rms level of the range sidelobes for the uncoded pulse duration are indicated in decibels with respect to the central response. Since the pulses were coded into 31 segments, the range sidelobes extend for 30 segments symmetrically on each side of the central response.

Column A of the table contains computed values for the first pulse train, which has uniformly spaced pulses. The ambiguity function along the doppler axis ($\tau = 0$) has the characteristic $\sin Nx/N \sin x$ shape, except for a small error due to deviations in the computer program which did not calculate in exact units of $1/2T_d$. The proximal range sidelobes for small doppler shifts are considerably reduced from the value at $\tau = 0$. This reduction indicates significant multiple-target resolution capability for range separations in excess of the segment length.

Column B shows the response when an uncoded pulse is added to the end of the previous pulse train to form a burst of 32 uniformly spaced pulses. Since the same units of doppler were used as in the computation

TABLE 14-2 Proximal Portion of Ambiguity Function for Bursts of 31-segment Phase-coded Pulses Normalized to 0 dB at the Origin

Approx. doppler shift in units of $1/T_d$	A 31-pulse uniform spacing	$\int\int$	B 32-pulse uniform spacing	$\int\int$	C 31-pulse staggered spacing	$\int\int$
4	Null	−39.2 dB rms	−25.0	−33.2	−28.1	−39.2
3½	−21.6	−38.4	−23.8	−32.8	−25.6	−38.5
3	Null	−37.6	−24.5	−32.7	−36.8	−37.0
2½	−18.4	−33.5	−19.7	−30.2	−19.3	−34.4
2	Null	−33.4	−24.3	−31.3	−25.4	−33.8
1½	−13.8	−31.0	−14.4	−29.5	−19.2	−29.5
1	Null	−27.6	−24.2	−27.2	−22.4	−27.6
½	− 4.2	−30.3	− 4.5	−29.2	− 3.6	−30.9
0	0 dB	−34.6 dB	0 dB	Null	0 dB	−34.6

$T_d = 30T$
T = interpulse spacing for 31 pulse train
All numerical values are ambiguity function levels in decibels.

for the 31 pulse train, the existing null structure along the doppler axis is not apparent in the table.[1] The clear region along the doppler axis increases into a more or less uniform sidelobe level for doppler shifts $\geq 1/2\,T_d$. This level is slightly higher than that obtained with the 31 pulse train.

Column C tabulates the response when the 31 pulses are staggered, but the total time T_d is kept the same as for the uniformly spaced train. The center pulses are spaced by 20 pulselengths, and the interpulse spacing increases by one unit towards the beginning and end of the pulse train. (The first and last pairs are pairs that are separated by 34 units.) The doppler again is computed in units of $1/2\,T_d$ of the 31-pulse uniform train. The ambiguity function for the first doppler lobes near the time axis is comparable to that for uniform spacing except that the central or zero doppler lobe is widened and the doppler nulls are filled in.

In each of the three cases, there is significant improvement due to coding in the rejection of discrete interference which is separated from the origin by at least one code segment in range and has a small doppler offset.

In uniform clutter which is within $cT/2$ range from the target and extends for at least twice the uncoded pulselength, the clutter contribution of the $2N - 2$ (60 in this case) range sidelobes must be added to the response at zero range offset. The average is shown in Table 14-3 for small doppler offsets. The clutter rejection reference level in this figure is that due to a *single* segment of one of the coded pulses. Unless the clutter can be placed in one of the doppler nulls, the coded pulse trains can provide an increased rejection of uniform clutter of not quite the compression ratio of one pulse. The addition of the uncoded 32d pulse reduces this increased rejection of uniform clutter by a few decibels.

The clutter rejection near one-half the ambiguous doppler[2] is more strongly affected by the coding and pulse spacing. The rejection for discrete interference is shown in Table 14-4. Remember that the clutter response at doppler frequencies near $(N - 1)/2T_d$ is minimized for both coded and uncoded pulses having uniform or nearly uniform spacing.

---

[1] Only one side is indicated on the table.

[2] The ambiguous doppler is located at the inverse of the average interpulse period $\approx 1/T$.

TABLE 14-3   Residues from Uniform Clutter with Coded Pulses and Small Doppler Offsets*

Approx. doppler in units of $1/T_d$	31 uncoded pulses	31 pulse uniform	32 pulse uniform	31 pulse staggered
0	+14.9 dB	0.0 dB	0.0 dB	0.0 dB
½		− 4.2	− 4.5	− 3.6
1	Null	− 9.8	− 9.4	− 9.8
1½	+ 1.8	−10.0	− 9.9	−11.0
2	Null	−15.6	−13.5	−16.0
2½		−13.8	−12.0	−14.8
3	Null	−19.9	−14.9	−19.2
3½	− 5.1	−18.0	−15.0	−19.5
4	Null	−21.4	−15.4	−20.4
4½		−20.5		−20.8
5	Null	−23.7		−23.7

*Residues are in decibels above or below clutter response from a single segment of the pulse. The clutter is assumed to be greater in extent than twice the 31-segment pulselength but does not extend for the entire interpulse period T. Negative values represent an improvement I.

The clutter at this offset frequency is reduced an average of 32.8 dB for a train of 31 uncoded pulses. In this *optimum* doppler region for range offsets exceeding one segment, the addition of the 32d uncoded pulse to a coded pulse train degrades the response about 11 or 12 dB. In the case of uniform clutter and coded pulse trains the energy from 60 range sidelobes must be added, hence these sidelobes will dominate the response at $\tau = 0$. For the uniformly spaced train of 31 coded pulses, the average rejection of uniform clutter is only about 10 dB greater for a compression ratio of 31:1 than for uncoded pulses. The uniform-clutter rejection with the added 32d uncoded pulse is actually about 1 dB poorer than an uncoded pulse train. Thus, the penalty for achieving a clear region on the doppler axis with the 32d (uncoded) pulse is poorer rejection of uniform clutter in the neighborhood of one-half the unambiguous doppler frequency. The overall effect is to create a *thumbtack*-type ambiguity function with better performance near the origin in exchange for some degradation at the optimum doppler offset frequency.

TABLE 14-4 Clutter Residues for Coded Pulses with Doppler Shifts near 2/NT, dB

Approximate doppler in units of $\approx 1/T_d$	31 pulse uniform		32 pulse uniform		31 pulse staggered*	
	at target range	rms side peaks	at target range	rms side peaks	at target range	rms side peaks
13	−43.3	−47.8	−36.6	−34.5	−30.0	−46.4
13½	−30.0	−45.6	−30.4	−34.6	−41.5	−48.7
14	−54.0	−47.4	−45.2	−35.0	−29.6	−48.5
14½	−29.8	−46.8	−30.1	−35.0	−38.5	−49.0
15					−29.8	−48.3
Average rejection of uniform clutter** with respect to single segment	~28 dB		~17 dB		~29 dB	
Average rejection of uniform clutter with respect to transmit pulse	~43 dB		~32 dB			
Average rejection of uncoded pulse train	~32.8 dB		~33.1 dB			

*Slightly longer period than for uniform spacing.
**Extended over the pulse duration but not at the ambiguous ranges, these values are the average improvements.

14.8 DIGITAL PULSE-TRAIN COHERENT PROCESSORS

As was the case for MTI, a coherent digital processor can be built for pulse-train waveforms. While it is possible to construct a digital pulse-doppler processor for a continuous train of pulses, it is simpler to discuss (and construct) a digital matched filter for a finite burst of pulses. This is often called a discrete or *batch* processor. This section will discuss a few of the many implementations for an N pulse train. If the target range is known to within one pulse length τ' it is only necessary to compute the discrete Fourier transform of the N echoes from the range gate of interest. As in previous discussions target acceleration effects are neglected and the radar's position is fixed. In addition, it is assumed that the target does not move an appreciable distance compared to $c\tau'/2$ during the pulse train. The echo from a moving target with constant radial velocity has a linear change of phase versus time if the transmit waveform consists of samples of a perfectly stable sine wave.

If the transmit pulse amplitudes and the interpulse period are constant, there are only N unambiguous doppler channels (including zero doppler) that need be computed. These doppler channels will overlap at their 4-dB points, and the 4-dB width is $1/T_d$ Hz where T_d is the duration of the entire pulse train. To eliminate all ambiguous velocities the PRF must be at least twice the highest target-doppler frequency.

The total phase shift for a target echo in the first unambiguous doppler channel is 2π radians during the time T_d. A target echo centered in the second doppler filter has a phase shift of 4π radians in time T_d, etc. Thus, to form the first doppler channel the required phase shift per interpulse period T is $-2\pi/N$ radians and $-4\pi/N$ radians in the second doppler channel, etc. The Nth doppler channel is formed with a phase shift of -2π radians per interpulse period which is ambiguous with zero doppler.

The digital processor in its basic form essentially performs these phase rotations. Since it is easier to *compute* with quantized video than signals on a carrier, the RF signals are mixed to zero carrier and divided into I and Q channels. (See Secs. 11.5, 12.6, and 14.3 and DiFranco and Rubin [96, sec. 4.11].) Each channel has a matched filter for the individual pulses. This is shown for rectangular pulses as the delay and subtract circuits on Fig. 14-17. The *pulse matched-filter* outputs in each quadrature channel are converted to digital form with a multibit A/D converter. In simple form the analog signals can be represented

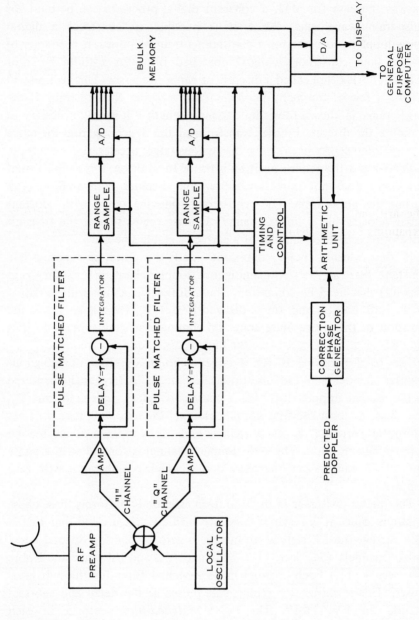

figure 14-17 *Homodyne pulse matched filter and digital pulse doppler processor.*

$$I = A \cos(\omega_d t + \phi)$$
$$Q = A \sin(\omega_d t + \phi) \tag{14-17}$$

where A is the input amplitude, ω_d is the radian doppler shift, and ϕ is the incident phase angle. The desired doppler *channel* is then formed by subtracting (or adding) the doppler frequency shift ω_c.

The *doppler-corrected* I and Q components, after filtering, become

$$I' = A \cos[(\omega_d - \omega_c)t + \phi]$$
$$Q' = A \sin[(\omega_d - \omega_c)t + \phi] \tag{14-18}$$

The arithmetic units of Fig. 14-17 perform the digital phase shifting. Expanding Eq. (14-18) we obtain in the *correct* doppler channel

$$I' = I \cos\omega_c t + Q \sin\omega_c t$$
$$Q' = I \sin\omega_c t + Q \sin\omega_c t \tag{14-19}$$

Since in the general case ω_d is not known, ω_c must be stepped through all possible values of ω_d. For a discrete number of pulses the equations for doppler channels for the rotated components become

$$I'_{n,d} = I_n \cos[(n\Delta\theta + \Delta_n)d] + Q_n \sin[(n\Delta\theta + \Delta_n)d]$$
$$Q'_{n,d} = Q_n \cos[(n\Delta\theta + \Delta_n)d] - I_n \sin[(n\Delta\theta + \Delta_n)d]$$

where n = pulse time index number $(n_{max} = N)$
d = doppler filter channel number
Δ_n = phase angle induced by transmit pulse variations from a constant interpulse period
$\Delta\theta$ = doppler phase increment between pulses for the first doppler channel

For negative doppler channels the sign of the sine terms is reversed. In the *correct* doppler channel and neglecting noise, the rotated components will all have the same polarity. That is, all of the I'_n signals will be positive, negative, or zero; and all Q'_n signals will also be the same. If signal is present, they both cannot be zero. The magnitude of the total signal is the vector sum of the N-rotated signals or

$$
E_d = \left[\left(\sum_{n=1}^{N} I'_{n,d} \right)^2 + \left(\sum_{n=1}^{N} Q'_{n,d} \right)^2 \right]^{1/2} \tag{14-20}
$$

In the correct doppler channel this amounts to coherent or voltage addition of the N-echo signal vectors. In the absence of clutter the mean value in the other channels is the power summation of the noise on each pulse plus the doppler *sidelobes* of the target echo. This processor can be thought of as a digital filter about zero frequency with a band pass of $1/TN = 1/T_d$ Hz.

To form all possible doppler channels, each of the N pulses must be phase shifted (multiplied) N times for both I and Q. Thus, there are $2N^2$ multiplications per range gate. It can be seen that many of these multiplications are redundant or are simply modulo 2π of another multiplication. If these redundancies can be eliminated by using a Fast Fourier Transform [72, 68], a considerable reduction in the number of multiplications is obtainable. If the number of pulses is $N = 2^K$ where K is any integer, only $2N \log_2 N$ multiplications are required instead of N^2 for the I or Q channel. For $N = 64$ the number of digital phase shifts is 768 rather than 4,096.

14.9 CORRELATORS FOR VELOCITY AND TURBULENCE MEASUREMENT

Most of the previous sections described matched filters or cross correlators. However, it is often adequate to study the autocorrelation function of echo signals from clutter or targets (see Chaps. 6 and 7). While a true correlator is generally difficult to construct because of the difficulties of multiplication and time storage of radar signals, a simple device that gives a good approximation is shown in Fig. 14-18. This hybrid correlator is based on the work of J. Gulick and R. Roll of the Applied Physics Laboratory, although other processors of this nature have also been built elsewhere.

The received echoes are mixed with the transmitted frequency, and the resulting bipolar video signal is divided into two channels. In the upper channel the signal is limited, its polarity sampled, and a *one* or a

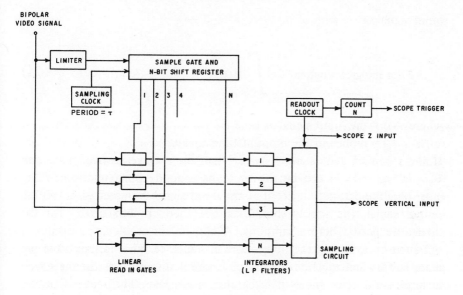

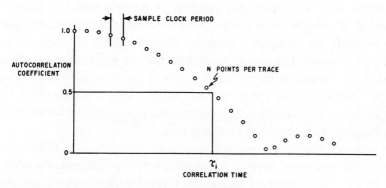

figure 14-18 *Sampled autocorrelator and typical display.*

zero fed into the shift-register circuit. The subsequent stages of the shift register comprise the set of successive time delays which are needed to compute the autocorrelation function. The lines from the stages of the shift register are connected to the *read-in* gates which pass either the input signal directly to the integrators or invert these signals depending upon the polarity in that stage for the duration of the sampling period. After a period of time, the integrators contain an approximation to the integral of the product of the input signal times a delayed replica of the input signal or

$$\text{First integrator output} = \int_0^T f(t)f(t + \tau_n)\, dt \approx R(\tau_n) \tag{14-21}$$

where $f(t)$ = the input signal at time t

$\quad f(t + \tau_n)$ = replica of the input at time τ_n later

If the mean of $f(t)$ is much larger than the average receiver noise and $T \gg \tau$, Eq. (14-21) can be shown to be a good approximation of the autocorrelation function $[R(\tau)]$ of the input signal evaluated at τ_n. With a pulsed signal, the inputs should be stretched or "boxcarred" for the interpulse period before sampling. A convenient oscilloscope display is obtained by sampling the integrators in sequence. The horizontal sweep is initiated when the first integrator is sampled (zero lag) and the sweep is terminated after the Nth integrator is sampled. The readout period should be long compared to the sampling clock interval, and the integrators can be dumped at that time if desired. The vertical deflection is obtained from the integrator outputs. A more pleasant display is obtained if the scope intensity is increased (Z modulation) at the time the integrators are sampled, thereby eliminating the display of the switching transients. The spectral width of the fluctuations can then be estimated by observing the value of τ where the sampled outputs are equal to one-half their value at the origin. This estimate of τ_i is substituted into Eq. (1-18). The sampling rate of the output need not be in synchronism with the input sampling rate.

One device which can extract combined velocity and turbulence measurements uses the *inphase I* and *quadrature Q* components of the radar echoes. The received signal is mixed with a replica of the transmitted signal and also with the same replica shifted by 90° (Sec. 14.3). The resulting video signals are the I and Q components of the received signal. For a single scatterer (raindrop, snowflake, etc.) in a given range cell of a radar beam, the received signal after mixing can be written [325]

$$I_i(t) = a_i \cos[(\omega_0 + \overline{\Delta\omega} + \Delta\omega_i)t + \beta_i] \cos\omega_0 t \tag{14-22}$$
$$Q_i(t) = a_i \cos[(\omega_0 + \overline{\Delta\omega} + \Delta\omega_i)t + \beta_i] \sin\omega_0 t$$

where a_i = measure of the received signal from the scatterer

β_i = the relative phase angle of the received signal

$\overline{\Delta\omega}$ = the average doppler frequency of *all* the scatterers in the cell

$\Delta\omega_i$ = the incremental doppler of the single scatterer

Expanding and eliminating double frequency terms,

$$I_i(t) = \frac{1}{2} a_i \cos[(\overline{\Delta\omega} + \Delta\omega_i)t + \beta_i]$$

$$Q_i(t) = \frac{1}{2} a_i \sin[(\overline{\Delta\omega} + \Delta\omega_i)t + \beta_i]$$

$$(14\text{-}23)$$

The autocorrelation function of each component is

$$R_{II}(\tau) = \overline{\Sigma I_i(t) \Sigma I_j(t + \tau)}$$

$$R_{QQ}(\tau) = \overline{\Sigma Q_i(t) \Sigma Q_j(t + \tau)}$$

$$(14\text{-}24)$$

where τ is the time lag between the signals, and short-pulsed signals are assumed. It was shown by Roll [325] that the expression for the cross correlation between these component signals for independent scatterers can be reduced to

$$R_{II}(\tau) = \cos(\Delta\omega\tau) \left[\frac{\overline{\Sigma a_i^2}}{8} \right] [\Sigma \overline{\cos \Delta\omega_i \tau}] \qquad (14\text{-}25)$$

$$R_{IQ}(\tau) = -\sin(\Delta\omega\tau) \left[\frac{\overline{\Sigma a_i^2}}{8} \right] [\Sigma \overline{\cos \Delta\omega_i \tau}] \qquad (14\text{-}26)$$

If Eqs. (14-25) and (14-26) are evaluated at many values of τ, the average doppler shift of *all* the scatterers in the radar cell is

$$\overline{\Delta\omega} = \frac{1}{\tau} \arctan \left[\frac{R_{IQ}(\tau)}{R_{QQ}(\tau)} \right] \qquad (14\text{-}27)$$

The autocorrelation function of the turbulence component can be obtained from

$$R(\tau)_{\text{turb}} = [R_{II}^2(\tau) + R_{IQ}^2(\tau)]^{1/2} \qquad (14\text{-}28)$$

which describes the correlation function which would have resulted had the received signal been mixed down to zero average doppler. The standard deviation of the turbulence spectrum can either be estimated by using Eqs. (14-28) and (1-18) or from the Fourier transform of $R_{II}(\tau)$ or $R_{QQ}(\tau)$. When taking the Fourier transform of R_{II} or R_{QQ}, the velocity resolution is determined by the maximum time lag τ_{max} over which the autocorrelation is computed. By varying τ_{max}, one also varies the resolution.

BIBLIOGRAPHY AND REFERENCES

1. Abramowitz, W., Analytical Signal Processing Techniques in a Reverberation-limited Environment, *Grumman Res. Dept. Rept.* RE-262, June, 1966 (AD485861).*
2. Adams, W. B., Phased Array Radar Performance with Wideband Signals, *EASCON Proc.,* IEEE, Washington, D. C., October, 1967; *IEEE Trans.,* vol. AES-3, no. 6, pp. 257-271, November, 1967.
3. Ament, W. S., J. A. Burkett, F. C. Macdonald, and D. L. Ringwalt, Characteristics of Radar Sea Clutter: Observations at 220 Mc, *Naval Res. Lab. Rept.* 5218, Washington, D. C., Nov. 19, 1958.
4. Annett, M. E., Engineering Report on Analysis of ASW Flight Data, *Philco Rept.* H-1084, July, 1952 (AD5707).
5. Ares, M., The Performance of Some Uniformly Spaced Burst Waveforms in Extended Clutter, *Gen. Elec. Heavy Military Electron. Dept. Rept.* R65EMH3, Syracuse, N. Y., Jan. 12, 1965.
6. Ares, M., The Performance of Burst Waveforms in Extended Clutter, *Gen. Elect. Heavy Military Electron. Dept.,* Syracuse, N.Y., Jan. 12, 1965.

*AD numbers in parentheses refer to Defense Documentation Center listings.

7. Ares, M., Some Anticlutter Waveforms Suitable for Phased Array Radars, *Gen. Elec. Heavy Military Electron. Dept. Rept.* R66EMH8, Syracuse, N. Y., November, 1965.

8. Ares, M., Optimum Burst Waveforms for Detection of Targets in Uniform Range-extended Clutter, *Gen. Elec. Heavy Military Electron. Dept. Rept.* R66EMH16, Syracuse, N. Y., March, 1966. Also *IEEE Trans.*, vol. AES-3, no. 1, pp. 138-141, January, 1967.

9. Ares, M., Postdetection Combining of Multiple Frequency Channel Radar Signals, *Gen. Elec. Heavy Military Electron. Dept. Rept.* R67EMH20, Syracuse, N.Y., June 27, 1967.

10. Armendariz, Manuel, and L. J. Rider, Wind Shear for Small Thickness Layers, *Atmos. Sci. Lab.*, ECOM-5040, White Sands Missile Range, N.M., March, 1966 (AD482329).

11. Ashby, R. M., V. Josephson, and S. Sydoriak, Signal Threshold Studies, *Naval Res. Lab. Rept.* R-3007, Dec. 1, 1946.

12. Atlas, David, Advances in Radar Meteorology, in Landsberg (ed.), "Advances in Geophysics," vol. 10, Academic Press, Inc., New York, 1964.

13. Atlas, David, K. R. Hardy, and T. G. Konrad, Radar Detection of the Tropopause and Clear Air Turbulence, *Proc. 12th Radar Meteorol. Conf.*, Norman, Okla., Oct. 19, 1966.

14. Baker, C. H., "Man and Radar Displays," Pergamon Press, Inc. and The Macmillan Company, New York, 1962.

15. Balch, H. T., J. C. Dale, T. W. Eddy, and R. M. Lauver, Estimation of the Mean of a Stationary Random Process by Periodic Sampling, *Bell Systems Tech. J.*, vol. 45, no. 5, pp. 733-741, May-June, 1966.

16. Bardash, M., Ambiguities in Radar Signal Measurements of Extended Targets, *RCA Damp Tech. Monograph* DTM-62-18, Moorestown, N.J., October, 1962.

17. Barker, R. H., "Group Synchronizing of Binary Digital Systems, Communication Theory," Academic Press, Inc., New York, 1953. (Papers read at a symposium on Applications of Communication Theory held at the Institution of Electrical Engineers, London, Sept. 22-26, 1952.)

18. Barlow, E. J., Doppler Radar, *Proc. IRE*, vol. 37, no. 4, pp. 340-355, April, 1949.

19. Barrick, D. E., Radar Signal Spectrum Distortions Produced by Volume and Surface Distributed Scatterers, USNC/USRI Spring Meeting, Apr. 9-12, 1968, National Academy of Sciences, Washington, D. C.

20. Barton, D. K., "Radar Systems Analysis," Prentice-Hall, Inc., Englewood Cliffs, N. J., 1964.

21. Barton, D. K., Radar System Performance Charts, *IEEE Trans.*, vol. MIL-9, nos. 3-4, pp. 255-263, July-October, 1965.

22. Barton, D. K., Detectability Factor and Integration Loss, *Raytheon Co. Intern. Memo.* DKB-66, Wayland, Mass., Aug. 6, 1965.

23. Barton, D. K., and W. M. Hall, Antenna Pattern Loss Factors for Scanning Radars, *Proc. IEEE (Corr.)*, vol. 53, no. 9, pp. 1257, 1258, September, 1965.

24. Barton, D. K., Radar Equations for Jamming and Clutter, *EASCON Proc.*, IEEE, Washington, D. C., October, 1967; *IEEE Trans.*, vol. AES-3, no. 6, pp. 340-355, November, 1967.

25. Battan, L. J., "Radar Meteorology," University of Chicago Press, Chicago, 1959.

26. Bean, B. R., The Radio Refractive Index of Air, *Proc. IRE,* vol. 50, pp. 260-273, March, 1962.
27. Beard, C. I., I. Katz, and L. M. Spetner, Phenomenological Vector Model of Microwave Reflection from the Ocean, *IRE Trans.,* vol. AP-4, no. 2, April, 1956.
28. Beard, C. I., and I. Katz, The Dependence of Microwave Radio Signal Spectra on Ocean Roughness and Wave Spectra, *IRE Trans.,* vol. AP-5, no. 2, pp. 183-191, April, 1957.
29. Beard, C. I., Coherent and Incoherent Scattering of Microwaves from the Ocean, *IRE Trans.,* vol. AP-9, no. 5, pp. 470-483, September, 1961.
30. Beard, C. I., Behavior of Non-Rayleigh Statistics of Microwave Forward Scatter from a Random Water Surface, *Boeing Res. Lab. Rept.* DI-82-0587, Seattle, Wash., December, 1966.
31. Beasley, E. W., Effect of Surface Reflections on Rain Cancellation in Radars Using Circular Polarization, *Proc. IEEE (Letters),* vol. 54, no. 12, December, 1966.
32. Bechtel, M. E., and H. W. Prensen, Errors in R-meter Measurements of the Velocity Spread of Meteorological Targets Resulting from Radar Frequency Instabilities, *12th Conf. Radar Meteorol.,* Norman, Okla.   Oct. 17-20, 1966.
33. Beckmann, P., and A. Spizzichino. "The Scattering of Electromagnetic Waves from Rough Surfaces," The Macmillan Company, New York, 1963.
34. Beckmann, P., Shadowing of Random Rough Surfaces, *IEEE-PGAP,* vol. AP-13, pp. 384-388, May, 1965.
35. Bell, J., Propagation Measurements at 3.6 and 11 Gc/s over a Line of Sight Path, *Proc. IEE,* vol. 114, no. 5, pp. 545-549, May, 1967.
36. Bello, P., and W. Higgins, Effect of Hard Limiting on the Probabilities of Incorrect Dismissal and False Alarm at the Output of an Envelope Detector, *IRE Trans. IT,* pp. 60-66, April 1961.
37. Bennett, W. R., "Spectra ot Quantized Signals," *Bell System Tech. J.,* vol. 27, July, 1948.
38. Berkowitz, R. S. (ed.), "Modern Radar—Analysis, Evaluation and System Design," John Wiley & Sons, Inc., New York, 1965.
39. Bernfeld, M., C. E. Cook, J. Paolillo, and C. A. Palmieri, Matched Filtering Pulse Compression and Waveform Design (four parts), *Microwave J.,* pt. 1, October, 1964, pp. 56-64; pt. 2, November, 1964, pp. 81-90; pt. 3, December, 1964, pp. 70-76; pt. 4, January, 1965, pp. 73-81.
40. Bickel, H. J., Spectrum Analysis with Delay-line Filters, *1959 WESCON Conv. Rec.,* pt. 8, pp. 59-67, IRE, 1959.
41. Biernson, G., and I. Jacobs, Clutter Spectra for an Airborne CW Doppler Radar, *Trans. 1959 Symp. Radar Return,* May 11-12, 1959, pp. 377-438.
42. Birkemeier, W. P., and N. D. Wallace, Radar Tracking Accuracy Improvement by means of Pulse-to-Pulse Frequency Modulation, *AIEE Trans. Commun. Electron.,* vol. 81, pp. 571-575, January, 1963.
43. Blackband, W. T., Radar Techniques for Detection, Tracking, and Navigation *8th Symp. AGARD Avionics Panel, AGARDograph* 100, Gordon and Breach, New York, 1966.
44. Blake, L. V., A Guide to Basic Pulse-radar Maximum-range Calculation Part 1— Equations, Definitions, and Aids to Calculation, *Naval Res. Lab. Rept.* 5868, Dec. 28, 1962. [Also Radio Ray (Radar) Range—Height-angle charts, *Naval Res. Lab. Rept.* 6650, Jan. 22, 1968.]

45. Blake, L. V., Recent Advances in Basic Radar Range Calculation Technique, *IRE Trans.*, vol. MIL-5, no. 2, pp. 154-164, April, 1961.

46. Blevis, B. C., R. M. Dohoo, and K. S. McCormick, Measurements of Rainfall Attenuation at 8 and 15 GHz, *IEEE Trans.*, vol. AP. 15, no. 3, pp. 394-403, May, 1967.

47. Bogotch, S. E., and C. E. Cook, The Effect of Limiting on the Detectability of Partially-time-coincident Pulse Compression Signals, *IEEE Trans.*, vol. MIL-9, no. 1, pp. 17-24, January, 1965.

48. Bongianni, W. L., and J. B. Harrington, Ultrawide Bandwidth Pulse Compression in YIG, *Proc. IEEE,* vol. 54, no. 8, pp. 1074-1075, August, 1966.

49. Boring, J. G., E. R. Flynt, M. W. Long, and V. R. Widerquist, Final Report Project No. 157-96, Sea Return Study, *Georgia Inst. Tech. Eng. Exp. Sta. Rept.,* Aug. 1, 1957 (AD246180).

50. Braasch, R. H., and A. Erteza, A Recursion for Determining Feedback Formulas for Maximal Length Linear Pseudo-random Sequences, *Proc. IEEE (Letters),* vol. 54, no. 7, pp. 999-1000, July, 1966.

51. Brennan, L. E., and F. S. Hill, Jr., A Two-step Sequential Procedure for Improving the Cumulative Probability of Detection in Radars, *IEEE Trans.*, vol. MIL-9, nos. 3-4, pp. 278-287, July-October, 1965.

52. Brennan, L. E., and I. S. Reed, Quantization Noise in Digital Moving Target Indication Systems, *IEEE Trans.*, vol. AES-2, no. 6, pp. 655-658, November, 1966.

53. Britt, C. O., et al. Back Scattering Cross-sections at 4.3 Millimeter Wavelengths of Moderate Sea Surfaces, *Univ. Texas Rept.* 95 [NONR 375(01)], Nov. 8, 1957.

54. Brookner, E., Signal Processing and Synthesis for Optimum Clutter Rejection, *Raytheon Space Inform. Syst. Rept.* FR-65-345, presented at the International Conference on Microwave, Circuits, and Information Theory in Tokyo, September, 1964, Oct. 30, 1964, and revised May 1, 1965.

55. Burroughs, H. H., Rain Intensity-Time Distributions, *Naval Ordnance Lab. Rept.* 729, Corona, Calif., June 15, 1967 (AD654709).

56. Bussgang, J. J., and D. Middleton, Optimum Sequential Detection of Signals in Noise, *IRE Trans. IT*-1, pp. 5-18, December, 1955.

57. Bussgang, J. J., P. Nesbeda, and H. Safran, A Unified Analysis of Range Performance of CW, Pulse and Pulse Doppler Radar, *Proc. IRE,* vol. 47, no. 10, pp. 1753-1762, October, 1959.

58. Cahn, C. R., A Note on Signal-to-Noise Ratio in Bandpass Limiters, *IRE Trans. IT,* pp. 39-43, January, 1961.

59. Campbell, J. P., Back-scattering Characteristics of Land and Sea at X-band, *Aeronautical Electron. 1958 Natl. Conf. Proc.,* Dayton, Ohio, May, 1958, pp. 12-14.

60. Capon, J., On the Properties of an Active Time-variable Network: The Coherent Memory Filter, *Proc. Symp. Active Networks Feedback Systems,* Apr. 19-20, 1960, pp. 561-581, Polytechnic Institute of Brooklyn.

61. Capon, J., High-speed Fourier Analysis with Recirculating Delay-line Heterodyner Feedback Loops, *IRE Trans. Instr.,* vol. I-10, no. 1, pp. 32-73, June 1, 1961.

62. Capon, J., Optimum Weighting Functions for the Detection of Sampled Signals in Noise, *IEEE Trans. Inform. Theory,* April, 1964.

63. Cassedy, E. S., and J. Fainberg, Back-scattering Cross Sections of Cylindrical Wires of Finite Conductivity, *IRE Trans. Antennas Propagation,* vol. AP-8, no. 1, pp. 1-7, January, 1960.

64. Chandler, J. P., An Introduction to Pseudo-Noise Modulation, *Harry Diamond Lab. Rept.* TM-64-4, HDL-16100, Washington, D. C., Jan. 30, 1964, (AD479308).

65. Cheston, T. C., and J. Frank, Antenna Arrays, in M. I. Skolnik (ed.), "Radar Handbook," chap. 11, McGraw-Hill Book Company, New York (to be published).

66. Clark, N., D. Nielson, G. Hagn, and L. Rorden, An Investigation of the Backscatter of High Frequency Radio Waves from Land, Sea Water, and Ice, *Stanford Res. Inst. Proj.* 2090, May, 1960.

67. Clarke, A. S., Target Noise Reduction by Pulse-to-Pulse Frequency Shifting (Non-Coherent Processing), *Hughes Aircraft Co. Aerospace Group Rept.* TM 893, Culver City, Calif., November, 1967.

68. Cochran, W. T., et al., What is the Fast Fourier Transform, *Proc. IEEE,* vol. 55, no. 10, pp. 1664-1674, October, 1967.

69. Coleman, S. D., and G. R. Hetrich, Ground Clutter and Its Calculation for Airborne Pulse Doppler Radar, *Proc. 5th Natl. Conv. Military Electron.,* Washington, D. C., 1961, pp. 409-415.

70. Conlon, J. R., High Resolution Radar, Part II—Sea Clutter Measurements, *Naval Res. Lab. Rept.* NRL-4951, Washington, D. C., Aug. 29, 1957.

71. Cook, C. E., and M. Bernfeld, "Radar Signals — An Introduction to Theory and Applications," Academic Press, Inc., New York, March, 1967.

72. Cooley, J. W., and J. W. Tukey, An Algorithm for the Machine Calculation of Complex Fourier Series, *Mathematics Computation,* vol. 19, pp. 297-301, April, 1965.

73. Cope, R., Tackle Noise Variations at Video, *Electronic Design,* no. 20, pp. 54-56, Sept. 26, 1968.

74. Corriher, H. A., Jr., and B. O. Pyron, Abstracts on Radar Reflectivity of Sea Targets, *Georgia Inst. Tech. Eng. Exp. Sta. Proj.* A914, vol. 1, Atlanta, Georgia, Dec. 15, 1966 (AD813955L).

75. Corriher, H. A., Jr., et al., "A Bibliography of Radar Reflection Characteristics," vol. 7, ECOM-03759-F, Georgia Institute of Technology, Apr. 14, 1967 (AD820157).

76. Cosgriff, R. L., W. H. Peake, and R. C. Taylor, "Terrain Scattering, Properties for Sensor System Design Terrain Handbook II," Ohio State University Antenna Laboratory, May, 1960.

77. Cox, Charles, The Relation of Backscattered Radiation to Wind-Stress at the Sea Surface, *Intern. Symp. Electromagnetic Sensing Earth Satellites,* Miami Beach, Fla., November, 1965.

78. Crawford, A. B., and D. C. Hogg, Measurement of Atmospheric Attenuation at Millimeter Wavelengths, *Bell Telephone Monograph* 2646, vol. 35, pp. 907-916, July, 1956.

79. Crispin, J. W., R. F. Goodrich, and K. M. Siegel, A Theoretical Method for the Calculation of the Radar Cross Sections of Aircraft and Missiles, *Univ. Michigan Radiation Lab. Rept.* 2591-1-H, July, 1959.

80. Crispin, J. W., and A. L. Maffett, Radar Cross Section Estimation for Simple Shapes, *Proc. IEEE,* vol. 53, no. 8, pp. 833-848, August, 1965.

81. Croney, J., Clutter on Radar Displays − Reduction by Use of Logarithmic Receivers, *Wireless Engr.*, vol. 33, pp. 83-96, April, 1956.

82. Croney, J., Improved Radar Visibility of Small Targets in Sea Clutter, *Radio Electron. Engr.*, September, 1966, pp. 135-148 (also *A.S.W.E. Lab. Note* XRA-65-1, June 1, 1965, Portsmouth, England).

83. Curry, G. R., A Study of Radar Clutter in Tradex, *M.I.T./Lincoln Lab., Group Rept.* 1964-29, May 25, 1964 (also *IRE Trans.*, vol. MIL 9, no. 1, pp. 39-44, January, 1965).

84. Daley, J. C., Sea Clutter Measurements at X and C Band, *Naval Res. Lab. Letter Rept.* 5720-18A:JCD, Aug. 17, 1966.

85. Daley, J. C., Airborne Radar Backscatter Study at Four Frequencies, *Naval Res. Lab. Letter Rept.* 5270-20A:JCD, Aug. 23, 1966.

86. Davenport, W. B., Jr., Signal-to-Noise Ratios in Band-pass Limiters, *J. Appl. Phys.*, vol. 24, no. 6, pp. 720-727, June, 1953.

87. Davenport, W. B., Jr., and W. L. Root, "Introduction to the Theory of Random Signals and Noise," McGraw-Hill Book Company, New York, 1958.

88. Davidson, M., Dolph-Tschebycheff Amplitude Tapering Factors for Waveforms, *Appl. Phys. Lab. Intern. Memo.* TWC-3-065, April, 1962.

89. Davies, H., and G. G. MacFarlane, Radar Echoes from the Sea Surface at Centimeter Wave-lengths, *Proc. Phys. Soc. London,* vol. 58, 1946.

90. Davies, H., The Reflection of Electromagnetic Waves from a Rough Surface, *Proc. IEE,* pt. 4, vol. 101, pp. 209-214, Jan. 15, 1954 (also *Oxford Univ. Inst. Monograph* no. 90).

91. Delano, R. H., A Theory of Target Glint or Angular Scintillation in Radar Tracking, *Proc. IRE,* vol. 41, no. 8, pp. 1778-1784, December, 1953.

92. Deley, G. W., Width Modulated Pulse-doppler Waveforms for Clutter Rejection, *Def. Res. Corp., Tech. Memo.* 364, June, 1966 (AD647195).

93. Delong, D. F., Jr., Experimental Autocorrelation of Binary Codes, *M.I.T./Lincoln Lab. Rept.* 47G-0006, Oct. 24, 1960 (AD245803).

94. Demin, I. D., The Effect of Sea Roughness on Radar Visibility of Small Ships, *Leningrad-Central Naval Sci.-Res. Inst. Navigation Commun.,* no. 20, issue 79 (translation), Air Force Systems Command/Foreign Technology Division, Aug. 8, 1966, pp. 56-63 (AD643974).

95. Dickey, F. R., Jr., Theoretical Performance of Airborne Moving Target Indicators, *IRE Trans.*, vol. ANE-8, pp. 12-23, June, 1953.

96. DiFranco, J. V., and W. L. Rubin, "Radar Detection," Prentice-Hall, Inc., Englewood Cliffs, N.J., 1968.

97. Di Toro, J. A., Clutter Model for AEW Radar Design, *U.S. Naval Air Development Center Rept.* NADC-AE-6638, Johnsville, Penn., Nov. 29, 1966 (AD644567).

98. Douce, J. L., Barker Sequences, *Electronics (Letters) (British),* School of Engineering Science/Univ. of Warwick Coventry, Worcester, April, 1966, p. 159.

99. Durlach, N. I., Influence of the Earth's Surface on Radar, *M.I.T./Lincoln Lab. Tech. Rept.* 373, Jan. 18, 1965 (AD627635).

100. Dye, R. A., Performance of the Delay-lock Tracking Discriminator with Binary Signals, *1965 IEEE Conf.,* vol. MIL-E-Con 9, pp. 151-158, September, 1965.

101. Easterbrook, B. J., and D. Turner, Prediction of Attenuation by Rainfall in the 10.7-11.7 GHz Communication Band, *Proc. IEE,* vol. 114, no. 5, pp. 557-565, May, 1967.

102. Edrington, T. S., Fluctuation of CW Radar Echoes from Aircraft, *Def. Res. Lab. Univ. Texas* DRL-448, April, 1960.

103. Edrington, T. S., The Amplitude Statistics of Aircraft Radar Echoes, *IEEE Trans.,* vol. MIL-9, no. 1, pp. 10-16, January, 1965.

104. Lieber, Robert, in E. H. Ehling (ed.), "Range Instrumentation," chap. 3, Prentice-Hall, Inc., Englewood Cliffs, N.J., 1967.

105. Elspas, B., A Radar System Based on Statistical Estimation and Resolution Considerations, *Appl. Electron. Lab. Stanford Univ. Tech. Rept.* 361-1, Aug. 1, 1955 (AD207896).

106. Erdmann, R. L., and R. D. Myers, The Effect of Number of Signal Pulses upon Signal Detectability with PPI Scopes, Symposium on Illumination and Visibility of Radar and Sonar Displays *(Proc. Symp. Vision Smithsonian Inst.,* publ. 595, National Academy of Sciences, National Research Council, Washington, D. C., Apr. 1, 1958.)

107. Erickson, C. W., Clutter Cancelling in Autocorrelation Functions by Binary Sequence Pairing, *U.S. Navy Electron. Lab. Rept.* 1047, San Diego, Calif., June 13, 1961 (AD01401).

108. Ericson, L. O., Terrain Return Measurements with an Airborne X-band Radar Station, *Proc. Sixth Conf. Swedish Natl. Comm. Sci. Radio,* Mar. 13, 1963, Research Institute of National Defense, Stockholm, 1963.

109. "Clutter Data Appendixes," vol. II, Evaluation of Airborne Overland Radar Techniques and Testing (also sec. V), Airborne Instrument Laboratories, RTD-TR-65, September, 1965 (AD487604).

110. Evans, G. C., Influence of Ground Reflections on Radar Tracking Accuracy, *Proc. IEE,* vol. 113, no. 8, pp. 1281-1286, August. 1966.

111. Eveleth, J. H., A Survey of Ultrasonic Delay Lines Operating Below 100 MHz, *Proc. IEEE,* vol. 53, no. 10, pp. 1406-1428, October, 1965.

112. Farrell, J. L., and R. L. Taylor, Doppler Radar Clutter, *IRE Trans.,* vol. ANE-11, no. 3, pp. 162-172, September, 1964.

113. Fehlner, L. F., Marcum's and Swerling's Data on Target Detection by a Pulsed Radar, *Appl. Phys. Lab./Johns Hopkins Univ. Rept.* TG-451, Silver Spring, Md., July, 1962 (also supplement TG-451A, September, 1964).

114. Fenwick, R. B., and G. H. Barry, Step-by-Step to a Linear Frequency Sweep, *Electronics,* July 16, 1965, pp. 66-70.

115. Ferrell, E. B., Plotting Experimental Data on Normal or Log-normal Probability Paper, *Industrial Quality Control,* vol. 15, no. 1, July, 1958.

116. Fine, T., On the Estimation of the Mean of a Random Process, *Proc. IEEE,* vol. 53, no. 2, pp. 187-188, February, 1965.

117. Finn, H. M., A New Approach to Sequential Detection in Phased Array Radar Systems, *Proc. IEEE Natl. Winter Conv. Military Electron.,* vol. 2, pp. 3-4, 1963.

118. Finn, H. M., and R. S. Johnson, Efficient Sequential Detection in the Presence of Strong Localized Signal Interference, *RCA Rev.,* September, 1966.

119. Finn, H. M., Adaptive Detection in Clutter, *Proc. NEC,* vol. 22, p. 562, 1966 (also see RCA APTM-1140-47, Dec. 1, 1966 and Adaptive Detection with Regulated Error Probabilities, *RCA Rev.,* vol. 28, no. 4, December, 1967).

120. Fishbein, W., S. W. Graveline, and O. E. Rittenbach, Clutter Attenuation Analysis, *USAECOM, Tech. Rept.* ECOM-2808, Fort Monmouth, N.J., March, 1967.

121. Flesher, G. T., and G. I. Cohn, The General Theory of Comb Filters, *Proc. NEC,* Chicago, 1958, pp. 282-295.

122. Floyd, W. L., and T. J. Lund, Scatterometer Program, *Proc. 1966 Aerospace Electron. Systems Conv. Rec.* (supplement to *IEEE Trans. Aerospace Electron. Sys.),* vol. AES-2, no. 6, November, 1966.

123. Fowle, E. N., E. J. Kelly, and J. A. Sheehan, Radar System Performance in a Dense-target Environment, *IRE Nat. Conv. Rec.,* pt. 4, 1961, pp. 136-145.

124. Fowle, E. N., E. J. Kelly, and J. A. Sheehan, Accuracy and Resolution, Unpublished notes at M.I.T./Lincoln Laboratory, June, 1961.

125. Fowle, E. N., The Design of Radar Signals, Mitre Corporation, SR-98, Bedford, Mass., Nov. 1, 1963 (AD617711).

126. Fowler, C. A., A. P. Uzzo Jr., and A. E. Ruvin, Signal Processing Techniques for Surveillance Radar Sets, *IRE Trans.,* vol. MIL-5, no. 2, pp. 103-108, April, 1961.

127. Frank, R. L., and S. A. Zadoff, Phase Shift Pulse Codes with Good Periodic Correlation Properties, *IRE Trans.,* vol. IT-8, pp. 381-382, October, 1962.

128. Frank, R. L., Polyphase Codes with Good Nonperiodic Correlation, *IEEE Trans.,* vol. IT-9, pp. 43-45, January, 1963.

129. Gaheen, A. F., J. McDonough, and D. P. Tice, "Frequency Diversity Radar Study," vol. 1, Experimental Verification of Frequency and Polarization Diversity on the Statistics of the Radar Cross Section of Satellite Targets, Westinghouse Electric Corp., Baltimore, Feb. 11, 1966 (AD480358).

130. Galejs, J., Enhancement of Pulse Train Signals by Comb Filters, *IRE Trans. Inform. Theory,* vol. IT-4, no. 3, pp. 114-125, September, 1958.

131. Gallagher, J. M., H. Krason, and W. Todd, Multifrequency Clutter Study, *Raytheon Co. Missile Sys. Div. BR*-3289, Bedford, Mass., Mar. 5, 1965.

132. Gardner, R. E., Doppler Spectral Characteristics of Aircraft Radar Targets at S-band, *Naval Res. Lab. Rept.* 5656, Aug. 3, 1961.

133. Gent, H., I. M. Hunter, and N. P. Robinson, Polarization of Radar Echoes, Including Aircraft Precipitation and Terrain, *Proc. IEE,* vol. 110, no. 12. pp. 2139-2148, December, 1963.

134. George, S. F., and A. S. Zamanakos, Comb Filters for Pulsed Radar Use, *Proc. IRE,* vol. 42, no. 7, pp. 1159-1165, July, 1954.

135. George, T. S., Fluctuations of Ground Clutter Return in Airborne Radar Equipment, *Proc. IEE (British),* vol. 99, pt. 4, no. 2, pp. 92-98, April, 1952.

136. Gill, W. J., A Comparison of Binary Delay-lock Tracking-loop Implementations, *IEEE Trans.,* vol. AES-2, no. 4, pp. 415-424, July, 1966.

137. Glover, K. M., D. R. Hardy, C. R. Landry, and T. Konrad, Radar Characteristics of Known Insects in Free Flight, *Proc. 12th Radar Meteorol. Conf.,* Norman, Okla., Oct. 19, 1966. (Also *Science,* vol. 154, no. 967, 1966.)

138. Godard, S., Properties of Attenuation of Radiowaves in the 0.86-cm Band by Rain, translated from *J. Rech. Atmos. (J. Atmos. Res.),* vol. 2, no. 4, pp. 121-167, 1965.

139. Goetz, L. P., and J. D. Albright, Airborne Pulse-Doppler Radar, *IRE Trans.,* vol. MIL-5, no. 2, pp. 116-126, April, 1954.

140. Goetz, L. P., and W. A. Skillman, Master Oscillator Requirements for Coherent Radar Sets, *IEEE-NASA Symp. Short Term Frequency Stability,* NASA SP-80, Nov. 23, 1964, pp. 19-27.

141. Goldstein, H., Frequency Dependence of the Properties of Sea Echo, *Phys. Rev.,* vol. 70, Dec. 1 and 15, 1956. (See also [211].)

142. Golomb, S. W., Sequences with Randomness Properties, MR-6193-K, Contract Req., no. 639498, Glenn L. Martin Co., June 14, 1955.

143. Golomb, S. W., et al., "Digital Communications," Prentice-Hall, Inc., Englewood Cliffs, N.J., 1964.

144. Golomb, S. W., and R. A. Scholtz, Generalized Barker Sequences, *IEEE Trans. Inform. Theory*, vol. IT-11, no. 4, pp. 533-537, October, 1965.

145. Grant, C. R., and B. S. Yaplee, Backscattering from Water and Land at Centimeter and Millimeter Wavelengths, *Proc. IRE*, vol. 45, no. 7, July, 1957.

146. Graves, C. D., Radar Polarization Scattering Matrix, *Proc. IRE,* vol. 44, no. 2, pp. 248-252, February, 1956.

147. Green, B. A., Jr., Radar Detection Probability with Logarithmic Detectors, *IRE Trans.*, vol. IT-4, no. 1, pp. 50-52, March, 1958.

148. Grisetti, R. S., M. M. Santa, and G. M. Kirkpatrick, Effect of Internal Fluctuation and Scanning on Clutter Attenuation in MTI Radar, *IRE Trans. — Aeronautical Navigational Electron.*, March, 1955, pp. 37-41.

149. Guinard, N. W., J. T. Ransone, Jr., M. B. Laing, and L. E. Hearton, NRL Terrain Clutter Study, Phase I, *Naval Res. Lab. Rept.* 6487, May 10, 1967 (AD653447).

150. Gunn, K. L. S., and T. W. R. East, The Microwave Properties of Precipitation Particles, *Quart. J. Roy. Meteorol. Soc.,* vol. 80, pp. 522-545, October, 1954.

151. Gustafson, B. G., and B. O. As, System Properties of Jumping-frequency Radars, *Phillips Telecommun. Rev.,* vol. 25, no. 1, pp. 70-76, July, 1964. (See also Lind, G., Reduction of Radar Tracking Errors With Frequency Agility, *IEEE Trans.,* vol. AES-4, no. 3, pp. 410-416, May, 1968.

152. Hadad, J. D., Basic Relation Between the Frequency Stability Specification and the Application, *IEE-NASA Symp. Short Term Frequency Stability,* NASA SP-80, Nov. 23, 1964.

153. Hagn, G., Investigation of Direct Backscatter of High Frequency Radio Waves from Land, Sea Water, and Ice, *Stanford Res. Final Rept.* 2, NONR 2917-(00), Menlo Park, Calif., May, 1962.

154. Hall, W. M., General Radar Equation, in "Space Aeronautics R&D Handbook," 1962-1963. (Also unpublished Appendix A.)

155. Hall, W. M., and H. R. Ward, Signal To Noise Loss in Moving Target Indicators, *Proc. IEEE (Letters),* vol. 56, no. 2, pp.233-234, February, 1968.

156. Hammond, D. L., and R. E. Morden, Signal Processing Techniques for a Lunar Radar System, *Suppl. IEEE Trans. Aerospace Electron. Sys.,* vol. AES-2, no. 6, pp. 395-399, November, 1966.

157. Hansen, R. C. (ed.), "Microwave Scanning Antennas," vol. I, Academic Press, Inc., New York, 1964.

158. Hansen, V. G., Studies of Logarithmic Radar Receiver Using Pulse-length Discrimination, *IEEE Trans.*, vol. AES-1, no. 3, pp. 246-253, December, 1965.

159. Harmer, J. D., and W. S. O'Hare, Some Advances in CW Radar Techniques, *IRE Nat. Conv. MIL Electron.,* 1961, pp. 311-323.

160. Harrington, J. V., An Analysis of the Detection of Repeated Signals in Noise by Binary Integration, *IRE Trans.,* vol. IT-1, no. 1, pp. 1-9, March 1955.

161. Harrison, A., Methods of Distinguishing Sea Targets from Clutter on a Civil Marine Radar, *Radio Electron. Engr.,* vol. 27, pp. 261-275, April, 1964.

162. Harrold, T. W., Attenuation of 8.6-mm Wavelength Radiation in Rain, *Proc. IEE,* vol. 114, no. 2, pp. 201-203, February, 1967.

163. Heffner, R. W., A Backscatter-Multipath Model for Ground Wave Pulse Communications Systems, *1964 IEEE Intern. Conv. Rec.,* vol. 6, no. 10, 1964.

164. Heidbreder, G. R., and R. L. Mitchell, Detection Probabilities for Log-normally Distributed Signals, *IEEE Trans.,* vol. AES-3, no. 1, pp. 5-13, January, 1967.

165. Heimiller, R. C., Phase Shift Pulse Codes with Good Periodic Correlation Properties, *IRE Trans.,* vol. IT-7, pp. 254-257, October, 1961.

166. Helgöstam, L. F., and B. Ronnerstam, Ground Clutter Calculation for Airborne Doppler Radars, *IEEE Trans.,* vol. MIL-9, nos. 3 and 4, pp. 294-297, July-October, 1965.

167. Helstrom, C. W., An Introduction to Sequential Detection in Radar, *Westinghouse Res. Lab. Rept.* 412 FF 512-R1, Pittsburgh, March, 1960. (Also see A Range Sampled Sequential Detection System, *IRE PGIT,* vol. IT-8, pp. 43-47, January, 1962.)

168. Helstrom, C. W., Analysis of Two-stage Signal Detection System, *Westinghouse Res. Lab. Rept.* 412 FF 512-R2, Pittsburgh, April, 1960.

169. Hicks, B. L., et al., Sea Clutter Spectrum Studies using Airborne Coherent Radar III, *Control Sys. Lab./Univ. Illinois Rept.* R-105, May, 1958.

170. Hicks, J. J., I. Katz, C. R. Landry, and K. R. Hardy, Clear Air Turbulence: Simultaneous Observations by Radar and Aircraft, *Science,* vol. 157, no. 3790, pp. 808, 809, Aug. 18, 1967.

171. Hicks, J. J., and J. K. Angell, Radar Observations of Braided Structures in the Visually Clear Atmosphere, *J. Appl. Meteorol.,* February, 1968.

172. Hill, R. T., Performance Characteristics of Radar Phased Arrays, *Assoc. Senior Engs.–Bureau Ships 3rd Ann. Tech. Symp.,* Mar. 25, 1966.

173. Hilst, G. R., Analysis of the Audio-frequency Fluctuations in Radar Storm Echoes: A Key to the Relative Velocities of the Precipitation Particles, *M.I.T. Dept. Meteorol. Weather Radar Res. Tech. Rept.* 9, pt. A, Nov. 1, 1949 (AD128649).

174. Hoffman, J., Relativistic and Classical Doppler Electronic Tracking Accuracies, *AIAA Space Flight Testing Conf.,* Cocoa Beach, Fla., Mar. 18, 1963.

175. Hoffman, L. A., H. J. Winthroub, and W. A. Garber, Propagation Observations at 3.2 Millimeters, *Proc. IEEE,* vol. 54, no. 4, pp. 449-454, April, 1966.

176. Holliday, E. M., W. E. Wood, D. E. Powell, and C. E. Basham, L-band Clutter Measurements, *U.S. Army Missile Command Rept.* RE-TR-65-1, Redstone Arsenal, Ala., November, 1964 (AD461590).

177. Hollis, E., Comparison of Combined Barker Codes for Coded Radar Use, *IEEE Trans. Aerospace Electron. Sys.,* vol. AES-3, no. 1, pp. 141-143, January, 1967.

178. Hollis, E. E., Predicting the Truncated Autocorrelation Functions of Combined Barker Sequences of any Length without Use of a Computer, *IEEE Trans.,* vol. AES-3, no. 2, pp. 368-369, March, 1967.

179. Hoover, R. M., and R. J. Urick, Sea Clutter in Radar and Sonar, *IRE Conv. Rec., 1957,* pt. 9, November, 1957, p. 17.

180. Howard, T. B., Application of Some Linear FM Results to Frequency-Diversity Waveforms, *RCA Rev.,* vol. 26, no. 1, pp. 75-105.

181. Hunter, I. M., and T. B. A. Senior, Experimental Studies of Sea-surface Effects on Low-angle Radars, *Proc. IEE,* vol. 113, no. 11, pp. 1731-1740, November, 1966.

182. Hynes, R., and R. E. Gardner, Doppler Spectra of S-band and X-band Signals, *IEEE Eastcon Proc.,* Washington, D. C., Oct. 16-18, 1967; *IEEE Trans.,* vol. AES-3, no. 6, pp. 356-365, November, 1967.

183. Delay Devices for Pulse Compression Radar, *IEE Conf. Publ.* 20, Feb. 21, 1966.

184. *Proc. IEEE (Radar Reflectivity* Issue), vol. 53, No. 8, August, 1965.

185. "Airborne Overland Radar Techniques and Testing Study," sec. V, Ground Clutter Model, Illinois Institute of Technology Research Institute, September, 1965.

186. Jacob, J. S., Graphical Comparison of a Doppler-shift Advantage for Three Pulse-compression Techniques, *1965 IEEE Conf.,* MIL-E-CON 9, September, 1965, pp. 182-187.

187. Jacobs, I., The Effects of Video Clipping on the Performance of an Active Satellite PSK Communication System, *IEEE Trans. Commun. Tech.,* June, 1965, pp. 195-201.

188. Jacobus, R. W., A Linear FM 1000:1 Pulse Compression System, *Tech. Documentary Rept.* ESD-TDR-63-237, Mitre Corp., Bedford, Mass., July, 1963, (AD411637).

189. James, W. J., The Effect of the Weather in Eastern England on the Performance of X-band Ground Radars, *Roy. Radar Establishment Tech. Note,* no. 655, July, 1961.

190. Janza, F. J., R. K. Moore, B. D. Warner, and A. R. Edison, Radar Cross Sections of Terrain Near-vertical Incidence at 415 Mc, 3800 Mc and Extension of Analysis to X-band, *Univ. New Mexico Eng. Expt. Sta. Tech. Rept.* EE-21, May, 1959.

191. Jiusto, J. E., and W. J. Eadie, Terminal Fall Velocity of Radar Chaff, *J. Geophysical Res.,* vol. 68, no. 9, pp. 2858-2861, May 1, 1963.

192. Johnson, C. M., S. P. Schlesinger, J. C. Wiltse, and C. W. Smith, Sea Scattering Measurements in the Region from 9.6 to 38 kMc, *Johns Hopkins Univ. Radiation Lab. Tech. Rept.* 27, September, 1955 (AD605545).

193. Johnson, C. M., S. P. Schlesinger, and J. C. Wiltse, Backscattering Characteristics of the Sea in the Region from 10 to 60 kMc, *Johns Hopkins Univ. Radiation Lab. Tech. Rept.* 35, May, 1956.

194. Johnson, C. M., S. P. Schlesinger, and J. C. Wiltse, Backscattering Characteristics of the Sea in the Region from 10 to 50 kMc, *Proc. IRE,* vol. 45, no. 2, pp. 220-228, February, 1957.

195. Johnson, Nicolas, Cumulative Detection Probability for Swerling III and IV Targets, *Proc. IEEE (Letters),* vol. 54, no. 11, pp. 1583, 1584, November, 1966.

196. Jones, J. J., Hard Limiting of Two Signals in Random Noise, *IEEE Trans.,* vol. IT-9, pp. 34-42, January, 1963.

197. Jonsen, G. L., On the Improvement of Detection Range Using Frequency Agile Techniques, *Boeing Airplane Co. Transport Div. Doc.* D6-6835, April, 1964.

198. K_A Radar Weather Performance Data and Analysis, *Emerson Elec.-Electron. Space Div. Rept.* 1840, St. Louis, Dec. 9, 1964.

199. Kaiteris, C., and W. L. Rubin, A Noncoherent Signal Design Technique for Achieving a Low Residue Ambiguity Function, *IEEE Trans. (Corr.),* vol. AES-2 no. 4, pp. 468-471, July, 1966.

200. Kaiteris, C. P., and W. L. Rubin, Radar Waveform Design for Detecting Targets in Clutter, *Proc. IEE,* vol. 114, no. 6., pp. 696-702, June, 1967.

201. Kaplan, E. L., Signal–Detection Studies with Applications, *Bell Sys. Tech. J.,* vol. 34, March, 1955.

202. Katz, Isadore, and L. M. Spetner, Two Statistical Models for Radar Terrain Return, *IRE Trans.,* vol. AP-8, no. 3, pp. 242-246, May, 1960.

203. Katz, Isadore, Radar Reflectivity of the Earth's Surface, *APL Tech. Digest,* vol. 2, no. 3, pp. 10-17, January-February, 1963.
204. Katz, Isadore, Radar Backscattering from Terrain at X-band, *Appl. Phys. Lab./Johns Hopkins Univ. Rept.* CLO-4-002, Silver Spring, Md., June 5, 1963.
205. Katz, Isadore, Wave Length Dependence of the Radar Reflectivity of the Earth and the Moon, *J. Geophysical Res.,* vol. 71, no. 2, pp. 361-366, Jan. 15, 1966.
206. Katz, I., A Polychromatic Radar, *12th Radar Meteorol. Conf.,* American Meteorological Society, Norman, Okla., October, 1966.
207. Katz, Isadore, Probing the Clear Atmosphere with Radar, *APL Tech. Digest,* vol. 6, no. 1, pp. 2-8, September-October, 1966.
208. Katzin, M., On the Mechanisms of Radar Sea Clutter, *Proc. IRE,* vol. 45, no. 1, pp. 44-54, January, 1957.
209. Kelly, E. J., and Wishner, Matched Filter Theory for High Velocity Accelerating Targets, *IEEE Trans.,* vol. MIL 9, pp. 56-59, January, 1965.
210. Kennedy, R. W., The Spatial and Spectral Characteristics of the Radar Cross Section of Satellite-type Targets, *Air Force Avionics Lab. Tech. Rept.* AFAL-TR-66-17, Research and Technology Division, Air Force Systems Command, Wright-Patterson AFB, Ohio, March, 1966.
211. Kerr, D. E. (ed.), "Propagation of Short Radio Waves," M.I.T. Radiation Laboratory Series, no. 13, McGraw-Hill Book Company, New York, 1951.
212. Key, E. L., E. N. Fowle, and R. C. Haggarty, A Method of Side-lobe Suppression in Phase-coded Pulse Compression Systems, *M.I.T./Lincoln Lab. Tech. Rept.* 209, Aug. 28, 1959.
213. Kiely, D. G., Rain Clutter Measurements with CW Radar Systems Operating in the 8mm Wavelength Band, *Proc. IEE (British),* vol. 101, pt. 3, no. 70, pp. 101-108, March, 1954.
214. Klauder, J. R., A. C. Price, S. Darlington, and W. J. Albersheim, The Theory and Design of Chirp Radars, *Bell System Tech. J.,* vol. 39, no. 4, pp. 745-808, July, 1960.
215. Klauder, J. R., The Design of Radar Signals Having Both High Range Resolution and High Velocity Resolution, *Bell System Tech. J.,* vol. 39, p. 808ff., July, 1960.
216. Kohlenstein, L. C., Effect of Quantization and Limiting on Signal-to-Noise Ratio and System Linearity, *Appl. Phys. Lab. Intern. Memo.* TWI-3-028, March, 1964 (not generally available).
217. Konrad, T. G., J. J. Hicks, and E. B. Dobson, Radar Characteristics of Known Single Birds in Flight, *Science,* vol. 159, no. 3812, pp. 274-280, Jan. 19, 1968.
218. Kovaly, J. J., et al., Sea Clutter Studies Using Airborne Coherent Radar, *Univ. Illinois Control Sys. Lab. Rept.* 37, June 26, 1953.
219. Krason, H., and G. Randig, Terrain Backscattering Characteristics at Low Grazing Angles for X and S Band, *Proc. IEEE (Letters)* (special issue on computers), vol. 54, no. 12, December, 1966.
220. Kraus, J. D., "Antennas," McGraw-Hill Book Company, New York, 1950.
221. Kroszezynski, J., On the Optimum MTI Reception, *IEEE Trans.,* vol. IT-11, no. 3, pp. 451-452, July, 1965.
222. Lane, P. E., and R. L. Robb, Sea Clutter Measurements at S and X Band, *A.S.R.E. Tech. Note* NX 55-6, Portsmouth, Hants, England, Aug. 19, 1955 (AD302300).
223. Lawson, J. L., and G. E. Uhlenbeck, "Threshold Signals," M.I.T. Radiation Laboratory Series, no. 24, McGraw-Hill Book Company, New York, 1950.

224. Lee, Y. W., "Statistical Theory of Communications," John Wiley & Sons, Inc., New York, September, 1964.

225. Lerner, R. M., R. Price, and R. Manasse, Loss of Detectability in Band-pass Limiters, *IRE Trans.*, vol. IT-4, pp. 34-38, March, 1958.

226. Lerner, R. M., A Matched Filter Detection System for Complicated Doppler Shifted Signals, *IRE Trans. Inform. Theory*, vol. IT-6, pp. 373-385, June, 1960.

227. Levine, Daniel, "Radargrammetry," McGraw-Hill Book Company, New York, 1960.

228. Lhermitte, R. M., Motions of Scatterers and the Variance of the Mean Intensity of Weather Radar Signals, *Sperry Rand Res. Center Program* 38310, SRRC-RR-63-57, Atmospheric Physics Department, Sudbury, Mass., November, 1963.

229. Lhermitte, R. M., Weather Echoes in Doppler and Conventional Radars, *Proc. 10th Weather Radar Conf.*, April, 1965.

230. Linder, R. A., and G. H. Kutz, Digital Moving Target Indicators, *IEEE Trans. (Suppl.)*, vol. AES-3, no. 6, pp. 374-385, November, 1967.

231. Lindgren, B. W., "Statistical Theory," The Macmillan Company, New York, 1962.

232. Linell, T., An Experimental Investigation of the Amplitude Distribution of Radar Terrain Return, *6th Conf. Swedish Natl. Comm. Sci. Radio*, Research Institute of National Defense, Stockholm, Mar. 13, 1963.

233. Locke, A. S., "Guidance," pp. 100-101, D. Van Nostrand Co., Inc., 1955.

234. Long, M. W., R. D. Wetherington, J. L. Edwards, and A. B. Abeling, Wavelength Dependence of Sea Echo, *Georgia Inst. Tech. Final Rept. Proj.* A-840, July 15, 1965 (AD477905).

235. Long, M. W., On the Polarization and the Wavelength Dependence of Sea Echo, *IEEE Trans.*, vol. AP-13, pp. 749-754, September, 1965. (See also Polarization and Sea Echo, *Electron. Letters*, February, 1967.)

236. Long, M. W., Backscattering for Circular Polarization, *Electron. Letters*, vol. 2, p. 341, September, 1966.

237. Luke, P. J., Rain Clutter for CW Radar, *Appl. Phys. Lab./Johns Hopkins Univ. Intern. Memo.* MRT-0-004, Silver Spring, Md., Nov. 5, 1965.

238. Luke, P. J., Detection of I Channel Only and of |I| and |Q|, *Appl. Phys. Lab./Johns Hopkins Univ. Intern. Memo.* MTR-7006, Oct. 5, 1967.

239. Macdonald, F. C., Correlation of Radar Sea Clutter on Vertical and Horizontal Polarizations with Wave Height and Slope, *Proc. IRE*, vol. 43, no. 14.2, 1955.

240. Mack, C. L., Jr., and B. Reiffen, RF Characteristics of Thin Dipoles, *Proc. IEEE*, vol. 52, no. 5, pp. 533-542, May, 1964.

241. Mack, R. B., and B. B. Gorr, Measured Radar Backscatter Cross Sections of the Project Mercury Capsule, *Air Force Cambridge Res. Lab. Rept.* AFCRL 38, Bedford, Mass., February, 1961.

242. Mallett, J. D., and L. E. Brennan, Cumulative Probability for Targets Approaching a Uniformly Scanning Search Radar, *Proc. IEEE*, vol. 51, pp. 596-601, April, 1963.

243. Manasse, R., The Use of Pulse Coding to Discriminate against Clutter, *M.I.T./Lincoln Lab. Group Rept.* 312-12 (rev. 1), June 7, 1961.

244. Marcum, J. I., and P. Swerling, Studies of Target Detection by Pulsed Radar, *IRE Trans.*, vol. IT-6, no. 2, pp. 59-267, April, 1960. (From *Rand Corp. Res. Memo.* RM 754, July, 1948.)

245. Marshall, J. S., and Walter Hitschfeld, Interpretation of the Fluctuating Echo from Randomly Distributed Scatters, *Can. J. Phys.*, vol. 31, pts. 1 and 2, pp. 962-994.

246. Max, J., Quantizing for Minimum Distortion, *IRE Trans.*, vol. IT-6, no. 1, pp. 7-12, March, 1960.

247. McGinn, J. W., and E. W. Pike, A Study of Sea Clutter Spectra, *Proc. Symp. Statistical Methods Radio Wave Propagation, UCLA,* June 18-20, 1958, Pergamon Press, Inc., New York, 1960.

248. Medhurst, R. G., Rainfall Attenuation of Centimeter Waves: Comparison of Theory and Measurement, *IEEE Trans.*, vol. AP-13, no. 4, pp. 550-564, July, 1965.

249. Menske, R. A., Detection Probability for a System with Instantaneous Automatic Gain Control, *1965 IEEE Conf. Military Electron.*, vol. MIL-E-Con 9, September, 1965, pp. 28-31.

250. Millman, Jacob, and Herbert Taub, "Pulse and Digital Circuits," McGraw-Hill Book Company, New York, 1956.

251. Mityashev, B. N., The Determination of the Time Position of Pulses in the Presence of Noise, translated by Scripta Technica, Inc., D. L. Jones (ed.), MacDonald, London, 1965.

252. Mooney, D., and G. Ralston, Performance in Clutter of Airborne Pulse, MTI, CW Doppler and Pulse Doppler Radar, *IRE Intern. Conv. Rec.*, 1961.

253. Moore, R. K., and W. J. Pierson, Measuring Sea State and Estimating Surface Winds from a Polar Orbiting Satellite, Center for Research in Engineering Science/University of Kansas and the Department of Meteorology and Oceanography/New York University, November, 1965.

254. Moore, R. K., Radar Scatterometry—An Active Remote Sensing Tool, *Proc. 4th Symp. Remote Sensing Environment,* Apr. 12, 13, and 14, 1966, University of Michigan 4864-11-x, June, 1966, pp. 339-374.

255. Muchmore, R. B., Aircraft Scintillation Spectra, *IRE Trans.*, vol. AP-8, pp. 201-212, March, 1960.

256. Muchmore, R. B., Reply to Comments by Leon Peters and F. C. Weiner, *IRE Trans. Antennas Propagation,* vol. AP-9, pp. 112-113, January, 1961.

257. Mueller, E. A., and A. L. Sims, Investigation of the Quantitative Determination Precipitation by Radar Echo Measurements, *Illinois State Water Survey, Tech. Rept.* ECOM-00032-F, Urbana, Ill., December, 1966 (AD645218).

258. Myers, G. F., High-resolution Radar, Part 3, Sea Clutter Analysis, *Naval Res. Lab. Rept.* 4952, Washington, D. C., Oct. 21, 1958.

259. Myers, G. F., High-resolution Radar, Part 4, Sea Clutter Analysis, *Naval Res. Lab. Rept.* 5191, Oct. 21, 1958.

260. Nathanson, F. E., and M. Davidson, A Coherent Frequency Waveform Synthesizer, *Proc. IEEE (Corr.),* vol. 51, no. 12, pp. 1773-1774, December, 1963.

261. Nathanson, F. E., Time Sidelobes in a Combined Phase-code and Pulse-doppler System, *Proc. IEEE (Corr.),* vol. 53, no. 11, pp. 1775-1776, November, 1965.

262. Nathanson, F. E., and J. P. Reilly, Radar Precipitation Echoes, *IEEE Trans.*, vol. AES-4, no. 4, pp. 505-514, July, 1968. (See also [263].)

263. Nathanson, F. E., and J. P. Reilly, Radar Precipitation Echoes—Experiments on Temporal, Spatial and Frequency Correlation, *Appl. Phys. Lab./Johns Hopkins Univ. Rept.* TG-899, April, 1967.

264. Nathanson, F. E., and J. P. Reilly, Clutter Statistics that Affect Radar Performance Analysis, *EASCON Proc.* (supplement to *IEEE Trans.*), vol. AES-3, no. 6, pp. 386-398, November, 1967.
265. Ament, W. S., F. C. Macdonald, H. J. Passerini, and R. O. Shrewbridge, Quantitative Measurements of Radar Echoes from Aircraft, series of 16 Naval Research Laboratory reports on RCS measurements made between October, 1950 and June, 1960, nos. 1-16.
266. Nichols, D. E. T., Determination of the Total Refraction of 3.6 cm Wavelength Radiation Passing through the Atmosphere, Royal Aircraft Establishment, RT 66191, June, 1966 (AD807747).
267. Nilssen, O. K., New Methods of Range Measuring Doppler Radar, *IRE Trans.*, vol. ANE-9, no. 4, pp. 255-265, December, 1962.
268. Nolen, J. C., J. Schneider, and J. Lacis, Statistics of Radar Detection, Bendix Corporation, Bendix Radio Division, Baltimore, February, 1966.
269. North, D.O., An Analysis of the Factors which Determine Signal Noise Discrimination in Pulsed Carrier Systems, *Proc. IEEE*, vol. 51, no. 7, pp. 1016-1027, July, 1963.
270. O'Hara, F. J., and G. M. Moore, A High Performance CW Receiver Using Feedthru Nulling, *Microwave J.*, vol. 6, no. 9, pp. 63-71, September, 1963.
271. Ohman, G. P., Getting High Range Resolution with Pulse Compression Radar, *Electronics*, vol. 33, no. 41, pp. 53-57, Oct. 7, 1960.
272. Olin, I. D., and F. E. Queen, Dynamic Measurement of Radar Cross Section, *Proc. IEEE*, vol. 53, no. 8, pp. 954-961, August, 1965.
273. O'Neill, H. J., A Wideband Stepped Frequency Pulse Compression System, *Appl. Electron. Lab. G.E.C. (Electron.) LTD Rept.* SLR 398, Stanmore, Middlesex, England, October, 1966.
274. Palmieri, C. A., Radar Waveform Design, *Sperry Eng. Rev. (Radar)*, pt. 1, Winter, 1962, pp. 32-43.
275. Peake, W. H., The Interaction of Electromagnetic Waves with Some Natural Surfaces, *Ohio State Univ. Rept.* 898-2, May 30, 1959.
276. Peake, W. H., Theory of Radar Return from Terrain, *IRE Natl. Conv. Rec.*, pt. 1, Antenna Laboratory, Ohio State University, 1959, pp. 27-41.
277. Pedersen, C., Detectability of Radar Echoes in Noise and Clutter Interference, *Columbia Univ. Tech. Rept.* T-2/318 CU-28-65-AF-1478-ERL, Sept. 30, 1965.
278. Peebles, P. Z., Jr., and G. H. Stevens, A Technique for the Generation of Highly Linear FM Pulse Radar Signals, *IEEE Trans.*, vol. MIL-9, no. 1, pp. 32-38, January, 1965.
279. Perlman, S. E., Staggered Rep Rate Fills Radar Blind Spots, *Electronics*, vol. 31, no. 47, pp. 82-85, Nov. 21, 1958.
280. Persons, C. E., Ambiguity Function of Pseudo-random Sequences, *Proc. IEEE (Corr.)*, vol. 54, no. 12, pp. 1946-1947, December, 1966.
281. Peters, L., and F. C. Weimer, Concerning the Assumption of Random Distribution of Scatterers as a Model of an Aircraft for Tracking Radars, *IRE Trans. Antennas Propagation*, vol. AP-9, pp. 110-111, January, 1961.
282. Petrushevskii, V. A., R. V. Ignatova, and E. Sal'man, Radar Echo Characteristics of Clouds (translation), American Meteorological Society, 1965 (AD630675).
283. Philippides, C., Degradation of a Pulse Compression System due to Random Phase Errors, *Appl. Phys. Lab./Johns Hopkins Univ. Rept.* TG-843, August, 1966.

284. Pidgeon, V. W., Bistatic Cross Section of the Sea, *IEEE Trans.*, vol. AP-14, no. 3, pp. 405-406, May, 1966.
285. Pidgeon, V. W., Bistatic Cross Section of the Sea for a Beaufort 5 Sea, *Space Sys. Planetary Geol. Geophys.*, American Astronautical Society, May, 1967.
286. Pidgeon, V. W., Time, Frequency, and Spatial Correlation of Radar Sea Return, *Space Sys. Planetary Geol. Geophys.*, American Astronautical Society, May, 1967.
287. Pidgeon, V. W., Radar Land Clutter for Small Grazing Angles at X and L Band, *Space Sys. Planetary Geol. Geophys.*, American Astronautical Society, May, 1967.
288. Pidgeon, V. W., The Doppler Dependence of Radar Sea Return, *J. Geophys. Res.*, Feb. 15, 1968.
289. Pidgeon, V. W., Private communication to Applied Physics Laboratory/Johns Hopkins University.
290. Pilie', R. J., J. E. Jiusto, and R. R. Rogers, Wind Velocity Measurement with Doppler Radar, *Proc. 10th Weather Radar Conf.*, American Meteorological Society, Washington, D. C., April, 1963.
291. Portner, E. M., Double Threshold Detection, *Appl. Phys. Lab. Intern. Memo.* MRT-0-037, Aug. 24, 1966.
292. Potts, B., Memorandum on the Average Radar Echo Area of Orbiting Satellites, *Antenna Lab. Ohio State Univ. Rept.* 1116-7, Columbus, Ohio, June 23, 1961 (AD264033).
293. Povejsil, D. J., R. S. Raven, and P. Waterman, "Airborne Radar," D. Van Nostrand Co., Inc., Princeton, N.J., 1961.
294. Preston, G. W., The Search Efficiency of the Probability Ratio Sequential Search Radar, *IRE Intern. Conv. Rec.*, vol. 8, pt. 4, pp. 116-124, 1960.
295. Price, R., and E. M. Hofstetter, Bounds on the Volume and Height Distributions of the Ambiguity Function, *IEEE Trans. Inform. Theory*, vol. IT-11, no. 2, pp. 207-214, April, 1965.
296. Probert-Jones, J. R., The Radar Equation in Meteorology, *Quart. J. Roy. Meteorol. Soc.*, vol. 88, pp. 485-495, 1962.
297. *Proc. 13th Radar Meteorol. Conf.*, McGill University, Montreal, July, 1968.
298. Queen, J. L., Effect of Time Sidelobes on Clutter Response of Zero-180° Phase Codes, *Appl. Phys. Lab. Intern. Memo.* MRT-0-057, May 23, 1967.
299. Queen, J. L., 4-Phase Code Correlation Functions, *Appl. Phys. Lab. Intern. Memo.* MRT-00-025, July 18, 1967.
300. Queen, J. L., Private communication on Detection Losses in Noise and Clutter to Applied Physics Laboratory/Johns Hopkins University.
301. Raduziner, D. M., and N. R. Gillespie, Coherort Radar FM Noise Limitations, *IEEE-NASA Symp. Short-term Frequency Stability*, Nov. 23-24, 1964, Greenbelt, Md., NASA SP-80, pp. 29-38.
302. Ramp, H. O., and E. R. Wingrove, Principles of Pulse Compression, *IRE Trans.*, vol. MIL-5, pp. 108-116, April, 1961.
303. Rappaport, S. S., On Optimum Dynamic Range Centering, *Proc. IEEE (Letters)*, vol. 54, no. 8, pp. 1067, 1068, August, 1966.
304. Ray, Howard, Improving Radar Range and Angle Detection with Frequency Agility, *Proc. 11th Ann. East Coast Conf. Aerospace Navigational Electron.*, pp. 1.3.6-1 to 1.3.6-7, IEEE, Baltimore, Oct. 21-23, 1964.
305. Optimum Waveform Study for Coherent Pulse Radar, Final Report on contract

NOnr-4649(00)(X), Radio Corporation of America, Moorestown, N.J., Feb. 28, 1965 (AD641391).

306. Reilly, J. P., "Clutter Reduction by Delay Line Cancellers," masters thesis, George Washington University, Washington, D. C., February, 1967.

307. Reitz, E. A., Radar Terrain Return Study, Final Report, Measurements of Terrain Backscattering Coefficients with an Airborne X-Band Radar, Goodyear Corporation, GERA-463, Sept. 30, 1959.

308. Remley, W. R., Doppler Dispersion Effects in Matched Filter Detection and Resolution, *Proc. IEEE*, vol. 54, no. 1, pp. 33-39, January, 1966.

309. Resnick, J. B., "High Resolution Waveforms Suitable for a Multiple Target Environment," masters thesis, Massachusetts Institute of Technology, June, 1962.

310. Rice, P., et al., Transmission Loss Predictions for Tropospheric Communication Circuits, *Natl. Bureau Std. Tech. Note* 101, vols. 1 and 2, 1965.

311. Rice, S. O., Mathematical Analysis of Random Noise, *Bell System Tech. J.*, vols. 23 and 24, pp. 282-332; 45-156, 1945.

312. Rice, S. O., Statistical Properties of a Sine Wave plus Random Noise, *Bell System Tech. J.*, vol. 27, pp. 109-157, January, 1948.

313. Rice, S. O., in N. Wax (ed.), "Mathematical Analysis of Random Noise" (selected papers on noise and stochastic processes), Dover Publications, Inc., New York, 1954.

314. Richman, D., Resolution of Multiple Targets in Clutter, *Inst. Defense Analyses Res. Paper* P-158 IDA/HQ 66-5086, June, 1966, revised April, 1967 (AD652578).

315. Riggs, R. F., The Angular Accuracy of Monopulse Radar in the Presence of Clutter, *Proc. NEC*, vol. 22, 1966.

316. Rihaczek, A. W., Radar Resolution Properties of Pulse Trains, *Proc. IEEE*, vol. 52, pp. 153-164, February, 1964.

317. Rihaczek, A. W., Radar Signal Design for Target Resolution, *Proc. IEEE*, vol. 53, pp. 116-128, February, 1965.

318. Rihaczek, A. W., Measurement Properties of the Chirp Signal, *Space Sys. Div. Air Force Sys. Command Rept.* TDR-469 (5230-43)2, SSD-TR-65-115, Aerospace Corporation, El Segundo, Calif., AFO4(695)-469, August, 1965.

319. Rihaczek, A. W., Optimum Filters for Signal Detection in Clutter, *IEEE Trans. Aerospace Electron. Sys.*, vol. AES-1, no. 3, pp. 297-299, December, 1965.

320. Rihaczek, A. W., Doppler-tolerant Signal Waveforms, *Proc. IEEE*, vol. 54, no. 6, pp. 849-857, June, 1966.

321. Rihaczek, A. W., "High Resolution Radar," McGraw-Hill Book Company, New York, 1969.

322. Rochefort, J. S., Matched Filters for Detecting Pulsed Signals in Noise, *IRE Natl. Conv. Rec.*, vol. 2, pp. 30-34, 1954.

323. Rogers, R. R., Radar Measurement of Gustiness, *Proc. 6th Weather Radar Conf.*, pp. 96-106, American Meteorological Society, Boston.

324. Rogers, R. R., and B. R. Tripp, Some Radar Measurements of Turbulence in Snow, *J. Appl. Meteorol.*, October, 1961, pp. 603-610.

325. Roll, R. G., Use of the Autocorrelation Function of the In-phase and Quadrature Power from a Turbulent Process. . . , *Appl. Phys. Lab./Johns Hopkins Univ. Rept.* BPD 65U-14, Silver Spring, Md., Nov. 10, 1965.

326. Ross, A. W., D. C. Fakley, and L. R. Palmer, Sea-Clutter Investigations, Admiralty Signal and Radar Establishment, England, R4/50/15, Sept. 7, 1950.

327. Roth, H. H., Linear Binary Shift Register Circuits Utilizing a Minimum Number of MOD-2 Adders, *IEEE Trans. Inform. Theory,* vol. IT-11, no. 2, pp. 215-220, April, 1965.

328. Rubin, W. L., and S. K. Kamen, S/N Ratios in a Two-channel Band-Pass Limiter, *Proc. IEEE,* pp. 389-390, February, 1963.

329. Rubin, W. L., and C. P. Kaiteris, Some Results Concerning Radar Waveform Design for Detecting Targets in Clutter, *Proc. IEEE (Letters),* vol. 54, no. 11, pp. 1609-1610, December, 1966.

330. Rummler, W. D., Clutter Suppression by Complex Weighting of Coherent Pulse Trains, *IEEE Trans.,* vol. AES-2, no. 6, pp. 689-699, November, 1966.

331. Rummler, W. D., A Technique for Improving the Clutter Performance of Coherent Pulse Train Signals, *IEEE Trans.,* vol. AES-3, no. 6, pp. 898-907, November, 1967.

332. Sakamoto, J., et al., Coded Pulse Radar System, *J. Faculty Eng. Univ. Tokyo,* vol. 27, no. 1, pp. 119-181, 1964.

333. Saxton, J. A., VHF and UHF Reception, Effects of Trees and Other Obstacles, *Wireless World,* May, 1955, pp. 229-232.

334. Scelling, Burrows, and Ferrell, Ultra-short-wave Propagation, *Proc. IRE,* vol. 21, pp. 440-442, 458-461, Mar. 3, 1933.

335. Schlesinger, R. J., "Principles of Electronic Warfare," Prentice-Hall Space Technology Series, Prentice-Hall, Inc., Englewood Cliffs, N.J., 1961.

336. Scholefield, P. H. R., Statistical Aspects of Ideal Radar Targets, *Proc. IEEE,* vol. 55, no. 4, pp. 587-590, April, 1967.

337. Schwartz, M., A Coincidence Procedure for Signal Detection, *IRE Trans.,* vol. IT-2, no. 4, pp. 135-139, December, 1956.

338. Schwartz, Mischa, "Information Transmission Modulation and Noise," McGraw-Hill Book Company, New York, 1959.

339. Scoggins, J. R., and M. Susko, FPS-16 Radar/Jimsphere Wind Data Measured at the Eastern Test Range, AGRO-Astrodynamics Laboratory, NASA TM X53290, July 9, 1965.

340. Senior, T. B. A., A Survey of Analytical Techniques for Cross-section Estimation, *Proc. IEEE,* vol. 53, no. 8, pp. 822-833, August, 1965.

341. Sevy, J. L., The Effect of Hard Limiting an Angle-modulated Signal Plus Noise, *IEEE Trans.,* vol. AES-4, no. 1, pp. 24-30, January, 1968. (See also pp. 125-128.)

342. Shannon, *Electronics,* vol. 35, no. 49, pp. 52-56, December, 1962.

343. Sherwood, E. M., and E. E. Ginston, Reflection Coefficients of Irregular Terrain at 10 cm, *Proc. IRE (Corr.),* vol. 43, no. 7, pp. 877,878, July, 1955.

344. Shrader, W., MTI Radar, in M.I. Skolnik (ed.), "Radar Handbook," chap. 17, McGraw-Hill Book Company, New York, 1970.

345. Siebert, W. McC., A Radar Detection Philosophy, *IRE Trans.,* vol. IT-2, no. 3, pp. 204-221, September, 1956.

346. Siegel, K. M., Far Field Scattering from Bodies of Revolution, *Appl. Sci. Res.,* sec. B, vol. 7, pp. 293-328, 1958.

347. Siegel, K. M., Low Frequency Radar Cross-section, *Proc. IEEE,* vol. 51, pp. 231-233, January, 1963.

348. Siegel, K. M., Radar Cross Section of a Cone Sphere, *Proc. IEEE (Corr.),* vol. 51, no. 1, pp. 231-232, January, 1963.

349. Silber, D., Probabilities of Detection and False Alarm for a Coherent Detector with Amplitude Limiting of Arbitrary Hardness, *1966 IEEE Intern. Conv. Rec.,* pt. 7, April, 1966, pp. 107-123.

350. Silver, S., "Microwave Antenna Theory and Design," M.I.T. Radiation Laboratory Series, vol. 12, chap. 6, McGraw-Hill Book Company, New York, 1949.

351. Simon, L. H., An Analysis of Frequency Step Pulse Compression, *RCA-Radar Sys. Memo.,* Moorestown, N. J., Apr. 12, 1961.

352. Sims, A. L., E. A. Mueller, and G. E. Stout, Investigation of the Quantitative Determination of Point and Areal Precipitation by Radar Echo Measurements, *Illinois State Water Survey Interim Rept.* 1, Oct. 1, 1964 - Mar. 31, 1965 (AD623409).

353. Sinsky, A. I., Private communication on Wide Bandwidth Signals in Phase Steered Arrays, 1967.

354. Skolnik, M. I., "Introduction to Radar Systems," McGraw-Hill Book Company, New York, 1962.

355. Slobodin, L., Private communication on Optical Signal Processor, Lockheed Electronics Co., December, 1967.

356. Smirnova, G. A., Experiments in Radar Measurement of Clear Sky-Turbulence with the Aid of Passive Reflectors, translated from *Tsentr. Aerologicheskaya Observatoriya, Trudy (Central Aerological Observatory Trans.),* no. 57, pp. 72-76, *Appl. Phys. Lab./Johns Hopkins Univ. Trans.* 1462, Silver Spring, Md., 1965.

357. Spafford, L. J., Optimum Radar Receive Waveforms in the Presence of Clutter, *Gen. Elec. Heavy Military Electron. Dept. Rept.* R65EMH14, Syracuse, N. Y., June 3, 1965.

358. Spafford, L. J., Ambiguity Function Catalog of Uniformly Spaced Envelope Recurrent Pulse Trains, *Gen. Elec. Heavy Military Electron. Dept. Rept.* R66EMH51, Syracuse, N.Y., November, 1966.

359. Spafford, L. J., Optimum Radar Signal Processing in Clutter, *Gen. Elec. Heavy Military Electron. Dept. Rept.* RG7EMH16, Syracuse, N.Y., June, 1967 (also PhD thesis Polytechnic Institute of Brooklyn, June, 1967 and *IEEE Trans.,* vol. IT-14, no. 2, pp. 734-743, September, 1968).

360. Spilker, J. J., Jr., Delay Lock Tracking of Binary Signals, *IEEE Trans.,* vol. SET-9, pp. 1-8, March, 1963.

361. Steele, J. G., Backscatter of 16 Mc/s Radio Waves from Land and Sea, *Austral. J. Phys.,* vol. 18, pp. 317-327, August, 1965. (See also *Proc. IEEE,* vol. 55, no. 9, pp. 1583-1590, September, 1967.)

362. Stewart, Dorathy A., Wind Shear and Baroclinity in Cross Sections along 80°W, *U.S. Army Missile Command Rept.* RR-TR-67-4, Redstone Arsenal, Ala., February, 1967 (AD652256).

363. Stutt, C. A., Preliminary Report on Ground-wave-radar Sea Clutter, *Lincoln Lab. Tech. Rept.* 134, Sept. 21, 1956.

364. Stutt, C. A., Integral Equation for Optimum Mismatched Filter for Suppressing Sidelobes over a Designated Region of the τ-Δ Plane, *GE Res. Lab. Memo.,* Schenectady, New York, December, 1964.

365. Stutt, C. A., A Mismatch Filter Approach to Clutter Discrimination, *GE Res. Lab. Memo.,* Schenectady, New York, December, 1964.

366. Stutt, C. A., and L. J. Spafford, A "Best" Mismatched Filter Response for Radar Clutter Discrimination, *IEEE Trans.,* vol. IT-4, no. 2, March, 1968.

367. Swerling, P., Probability of Detection for Fluctuating Targets, Rand Corporation, RM-1217, Mar. 17, 1954.
368. Swerling, P., Probability of Detection for Some Additional Fluctuating Target Cases, *Aerospace Corp. Rept.* TOR-699 (9990)-14, El Segundo, Calif., March, 1966.
369. Swerling, P., Radar Target Signatures, Intensive Lecture Series, Technology Service Corporation, Santa Monica, Calif., Aug. 26-30, 1968.
370. Tamir, T., On Radio Wave Propagation in Forest Environments, *IEEE Trans.*, vol. AP-15, no. 6, pp. 806-817, November, 1967.
371. Taylor, R. C., Terrain Measurements at X-, K_u-, and K_a-Band, *IRE Natl. Conv. Rec.*, vol. 7, pt. 1, pp. 27-41, 1959 on Antennas and Propagation.
372. Taylor, S. A., and J. L. MacArthur, Digital Pulse Compression Radar Receiver, *Appl. Phys. Lab. Tech. Digest,* vol. 6, no. 4, pp. 2-10, 1967.
373. Temes, C. L., Sidelobe Suppression in a Range-channel Pulse Compression Radar, *IRE Trans.*, vol. MIL-6, no. 2, pp. 162-169, April, 1962.
374. Tolbert, C. W., et al., Back Scattering Cross Sections of Water Droplets, Rain, and Foliage at 4.3 Millimeter Radio Wavelengths, *Univ. Texas Elec. Eng. Res. Lab. Rept.* 91, Apr. 30, 1957 (AD200368).
375. Tompkins, D. N., Codes with Zero Correlation, *Hughes Aircraft Tech. Memo.* 651, June, 1960.
376. Totty, R. E., et al., Investigation of Chaff Communications at 4,650 MHz, *Tech. Rept.* ECOM-0202-3, Radiation Inc., Melbourne, Fla., August, 1967 (AD664124).
377. Turin, G. L., A Review of Correlation, Matched-filter and Signal Coding Techniques with Emphasis on Radar Applications, *Hughes Aircraft Sys. Devel. Lab. Tech. Memo.* 559, vol. 1, April, 1957.
378. Turin, G. L., An Introduction to Matched Filters, *IRE Trans.*, vol. IT-6, pp. 311-329, June, 1960.
379. Turyn, R., On Barker Codes of Even Length, *Proc. IEEE (Corr.),* vol. 51, p. 1256, September, 1963.
380. Urkowitz, H., Filters for Detection of Small Radar Signals in Clutter, *J. Appl. Phys.,* vol. 24, no. 8, pp. 1024-1031, August, 1953.
381. Urkowitz, H., Analysis and Synthesis of Delay Line Periodic Filters, *IRE Trans.,* vol. CT-4, pp. 41-53, June, 1957.
382. Urkowitz, H., Analysis of Periodic Filters with Stationary Random Inputs, *IRE Trans.,* vol. CT-6, no. 4, pp. 330-334, December, 1959.
383. Urkowitz, H., Some Properties and Effects of Reverberation in Acoustic Surveillance, *Gen. Atronics Corp. Rept.* 1594-2041-2, February, 1967.
384. Valley, S. L. (ed.), "Handbook of Geophysics and Space Environments," McGraw-Hill Book Company, New York, April, 1965.
385. Van Trees, H. L., Optimum Signal Design and Processing for Reverberation-limited Environments, *IEEE Trans.,* vol. MIL-9, pp. 212-229, July-October, 1965.
386. Van Vleck, J. H., and D. Middleton, The Spectrum of Clipped Noise, *Proc. IEEE,* vol. 54, no. 1, January, 1966 (reprint from 1943 report).
387. Vinitskiy, A. S., Principles of Continuous Wave Radar, translated from the Russian (Leningrad, 1961), English translation OTS-64-21507, Feb. 6, 1964.
388. Vogel, M., Uber das Statistische Verhalten der Radarechoes von Flugzeugen, *Wiss. Ges. Tuftfart E. V. Berlin,* January, 1962.

389. Voles, R., Frequency Correlation of Clutter, *Proc. IEEE (Letters)*, vol. 54, no. 6, pp. 881, 882, June, 1966.

390. Waddell, M. C., "Air Battle Analyzer Handbook," Applied Physics Laboratory/ Johns Hopkins University, TG-421, Silver Spring, Md., April, 1963.

391. Wagner, T. C., "Analytical Transients," John Wiley & Sons, Inc., New York, 1959.

392. Wainstein, L. A. and V. D. Zubakov, "Extraction of Signals from Noise," (translation from Russian edited by R. A. Silverman), Prentice-Hall, Inc. Englewood Cliffs, N.J., 1962.

393. Walcoff, P., Duty Cycle Optimization for a High PRF Radar, *Radar Comm., Dev. Engr. Tech. Memo.* 39, Westinghouse Air Arm, Baltimore, Feb. 25, 1964.

394. Wald, A., "Sequential Analysis," John Wiley & Sons, Inc., New York.

395. Wallace, P. R., Interpretation of the Fluctuating Echo from Randomly Distributed Scatterers, *Can. J. Phys.*, vol. 31, pt. 2, pp. 995-1009, September, 1952.

396. Ward, H. R., Dynamic Range Centering for Minimum Probability of Excluding a Rayleigh Distributed Signal, *Proc. IEEE (Letters)*, vol. 54, no. 1, pp. 59, 60, January, 1966.

397. Ward, H. R., and W. W. Shrader, MTI Performance Degradation Caused by Limiting, *IEEE EASCON Rec.*, vol. 68-C3 AES, pp. 168-174, September, 1968.

398. Waters, P. L., Frequency Diversity Performance of a Ground Surveillance Radar, *Proc. IEE (Letters)*, vol. 1, no. 10, pp. 282-283, December, 1965.

399. Weiner, S. D., A Model of Radar Scattering from the Cone Sphere, *M.I.T./Lincoln Lab. Tech. Note* 1966-47, Lexington, Mass., Oct. 20, 1966 (AD646853).

400. Weinstock, W., "Target Cross Section Models for Radar Systems Analysis," doctoral dissertation, University of Pennsylvania, Philadelphia, 1964.

401. Westerfield, E. C., R. H. Prager, and J. L. Stewart, Processing Gains Against Reverberation (Clutter) Using Matched Filters, *IRE Trans. Inform. Theory*, vol. IT-6, pp. 342-348, June, 1960.

402. White, D. M., Synthesis of Pulse Compression Waveforms with Weighted Finite Frequency Combs, *Appl. Phys. Lab. Rept.* TG-934, August, 1967.

403. White, W. O., and A. E. Ruvin, Recent Advances in the Synthesis of Comb Filters, *IRE Nat. Conv. Rec.*, vol. 5, pt. 2, pp. 186-199, 1957.

404. Widrow, B., Statistical Analysis of Amplitude Quantized Sampled-data Systems, *AIEE Trans.*, vol. 79, pp. 555-568, January, 1961.

405. Wild, T. A., Private communication on Calculated Return from Total Rainfall in a Collimated Bistatic Radar to Applied Physics Laboratory.

406. Wilkes, M. V., and J. A. Ramsay, A Theory of the Performance of Radar on Ship Targets, *Proc. Cambridge Phil. Soc.*, vol. 43, pp. 202-231, 1947.

407. Woodward, P. M., "Probability and Information Theory, with Applications to Radar," Pergamon Press, Ltd., London, McGraw-Hill Book Company, New York, 1953.

408. Woodward, P. M., Radar Ambiguity Analysis, *RRE Tech. Note* 731, Royal Radar Establishment, Malvern Worcs, England, February, 1967 (AD653404).

409. Worley, R., Optimum Thresholds for Binary Integration, *IEEE Trans.*, vol. IT-14, no. 2, pp. 349-352, March, 1968.

410. Young, C. A., Signal-to-Noise Ratios, *Appl. Phys. Lab./Johns Hopkins Univ. Intern. Memo.* MRD-0-013, Silver Spring, Md., Nov. 9, 1965. (Not generally available.)
411. Zierler, N. Linear Recurring Sequences, *J Soc. Ind. and Appl. Math,* vol. 7, no. 1, pp. 31-48, March, 1959.

INDEX